Undergraduate Texts

S. Axler
K.A. Ribet

For other titles published in this series, go to
http://www.springer.com/series/666

John Stillwell

Mathematics
and Its History

Third Edition

 Springer

John Stillwell
Department of Mathematics
University of San Francisco
San Francisco, CA 94117-1080
USA
stillwell@usfca.edu

ISSN 0172-6056
ISBN 978-1-4614-2632-5   ISBN 978-1-4419-6053-5 (eBook)
DOI 10.1007/978-1-4419-6053-5
Springer New York Dordrecht Heidelberg London

Mathematics Subject Classification (2010): 01-xx, 01Axx

Printed on acid-free paper

Springer is part of Springer Science+Business Media (www.springer.com)

To Elaine, Michael, and Robert

Preface to the Third Edition

The aim of this book, announced in the first edition, is to give a bird's-eye view of undergraduate mathematics and a glimpse of wider horizons. The second edition aimed to broaden this view by including new chapters on number theory and algebra, and to engage readers better by including many more exercises. This third (and possibly last) edition aims to increase breadth and depth, but also *cohesion*, by connecting topics that were previously strangers to each other, such as projective geometry and finite groups, and analysis and combinatorics.

There are two new chapters, on simple groups and combinatorics, and several new sections in old chapters. The new sections fill gaps and update areas where there has been recent progress, such as the Poincaré conjecture. The simple groups chapter includes some material on Lie groups, thus redressing one of the omissions I regretted in the first edition of this book. The coverage of group theory has now grown from 17 pages and 10 exercises in the first edition to 61 pages and 85 exercises in this one. As in the second edition, exercises often amount to proofs of big theorems, broken down into small steps. In this way we are able to cover some famous theorems, such as the Brouwer fixed point theorem and the simplicity of A_5, that would otherwise consume too much space.

Each chapter now begins with a "Preview" intended to orient the reader with motivation, an outline of its contents and, where relevant, connections to chapters that come before and after. I hope this will assist readers who like to have an overview before plunging into the details, and also instructors looking for a path through the book that is short enough for a one-semester course. Many different paths exist, at many different levels. Up to Chapter 10, the level should be comfortable for most junior or senior undergraduates; after that, the topics become more challenging, but also of greater current interest.

All the figures have now been converted to electronic form, which has enabled me to reduce some that were excessively large, and hence mitigate the bloating that tends to occur in new editions.

Some of the new material on mechanics in Section 13.2 originally appeared (in Italian) in a chapter I wrote for Volume II of *La Matematica*, edited by Claudio Bartocci and Piergiorgio Odifreddi (Einaudi, Torino, 2008). Likewise, the new Section 8.6 contains material that appeared in my book *The Four Pillars of Geometry* (Springer, 2005).

Finally, there are many improvements and corrections suggested to me by readers. Special thanks go to France Dacar, Didier Henrion, David Kramer, Nat Kuhn, Tristan Needham, Peter Ross, John Snygg, Paul Stanford, Roland van der Veen, and Hung-Hsi Wu for these, and to my son Robert and my wife, Elaine, for their tireless proofreading.

I also thank the University of San Francisco for giving me the opportunity to teach the courses on which much of this book is based, and Monash University for the use of their facilities while revising it.

JOHN STILLWELL

Monash University and the University of San Francisco

March 2010

Preface to the Second Edition

This edition has been completely retyped in LaTeX, and many of the figures redone using the PSTricks package, to improve accuracy and make revision easier in the future. In the process, several substantial additions have been made.

- There are three new chapters, on Chinese and Indian number theory, on hypercomplex numbers, and on algebraic number theory. These fill some gaps in the first edition and give more insight into later developments.

- There are many more exercises. This, I hope, corrects a weakness of the first edition, which had too few exercises, and some that were too hard. Some of the monster exercises in the first edition, such as the one in Section 2.2 comparing volume and surface area of the icosahedron and dodecahedron, have now been broken into manageable parts. Nevertheless, there are still a few challenging questions for those who want them.

- Commentary has been added to the exercises to explain how they relate to the preceding section, and also (when relevant) how they foreshadow later topics.

- The index has been given extra structure to make searching easier. To find Euler's work on Fermat's last theorem, for example, one no longer has to look at 41 different pages under "Euler." Instead, one can find the entry "Euler, and Fermat's last theorem" in the index.

- The bibliography has been redone, giving more complete publication data for many works previously listed with little or none. I have found the online catalogue of the Burndy Library of the Dibner Institute at MIT helpful in finding this information, particularly for early printed works. For recent works I have made extensive use of MathSciNet, the online version of *Mathematical Reviews*.

There are also many small changes, some prompted by recent mathematical events, such as the proof of Fermat's last theorem. (Fortunately, this one did not force a major rewrite, because the background theory of elliptic curves was covered in the first edition.)

I thank the many friends, colleagues, and reviewers who drew my attention to faults in the first edition, and helped me in the process of revision. Special thanks go to the following people.

- My sons, Michael and Robert, who did most of the typing, and my wife, Elaine, who did a great deal of the proofreading.

- My students in Math 310 at the University of San Francisco, who tried out many of the exercises, and to Tristan Needham, who invited me to USF in the first place.

- Mark Aarons, David Cox, Duane DeTemple, Wes Hughes, Christine Muldoon, Martin Muldoon, and Abe Shenitzer, for corrections and suggestions.

JOHN STILLWELL
Monash University
Victoria, Australia
2001

Preface to the First Edition

One of the disappointments experienced by most mathematics students is that they never get a course on mathematics. They get courses in calculus, algebra, topology, and so on, but the division of labor in teaching seems to prevent these different topics from being combined into a whole. In fact, some of the most important and natural questions are stifled because they fall on the wrong side of topic boundary lines. Algebraists do not discuss the fundamental theorem of algebra because "that's analysis" and analysts do not discuss Riemann surfaces because "that's topology," for example. Thus if students are to feel they really know mathematics by the time they graduate, there is a need to unify the subject.

This book aims to give a unified view of undergraduate mathematics by approaching the subject through its history. Since readers should have had some mathematical experience, certain basics are assumed and the mathematics is not developed formally as in a standard text. On the other hand, the mathematics is pursued more thoroughly than in most general histories of mathematics, because mathematics is our main goal and history only the means of approaching it. Readers are assumed to know basic calculus, algebra, and geometry, to understand the language of set theory, and to have met some more advanced topics such as group theory, topology, and differential equations. I have tried to pick out the dominant themes of this body of mathematics, and to weave them together as strongly as possible by tracing their historical development.

In doing so, I have also tried to tie up some traditional loose ends. For example, undergraduates can solve quadratic equations. Why not cubics? They can integrate $1/\sqrt{1 - x^2}$ but are told not to worry about $1/\sqrt{1 - x^4}$. Why? Pursuing the history of these questions turns out to be very fruitful, leading to a deeper understanding of complex analysis and algebraic geometry, among other things. Thus I hope that the book will be not only a bird's-eye view of undergraduate mathematics but also a glimpse of wider horizons.

Some historians of mathematics may object to my anachronistic use of modern notation and (fairly) modern interpretations of classical mathematics. This has certain risks, such as making the mathematics look simpler than it really was in its time, but the risk of obscuring ideas by cumbersome, unfamiliar notation is greater, in my opinion. Indeed, it is practically a truism that mathematical ideas generally arise before there is notation or language to express them clearly, and that ideas are implicit before they become explicit. Thus the historian, who is presumably trying to be both clear and explicit, often has no choice but to be anachronistic when tracing the origins of ideas.

Mathematicians may object to my choice of topics, since a book of this size is necessarily incomplete. My preference has been for topics with elementary roots and strong interconnections. The major themes are the concepts of number and space: their initial separation in Greek mathematics, their union in the geometry of Fermat and Descartes, and the fruits of this union in calculus and analytic geometry. Certain important topics of today, such as Lie groups and functional analysis, are omitted on the grounds of their comparative remoteness from elementary roots. Others, such as probability theory, are mentioned only briefly, as most of their development seems to have occurred outside the mainstream. For any other omissions or slights I can only plead personal taste and a desire to keep the book within the bounds of a one- or two-semester course.

The book has grown from notes for a course given to senior undergraduates at Monash University over the past few years. The course was of half-semester length and a little over half the book was covered (Chapters 1–11 one year and Chapters 5–15 another year). Naturally I will be delighted if other universities decide to base a course on the book. There is plenty of scope for custom course design by varying the periods or topics discussed. However, the book should serve equally well as general reading for the student or professional mathematician.

Biographical notes have been inserted at the end of each chapter, partly to add human interest but also to help trace the transmission of ideas from one mathematician to another. These notes have been distilled mainly from secondary sources, the *Dictionary of Scientific Biography* (DSB) normally being used in addition to the sources cited explicitly. I have followed the DSB's practice of describing the subject's mother by her maiden name.

References are cited in the name (year) form, for example, Newton (1687) refers to the *Principia*, and the references are collected at the end of the book.

The manuscript has been read carefully and critically by John Crossley, Jeremy Gray, George Odifreddi, and Abe Shenitzer. Their comments have resulted in innumerable improvements, and any flaws remaining may be due to my failure to follow all their advice. To them, and to Anne-Marie Vandenberg for her usual excellent typing, I offer my sincere thanks.

JOHN STILLWELL
Monash University
Victoria, Australia
1989

Contents

1

The Theorem of Pythagoras

PREVIEW

The Pythagorean theorem is the most appropriate starting point for a book on mathematics and its history. It is not only the oldest mathematical theorem, but also the source of three great streams of mathematical thought: numbers, geometry, and infinity.

The number stream begins with *Pythagorean triples*; triples of integers (a, b, c) such that $a^2 + b^2 = c^2$. The geometry stream begins with the interpretation of a^2, b^2, and c^2 as squares on the sides of a right-angled triangle with sides a, b, and hypotenuse c. The infinity stream begins with the discovery that $\sqrt{2}$, the hypotenuse of the right-angled triangle whose other sides are of length 1, is an *irrational* number.

These three streams are followed separately through Greek mathematics in Chapters 2, 3, and 4. The geometry stream resurfaces in Chapter 7, where it takes an *algebraic* turn. The basis of algebraic geometry is the possibility of describing points by numbers—their *coordinates*—and describing each curve by an equation satisfied by the coordinates of its points.

This fusion of numbers with geometry is briefly explored at the end of this chapter, where we use the formula $a^2 + b^2 = c^2$ to define the concept of *distance* in terms of coordinates.

J. Stillwell, *Mathematics and Its History*, Undergraduate Texts in Mathematics, DOI 10.1007/978-1-4419-6053-5_1, © Springer Science+Business Media, LLC 2010

1.1 Arithmetic and Geometry

If there is one theorem that is known to all mathematically educated people, it is surely the theorem of Pythagoras. It will be recalled as a property of right-angled triangles: the square of the hypotenuse equals the sum of the squares of the other two sides (Figure 1.1). The "sum" is of course the sum of areas and the area of a square of side l is l^2, which is why we call it "l squared." Thus the Pythagorean theorem can also be expressed by

$$a^2 + b^2 = c^2, \tag{1}$$

where a, b, c are the lengths shown in Figure 1.1.

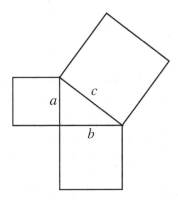

Figure 1.1: The Pythagorean theorem

Conversely, a solution of (1) by positive numbers a, b, c can be realized by a right-angled triangle with sides a, b and hypotenuse c. It is clear that we can draw perpendicular sides a, b for any given positive numbers a, b, and then the hypotenuse c must be a solution of (1) to satisfy the Pythagorean theorem. This converse view of the theorem becomes interesting when we notice that (1) has some very simple solutions. For example,

$$(a, b, c) = (3, 4, 5), \qquad (3^2 + 4^2 = 9 + 16 = 25 = 5^2),$$
$$(a, b, c) = (5, 12, 13), \quad (5^2 + 12^2 = 25 + 144 = 169 = 13^2).$$

It is thought that in ancient times such solutions may have been used for the construction of right angles. For example, by stretching a closed rope with 12 equally spaced knots one can obtain a $(3, 4, 5)$ triangle with right angle between the sides 3, 4, as seen in Figure 1.2.

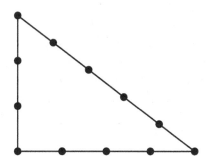

Figure 1.2: Right angle by rope stretching

Whether or not this is a practical method for constructing right angles, the very existence of a geometrical interpretation of a purely arithmetical fact like

$$3^2 + 4^2 = 5^2$$

is quite wonderful. At first sight, arithmetic and geometry seem to be completely unrelated realms. Arithmetic is based on counting, the epitome of a *discrete* (or *digital*) process. The facts of arithmetic can be clearly understood as outcomes of certain counting processes, and one does not expect them to have any meaning beyond this. Geometry, on the other hand, involves *continuous* rather than discrete objects, such as lines, curves, and surfaces. Continuous objects cannot be built from simple elements by discrete processes, and one expects to *see* geometrical facts rather than arrive at them by calculation.

The Pythagorean theorem was the first hint of a hidden, deeper relationship between arithmetic and geometry, and it has continued to hold a key position between these two realms throughout the history of mathematics. This has sometimes been a position of cooperation and sometimes one of conflict, as followed the discovery that $\sqrt{2}$ is irrational (see Section 1.5). It is often the case that new ideas emerge from such areas of tension, resolving the conflict and allowing previously irreconcilable ideas to interact fruitfully. The tension between arithmetic and geometry is, without doubt, the most profound in mathematics, and it has led to the most profound theorems. Since the Pythagorean theorem is the first of these, and the most influential, it is a fitting subject for our first chapter.

1.2   Pythagorean Triples

Pythagoras lived around 500 BCE (see Section 1.7), but the story of the Pythagorean theorem begins long before that, at least as far back as 1800 BCE in Babylonia. The evidence is a clay tablet, known as Plimpton 322, which systematically lists a large number of integer pairs (a, c) for which there is an integer b satisfying

$$a^2 + b^2 = c^2. \tag{1}$$

A translation of this tablet, together with its interpretation and historical background, was first published by Neugebauer and Sachs (1945) (for a more recent discussion, see van der Waerden (1983), p. 2). Integer triples (a, b, c) satisfying (1)—for example, $(3, 4, 5)$, $(5, 12, 13)$, $(8, 15, 17)$—are now known as *Pythagorean triples*. Presumably the Babylonians were interested in them because of their interpretation as sides of right-angled triangles, though this is not known for certain. At any rate, the problem of finding Pythagorean triples was considered interesting in other ancient civilizations that are known to have possessed the Pythagorean theorem; van der Waerden (1983) gives examples from China (between 200 BCE and 220 CE) and India (between 500 and 200 BCE). The most complete understanding of the problem in ancient times was achieved in Greek mathematics, between Euclid (around 300 BCE) and Diophantus (around 250 CE).

We now know that the general formula for generating Pythagorean triples is

$$a = (p^2 - q^2)r, \qquad b = 2qpr, \qquad c = (p^2 + q^2)r.$$

It is easy to see that $a^2 + b^2 = c^2$ when a, b, c are given by these formulas, and of course a, b, c will be integers if p, q, r are. Even though the Babylonians did not have the advantage of our algebraic notation, it is plausible that this formula, or the special case

$$a = p^2 - q^2, \qquad b = 2pq, \qquad c = p^2 + q^2$$

(which gives all solutions a, b, c, without common divisor) was the basis for the triples they listed. Less general formulas have been attributed to Pythagoras himself (around 500 BCE) and Plato (see Heath (1921), Vol. 1, pp. 80–81); a solution equivalent to the general formula is given in Euclid's *Elements*, Book X (lemma following Prop. 28). As far as we know, this is the first statement of the general solution and the first proof that it is general. Euclid's proof is essentially arithmetical, as one would expect since the problem seems to belong to arithmetic.

However, there is a far more striking solution, which uses the geometric interpretation of Pythagorean triples. This emerges from the work of Diophantus, and it is described in the next section.

EXERCISES

The integer pairs (a, c) in Plimpton 322 are

a	c
119	169
3367	4825
4601	6649
12709	18541
65	97
319	481
2291	3541
799	1249
481	769
4961	8161
45	75
1679	2929
161	289
1771	3229
56	106

Figure 1.3: Pairs in Plimpton 322

1.2.1 For each pair (a, c) in the table, compute $c^2 - a^2$, and confirm that it is a perfect square, b^2. (Computer assistance is recommended.)

You should notice that in most cases b is a "rounder" number than a or c.

1.2.2 Show that most of the numbers b are divisible by 60, and that the rest are divisible by 30 or 12.

Such numbers were in fact exceptionally "round" for the Babylonians, because 60 was the base for their system of numerals. It looks like they computed Pythagorean triples starting with the "round" numbers b and that the column of b values later broke off the tablet.

Euclid's formula for Pythagorean triples comes out of his theory of divisibility, which we shall take up in Section 3.3. Divisibility is also involved in some basic properties of Pythagorean triples, such as their evenness or oddness.

1.2.3 Show that any integer square leaves remainder 0 or 1 on division by 4.

1.2.4 Deduce from Exercise 1.2.3 that if (a, b, c) is a Pythagorean triple then a and b cannot both be odd.

1.3   Rational Points on the Circle

We know from Section 1.1 that a Pythagorean triple (a, b, c) can be realized by a triangle with sides a, b and hypotenuse c. This in turn yields a triangle with fractional (or *rational*) number sides $x = a/c$, $y = b/c$ and hypotenuse 1. All such triangles can be fitted inside the circle of radius 1 as shown in Figure 1.4. The sides x and y become what we now call the *coordinates* of

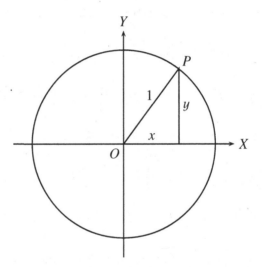

Figure 1.4: The unit circle

the point P on the circle. The Greeks did not use this language; however, they could derive the relationship between x and y we call the *equation of the circle*. Since

$$a^2 + b^2 = c^2 \tag{1}$$

we have

$$\left(\frac{a}{c}\right)^2 + \left(\frac{b}{c}\right)^2 = 1,$$

so the relationship between $x = a/c$ and $y = b/c$ is

$$x^2 + y^2 = 1. \tag{2}$$

Consequently, finding integer solutions of (1) is equivalent to finding rational solutions of (2), or finding *rational points* on the curve (2).

Such problems are now called *Diophantine*, after Diophantus, who was the first to deal with them seriously and successfully. *Diophantine equations* have acquired the more special connotation of equations for which integer solutions are sought, although Diophantus himself sought only rational solutions. (There is an interesting open problem that turns on this distinction. Matiyasevich (1970) proved that there is no algorithm for deciding which polynomial equations have integer solutions. It is not known whether there is an algorithm for deciding which polynomial equations have *rational* solutions.)

Most of the problems solved by Diophantus involve quadratic or cubic equations, usually with one obvious trivial solution. Diophantus used the obvious solution as a stepping stone to the nonobvious, but no account of his method survived. It was ultimately reconstructed by Fermat and Newton in the 17th century, and this so-called *chord–tangent construction* will be considered later. Here, we need it only for the equation $x^2 + y^2 = 1$, which is an ideal showcase for the method in its simplest form.

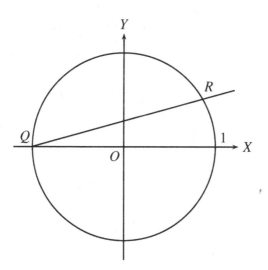

Figure 1.5: Construction of rational points

A trivial solution of this equation is $x = -1$, $y = 0$, which is the point Q on the unit circle (Figure 1.5). After a moment's thought, one realizes that a line through Q, with rational slope t,

$$y = t(x + 1) \tag{3}$$

will meet the circle at a second rational point R. This is because substitution of $y = t(x + 1)$ in $x^2 + y^2 = 1$ gives a quadratic equation with rational coefficients and one rational solution $(x = -1)$; hence the second solution must also be a rational value of x. But then the y value of this point will also be rational, since t and x will be rational in (3). Conversely, the chord joining Q to any other rational point R on the circle will have a rational slope. Thus by letting t run through all rational values, we find all rational points $R \neq Q$ on the unit circle.

What are these points? We find them by solving the equations just discussed. Substituting $y = t(x + 1)$ in $x^2 + y^2 = 1$ gives

$$x^2 + t^2(x + 1)^2 = 1,$$

or

$$x^2(1 + t^2) + 2t^2x + (t^2 - 1) = 0.$$

This quadratic equation in x has solutions -1 and $(1 - t^2)/(1 + t^2)$. The nontrivial solution $x = (1 - t^2)/(1 + t^2)$, when substituted in (3), gives $y = 2t/(1 + t^2)$.

EXERCISES

The parameter t in the pair $\left(\frac{1-t^2}{1+t^2}, \frac{2t}{1+t^2}\right)$ runs through all rational numbers if $t = q/p$ and p, q run through all pairs of integers.

1.3.1 Deduce that if (a, b, c) is any Pythagorean triple then

$$\frac{a}{c} = \frac{p^2 - q^2}{p^2 + q^2}, \quad \frac{b}{c} = \frac{2pq}{p^2 + q^2}$$

for some integers p and q.

1.3.2 Use Exercise 1.3.1 to prove Euclid's formula for Pythagorean triples.

The triples (a, b, c) in Plimpton 322 seem to have been computed to provide right-angled triangles covering a range of shapes—their angles actually follow an increasing sequence in roughly equal steps. This raises the question, can the shape of any right-angled triangle be approximated by a Pythagorean triple?

1.3.3 Show that any right-angled triangle with hypotenuse 1 may be approximated arbitrarily closely by one with rational sides.

Some important information may be gleaned from Diophantus's method if we compare the angle at O in Figure 1.4 with the angle at Q in Figure 1.5. The two angles are shown in Figure 1.6, and hopefully you know from high school geometry the relation between them.

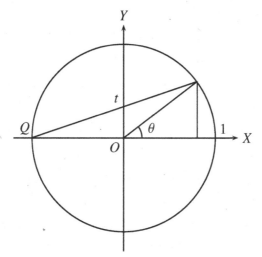

Figure 1.6: Angles in a circle

1.3.4 Use Figure 1.6 to show that $t = \tan \frac{\theta}{2}$ and

$$\cos \theta = \frac{1 - t^2}{1 + t^2}, \quad \sin \theta = \frac{2t}{1 + t^2}.$$

1.4   Right-Angled Triangles

It is high time we looked at the Pythagorean theorem from the traditional point of view, as a theorem about right-angled triangles; however, we shall be rather brief about its proof. It is not known how the theorem was first proved, but probably it was by simple manipulations of area, perhaps suggested by rearrangement of floor tiles. Just how easy it can be to prove the Pythagorean theorem is shown by Figure 1.7, given by Heath (1925) in his edition of Euclid's *Elements*, Vol. 1, p. 354. Each large square contains four copies of the given right-angled triangle. Subtracting these four triangles from the large square leaves, on the one hand (Figure 1.7, *left*), the sum of the squares on the two sides of the triangle. On the other hand (*right*), it also leaves the square on the hypotenuse. This proof, like the hundreds of others that have been given for the Pythagorean theorem, rests on certain geometric assumptions. It is in fact possible to transcend geometric assumptions by using numbers as the foundation for geometry, and the Pythagorean theorem then becomes true almost by definition, as an immediate consequence of the definition of distance (see Section 1.6).

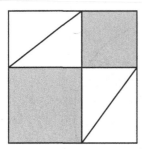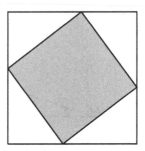

Figure 1.7: Proof of the Pythagorean theorem

To the Greeks, however, it did not seem possible to build geometry on the basis of numbers, due to a conflict between their notions of number and length. In the next section we shall see how this conflict arose.

EXERCISES

A way to see the Pythagorean theorem in a tiled floor was suggested by Magnus (1974), p. 159, and it is shown in Figure 1.8. (The dotted squares are not tiles; they are a hint.)

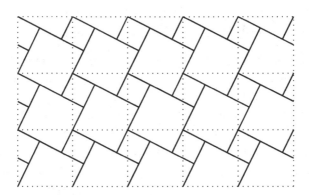

Figure 1.8: Pythagorean theorem in a tiled floor

1.4.1 What has this figure to do with the Pythagorean theorem?

Euclid's first proof of the Pythagorean theorem, in Book I of the *Elements*, is also based on area. It depends only on the fact that triangles with the same base and height have equal area, though it involves a rather complicated figure. In Book VI, Proposition 31, he gives another proof, based on similar triangles (Figure 1.9).

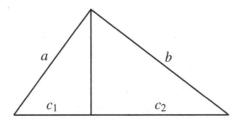

Figure 1.9: Another proof of the Pythagorean theorem

1.4.2 Show that the three triangles in Figure 1.9 are similar, and hence prove the Pythagorean theorem by equating ratios of corresponding sides.

1.5   Irrational Numbers

We have mentioned that the Babylonians, although probably aware of the geometric meaning of the Pythagorean theorem, devoted most of their attention to the whole-number triples it had brought to light, the Pythagorean triples. Pythagoras and his followers were even more devoted to whole numbers. It was they who discovered the role of numbers in musical harmony: dividing a vibrating string in two raises its pitch by an octave, dividing in three raises the pitch another fifth, and so on. This great discovery, the first clue that the physical world might have an underlying mathematical structure, inspired them to seek numerical patterns, which to them meant *whole-number* patterns, everywhere. Imagine their consternation when they found that the Pythagorean theorem led to quantities that were not numerically computable. They found lengths that were *incommensurable*, that is, not measurable as integer multiples of the same unit. The ratio between such lengths is therefore not a ratio of whole numbers, hence in the Greek view not a ratio at all, or *irrational*.

The incommensurable lengths discovered by the Pythagoreans were the side and diagonal of the unit square. It follows immediately from the Pythagorean theorem that

$$(\text{diagonal})^2 = 1 + 1 = 2.$$

Hence if the diagonal and side are in the ratio m/n (where m and n can be assumed to have no common divisor), we have

$$m^2/n^2 = 2,$$

whence

$$m^2 = 2n^2.$$

The Pythagoreans were interested in odd and even numbers, so they probably observed that the latter equation, which says that m^2 is even, also implies that m is even, say $m = 2p$. But if

$$m = 2p,$$

then

$$2n^2 = m^2 = 4p^2;$$

hence

$$n^2 = 2p^2,$$

which similarly implies that n is even, contrary to the hypothesis that m and n have no common divisor. (This proof is in Aristotle's *Prior Analytics*. An alternative, more geometric, proof is mentioned in Section 3.4.)

This discovery had profound consequences. Legend has it that the first Pythagorean to make the result public was drowned at sea (see Heath (1921), Vol. 1, pp. 65, 154). It led to a split between the theories of number and space that was not healed until the 19th century (if then, some believe). The Pythagoreans could not accept $\sqrt{2}$ as a number, but no one could deny that it was the diagonal of the unit square. Consequently, geometrical quantities had to be treated separately from numbers or, rather, without mentioning any numbers except rationals. Greek geometers thus developed ingenious techniques for precise handling of arbitrary lengths in terms of rationals, known as the *theory of proportions* and the *method of exhaustion*.

When Dedekind reconsidered these techniques in the 19th century, he realized that they provided an arithmetical interpretation of irrational quantities after all (Chapter 4). It was then possible, as Hilbert (1899) showed, to reconcile arithmetic with geometry. The key role of the Pythagorean theorem in this reconciliation is described in the next section.

The crucial step in the proof that $\sqrt{2}$ is irrational is showing that m^2 even implies m is even or, equivalently, that m odd implies m^2 odd. It is worth taking a closer look at why this is true.

1.5.1 Writing an arbitrary odd number m in the form $2q + 1$, for some integer q, show that m^2 also has the form $2r + 1$, which shows that m^2 is also odd.

You probably did some algebra like this in Exercise 1.2.3, but if not, here is your chance:

1.5.2 Show that the square of $2q + 1$ is in fact of the form $4s + 1$, and hence explain why every integer square leaves remainder 0 or 1 on division by 4.

1.6   The Definition of Distance

The numerical interpretation of irrationals gave each length a numerical measure and hence made it possible to give coordinates x, y to each point P on the plane. The simplest way is to take a pair of perpendicular lines (*axes*) OX, OY and let x, y be the lengths of the perpendiculars from P to OX and OY respectively (Figure 1.10). Geometric properties of P are then reflected by arithmetical relations between x and y. This opens up the possibility of *analytic geometry*, whose development is discussed in Chapter 7. Here we want only to see how coordinates give a precise meaning to the basic geometric notion of *distance*.

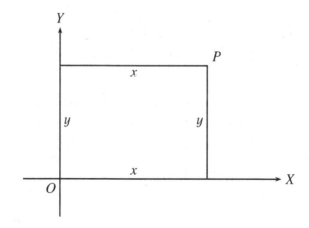

Figure 1.10: Perpendicular axes

We have already said that the perpendicular distances from P to the axes are the numbers x, y. The distance between points on the same perpendicular to an axis should therefore be defined as the difference between the appropriate coordinates. In Figure 1.11 this is $x_2 - x_1$ for RQ and $y_2 - y_1$ for PQ. But then the Pythagorean theorem tells us that the distance PR is

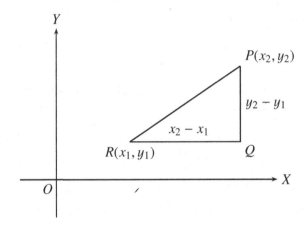

Figure 1.11: Defining distance

given by

$$PR^2 = RQ^2 + PQ^2 = (x_2 - x_1)^2 + (y_2 - y_1)^2.$$

That is,

$$PR = \sqrt{(x_2 - x_1)^2 + (y_2 - y_1)^2}. \tag{1}$$

Since this construction applies to arbitrary points P, R in the plane, we now have a general formula for the distance between two points.

We derived this formula as a consequence of geometric assumptions, in particular the Pythagorean theorem. Although this makes geometry amenable to arithmetical calculation—a very useful situation, to be sure—it does not say that geometry *is* arithmetic. In the early days of analytic geometry, the latter was a very heretical view (see Section 7.6). Eventually, however, Hilbert (1899) realized it could be made a fact by taking (1) as a *definition* of distance. Of course, all other geometric concepts have to be defined in terms of numbers, too, but this boils down to defining a point, which is simply an *ordered pair* (x, y) of numbers. Equation (1) then gives the distance between the points (x_1, y_1) and (x_2, y_2).

When geometry is reconstructed in this way, all geometric facts become facts about numbers (though they do not necessarily become easier to see). In particular, the Pythagorean theorem becomes true by definition since it has been built into the definition of distance. This is not to say that the Pythagorean theorem ultimately is trivial. Rather, it shows that the Pythagorean theorem is precisely what is needed to interpret arithmetical facts as geometry.

I mention these more recent ideas only to update the Pythagorean theorem and to give a precise statement of its power to transform arithmetic into geometry. In ancient Greek times, geometry was based much more on seeing than on calculation. We shall see in the next chapter how the Greeks managed to build geometry on the basis of visually evident facts.

EXERCISES

Most mathematicians today are more familiar with coordinates than traditional geometry, yet certain theorems of analytic geometry are seldom proved, because they seem visually obvious. A good example is what Hilbert (1899) calls *additivity of segments*: if A, B, C are points in that order on a line, then $AB + BC = AC$.

1.6.1   By suitably naming the coordinates for A, B, and C, show that the equation $AB + BC = AC$ is equivalent to

$$\sqrt{x_1^2 + y_1^2} + \sqrt{x_2^2 + y_2^2} = \sqrt{(x_1 + x_2)^2 + (y_1 + y_2)^2}, \qquad (*)$$

where $x_1 y_2 = y_1 x_2$. *Hint*: It is convenient to let B be the origin.

1.6.2   Prove (*) by proving an equivalent rational equation obtained by squaring twice and using $x_1 y_2 = y_1 x_2$.

It should be stressed that Hilbert (1899) is concerned not only with defining geometric concepts in terms of coordinates, but also with the reverse process: setting up geometric assumptions from which coordinates may be rigorously derived. There is more about this in Sections 2.1 and 20.7.

1.7   Biographical Notes: Pythagoras

Very little is known for certain about Pythagoras, although he figures in many legends. No documents have survived from the period in which he lived, so we have to rely on stories that were passed down for several centuries before being recorded. It appears that he was born on Samos, a Greek island near the coast of what is now Turkey, around 580 BCE. He traveled to the nearby mainland town of Miletus, where he learned mathematics from Thales (624–547 BCE), traditionally regarded as the founder of Greek mathematics. Pythagoras also traveled to Egypt and Babylon, where he presumably picked up additional mathematical ideas. Around 540 BCE he settled in Croton, a Greek colony in what is now southern Italy.

There he founded a school whose members later became known as the Pythagoreans. The school's motto was "All is number," and the Pythagoreans tried to bring the realms of science, religion, and philosophy all under the rule of number. The very word *mathematics* ("that which is learned") is said to be a Pythagorean invention. The school imposed a strict code of conduct on its members, which included secrecy, vegetarianism, and a curious taboo on the eating of beans. The code of secrecy meant that mathematical results were considered to be the property of the school, and their individual discoverers were not identified to outsiders. Because of this, we do not know who discovered the Pythagorean theorem, the irrationality of $\sqrt{2}$, or other arithmetical results that will be mentioned in Chapter 3.

As mentioned in Section 1.5, the most notable scientific success of the Pythagorean school was the explanation of musical harmony in terms of whole-number ratios. This success inspired the search for a numerical law governing the motions of planets, a "harmony of the spheres." Such a law probably cannot be expressed in terms that the Pythagoreans would have accepted; nevertheless, it seems reasonable to view the expansion of the number concept to meet the needs of geometry (and hence mechanics) as a natural extension of the Pythagorean program. In this sense, Newton's law of gravitation (Section 13.3) expresses the harmony that the Pythagoreans were looking for. Even in the strictest sense, Pythagoreanism is very much alive today. With the digital computer, digital audio, and digital video coding everything, at least approximately, into sequences of whole numbers, we are closer than ever to a world in which "all is number."

Whether the complete rule of number is wise remains to be seen. It is said that when the Pythagoreans tried to extend their influence into politics they met with popular resistance. Pythagoras fled, but he was murdered in nearby Metapontum in 497 BCE.

2

Greek Geometry

PREVIEW

Geometry was the first branch of mathematics to become highly developed. The concepts of "theorem" and "proof" originated in geometry, and most mathematicians until recent times were introduced to their subject through the geometry in Euclid's *Elements*.

In the *Elements* one finds the first attempt to derive theorems from supposedly self-evident statements called *axioms*. Euclid's axioms are incomplete and one of them, the so-called *parallel* axiom, is not as obvious as the others. Nevertheless, it took over 2000 years to produce a clearer foundation for geometry.

The climax of the *Elements* is the investigation of the regular polyhedra, five symmetric figures in three-dimensional space. The five regular polyhedra make several appearances in mathematical history, most importantly in the theory of symmetry—*group theory*—discussed in Chapters 19 and 23.

The *Elements* contains not only proofs but also many *constructions*, by ruler and compass. However, three constructions are conspicuous by their absence: duplication of the cube, trisection of the angle, and squaring the circle. These problems were not properly understood until the 19th century, when they were resolved (in the negative) by algebra and analysis.

The only curves in the *Elements* are circles, but the Greeks studied many other curves, such as the conic sections. Again, many problems that the Greeks could not solve were later clarified by algebra. In particular, curves can be classified by *degree*, and the conic sections are the curves of degree 2, as we will see in Chapter 7.

J. Stillwell, *Mathematics and Its History*, Undergraduate Texts in Mathematics,   17
DOI 10.1007/978-1-4419-6053-5_2, © Springer Science+Business Media, LLC 2010

2.1 The Deductive Method

> He was 40 years old before he looked on Geometry; which
> happened accidentally. Being in a Gentleman's Library, Eu-
> clid's Elements lay open, and 'twas the 47 El. libri I. He read
> the Proposition. *By* G——sayd he (he would now and then
> sweare an emphaticall Oath by way of emphasis) *this is im-
> possible*! So he reads the Demonstration of it, which referred
> him back to such a Proposition; which proposition he read.
> That referred him back to another, which he also read ... that
> at last he was demonstratively convinced of that trueth. This
> made him in love with Geometry.

This quotation about the philosopher Thomas Hobbes (1588–1679),
from Aubrey's *Brief Lives*, beautifully captures the force of Greece's most
important contribution to mathematics, the deductive method. (The proposition mentioned, incidentally, is the Pythagorean theorem.)

We have already seen that significant results were *known* before the period of classical Greece, but the Greeks were the first to construct mathematics by deduction from previously established results, resting ultimately
on the most evident possible statements, called *axioms*. Thales (624–547
BCE) is thought to be the originator of this method (see Heath (1921),
p. 128), and by 300 BCE it had become so sophisticated that Euclid's *Elements* set the standard for mathematical rigor until the 19th century. The
Elements was in fact too subtle for most mathematicians, let alone their students, so that in time Euclid's geometry was boiled down to the simplest
and driest propositions about straight lines, triangles, and circles. This
part of the *Elements* is based on the following axioms (in the translation of
Heath (1925), p. 154), which Euclid called *postulates* and *common notions*.

Postulates

Let the following be postulated:

1. To draw a straight line from any point to any point.

2. To produce a finite straight line continuously in a straight line.

3. To describe a circle with any center and distance.

4. That all right angles are equal to one another.

5. That, if a straight line falling on two straight lines make the interior angles on the same side less than two right angles, the two straight lines, if produced indefinitely, meet on that side on which are the angles less than the two right angles.

Common Notions

1. Things which are equal to the same thing are also equal to one another.

2. If equals be added to equals, the wholes are equal.

3. If equals be subtracted from equals, the remainders are equal.

4. Things which coincide with one another are equal to one another.

5. The whole is greater than the part.

It appears that Euclid's intention was to deduce geometric propositions from visually evident statements (the postulates) using evident principles of logic (the common notions). Actually, he often made unconscious use of visually plausible assumptions that are not among his postulates. His very first proposition used the unstated assumption that two circles meet if the center of each is on the circumference of the other (Heath (1925), p. 242). Nevertheless, such flaws were not noticed until the 19th century, and they were rectified by Hilbert (1899). By themselves, they probably would not have been enough to end the *Elements'* run of 22 centuries as a leading textbook. The *Elements* was overthrown by more serious mathematical upheavals in the 19th century. The so-called non-Euclidean geometries, using alternatives to Euclid's fifth postulate (the *parallel axiom*), developed to the point where the old axioms could no longer be considered self-evident (see Chapter 18). At the same time, the concept of number matured to the point where irrational numbers became acceptable, and indeed preferable to intuitive geometric concepts, in view of the doubts about what the self-evident truths of geometry really were.

The outcome was a more adaptable language for geometry in which "points," "lines," and so on, could be defined, usually in terms of numbers, so as to suit the type of geometry under investigation. Such a development was long overdue, because even in Euclid's time the Greeks were investigating curves more complicated than circles, which did not fit conveniently in Euclid's system. Descartes (1637) introduced the coordinate method, which gives a single framework for handling both Euclid's geometry and higher curves (see Chapter 7), but it was not at first realized that coordinates allowed geometry to be entirely rebuilt on numerical foundations.

The comparatively trivial step (for us) of passing to axioms about numbers from axioms about points had to wait until the 19th century, when geometric axioms about points lost authority and number-theoretic axioms gained it. We shall say more about these developments later (and of problems with the authority of axioms in general, which arose in the 20th century). For the remainder of this chapter we shall look at some important nonelementary topics in Greek geometry, using the coordinate framework where convenient.

EXERCISES

Euclid's Common Notions 1 and 4 define what we now call an *equivalence relation*, which is not necessarily the equality relation. In fact, the kind of relation Euclid had in mind was equality in *some* geometric quantity such as length or angle (but not necessarily equality in all respects—the latter is what he meant by "coinciding"). An equivalence relation ≅ is normally defined by three properties. For any a, b and c:

$$a \cong a, \qquad \text{(reflexive)}$$
$$a \cong b \implies b \cong a, \qquad \text{(symmetric)}$$
$$a \cong b \text{ and } b \cong c \implies a \cong c. \qquad \text{(transitive)}$$

2.1.1 Explain how Common Notions 1 and 4 may be interpreted as the transitive and reflexive properties. Note that the natural way to write Common Notion 1 symbolically is slightly different from the statement of transitivity above.

2.1.2 Show that the symmetric property follows from Euclid's Common Notions 1 and 4.

Hilbert (1899) took advantage of Euclid's Common Notions 1 and 4 in his rectification of Euclid's axiom system. He *defined* equality of length by postulating a transitive and reflexive relation on line segments, and stated transitivity in the style of Euclid, so that the symmetric property was a consequence.

2.2   The Regular Polyhedra

Greek geometry is virtually complete as far as the elementary properties of plane figures are concerned. It is fair to say that only a handful of interesting elementary propositions about triangles and circles have been discovered since Euclid's time. Solid geometry is much more challenging, even today, so it is understandable that it was left in a less complete state by the Greeks. Nevertheless, they made some very impressive discoveries and managed to complete one of the most beautiful chapters in solid geometry, the enumeration of the regular polyhedra. The five possible regular polyhedra are shown in Figure 2.1.

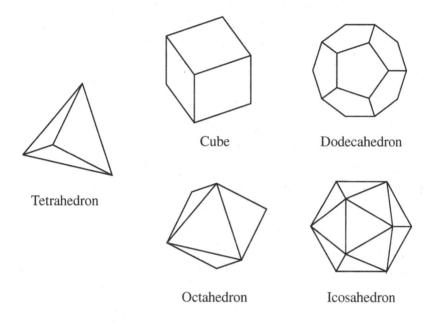

Figure 2.1: The regular polyhedra

Each polyhedron is convex and is bounded by a number of congruent polygonal faces, the same number of faces meet at each vertex, and in each face all the sides and angles are equal, hence the term *regular polyhedron*. A regular polyhedron is a spatial figure analogous to a regular polygon in the plane. But whereas there are regular polygons with any number $n \geq 3$ of sides, there are only five regular polyhedra.

This fact is easily proved and may go back to the Pythagoreans (see, for example Heath (1921), p. 159). One considers the possible polygons that can occur as faces, their angles, and the numbers of them that can occur at a vertex. For a 3-gon (triangle) the angle is $\pi/3$, so three, four, or five can occur at a vertex, but six cannot, as this would give a total angle 2π and the vertex would be flat. For a 4-gon the angle is $\pi/2$, so three can occur at a vertex, but not four. For a 5-gon the angle is $3\pi/5$, so three can occur at a vertex, but not four. For a 6-gon the angle is $2\pi/3$, so not even three can occur at a vertex. But at least three faces must meet at each vertex of a polyhedron, so 6-gons (and, similarly, 7-gons, 8-gons, ...) cannot occur as faces of a regular polyhedron. This leaves only the five possibilities just listed, which correspond to the five known regular polyhedra.

But do we really know that these five exist? There is no difficulty with the tetrahedron, cube, or octahedron, but it is not clear that, say, 20 equilateral triangles will fit together to form a closed surface. Euclid found this problem difficult enough to be placed near the end of the *Elements*, and few of his readers ever mastered his solution. A beautiful direct construction was given by Luca Pacioli, a friend of Leonardo da Vinci's, in his book *De divina proportione* (1509). Pacioli's construction uses three copies of the *golden rectangle*, with sides 1 and $(1 + \sqrt{5})/2$, interlocking as in Figure 2.2. The 12 vertices define 20 triangles such as ABC, and it suffices to show that these are equilateral, that is, $AB = 1$. This is a straightforward exercise in the Pythagorean theorem (Exercise 2.2.2).

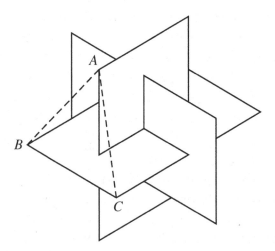

Figure 2.2: Pacioli's construction of the icosahedron

The regular polyhedra will make another important appearance in connection with yet another 19th-century development, the theory of finite groups and Galois theory. Before the regular polyhedra made this triumphant comeback, they also took part in a famous fiasco: the Kepler (1596) theory of planetary distances. Kepler's theory is summarized by his famous diagram (Figure 2.3) of the five polyhedra, nested in such a way as to produce six spheres of radii proportional to the distances of the six planets then known. Unfortunately, although mathematics could not permit any more regular polyhedra, nature could permit more planets, and Kepler's theory was ruined when Uranus was discovered in 1781.

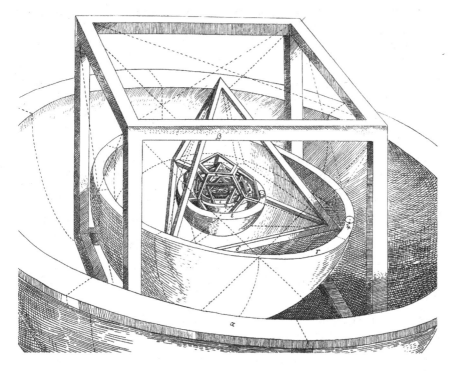

Figure 2.3: Kepler's diagram of the polyhedra

EXERCISES

The ratios between successive radii in Kepler's construction depend on what may be called the *inradius* and *circumradius* of each polyhedron—the radii of the spheres that touch it on the inside and the outside. It happens that the ratio

$$\frac{\text{circumradius}}{\text{inradius}}$$

is the same for the cube and the octahedron, and it is also the same for the dodecahedron and the icosahedron. This implies that the cube and octahedron can be exchanged in Kepler's construction, as can the dodecahedron and the icosahedron. Thus there are at least four different arrangements of the regular polyhedra that yield the same sequence of radii.

It is easy to see why the cube and the octahedron are interchangeable.

2.2.1 Show that $\dfrac{\text{circumradius}}{\text{inradius}} = \sqrt{3}$ for both the cube and the octahedron.

To compute circumradius/inradius for the icosahedron and the dodecahedron, we pursue Pacioli's construction a little further, with the help of vector addition.

2.2.2 First check Pacioli's construction: use the Pythagorean theorem to show that $AB = BC = CA$ in Figure 2.2. (It may help to use the additional fact that $\tau = (1 + \sqrt{5})/2$ satisfies $\tau^2 = \tau + 1$. This is also useful in the exercises below.)

Now, to simplify coordinates, we take golden rectangles that are twice the normal size—length 2τ and width 2—and place them in the three coordinate planes in the relative positions shown in Figure 2.2, so $O = (0, 0, 0)$ is at the center of each rectangle.

2.2.3 Show that the coordinates of the vertices of the icosahedron are $(\pm 1, 0, \pm\tau)$, $(\pm\tau, \pm 1, 0)$, and $(0, \pm\tau, \pm 1)$, for all possible combinations of $+$ and $-$ signs.

2.2.4 In particular, show that suitably chosen axes give $A = (1, 0, \tau)$, $B = (\tau, -1, 0)$, and $C = (\tau, 1, 0)$ in Figure 2.2. Deduce that

$$\text{circumradius} = \sqrt{\tau + 2} \quad \text{for this icosahedron.}$$

To find the inradius, we find the center of the triangle ABC, then compute its distance from O.

2.2.5 Show that the center of the triangle ABC is $\frac{1}{3}(2\tau + 1, 0, \tau)$, and hence that

$$\text{inradius} = \frac{1}{3}\sqrt{9\tau + 6} \quad \text{for this icosahedron.}$$

It follows that

$$\frac{\text{circumradius}}{\text{inradius}} = \frac{3\sqrt{\tau + 2}}{\sqrt{9\tau + 6}} \quad \text{for any icosahedron,}$$

but it will be helpful to have this number in a simpler form.

2.2.6 Show that $\frac{3\sqrt{\tau+2}}{\sqrt{9\tau+6}} = \sqrt{3(7 - 4\tau)} = \sqrt{\frac{15}{4\tau+3}}$.

Now to compute the ratio circumradius/inradius for the dodecahedron, we use the *dual dodecahedron*, whose vertices are the face centers, such as $\frac{1}{3}(A + B + C)$, of the icosahedron above. This immediately gives

$$\text{circumradius of dual dodecahedron} = \text{inradius of icosahedron} = \frac{1}{3}\sqrt{9\tau + 6}.$$

Thus it remains to find the inradius of the dual dodecahedron, which is the distance from O to one of *its* face centers. A face of the dual dodecahedron is a pentagon, with vertices, for example,

$$\frac{1}{3}(A+B+C), \quad \frac{1}{3}(A+C+D), \quad \frac{1}{3}(A+D+E), \quad \frac{1}{3}(A+E+F), \quad \frac{1}{3}(A+F+B),$$

where B, C, D, E, F are the five vertices of the icosahedron equidistant from A.

2.2.7 Using $A = (1, 0, \tau)$, $B = (\tau, -1, 0)$, $C = (\tau, 1, 0)$, $D = (0, \tau, 1)$, $E = (-1, 0, \tau)$, and $F = (0, -\tau, 1)$, show that the face center of the pentagon with the above vertices is

$$\frac{1}{15}(5A+2B+2C+2D+2E+2F) = \frac{1}{15}(4\tau+3, 0, 7\tau+4) = \frac{4\tau+3}{15}(1, 0, \tau),$$

and hence that

$$\text{inradius of the dual dodecahedron} = \frac{4\tau+3}{15}\sqrt{\tau+2}.$$

2.2.8 Deduce from Exercises 2.2.7 and 2.2.6 that

$$\frac{\text{circumradius}}{\text{inradius}} \text{ for dodecahedron} = \sqrt{\frac{15}{4\tau+3}} = \frac{\text{circumradius}}{\text{inradius}} \text{ for icosahedron.}$$

Another remarkable result follows from this, using the fact that the volume of a pyramid $= 1/3$ base area $\times$ height. The result is attributed to Apollonius.

2.2.9 By dividing the polyhedra into pyramids with bases equal to the faces, and height equal to the inradius, establish the following relationship between the dodecahedron D and the icosahedron I of the same circumradius:

$$\frac{\text{surface area } D}{\text{surface area } I} = \frac{\text{volume } D}{\text{volume } I}.$$

2.3 Ruler and Compass Constructions

Greek geometers prided themselves on their logical purity; nevertheless, they were guided by intuition about physical space. One aspect of Greek geometry that was peculiarly influenced by physical considerations was the theory of constructions. Much of the elementary geometry of straight lines and circles can be viewed as the theory of constructions by ruler and compass. The very subject matter, lines and circles, reflects the instruments used to draw them. And many of the elementary problems of geometry— for example, to bisect a line segment or angle, construct a perpendicular, or draw a circle through three given points—can be solved by ruler and compass constructions.

When coordinates are introduced, it is not hard to show that the points constructible from points $P_1, \ldots, P_n$ have coordinates in the set of numbers generated from the coordinates of $P_1, \ldots, P_n$ by the operations $+, -, \times, \div$, and $\sqrt{}$ (see Moise (1963) or the exercises to Section 6.3). Square roots arise, of course, because of the Pythagorean theorem: if points (a, b) and (c, d) have been constructed, then so has the distance $\sqrt{(c - a)^2 + (d - b)^2}$ between them (Section 1.6 and Figure 2.4). Conversely, it is possible to construct $\sqrt{l}$ for any given length l (Exercise 2.3.2).

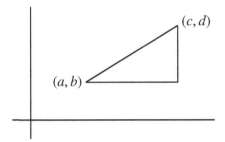

Figure 2.4: Construction of a distance

Looked at from this point of view, ruler and compass constructions look very special and unlikely to yield numbers such as $\sqrt[3]{2}$, for example. However, the Greeks tried very hard to solve just this problem, which was known as *duplication of the cube* (so-called because to double the volume of a cube one must multiply the side by $\sqrt[3]{2}$). Other notorious problems were *trisection of the angle* and *squaring the circle*. The latter problem was to construct a square equal in area to a given circle or to construct the number π, which amounts to the same thing. They never seem to have given up these goals, though the possibility of a negative solution was admitted and solutions by less elementary means were tolerated. We shall see some of these in the next sections.

The impossibility of solving these problems by ruler and compass constructions was not proved until the 19th century. For the duplication of the cube and trisection of the angle, impossibility was shown by Wantzel (1837). Wantzel seldom receives credit for settling these problems, which had baffled the best mathematicians for 2000 years, perhaps because his methods were superseded by the more powerful theory of Galois.

The impossibility of squaring the circle was proved by Lindemann (1882), in a very strong way. Not only is π undefinable by rational operations and square roots; it is also *transcendental*, that is, not the root of any polynomial equation with rational coefficients. Like Wantzel's work, this was a rare example of a major result being proved by a minor mathematician. In Lindemann's case the explanation is perhaps that a major step had already been taken when Hermite (1873) proved the transcendence of e. Accessible proofs of both these results can be found in Klein (1924). Lindemann's subsequent career was mathematically undistinguished, even embarrassing. In response to skeptics who thought his success with π had been a fluke, he took aim at the most famous unsolved problem in mathematics, "Fermat's last theorem" (see Chapter 11 for the origin of this problem). His efforts fizzled out in a series of inconclusive papers, each one correcting an error in the one before. Fritsch (1984) has written an interesting biographical article on Lindemann.

One ruler and compass problem is still open: which regular n-gons are constructible? Gauss discovered in 1796 that the 17-gon is constructible and then showed that a regular n-gon is constructible if and only if $n = 2^m p_1 p_2 \cdots p_k$, where the p_i are distinct primes of the form $2^{2^h} + 1$. (This problem is also known as *circle division*, because it is equivalent to dividing the circumference of a circle, or the angle 2π, into n equal parts.) The proof of necessity was actually completed by Wantzel (1837). However, it is still not explicitly known what these primes are, or even whether there are infinitely many of them. The only ones known are for $h = 0, 1, 2, 3, 4$.

EXERCISES

Many of the constructions made by the Greeks can be simplified by translating them into algebra, where it turns out that constructible lengths are those that can be built from known lengths by the operations of $+$, $-$, $\times$, $\div$, and $\sqrt{\ }$. It is therefore enough to know constructions for these five basic operations. Addition and subtraction are obvious, and the other operations are covered in the following exercises, together with an example in which algebra is a distinct advantage.

2.3.1 Show, using similar triangles, that if lengths l_1 and l_2 are constructible, then so are $l_1 l_2$ and l_1 / l_2.

2.3.2 Use similar triangles to explain why $\sqrt{l}$ is the length shown in Figure 2.5, and hence show that $\sqrt{l}$ is constructible from l.

One of the finest ruler and compass constructions from ancient times is that of the regular pentagon, which includes, yet again, the golden ratio $\tau = (1 + \sqrt{5})/2$.

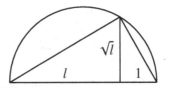

Figure 2.5: Square root construction

Knowing (from the questions above) that this number is constructible, it becomes easy for us to construct the pentagon itself.

2.3.3 By finding some parallels and similar triangles in Figure 2.6, show that the diagonal x of the regular pentagon of side 1 satisfies $x/1 = 1/(x - 1)$.

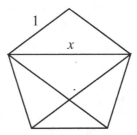

Figure 2.6: The regular pentagon

2.3.4 Deduce from Exercise 2.3.3 that the diagonal of the pentagon is $(1 + \sqrt{5})/2$ and hence that the regular pentagon is constructible.

2.4   Conic Sections

Conic sections are the curves obtained by intersecting a circular cone by a plane: hyperbolas, ellipses (including circles), and parabolas (Figure 2.7, left to right). Today we know the conic sections better in terms of their equations in cartesian coordinates:

$$\frac{x^2}{a^2} - \frac{y^2}{b^2} = 1, \qquad \text{(hyperbola)}$$

$$\frac{x^2}{a^2} + \frac{y^2}{b^2} = 1, \qquad \text{(ellipse)}$$

$$y = ax^2. \qquad \text{(parabola)}$$

More generally, any second-degree equation represents a conic section or a pair of straight lines, a result that was proved by Descartes (1637).

Figure 2.7: The conic sections

The invention of conic sections is attributed to Menaechmus (fourth century BCE), a contemporary of Alexander the Great. Alexander is said to have asked Menaechmus for a crash course in geometry, but Menaechmus refused, saying, "There is no royal road to geometry." Menaechmus used conic sections to give a very simple solution to the problem of duplicating the cube. In analytic notation, this can be described as finding the intersection of the parabola $y = \frac{1}{2}x^2$ with the hyperbola $xy = 1$. This yields

$$x\frac{1}{2}x^2 = 1 \qquad \text{or} \qquad x^3 = 2.$$

Although the Greeks accepted this as a "construction" for duplicating the cube, they apparently never discussed instruments for actually drawing conic sections. This is very puzzling since a natural generalization of the compass immediately suggests itself (Figure 2.8). The arm A is set at a fixed position relative to a plane P, while the other arm rotates about it at a fixed angle θ, generating a cone with A as its axis of symmetry. The pencil, which is free to slide in a sleeve on this second arm, traces the section of the cone lying in the plane P. According to Coolidge (1945), p. 149, this

instrument for drawing conic sections was first described as late as 1000 CE
by the Arab mathematician al-Kuji. Yet nearly all the *theoretical* facts one
could wish to know about conic sections had already been worked out by
Apollonius (around 250–200 BCE)!

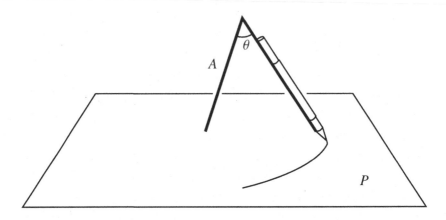

Figure 2.8: Generalized compass

The theory and practice of conic sections finally met when Kepler
(1609) discovered the orbits of the planets to be ellipses, and Newton
(1687) explained this fact by his law of gravitation. This wonderful vin-
dication of the theory of conic sections has often been described in terms
of basic research receiving its long overdue reward, but perhaps one can
also see it as a rebuke to Greek disdain for applications. Kepler would not
have been sure which it was. To the end of his days he was proudest of
his theory explaining the distances of the planets in terms of the five regular polyhedra (Section 2.2). The fascinating and paradoxical character of
Kepler has been warmly described in two excellent books, Koestler (1959)
and Banville (1981).

EXERCISES

A key feature of the ellipse for both geometry and astronomy is a point called
the *focus*. The term is the Latin word for fireplace, and it was introduced by
Kepler. The ellipse actually has two foci, and they have the geometric property
that the sum of the distances from the foci F_1, F_2 to any point P on the ellipse is
constant.

2.4.1 This property gives a way to draw an ellipse using two pins and piece of
string. Explain how.

2.4.2 By introducing suitable coordinate axes, show that a curve with the above "constant sum" property indeed has an equation of the form

$$\frac{x^2}{a^2} + \frac{y^2}{b^2} = 1.$$

(It is a good idea to start with the two square root terms, representing the distances F_1P and F_2P, on opposite sides of the equation.) Show also that *any* equation of this form is obtainable by suitable choice of F_1, F_2, and $F_1P + F_2P$.

Another interesting property of the lines from the foci to a point P on the ellipse is that they make equal angles with the tangent at P. It follows that a light ray from F_1 to P is reflected through F_2. A simple proof of this can be based on the *shortest-path property of reflection*, shown in Figure 2.9 and discovered by the Greek scientist Heron around 100 CE.

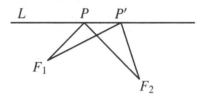

Figure 2.9: The shortest-path property

Shortest-path property. The path F_1PF_2 of reflection in the line L from F_1 to F_2 is shorter than any other path $F_1P'F_2$ from F_1 to L to F_2.

2.4.3 Prove the shortest-path property, by considering the two paths $F_1P\overline{F_2}$ and $F_1P'\overline{F_2}$, where $\overline{F_2}$ is the reflection of the point F_2 in the line L.

Thus to prove that the lines F_1P and F_2P make equal angles with the tangent, it is enough to show that F_1PF_2 is shorter than $F_1P'F_2$ for any other point P' on the tangent at P.

2.4.4 Prove this, using the fact that F_1PF_2 has the same length for all points P on the ellipse.

Kepler's great discovery was that the focus is also significant in astronomy. It is the point occupied by the sun as the planet moves along its ellipse.

2.5   Higher-Degree Curves

The Greeks lacked a systematic theory of higher-degree curves, because they lacked a systematic algebra. They could find what amounted to cartesian equations of individual curves—"symptoms," as they called them; see van der Waerden (1954), p. 241—but they did not consider equations in general or notice any of their properties relevant to the study of curves, for example, the degree. Nevertheless, they studied many interesting special curves, which Descartes and his followers cut their teeth on when algebraic geometry finally emerged in the 17th century. An excellent and well-illustrated account of these early investigations may be found in Brieskorn and Knörrer (1981), Chapter 1.

In this section we must confine ourselves to brief remarks on a few examples.

The Cissoid of Diocles (around 100 BCE)

This curve is defined using an auxiliary circle, which for convenience we take to be the unit circle, and vertical lines through x and $-x$. It is the set of all points P seen in Figure 2.10.

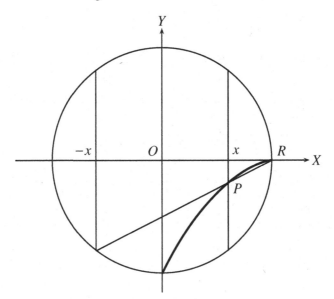

Figure 2.10: Construction of the cissoid

The portion shown results from varying x between 0 and 1. It is a cubic curve with cartesian equation

$$y^2(1 + x) = (1 - x)^3.$$

This equation shows that if (x, y) is a point on the curve, then so is $(x, -y)$. Hence one gets the complete picture of it by reflecting the portion shown in Figure 2.10 in the x-axis. The result is a sharp point at R, a *cusp*, a phenomenon that first arises with cubic curves. Diocles showed that the cissoid could be used to duplicate the cube, which is plausible (though still not obvious!) once one knows that this curve is cubic.

The Spiric Sections of Perseus (around 150 BCE)

Apart from the sphere, cylinder, and cone—whose sections are all conic sections—one of the few surfaces studied by the Greeks was the *torus*. This surface, generated by rotating a circle about an axis outside the circle, but in the same plane, was called a *spira* by the Greeks—hence the name spiric sections for the sections by planes parallel to the axis. These sections, which were first studied by Perseus, have four qualitatively distinct forms (see Figure 2.11, which is adapted from Brieskorn and Knörrer (1981), p. 20).

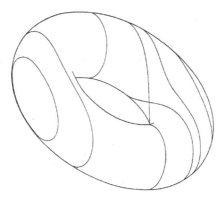

Figure 2.11: Spiric sections

These forms—convex ovals, "squeezed" ovals, the figure 8, and pairs of ovals—were rediscovered in the 17th century when analytic geometers looked at curves of degree 4, of which the spiric sections are examples. For suitable choice of torus, the figure 8 curve becomes the *lemniscate of Bernoulli* and the convex ovals become *Cassini ovals*. Cassini (1625–1712) was a distinguished astronomer but an opponent of Newton's theory of gravitation. He rejected Kepler's ellipses and instead proposed Cassini ovals as orbits for the planets.

The Epicycles of Ptolemy (140 CE)

These curves are known from a famous astronomical work, the *Almagest* of Claudius Ptolemy. Ptolemy himself attributes the idea to Apollonius. It seems almost certain that this is the Apollonius who mastered conic sections, which is ironic, because epicycles were his candidates for the planetary orbits, destined to be defeated by those very same conic sections.

An epicycle, in its simplest form, is the path traced by a point on a circle that rolls on another circle (Figure 2.12). More complicated epicycles can be defined by having a third circle roll on the second, and so on. The Greeks introduced these curves to try to reconcile the complicated movements of the planets, relative to the fixed stars, with a geometry based on the circle. In principle, this is possible! Lagrange (1772) showed that *any* motion along the celestial equator can be approximated arbitrarily closely by epicylic motion, and a more modern version of the result may be found in Sternberg (1969). But Ptolemy's mistake was to accept the apparent complexity of the motions of the planets as actual in the first place. As we now know, the motion becomes simple when one considers motion relative to the sun rather than to the earth and allows orbits to be ellipses.

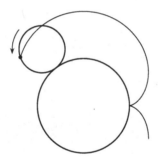

Figure 2.12: Generating an epicycle

Epicycles still have a role to play in engineering, and their mathematical properties are interesting. Some of them are closed curves and turn out to be algebraic, that is, of the form $p(x, y) = 0$ for a polynomial p. Others, such as those that result from rolling circles whose radii have an irrational ratio, lie densely in a certain region of the plane and hence cannot be algebraic; an algebraic curve $p(x, y) = 0$ can meet a straight line $y = mx + c$ in only a finite number of points, corresponding to roots of the polynomial equation $p(x, mx + c) = 0$, and the dense epicycles meet some lines infinitely often.

EXERCISES

The equation of the cissoid is derivable as follows.

2.5.1 Using X and Y for the horizontal and vertical coordinates, show that the straight line RP in Figure 2.10 has equation

$$Y = \frac{\sqrt{1 - x^2}}{1 + x}(X - 1).$$

2.5.2 Deduce the equation of the cissoid from Exercise 2.5.1.

The simplest epicyclic curve is the *cardioid* ("heart-shape"), which results from a circle rolling on a fixed circle of the same size.

2.5.3 Sketch a picture of the cardioid, confirming that it is heart-shaped (sort of).

2.5.4 Show that if both circles have radius 1, and we follow the point on the rolling circle initially at $(1, 0)$, then the cardioid it traces out has parametric equations

$$x = 2 \cos \theta - \cos 2\theta,$$
$$y = 2 \sin \theta - \sin 2\theta.$$

The cardioid is an algebraic curve. Its cartesian equation may be hard to discover, but it is easy to verify, especially if one has a computer algebra system.

2.5.5 Check that the point (x, y) on the cardioid satisfies

$$(x^2 + y^2 - 1)^2 = 4((x - 1)^2 + y^2).$$

2.6 Biographical Notes: Euclid

Even less is known about Euclid than about Pythagoras. We know only that he flourished around 300 BCE and taught in Alexandria, the Greek city in Egypt founded by Alexander the Great in 322 BCE. Two stories are told about him. The first—the same that is told about Menaechmus and Alexander—has Euclid telling King Ptolemy I, "There is no royal road to geometry." The second concerns a student who asked the perennial question, "What shall I gain from learning mathematics?" Euclid called his slave and said, "Give him a coin if he must profit from what he learns."

The most important fact of Euclid's life was undoubtedly his writing of the *Elements*, though we do not know how much of the mathematics in it was actually his own work. Certainly the elementary geometry of triangles and circles was known before Euclid's time. Some of the most sophisticated parts of the *Elements*, too, are due to earlier mathematicians. The theory of irrationals in Book V is due to Eudoxus (around 400–347 BCE), as is the "method of exhaustion" of Book XII (see Chapter 4). The theory of regular polyhedra of Book XIII is due, at least partly, to Theaetetus (around 415–369 BCE).

But whatever Euclid's "research" contribution may have been, it was dwarfed by his contribution to the organization and dissemination of mathematical knowledge. For 2000 years the *Elements* was not only the core of mathematical education but at the heart of Western culture. The most glowing tributes to the *Elements* do not, in fact, come from mathematicians but from philosophers, politicians, and others. We saw Hobbes's response to Euclid in Section 2.1. Here are some others:

> He studied and nearly mastered the six books of Euclid since he was a member of Congress. He regrets his want of education, and does what he can to supply the want.
>
> Abraham Lincoln (writing of himself), *Short Autobiography*
>
> ...he studied Euclid until he could demonstrate with ease all the propositions in the six books.
>
> Herndon's *Life of Lincoln*
>
> At the age of eleven, I began Euclid. ...This was one of the great events of my life, as dazzling as first love. I had not imagined there was anything so delicious in the world.
>
> Bertrand Russell, *Autobiography*, vol. 1

Perhaps the low cultural status of mathematics today, not to mention the mathematical ignorance of politicians and philosophers, reflects the lack of an *Elements* suitable for the modern world.

3

Greek Number Theory

PREVIEW

Number theory is the second large field of mathematics that comes to us
from the Pythagoreans via Euclid. The Pythagorean theorem led mathematicians to the study of squares and sums of squares; Euclid drew attention to the *primes* by proving that there are infinitely many of them.

Euclid's investigations were based on the so-called *Euclidean algorithm*, a method for finding the greatest common divisor of two natural
numbers. Common divisors are the key to basic results about prime numbers, in particular *unique prime factorization*, which says that each natural
number factors into primes in exactly one way.

Another discovery of the Pythagoreans, the irrationality of $\sqrt{2}$, has
repercussions in the world of natural numbers. Since $\sqrt{2} \neq m/n$ for any
natural numbers m, n, there is no solution of the equation $x^2 - 2y^2 = 0$ in
the natural numbers. But, surprisingly, there are natural number solutions
of $x^2 - 2y^2 = 1$, and in fact infinitely many of them. The same is true of
the equation $x^2 - Ny^2 = 1$ for any nonsquare natural number N.

The latter equation, called *Pell's equation*, is perhaps second in fame
only to the Pythagorean equation $x^2 + y^2 = z^2$, among equations for which
integer solutions are sought. Methods for solving the Pell equation for
general N were first discovered by Indian mathematicians, whose work we
study in Chapter 5.

Equations for which integer or rational solutions are sought are called
Diophantine, after Diophantus. The methods he used to solve quadratic
and cubic Diophantine equations are still of interest. We study his method
for cubics in this chapter, and take it up again in Chapters 11 and 16.

J. Stillwell, *Mathematics and Its History*, Undergraduate Texts in Mathematics,   37
DOI 10.1007/978-1-4419-6053-5_3, © Springer Science+Business Media, LLC 2010

3.1   The Role of Number Theory

In Chapter 1 we saw that number theory has been important in mathematics for at least as long as geometry, and from a foundational point of view it may be more important. Despite this, number theory has never submitted to a systematic treatment like that undergone by elementary geometry in Euclid's *Elements*. At all stages in its development, number theory has had glaring gaps because of the intractability of elementary problems. Most of the really old unsolved problems in mathematics, in fact, are simple questions about the natural numbers $1, 2, 3, \ldots$. The nonexistence of a general method for solving Diophantine equations (Section 1.3) and the problem of identifying the primes of the form $2^{2^h} + 1$ (Section 2.3) have been noted. Other unsolved number theory problems will be mentioned in the sections that follow.

As a consequence, the role of number theory in the history of mathematics has been quite different from that of geometry. Geometry has played a stabilizing and unifying role, to the point of retarding further development at times and creating the popular impression that mathematics is a static subject. For those able to understand it, number theory has been a spur to progress and change. Before 1800, only a handful of mathematicians contributed to advances in number theory, but they include some of the greats—Diophantus, Fermat, Euler, Lagrange, and Gauss. This book stresses those advances in number theory that sprang from its deep connections with other parts of mathematics, particularly geometry, since these were the most significant for mathematics as a whole. Nevertheless, there are topics in number theory that are too interesting to ignore, even though they seem (at present) to be outside the mainstream. We discuss a few of them in the next section.

3.2   Polygonal, Prime, and Perfect Numbers

The *polygonal numbers*, which were studied by the Pythagoreans, result from a naive transfer of geometric ideas to number theory. From Figure 3.1 it is easy to calculate an expression for the *m*th *n*-gonal number as the sum of a certain arithmetic series (Exercise 3.2.3) and to show, for example, that a square is the sum of two triangular numbers. Apart from Diophantus's work, which contains impressive results on sums of squares, Greek results on polygonal numbers were of this elementary type.

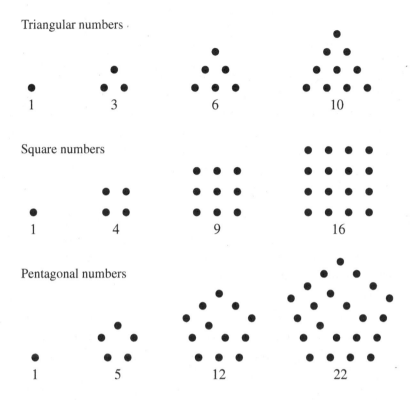

Figure 3.1: Polygonal numbers

On the whole, the Greeks seem to have been mistaken in attaching much importance to polygonal numbers. There are no major theorems about them, except perhaps the following two. The first is the theorem conjectured by Bachet de Méziriac (1621) (in his edition of Diophantus) that every positive integer is the sum of four integer squares. This was proved by Lagrange (1770). A generalization, which Fermat (1670) stated without proof, is that every positive integer is the sum of n n-agonal numbers. This was proved by Cauchy (1813a), though the proof is a bit of a letdown because all but four of the numbers can be 0 or 1. A short proof of Cauchy's theorem has been given by Nathanson (1987). The other remarkable theorem about polygonal numbers is the formula

$$\prod_{n=1}^{\infty}(1 - x^n) = 1 + \sum_{k=1}^{\infty}(-1)^k(x^{(3k^2-k)/2} + x^{(3k^2+k)/2})$$

proved by Euler (1750) and known as Euler's pentagonal number theorem, since the exponents $(3k^2 - k)/2$ are pentagonal numbers. For a proof see Hall (1967), p. 33.

The four-square theorem and the pentagonal number theorem were both absorbed around 1830 into Jacobi's theory of theta functions, a much larger theory. Theta functions are related to the *elliptic functions* that we study in Chapters 12 and 16.

The *prime numbers* were also considered within the geometric framework, as the numbers with no rectangular representation. A prime number, having no divisors apart from itself and 1, has only a "linear" representation. Of course this is merely a restatement of the definition of prime, and most theorems about prime numbers require much more powerful ideas; however, the Greeks did come up with one gem. This is the proof that there are infinitely many primes, in Book IX of Euclid's *Elements*.

Given any finite collection of primes $p_1, p_2, \ldots, p_n$, we can find another by considering

$$p = p_1 p_2 \cdots p_n + 1.$$

This number is not divisible by $p_1, p_2, \ldots, p_n$ (each leaves remainder 1). Hence either p itself is a prime, and $p > p_1, p_2, \ldots, p_n$, or else it has a prime divisor $\neq p_1, p_2, \ldots, p_n$.

A *perfect number* is one that equals the sum of its divisors (including 1 but excluding itself). For example, $6 = 1 + 2 + 3$ is a perfect number, as is $28 = 1+2+4+7+14$. Although this concept goes back to the Pythagoreans, only two noteworthy theorems about perfect numbers are known. Euclid concludes Book IX of the *Elements* by proving that if $2^n - 1$ is prime, then $2^{n-1}(2^n-1)$ is perfect (Exercise 3.2.5). These perfect numbers are of course even, and Euler (1849) (a posthumous publication) proved that every even perfect number is of Euclid's form. Euler's surprisingly simple proof may be found in Burton (1985), p. 504. It is not known whether there are any odd perfect numbers; this may be the oldest open problem in mathematics.

In view of Euler's theorem, the existence of even perfect numbers depends on the existence of primes of the form $2^n - 1$. These are known as Mersenne primes, after Marin Mersenne (1588–1648), who first drew attention to the problem of recognizing primes of this form. It is not known whether there are infinitely many Mersenne primes, though larger and larger ones seem to be found quite regularly. In recent years each new world-record prime has been a Mersenne prime, giving a corresponding world-record perfect number.

Infinitely many natural numbers are not sums of three (or fewer) squares. The smallest of them is 7, and it can be shown as follows that no number of the form $8n + 7$ is a sum of three squares.

3.2.1 Show that any square leaves remainder 0, 1, or 4 on division by 8.

3.2.2 Deduce that a sum of three squares leaves remainder 0, 1, 2, 3, 4, 5, or 6 on division by 8.

One reason polygonal numbers play only a small role in mathematics is that questions about them are basically questions about squares—hence the focus is on problems about squares.

3.2.3 Show that the kth pentagonal number is $(3k^2 - k)/2$.

3.2.4 Show that each square is the sum of two consecutive triangular numbers.

Euclid's theorem about perfect numbers depends on the prime divisor property, which will be proved in the next section. Assuming this for the moment, it follows that if $2^n - 1$ is a prime p, then the proper divisors of $2^{n-1}p$ (those unequal to $2^{n-1}p$ itself) are

$$1, 2, 2^2, \ldots, 2^{n-1} \quad \text{and} \quad p, 2p, 2^2 p \ldots, 2^{n-2} p.$$

3.2.5 Given that the divisors of $2^{n-1}p$ are those just listed, show that $2^{n-1}p$ is perfect when $p = 2^n - 1$ is prime.

3.3 The Euclidean Algorithm

This algorithm is named after Euclid because its earliest known appearance is in Book VII of the *Elements*. However, in the opinion of many historians (for example, Heath (1921), p. 399) the algorithm and some of its consequences were probably known earlier. At the very least, Euclid deserves credit for a masterly presentation of the fundamentals of number theory, based on this algorithm.

The Euclidean algorithm is used to find the greatest common divisor (gcd) of two positive integers a, b. The first step is to construct the pair (a_1, b_1), where

$$a_1 = \max(a, b) - \min(a, b),$$
$$b_1 = \min(a, b),$$

and then one simply repeats this operation of subtracting the smaller number from the larger. That is, if the pair constructed at step i is (a_i, b_i), then the pair constructed at step $i + 1$ is

$$a_{i+1} = \max(a_i, b_i) - \min(a_i, b_i),$$
$$b_{i+1} = \min(a_i, b_i).$$

The algorithm terminates at the first stage when $a_{i+1} = b_{i+1}$, and this common value is $\gcd(a, b)$. This is because taking differences preserves any common divisors; hence when $a_{i+1} = b_{i+1}$ we have

$$\gcd(a, b) = \gcd(a_1, b_1) = \cdots = \gcd(a_{i+1}, b_{i+1}) = a_{i+1} = b_{i+1}.$$

The sheer simplicity of the algorithm makes it easy to draw some important consequences. Euclid of course did not use our notation, but nevertheless he had results close to the following.

1. If $\gcd(a, b) = 1$, then there are integers m, n such that $ma + nb = 1$.

    The equations

    $$a_1 = \max(a, b) - \min(a, b),$$
    $$b_1 = \min(a, b),$$

    $$\vdots$$

    $$a_{i+1} = \max(a_i, b_i) - \min(a_i, b_i),$$
    $$b_{i+1} = \min(a_i, b_i)$$

    show first that a_1, b_1 are integral linear combinations, $ma + nb$, of a and b, hence so are a_2, b_2, hence so are $a_3, b_3, \ldots$, and finally this is true of $a_{i+1} = b_{i+1}$. But $a_{i+1} = b_{i+1} = 1$, since $\gcd(a, b) = 1$; hence $1 = ma + nb$ for some integers m, n.

2. If p is a prime number that divides ab, then p divides a or b (the *prime divisor property*).

    To see this, suppose p does *not* divide a. Then since p has no other divisors except 1, we have $\gcd(p, a) = 1$. Hence by the previous result we get integers m, n such that

    $$ma + np = 1.$$

Multiplying each side by b gives

$$mab + nbp = b.$$

By hypothesis, p divides ab; hence p divides *both* terms on the left-hand side, and therefore p divides the right-hand side b.

3. Each positive integer has a unique factorization into primes (the *fundamental theorem of arithmetic*).

   Suppose on the contrary that some integer n has two different prime factorizations:

   $$n = p_1 p_2 \cdots p_j = q_1 q_2 \cdots q_k.$$

   By removing common factors, if necessary, we can assume that there is a p_i that is not among the q's. But this contradicts the previous result, because p_i divides $n = q_1 q_2 \cdots q_k$, yet it does not divide any of $q_1, q_2, \ldots, q_k$ individually, since these are prime numbers $\neq p_i$.

EXERCISES

We can now fill the gap in the proof of Euclid's theorem on perfect numbers (previous exercise set), using the prime divisor property.

3.3.1 Use the prime divisor property to show that the proper divisors of $2^{n-1}p$, for any odd prime p, are $1, 2, 2^2, \ldots, 2^{n-1}$ and $p, 2p, 2^2 p \ldots, 2^{n-2}p$.

The result that if $\gcd(a, b) = 1$ then $1 = ma + nb$ for some integers m and n is a special case of the following way to represent the gcd.

3.3.2 Show that, for any integers a and b, there are integers m and n such that $\gcd(a, b) = ma + nb$.

This in turn gives a general way to find integer solutions of linear equations.

3.3.3 Deduce from Exercise 3.3.2 that the equation $ax + by = c$ with integer coefficients a, b, and c has an integer solution x, y if $\gcd(a, b)$ divides c.

The converse of this result is also valid, as one discovers when considering a *necessary* condition for $ax + by = c$ to have an integer solution.

3.3.4 The equation $12x + 15y = 1$ has no integer solution. Why?

3.3.5 (Solution of linear Diophantine equations) Give a test to decide, for any given integers a, b, c, whether there are integers x, y such that

$$ax + by = c.$$

3.4   Pell's Equation

The Diophantine equation $x^2 - Ny^2 = 1$, where N is a nonsquare integer, is known as Pell's equation because Euler mistakenly attributed a solution of it to the 17th-century English mathematician Pell (it should have been attributed to Brouncker). Pell's equation is probably the best-known Diophantine equation after the equation $a^2 + b^2 = c^2$ for Pythagorean triples, and in some ways it is more important. Solving Pell's equation is the main step in the solution of the general quadratic Diophantine equation in two variables (see, for example, Gelfond (1961)), and also a key tool in proving the theorem of Matiyasevich mentioned in Section 1.3 that there is no algorithm for solving all Diophantine equations (see, for example, Davis (1973) or Jones and Matiyasevich (1991)). In view of this, it is fitting that Pell's equation should make its first appearance in the foundations of Greek mathematics, and it is impressive to see how well the Greeks understood it.

The simplest instance of Pell's equation,

$$x^2 - 2y^2 = 1,$$

was studied by the Pythagoreans in connection with $\sqrt{2}$. If x, y are large solutions to this equation, then $x/y \approx \sqrt{2}$, and in fact the Pythagoreans found a way of generating larger and larger solutions by means of the recurrence relations

$$x_{n+1} = x_n + 2y_n,$$
$$y_{n+1} = x_n + y_n.$$

A short calculation shows that

$$x_{n+1}^2 - 2y_{n+1}^2 = -(x_n^2 - 2y_n^2),$$

so if (x_n, y_n) satisfies $x^2 - 2y^2 = \pm 1$, then (x_{n+1}, y_{n+1}) satisfies $x^2 - 2y^2 = \mp 1$. Starting with the trivial solution $(x_0, y_0) = (1, 0)$ of $x^2 - 2y^2 = 1$, we get successively larger solutions $(x_2, y_2), (x_4, y_4), \ldots$ of the equation $x^2 - 2y^2 = 1$. (The pairs (x_n, y_n) were known as *side and diagonal numbers* because the ratio y_n/x_n tends to that of the side and diagonal in a square.)

But how might these recurrence relations have been discovered in the first place? Van der Waerden (1976) and Fowler (1980, 1982) suggest that the key is the Euclidean algorithm applied to line segments, an operation

the Greeks called *anthyphairesis*. Given any two lengths a, b, one can define the sequence $(a_1, b_1), (a_2, b_2), \ldots$, as in Section 3.2, by repeated subtraction of the smaller length from the larger. If a, b are integer multiples
of some unit, then the process terminates as in Section 3.3, but if b/a is
irrational, it continues forever. We can well imagine that the Pythagoreans
would have been interested in anthyphairesis applied to $a = 1$, $b = \sqrt{2}$.
Here is what happens. We represent a, b by sides of a rectangle, and each
subtraction of the smaller number from the larger is represented by cutting
off the square on the shorter side (Figure 3.2). We notice that the rectangle
remaining after step 2, with sides $\sqrt{2} - 1$ and $2 - \sqrt{2} = \sqrt{2}(\sqrt{2} - 1)$, is the
same shape as the original, though the long side is now vertical instead of
horizontal. It follows that similar steps will recur forever, which is another
proof that $\sqrt{2}$ is irrational, incidentally.

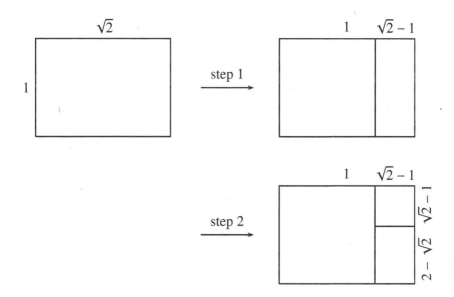

Figure 3.2: The Euclidean algorithm on $\sqrt{2}$ and 1

Our present interest, however, is in the relation between successive
similar rectangles. If we let the long and short sides of successive similar rectangles be x_{n+1}, y_{n+1} and x_n, y_n, we can derive a recurrence relations
for x_{n+1}, y_{n+1} from Figure 3.3:

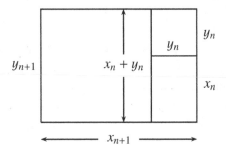

Figure 3.3: The recurrence relation

$$x_{n+1} = x_n + 2y_n,$$
$$y_{n+1} = x_n + y_n,$$

exactly the relations of the Pythagoreans! The difference is that our x_n, y_n are not integers, and they satisfy $x^2 - 2y^2 = 0$, not $x^2 - 2y^2 = 1$. Nevertheless, one feels that Figure 3.3 gives the most natural interpretation of these relations. The discovery that the same relations generate solutions of $x^2 - 2y^2 = 1$ possibly arose from wishing that the Euclidean algorithm terminated with $x_1 = y_1 = 1$. If the Pythagoreans started with $x_1 = y_1 = 1$ and applied the recurrence relations, then they may well have found that (x_n, y_n) satisfies $x^2 - 2y^2 = (-1)^n$, as we did earlier.

Many other instances of the Pell equation $x^2 - Ny^2 = 1$ occur in Greek mathematics, and these can be understood in a similar way by applying anthyphairesis to the rectangle with sides 1, $\sqrt{N}$. In the seventh century CE the Indian mathematician Brahmagupta gave a recurrence relation for generating solutions of $x^2 - Ny^2 = 1$, as we shall see in Chapter 5. The Indians called the Euclidean algorithm the "pulverizer" because it breaks numbers down to smaller and smaller pieces. To obtain a recurrence one has to know that a rectangle proportional to the original eventually recurs, a fact that was rigorously proved only in 1768 by Lagrange. The later European work on Pell's equation, which began in the 17th century with Brouncker and others, was based on the continued fraction for $\sqrt{N}$, though this amounts to the same thing as anthyphairesis (see exercises). For a condensed but detailed history of Pell's equation, see Dickson (1920), pp. 341–400.

An interesting aspect of the theory is the very irregular relationship between N and the number of steps of anthyphairesis before a rectangle proportional to the original recurs. If the number of steps is large, the smallest nontrivial solution of $x^2 - Ny^2 = 1$ is enormous. A famous example is the so-called *cattle problem* of Archimedes (287–212 BCE). This problem leads to the equation

$$x^2 - 4729494y^2 = 1,$$

the smallest solution of which was found by Krummbiegel and Amthor (1880) to have 206,545 digits!

A recent paper on the cattle problem, Lenstra (2002), gives a strikingly condensed form of solution: "for the first time in history, *all* infinitely many solutions to the cattle problem are displayed in a handy little table."

EXERCISES

The continued fraction of a real number $\alpha > 0$ is written

$$\alpha = n_1 + \cfrac{1}{n_2 + \cfrac{1}{n_3 + \cfrac{1}{n_4 + \cfrac{1}{\ddots}}}},$$

where $n_1, n_2, n_3, n_4, \ldots$ are integers obtained by the following algorithm. Let

n_1 = integer part of α.

Then $\alpha - n_1 < 1$ and $\alpha_1 = 1/(\alpha - n_1) > 1$, so we can take

n_2 = integer part of α_1.

Then $\alpha_1 - n_2 < 1$ and $\alpha_2 = 1/(\alpha_1 - n_2) > 1$, so we can take

n_3 = integer part of α_2, and so on.

3.4.1 Apply the above algorithm to the number $\alpha = 157/68$, and hence show that

$$\frac{157}{68} = 2 + \cfrac{1}{3 + \cfrac{1}{4 + \cfrac{1}{5}}}.$$

You may notice that what happens is essentially the Euclidean algorithm applied to the pair $(157, 68)$, except that repeated operations of subtraction are replaced by division with remainder. The integers $2, 3, 4, 5$ are the successive quotients obtained in these divisions: 157 divided by 68 gives quotient 2 and remainder 21, 68 divided by 21 gives quotient 3 and remainder 5, and so on.

Thus the Euclidean algorithm on integers a, b yields results that may be encoded by the (finite) continued fraction for a/b. This idea was introduced by Euler, and it became the preferred approach to the Euclidean algorithm for some mathematicians. Gauss (1801), in particular, always speaks of the Euclidean algorithm as the "continued fraction algorithm."

The Euclidean algorithm on a pair $(\alpha, 1)$, where α is irrational, is in fact better known as the continued fraction algorithm.

3.4.2 Interpret the operations in the continued fraction algorithm—detaching the integer part and taking the reciprocal of the remainder—in terms of anthyphairesis.

3.4.3 Show that

$$\sqrt{2} = 1 + \cfrac{1}{2 + \cfrac{1}{2 + \cfrac{1}{2 + \cfrac{1}{\ddots}}}} \; .$$

Exercise 3.4.3 implies that $\sqrt{2} + 1$ is the *periodic* continued fraction

$$2 + \cfrac{1}{2 + \cfrac{1}{2 + \cfrac{1}{2 + \cfrac{1}{\ddots}}}} \; .$$

3.4.4 Show that $\sqrt{3} + 1$ also has a periodic continued fraction, and hence derive the continued fraction for $\sqrt{3}$.

3.5   The Chord and Tangent Methods

In Section 1.3 we used a method of Diophantus to find all rational points on the circle. If $p(x, y) = 0$ is any quadratic equation in x and y with rational coefficients, and if the equation has one rational solution $x = r_1, y = s_1$, then we can find any rational solution by drawing a rational line $y = mx + c$ through the point r_1, s_1 and finding its other intersection with the curve $p(x, y) = 0$. The two intersections with the curve, $x = r_1, r_2$, say, are given by the roots r_1, r_2 of the equation

$$p(x, mx + c) = 0.$$

This means that $p(x, mx + c) = k(x - r_1)(x - r_2)$, and since all coefficients on the left-hand side are rational and r_1 is rational, then k and r_2 must also be rational. The y value when $x = r_2$, $y = s_2 = mr_2 + c$, is rational since m and c are; hence (r_2, s_2) is another rational point on $p(x, y) = 0$. Conversely, any line through two rational points is rational, and hence all rational points are found in this way.

Now if $p(x, y) = 0$ is a curve of degree 3, its intersections with a line $y = mx + c$ are given by the roots of the cubic equation $p(x, mx + c) = 0$. If we know two rational points on the curve, then the line through them will be rational, and its third intersection with the curve will also be rational, by an argument like the preceding one. This fact becomes more useful when one realizes that the two known rational points can be taken to coincide, in which case the line is the tangent through the known rational point. Thus from one rational solution we can generate another by the tangent construction, and from two we can construct a third by taking the chord between the two.

Diophantus found rational solutions to cubic equations in what seems to have been essentially this way. The surviving works of Diophantus reveal little of his methods, but a plausible reconstruction—an algebraic version of the tangent and chord constructions—has been given by Bashmakova (1981). Probably the first to understand Diophantus's methods was Fermat, in the 17th century, and the first to give the tangent and chord interpretation was Newton (1670s).

In contrast to the quadratic case, we have no choice in the slope of the rational line for cubics. Thus it is by no means clear that this method will give us *all* rational points on a cubic. A remarkable theorem, conjectured by Poincaré (1901) and proved by Mordell (1922), says that all rational points can be generated by tangent and chord constructions applied to finitely many points. However, it is still not known whether there is an algorithm for finding a finite set of such rational generators on each cubic curve.

EXERCISES

3.5.1 Explain the solution $x = 21/4$, $y = 71/8$ to $x^3 - 3x^2 + 3x + 1 = y^2$ given by Diophantus (Heath (1910), p. 242) by constructing the tangent through the obvious rational point on this curve.

3.5.2 Rederive the following rational point construction of Viète (1593), p. 145. Given the rational point (a, b) on $x^3 - y^3 = a^3 - b^3$, show that the tangent at (a, b) is

$$y = \frac{a^2}{b^2}(x - a) + b,$$

and that the other intersection of the tangent with the curve is the rational point

$$x = a\frac{a^3 - 2b^3}{a^3 + b^3}, \qquad y = b\frac{b^3 - 2a^3}{a^3 + b^3}.$$

3.6 Biographical Notes: Diophantus

Diophantus lived in Alexandria during the period when Greek mathematics, along with the rest of Western civilization, was generally in decline. The catastrophes that engulfed the West with the fall of Rome and the rise of Islam, culminating in the burning of the library in Alexandria in 640 CE, buried almost all details of Diophantus's life. His dates can be placed with certainty only between 150 and 350 CE, since he mentions Hypsicles (known to be around 150) and is mentioned by Theon of Alexandria (around 350). One other scrap of evidence, a letter of Michael Psellus (11th century), suggests 250 CE as the most likely time when Diophantus flourished. Apart from this, the only clue to Diophantus's life is a conundrum in the *Greek Anthology* (around 600 CE):

> God granted him to be a boy for the sixth part of his life, and adding a twelfth part to this, He clothed his cheeks with down. He lit him the light of wedlock after a seventh part, and five years after his marriage He granted him a son. Alas! late-born wretched child; after attaining the measure of half his father's life, chill Fate took him. After consoling his grief by this science of numbers for four years he ended his life.
>
> Cohen and Drabkin (1958), p. 27

If this information is correct, then Diophantus married at 33 and had a son who died at 42, four years before Diophantus himself died by his own hand at 84.

Diophantus's work went almost unnoticed for many centuries, and only parts of it survive. The first stirrings of interest in Diophantus occurred in the Middle Ages, but much of the credit for the eventual revival of Diophantus belongs to Rafael Bombelli (1526–1572) and Wilhelm Holtzmann

(known as Xylander, 1532–1576). Bombelli discovered a copy of Diophantus's *Arithmetic* in the Vatican library and published 143 problems
from it in his *Algebra* (1572). The most famous edition of the *Arithmetic*
was that of Bachet de Méziriac (1621). Bachet glimpsed the possibility of
general principles behind the special problems of the *Arithmetic* and, in his
commentary on the book, alerted his contemporaries to the challenge of
properly understanding Diophantus and carrying his ideas further. It was
Fermat who took up this challenge and made the first significant advances
in number theory since the classical era (see Chapter 11).

4

Infinity in Greek Mathematics

PREVIEW

Perhaps the most interesting—and most modern—feature of Greek mathematics is its treatment of infinity. The Greeks feared infinity and tried to avoid it, but in doing so they laid the foundations for a rigorous treatment of infinite processes in 19th century calculus.

The most original contributions to the theory of infinity in ancient times were the *theory of proportions* and the *method of exhaustion.* Both were devised by Eudoxus and expounded in Book V of Euclid's *Elements.*

The theory of proportions develops the idea that a "quantity" λ (what we would now call a real number) can be known by its position among the rational numbers. That is, λ is known if we know the rational numbers less than λ and the rational numbers greater than λ.

The method of exhaustion generalizes this idea from "quantities" to regions of the plane or space. A region becomes "known" (in area or volume) when its position among known areas or volumes is known. For example, we know the area of a circle when we know the areas of the polygons inside it and the areas of polygons outside it; we know the volume of a pyramid when we know the volumes of stacks of prisms inside it and outside it.

Using this method, Euclid found that the volume of a tetrahedron equals 1/3 of its base area times its height, and Archimedes found the area of a parabolic segment. Both of them relied on an infinite process that is fundamental to many calculations of area and volume: the summation of an infinite geometric series.

J. Stillwell, *Mathematics and Its History*, Undergraduate Texts in Mathematics, DOI 10.1007/978-1-4419-6053-5_4, © Springer Science+Business Media, LLC 2010

4.1   Fear of Infinity

Reasoning about infinity is one of the characteristic features of mathematics as well as its main source of conflict. We saw, in Chapter 1, the conflict that arose from the discovery of irrationals, and in this chapter we shall see that the Greeks' rejection of irrational numbers was just part of a general rejection of infinite processes. In fact, until the late 19th century most mathematicians were reluctant to accept infinity as more than "potential." The infinitude of a process, collection, or magnitude was understood as the possibility of its indefinite continuation, and no more—certainly not the possibility of eventual completion. For example, the natural numbers $1, 2, 3, \ldots$, can be accepted as a potential infinity—generated from 1 by the process of adding 1—without accepting that there is a completed totality $\{1, 2, 3, \ldots\}$. The same applies to any sequence $x_1, x_2, x_3, \ldots$ (of rational numbers, say), where x_{n+1} is obtained from x_n by a definite rule.

And yet a beguiling possibility arises when x_n tends to a limit x. If x is something we already accept—for geometric reasons, say—then it is very tempting to view x as somehow the "completion" of the sequence $x_1, x_2, x_3, \ldots$. It seems that the Greeks were afraid to draw such conclusions. According to tradition, they were frightened off by the paradoxes of Zeno, around 450 BCE.

We know of Zeno's arguments only through Aristotle, who quotes them in his *Physics* in order to refute them, and it is not clear what Zeno himself wished to achieve. Was there, for example, a tendency toward speculation about infinity that he disapproved of? His arguments are so extreme they could almost be parodies of loose arguments about infinity he heard among his contemporaries. Consider his first paradox, the *dichotomy*:

> There is no motion because that which is moved must arrive
> at the middle (of its course) before it arrives at the end.
>
> Aristotle, *Physics*, Book VI, Ch. 9

The full argument presumably is that before getting anywhere one must first get half way, and before that a quarter of the way, and before that one eighth of the way, ad infinitum. The completion of this infinite sequence of steps no longer seems impossible to most mathematicians, since it represents nothing more than an infinite set of points within a finite interval. It must have frightened the Greeks though, because in all their proofs they were very careful to avoid completed infinities and limits.

The first mathematical processes we would recognize as infinite were probably devised by the Pythagoreans, for example, the recurrence relations

$$x_{n+1} = x_n + 2y_n,$$
$$y_{n+1} = x_n + y_n$$

for generating integer solutions of the equations $x^2 - 2y^2 = \pm 1$. We saw in Section 3.4 why it is likely that these relations arose from an attempt to understand $\sqrt{2}$, and it is easy for us to see that $x_n/y_n \rightarrow \sqrt{2}$ as $n \rightarrow \infty$.

However, it is unlikely that the Pythagoreans would have viewed $\sqrt{2}$ as a "limit" or seen the sequence as a meaningful object at all. The most we can say is that, by stating a recurrence, the Pythagoreans *implied* a sequence with limit $\sqrt{2}$, but only a much later generation of mathematicians could accept the infinite sequence as such and appreciate its importance in defining the limit.

In a problem where we would find it natural to reach a solution α by a limiting process, the Greeks would instead eliminate any solution *but* α. They would show that any number $<\alpha$ was too small and any number $>\alpha$ was too large to be the solution. In the following sections we shall study some examples of this style of proof and see how it ultimately bore fruit in the foundations of mathematics. As a method of finding solutions to problems, however, it was sterile: how does one guess the number α in the first place? When mathematicians returned to problems of finding limits in the 17th century, they found no use for the rigorous methods of the Greeks. The dubious 17th-century methods of infinitesimals were criticized by the Zeno of the time, Bishop Berkeley, but little was done to meet his objections until much later, since infinitesimals did not seem to lead to incorrect results. It was Dedekind, Weierstrass, and others in the 19th century who eventually restored Greek standards of rigor.

The story of rigor lost and rigor regained took an amazing turn when a previously unknown manuscript of Archimedes, *The Method*, was discovered in 1906. In it he reveals that his deepest results were found using dubious infinitary arguments, and only later proved rigorously. Because, as he says, "It is of course easier to supply the proof when we have previously acquired some knowledge of the questions by the method, than it is to find it without any previous knowledge."

The importance of this statement goes beyond its revelation that infinity can be used to discover results that are not initially accessible to logic.

Archimedes was probably the first mathematician candid enough to explain that there is a difference between the way theorems are discovered and the way they are proved.

4.2   Eudoxus's Theory of Proportions

The theory of proportions is credited to Eudoxus (around 400–350 BCE) and is expounded in Book V of Euclid's *Elements*. The purpose of the theory is to enable lengths (and other geometric quantities) to be treated as precisely as numbers, while admitting only the use of rational numbers. We saw the motivation for this in Section 1.5: the Greeks could not accept irrational numbers, but they accepted irrational geometric quantities such as the diagonal of the unit square. To simplify the exposition of the theory, let us call lengths *rational* if they are rational multiples of a fixed length.

The idea of Eudoxus was to say that a length λ is determined by those rational lengths less than it and those greater than it. To be precise, he says $\lambda_1 = \lambda_2$ if any rational length $<\lambda_1$ is also $<\lambda_2$, and vice versa. Likewise $\lambda_1 < \lambda_2$ if there is a rational length $>\lambda_1$ but $<\lambda_2$. This definition uses the rationals to give an infinitely sharp notion of length while avoiding any overt use of infinity. Of course the infinite set of rational lengths $<\lambda$ is present in spirit, but Eudoxus avoids mentioning it by speaking of an arbitrary rational length $<\lambda$.

The theory of proportions was so successful that it delayed the development of a theory of real numbers for 2000 years. This was ironic, because the theory of proportions can be used to define irrational numbers just as well as lengths. It was understandable though, because the common irrational lengths, such as the diagonal of the unit square, arise from constructions that are intuitively clear and finite from the geometric point of view. Any *arithmetic* approach to $\sqrt{2}$, whether by sequences, decimals, or continued fractions, is infinite and therefore less intuitive. Until the 19th century this seemed a good reason for considering geometry to be a better foundation for mathematics than arithmetic. Then the problems of geometry came to a head, and mathematicians began to fear geometric intuition as much as they had previously feared infinity. There was a purge of geometric reasoning from the textbooks and industrious reconstruction of mathematics on the basis of numbers and sets of numbers. Set theory is discussed further in Chapter 24. Suffice to say, for the moment, that set theory depends on the acceptance of completed infinities.

The beauty of the theory of proportion was its adaptability to this new climate. Instead of rational lengths, take rational numbers. Instead of comparing existing irrational lengths by means of rational lengths, construct irrational numbers from scratch using sets of rationals! The length $\sqrt{2}$ is determined by the two sets of positive rationals

$$L_{\sqrt{2}} = \{r : r^2 < 2\}, \qquad U_{\sqrt{2}} = \{r : r^2 > 2\}.$$

Dedekind (1872) decided to let $\sqrt{2}$ *be* this pair of sets! In general, let any partition of the positive rationals into sets L, U such that any member of L is less than any member of U *be* a positive real number. This idea, now known as a *Dedekind cut*, is more than just a twist of Eudoxus; it gives a complete and uniform construction of all real numbers, or points on the line, using just the rationals. In short, it is an explanation of the *continuous* in terms of the *discrete*, finally resolving the fundamental conflict in Greek mathematics. Dedekind was understandably pleased with his achievement. He wrote

> The statement is so frequently made that the differential calculus deals with continuous magnitude, and yet an explanation of this continuity is nowhere given.... It then only remained to discover its true origin in the elements of arithmetic and thus at the same time secure a real definition of the essence of continuity. I succeeded Nov. 24 1858.
>
> Dedekind (1872), p. 2

EXERCISES

There is only one Dedekind cut (L, U) corresponding to an irrational number α, but there are two cuts corresponding to a rational number a:

$$L = \{r : r \leq a\}, \quad U = \{r : r > a\}$$

and

$$L = \{r : r < a\}, \quad U = \{r : r \geq a\}.$$

To unify the theory of all reals we choose the latter cut, call it

$$L_a = \{r : r < a\}, \quad U_a = \{r : r \geq a\},$$

as the standard way to represent a rational a. We can then say, whether x is rational or irrational, that the lower set for x is

$$L_x = \{r : r < x\}.$$

Now we use lower sets to define $x + y$ and xy for positive reals x and y as follows:

$$L_{x+y} = \{r + s : r < x \text{ and } s < y, \text{ where } r, s \text{ are rational}\}$$
$$L_{xy} = \{rs : r < x \text{ and } s < y, \text{ where } r, s \text{ are rational}\}.$$

4.2.1 Show that these are valid definitions of $x + y$ and xy when x and y are rational.

The true power of these definitions, as Dedekind realized, is that they allow rigorous proofs of results like $\sqrt{2}\,\sqrt{3} = \sqrt{6}$ that (in Dedekind's opinion) had never been rigorously proved before. Such proofs are possible, but still not trivial. Even to prove that $\sqrt{2}\,\sqrt{2} = 2$ one still has to prove the next two results.

4.2.2 If $r^2 < 2$ and $s^2 < 2$, show that $rs < 2$.

4.2.3 If a rational $t < 2$, show that $t = rs$ for some rational r, s with $r^2 < 2$, $s^2 < 2$.

4.2.4 Why do Exercises 4.2.2 and 4.2.3 show that $\sqrt{2}\,\sqrt{2} = 2$?

4.2.5 Give a similar proof that $\sqrt{2}\,\sqrt{3} = \sqrt{6}$.

4.3   The Method of Exhaustion

The method of exhaustion, also credited to Eudoxus, is a generalization of his theory of proportions. Just as an irrational length is determined by the rational lengths on either side of it, more general unknown quantities become determined by arbitrarily close approximations using known figures. Examples given by Eudoxus (and expounded in Book XII of Euclid's *Elements*) are an approximation of the circle by inner and outer polygons (Figure 4.1) and an approximation of a pyramid by stacks of prisms (Figure 4.2, which shows the most obvious approximation, not the cunning one actually used by Euclid). In both cases the approximating figures are known quantities, on the basis of the theory of proportions and the theorem that area of triangle $= 1/2$ base $\times$ height.

The polygonal approximations are used to show that the area of any circle is proportional to the square on its radius, as follows. Suppose $P_1 \subset P_2 \subset P_3 \subset \cdots$ are the inner polygons and $Q_1 \supset Q_2 \supset Q_3 \supset \cdots$ are the outer polygons. Each polygon is obtained from its predecessor by bisecting the arcs between its vertices, as shown in Figure 4.1. It can then be shown, by elementary geometry, that the area difference $Q_i - P_i$ can be made arbitrarily small, and hence P_i approximates the area C of the circle arbitrarily closely.

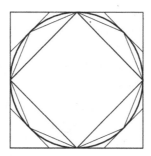

Figure 4.1: Approximating the circle

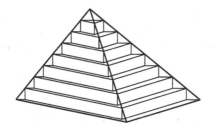

Figure 4.2: Approximating the pyramid

On the other hand, elementary geometry also shows that the area P_i is proportional to the square, R^2, of the radius. Writing the area as $P_i(R)$ and using the theory of proportions to handle ratios of areas, we have

$$P_i(R) : P_i(R') = R^2 : R'^2. \tag{1}$$

Now let $C(R)$ denote the area of the circle of radius R, and suppose

$$C(R) : C(R') < R^2 : R'^2. \tag{2}$$

By choosing a P_i that approximates C sufficiently closely we also get

$$P_i(R) : P_i(R') < R^2 : R'^2,$$

which contradicts (1). Hence the < sign in (2) is incorrect, and we can similarly show that > is incorrect. Thus the only possibility is

$$C(R) : C(R') = R^2 : R'^2,$$

that is, the area of a circle is proportional to the square of its radius.

Notice that "exhaustion" does not mean using an infinite sequence of steps to show that area is proportional to the square of the radius. Rather, one shows that any *disproportionality* can be refuted in a *finite* number of steps (by going to a suitable P_i). This is typical of the way in which exhaustion arguments avoid mention of limits and infinity.

In the case of the pyramid, one uses elementary geometry again to show that stacks of prisms approximate the pyramid arbitrarily closely. Then exhaustion shows that the volume of a pyramid, like that of a prism, is proportional to base × height (see exercises below). Finally, there is a clever argument to show that the constant of proportionality is 1/3. We can restrict to the case of triangular pyramids (since any pyramid can be cut into these), and Figure 4.3 shows how a triangular prism is cut into three triangular pyramids. Any two of these pyramids can be seen to have equal base and height—although which face is taken to be the base depends on which pyramids are being compared—hence all three are equal in volume. Each is therefore one-third of the prism, that is, 1/3 base × height.

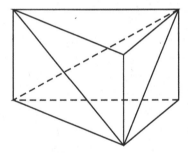

Figure 4.3: Cutting a prism into pyramids

It is interesting that Euclid does *not* need the method of exhaustion in the theory of area for polygons. All this can be done by dissection arguments such as that showing area of triangle = 1/2 base×height (Figure 4.4). In fact, it was shown by Farkas Bolyai (1832a) that any polygons P, Q of equal area can be cut into polygonal pieces $P_1, \ldots, P_n$ and $Q_1, \ldots, Q_n$ such that P_i is congruent to Q_i. Thus we can *define* polygons to be equal in area if they possess dissections into such correspondingly congruent pieces.

In Hilbert's famous list of mathematical problems, Hilbert (1900a), the third was to decide whether an analogous definition was possible for polyhedra. Dehn (1900) showed that it was not; in fact, a tetrahedron and a cube of equal volume cannot be dissected into corresponding congruent polyhe-

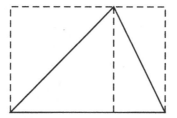

Figure 4.4: Area of a triangle

dral pieces. Hence infinite processes of some kind, such as the method of exhaustion, are needed to define equality of volume. A readable account of Dehn's theorem and related results may be found in Boltyansky (1978).

EXERCISES

Although the method of exhaustion is not needed for the area theory of polygons, it is nevertheless a helpful stepping stone toward cases in which exhaustion *is* necessary, such as volumes of polyhedra or areas of curved regions.

4.3.1 Show that the area of two triangles with the same base and height can be approximated arbitrarily closely by the same set of rectangles, differently stacked (Figure 4.5).

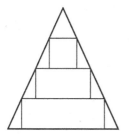 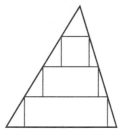

Figure 4.5: Approximations to triangles

4.3.2 Show similarly that any two tetrahedra with the same base and height can be approximated arbitrarily closely by the same prisms, differently stacked (Figure 4.6).

Around 1800, Legendre used the result of Exercise 4.3.2 to give another proof that the volume of a pyramid is 1/3 that of a prism with the same base and height (see Heath (1925), Book XII, Proposition 5). He used the above dissection of a prism into three tetrahedra, pairwise of the same base and height, so he only had to do the following.

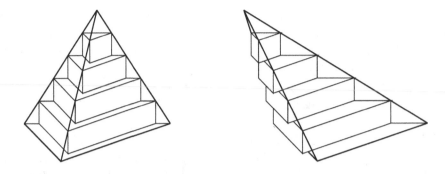

Figure 4.6: Approximations to tetrahedra

4.3.3 Deduce from Exercise 4.3.2 that pyramids of equal base and height have equal volume.

Another interesting approach to the volume of the tetrahedron by exhaustion was given by Euclid (see Heath (1925), Book XII, Proposition 4). He dissected the tetrahedron into two smaller tetrahedra and two prisms as shown in Figure 4.7, with vertices at the edge midpoints of the original tetrahedron.

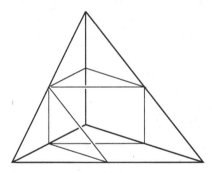

Figure 4.7: Euclid's dissection of the tetrahedron

4.3.4 Show that the two prisms occupy more than half the volume of the tetrahedron. (Hence, by iterating the construction in the smaller tetrahedra, the volume of the tetrahedron may be approximated arbitrarily closely by prisms.)

4.3.5 Show that the volume of the two prisms in Figure 4.7 is 1/4 base × height (the base and height of the tetrahedron, that is).

By computing the volumes of the corresponding prisms in the smaller tetra-hedra (Figure 4.8), and repeating, we find the volume of the original tetrahedron as a sum of a geometric series.

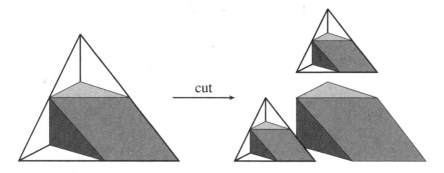

Figure 4.8: Repeated dissection of the tetrahedron

4.3.6 Show that the total volume of the prisms is

$$\left(\frac{1}{4} + \frac{1}{4^2} + \frac{1}{4^3} + \cdots\right) \text{base} \times \text{height} = 1/3 \text{ base} \times \text{height}.$$

In the next section we study a construction of Archimedes that is curiously similar to this one of Euclid. Each step cuts pieces out of the leftovers from the previous step and leads to a similar geometric series.

4.4   The Area of a Parabolic Segment

The method of exhaustion was brought to full maturity by Archimedes (287–212 BCE). Among his most famous results are the volume and surface area of the sphere and the area of a parabolic segment. As mentioned in Section 4.1, Archimedes first discovered these results by nonrigorous methods, later confirming them by the method of exhaustion. Perhaps the most interesting and natural of his exhaustion proofs is the one for the area of the parabolic segment. The segment is exhausted by polygons similarly to Eudoxus' exhaustion of the circle, but the area is obtained outright and not merely in proportion to another figure.

To simplify the construction slightly we assume that the segment is cut off by a chord perpendicular to the axis of symmetry of the parabola. Archimedes divides the parabolic segment into triangles $\Delta_1, \Delta_2, \Delta_3, \ldots$, as shown in Figure 4.9 (labeled by their subscripts). The middle vertex of each triangle lies on the parabola halfway between the other two (measured horizontally). These triangles clearly exhaust the parabolic segment, and so it remains to compute their area. Quite surprisingly, this turns into a geometric series.

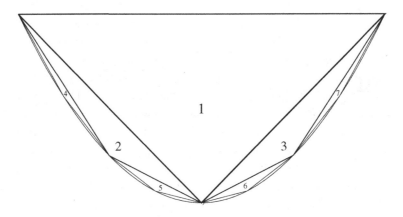

Figure 4.9: The parabolic segment

We indicate how this comes about by looking at Δ_3 (Figure 4.10).

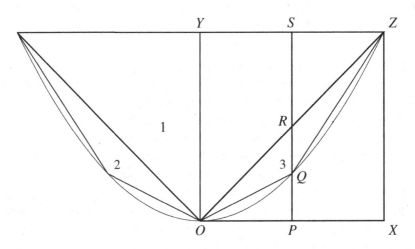

Figure 4.10: A triangle in the segment

Since $OP = \frac{1}{2}OX$, $PQ = \frac{1}{4}PS$ by definition of the parabola. On the other hand, $SR = \frac{1}{2}PS$, so $QR = \frac{1}{4}PS$. Now Δ_3 is the sum of the triangles RQZ and OQR, which have the same base RQ and "height" $OP = PX$, hence equal area. We have just seen that RQZ has half the base of SRZ and it has the same height; hence (calling figures equal when they have the same area)

$$\Delta_3 = SRZ = \frac{1}{4}OYZ = \frac{1}{8}\Delta_1.$$

By symmetry, $\Delta_2 = \Delta_3$, so $\Delta_2 + \Delta_3 = \frac{1}{4}\Delta_1$.

A similar argument shows that

$$\Delta_4 + \Delta_5 + \Delta_6 + \Delta_7 = \frac{1}{16}\Delta_1$$

and so on, each new chain of triangles having one-fourth the area of the previous chain. Consequently,

$$\text{area of parabolic segment} = \Delta_1\left(1 + \frac{1}{4} + \left(\frac{1}{4}\right)^2 + \cdots\right)$$

$$= \frac{4}{3}\Delta_1.$$

Of course, Archimedes does not use the infinite series but uses exhaustion, showing that any area $< \frac{4}{3}\Delta_1$ can be exceeded by taking sufficiently many of the triangles Δ_i. The sum of the *finite* geometric series needed for this was known from Euclid's *Elements*, Book IX, where Euclid used it for the theorem about perfect numbers (Section 3.2).

EXERCISES

Archimedes' method of approximation by triangles was a brilliant success on the parabolic segment, but not suited to many other curves. A more generally useful method is approximation by rectangles, probably known to you from calculus. The·area of a parabolic segment can also be computed in this way, though less gracefully, and indeed Archimedes did this too. We look at other curved areas that can be evaluated by rectangle approximation in Section 9.2.

Probably the simplest area that *cannot* be found by this method is the area under the hyperbola $y = 1/x$, from $x = 1$ to $x = t$. This is because the area in question is $\log t$, and the logarithm function cannot be defined by elementary means. But if instead one takes the area to be $\log t$ by *definition*, then it is possible to derive the basic *property* of the logarithm—

$$\log ab = \log a + \log b$$

—and by means Archimedes would have understood.

4.4.1 Suppose we approximate the area $\log a$ under $y = 1/x$ from 1 to a by n rectangles of equal width, as shown in Figure 4.11.

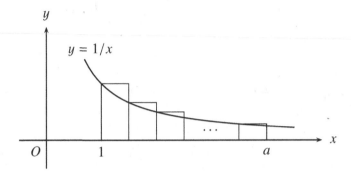

Figure 4.11: Rectangle approximation to $\log a$

Show that the corresponding approximation to the area under $y = 1/x$ from b to ab by n rectangles has exactly the same area. (In fact, corresponding rectangles have equal area.)

4.4.2 Deduce from Exercise 4.4.1, by the method of exhaustion, that the areas under $y = 1/x$ from 1 to a and from b to ab are equal.

4.4.3 Deduce from Exercise 4.4.2, and the above definition of log, that

$$\log ab = \log a + \log b.$$

4.5   Biographical Notes: Archimedes

Archimedes is one of the few ancient mathematicians whose life is known in any detail, thanks to the attention he received from classical authors such as Plutarch, Livy, and Cicero and his participation in the historically significant siege of Syracuse in 212 BCE. He was born in Syracuse (a Greek city in what is now Sicily) around 287 BCE and did most of his important work there, though he may have studied for a time in Alexandria. He seems to have been related to the ruler of Syracuse, King Hieron II, or at least on good terms with him. There are many stories of mechanisms invented by Archimedes for the benefit of Hieron: compound pulleys for moving ships, ballistic devices for the defense of Syracuse, and a model planetarium.

The most famous story about Archimedes is the one told by Vitruvius (*De architectura*, Book IX, Ch. 3), which has Archimedes leaping from his bath with a shout of "Eureka!" when he realized that weighing a crown immersed in water would give a means of testing whether it was pure gold. Historians doubt the authenticity of this story, but it does at least recognize Archimedes' understanding of hydrostatics.

In ancient times Archimedes' reputation rested on his mechanical inventions, which no doubt were more understandable to most people than his pure mathematics. However, it can also be argued that his theoretical mechanics (including the law of the lever, centers of mass, equilibrium, and hydrostatic pressure) was his most original contribution to science. Before Archimedes there was no mathematical theory of mechanics at all, only the thoroughly incorrect mechanics of Aristotle. In pure mathematics, Archimedes did not make any comparable *conceptual* advances, except perhaps in his *Method*, which uses his ideas from statics as a means of discovering results on areas and volumes. The concepts that Archimedes needed for proofs in geometry—the theory of proportions and the method of exhaustion—had already been supplied by Eudoxus, and it was Archimedes' phenomenal insight and technique that lifted him head and shoulders above his contemporaries.

The story of Archimedes' death has often been told, though with varying details. He was killed by a Roman soldier when Syracuse fell to the Romans under Marcellus in 212 BCE. Probably he was doing mathematics at the time of his death, but whether he enraged a soldier by saying "Stand away from my diagram!" is conjectural. This story has come down to us from Tzetzes (*Chiliad*, Book II). Another version of the death of Archimedes is given in Plutarch's *Lives*, in the chapter on Marcellus. Plutarch also tells us that Archimedes asked that his gravestone be inscribed with a figure and description of his favorite result, the relation between the volumes of the sphere and the cylinder. (He showed that the volume of the sphere is two-thirds that of the enveloping cylinder. See Heath (1897), p. 43, and Exercise 9.2.5.) A century and a half later, Cicero (*Tusculan Disputations*, Book V) reported finding the gravestone when he was a quaestor in Sicily in 75 BCE. The grave had been neglected, but the figure of the sphere and cylinder was still recognizable.

5

Number Theory in Asia

Preview

In the next three chapters we see algebra, in the form of techniques for manipulating equations, becoming firmly established in mathematics. The present chapter shows equations applied to number theory, Chapter 6 shows equations studied for their own sake, and Chapter 7 shows equations applied to geometry.

As we saw in Chapter 3, Diophantus had methods for finding rational solutions of quadratic and cubic equations. But when *integer* solutions are sought, even linear equations are not trivial. The first general solutions of linear equations in integers were found in China and India, along with independent discoveries of the Euclidean algorithm.

The Indians also rediscovered Pell's equation $x^2 - Ny^2 = 1$, and found methods of solving it for general natural number values of N.

The first advance on Pell's equation was made by Brahmagupta, who in 628 CE found a way of "composing" solutions of $x^2 - Ny^2 = k_1$ and $x^2 - Ny^2 = k_2$ to produce a solution of $x^2 - Ny^2 = k_1 k_2$. (We also touch on a curious formula of Brahmagupta that gives all triangles with rational sides and rational area.)

In 1150 CE, Bhâskara II found an extension of Brahmagupta's method that finds a solution of $x^2 - Ny^2 = 1$ for any nonsquare natural number N. He illustrated it with the case $N = 61$, for which the least nontrivial solution is extraordinarily large.

J. Stillwell, *Mathematics and Its History*, Undergraduate Texts in Mathematics,
DOI 10.1007/978-1-4419-6053-5_5, © Springer Science+Business Media, LLC 2010

5.1   The Euclidean Algorithm

It is clear from the preceding chapters of this book that ancient Greece had an enormous influence on world mathematics and that most of the fundamental concepts of mathematics can be found there. This does not mean, however, that the Greeks discovered everything first, or that they did everything best. We have already seen that the Pythagorean theorem was known in Babylon earlier than in Greece, and that Pythagorean triples were understood better there than they ever were in Greece, at least until the time of Diophantus.

In fact, the Pythagorean theorem and Pythagorean triples were also known in ancient China and India. As far as we know these were independent discoveries, so it rather seems that the Pythagorean theorem is mathematically universal, likely to arise in any sufficiently advanced civilization. Other such cultural universals are the concept of π—the ratio of diameter to circumference in the circle—and the Euclidean algorithm. As we shall see in this chapter, the Euclidean algorithm seems to arise whenever there is an interest in multiples, divisors, or integer solutions of linear and quadratic equations.

For Euclid, there were two quite separate applications of the Euclidean algorithm. In the first, the algorithm was applied to integers and used to draw conclusions about divisibility and primes. In the second, the algorithm was applied to line segments and was used as a *criterion for irrationality*: if the algorithm does not terminate, then the ratio of the segments is irrational. As we saw in Section 3.4, it is possible that the Greeks pushed the nonterminating Euclidean algorithm far enough to see that it becomes *periodic* in certain cases; for example, when the two line segments have lengths 1 and $\sqrt{2}$.

Independently of these developments, the first form of the Euclidean algorithm arose in China in the Han dynasty, between 200 BCE and 200 CE. It was used by the Chinese to simplify fractions—dividing numerator and denominator by their gcd—and also to find integer solutions of linear equations.

A typical "application" of such an equation is the following. Suppose there are $365\frac{1}{4}$ days in a year and $29\frac{1}{2}$ days in a lunar month. If we go to units of 1/4 day, the year and lunar month are then measured by the integers 1461 and 118. Now suppose there is a full moon on the first day of the year. How long will it be before there is a full moon on the *second* day of the year? This will happen in x years (and y months), where

$$1461x = 118y - 4.$$

We therefore seek the least integer solution of this equation and, as we saw in Section 3.3, this depends on expressing $1 = \gcd(1461, 118)$ as a combination of the form $118y - 1461x$, which can be done with the help of the Euclidean algorithm. In the equation, of course, we are interested only in part of the solution—the number x—because we only want to know *a* multiple of 1461 that is 4 less than *some* multiple of 118 (we don't care which one). Such a problem would later be described as a *congruence* problem: we seek an x such that $1461x$ is *congruent to* −4, *mod* 118. The Chinese became highly skilled in such problems, extending their methods to multiple congruences, as the next section explains. This led to an important theorem, known today as the *Chinese remainder theorem*.

Around the fifth and sixth centuries CE, similar linear Diophantine equations were solved in India, and perhaps with similar calendar problems in mind. However, the Indians took the idea in a different direction. They independently discovered the Pell equation $x^2 - Ny^2 = 1$, found by the Greeks in trying to understand $\sqrt{N}$, and also rediscovered the periodicity in it. Most remarkable of all, they did this without any split between rational and irrational. Their treatment of the Pell equation is completely based on integer operations, and it blends smoothly with their treatment of linear equations.

5.2   The Chinese Remainder Theorem

The origin of this theorem is the following problem, occurring in the *Mathematical Manual* of Sun Zi, late in the third century CE. It is required to find a number that leaves remainder 2 on division by 3, remainder 3 on division by 5, and remainder 2 on division by 7. The answer can easily be found by experiment to be 23, but Sun Zi offers the following explanation, presumably to indicate a general method.

> If we count by threes and there is a remainder 2, put down 140.
> If we count by fives and there is a remainder 3, put down 63.
> If we count by sevens and there is a remainder 2, put down 30.
> Add them to obtain 233 and subtract 210 to get the answer.
>
> From the translation of Sun Zi's *Mathematical Manual* in Lam and Ang (1992), p. 178

The numbers 140, 63, and 30 were chosen because of the following properties:

- $140 = 4 \times (5 \times 7)$
  leaves remainder 2 on division by 3,
  and remainder 0 on division by 5, 7.

- $63 = 3 \times (3 \times 7)$
  leaves remainder 3 on division by 5,
  and remainder 0 on division by 3, 7.

- $30 = 2 \times (3 \times 5)$
  leaves remainder 2 on division by 7,
  and remainder 0 on division by 3, 5.

Hence their sum 233 necessarily leaves remainders 2, 3, 2 on division by 3, 5, 7, respectively. Since $3 \times 5 \times 7 = 105$ leaves remainder 0 on division by 3, 5, and 7, we can subtract 105 from 233 and obtain a smaller number that leaves the same remainders on division by 3, 5, and 7. Subtracting 105 twice gives the smallest solution, 23.

But why choose 140, 63, and 30, in particular? It would be simpler to choose 35 in place of 140, because

- $35 = 5 \times 7$
  leaves remainder 2 on division by 3,
  and remainder 0 on division by 5, 7.

Sun Zi's explanation continues:

> If we count by threes and there is a remainder 1, put down 70.
> If we count by fives and there is a remainder 1, put down 21.
> If we count by sevens and there is a remainder 1, put down 15.

Apparently he began with $70 = 2 \times (5 \times 7)$ because it is the smallest multiple of 5 and 7 leaving remainder 1 on division by 3, then multiplied by 2 to get a number leaving remainder 2 on division by 3.

The numbers 63 and 30 can also be explained this way. The smallest multiple of 3 and 7 that leaves remainder 1 on division by 5 is $21 = 3 \times 7$. Therefore, $63 = 3 \times (3 \times 7)$ is a multiple of 3 and 7 that leaves remainder 3 on division by 5. Similarly, $15 = 3 \times 5$ is the smallest multiple of 3 and 5 that leaves remainder 1 on division by 7, so 30 is the smallest multiple of 3 and 5 that leaves remainder 2 on division by 7.

An interesting question arises at this point. If Sun Zi intended this to be a general method, with integers p, q, r in place of 3, 5, 7, he needed to know that there is a multiple $m(qr)$ of qr that leaves remainder 1 on division by p. Did he know this? Such a number m is what we now call an *inverse* of qr, mod p, and Sun Zi's problem is probably the first occasion in the history of mathematics where these inverses are called for.

A method for solving Sun Zi's problem in full generality was first given in the *Mathematical Treatise in Nine Sections* by Qin Jiushao in 1247. He solved the crucial problem of finding inverses by the Euclidean algorithm. Given integers p and a with $\gcd(p, a) = 1$, we know from Section 2.4 that there are integers m and n such that

$$mp + na = 1.$$

But then

$$mp = 1 - na,$$

so mp leaves remainder 1 on division by a, and m is an inverse of p, mod a. Qin Jiushao found m by running the Euclidean algorithm on p and a, then substituting back to find m and n with $mp + na = 1$. He called it the "method of finding 1."

It is not hard to show (Exercise 5.2.1) that p has an inverse mod a *only* if $\gcd(p, a) = 1$. Thus in a Chinese remainder problem we generally need the divisors to be relatively prime. The method of inverses then gives the following.

Chinese remainder theorem. *If $p_1, \ldots, p_k$ are relatively prime integers and $r_1 < p_1, \ldots, r_k < p_k$ are any integers ≥ 0, then there is an integer n such that, for each i, n leaves remainder r_i on division by p_i.* □

This theorem has made many appearances in the history of number theory, and has often been the vehicle for important new concepts and results. Its later development in China is described in Libbrecht (1973). When it was eventually discovered in Europe, Euler and Gauss made excellent use of it.

EXERCISES

5.2.1 Prove that if mp leaves remainder 1 on division by a, then $\gcd(p, a) = 1$.

5.2.2 Give a proof of the Chinese remainder theorem, using the existence of inverses mod p_i to justify Sun Zi's method.

5.3   Linear Diophantine Equations

We have seen how the Chinese came to use the Euclidean algorithm for
remainder problems, somewhere between the third century CE and Qin
Jiushao's *Mathematical Treatise* of 1247. The algorithm was also used ex-
tensively in India during the same period, beginning with the *Āryabhaṭīya*
of Āryabhaṭa in 499 CE. Āryabhaṭa was born in 476 CE and is also known
as Āryabhaṭa I, to distinguish him from another mathematician of the same
name who lived around 950 CE.

His most important contribution was a method for finding integer solutions of equations of the form $ax + by = c$, where a, b, and c are integers. Like the Chinese remainder problem, which it closely resembles, this
problem cries out for the Euclidean algorithm. Both problems boil down
to expressing $\gcd(a, b)$ in the form $ma + nb$, and in the case of the equation
$ax + by = c$ the underlying reason is the following:

Criterion for integer solution of $ax + by = c$. *The equation $ax + by = c$,
where a, b, c are integers, has an integer solution $\Leftrightarrow \gcd(a, b)$ divides c.*

Proof. If x and y are integers, then $\gcd(a, b)$ divides $ax + by$; hence if
$ax + by = c$, then $\gcd(a, b)$ divides c. Conversely, we know from Section
3.3 that there are integers m and n such that $\gcd(a, b) = ma + nb$. Hence, if
$\gcd(a, b)$ divides c, we have $ma + nb$ divides c, say $(ma + nb)d = c$. Then
$x = md$, $y = nd$ is a solution of the equation $ax + by = c$. $\square$

As mentioned in Section 3.3, $\gcd(a, b) = ma + nb$ is an easy consequence of the Euclidean algorithm, though Euclid apparently missed it.
We also cannot be sure that Āryabhaṭa noticed it, since his book contains
only a few lines on the problem of solving $ax + by = c$, and these were
made intelligible only by the efforts of later commentators. The first of
these was Bhâskara I, who observed in 522 CE that, by dividing a and b by
their gcd, the problem reduces to solving

$$a'x + b'y = 1,$$

where $\gcd(a', b') = 1$, and that the latter problem can always be solved.
Thus Bhâskara I assumed that $1 = \gcd(a', b') = m'a' + n'b'$ for some
integers m' and n', and it follows that $\gcd(a, b) = ma + nb$, after multiplying
both sides by $\gcd(a, b)$.

Bhâskara I also introduced the vivid term *kuṭṭaka*, meaning *pulverizer*,
for the Euclidean algorithm. The numbers a and b are "pulverized" by the

algorithm into smaller and smaller parts, with the smallest part being their gcd. The Indian pulverizer was the division-with-remainder form of the algorithm, though of course the word applies equally well to the subtractive form. To solve the equation $ax + by = c$, where $\gcd(a, b)$ divides c, the pulverizer was combined with substitution to find coefficients m and n such that $ma + nb = \gcd(a, b)$, and multiplication by a suitable factor to obtain x and y such that $ax + by = c$. Examples may be seen in Srinivasiengar (1967).

EXERCISES

Finding m and n such that $\gcd(a, b) = ma + nb$ can be done by running the Euclidean algorithm on the numbers a and b, in parallel with the algorithm on the literal *symbols* a and b. The symbols a and b hold the numbers m and n as their coefficients. For example, to find m and n such that $1 = 21m + 17n$, one runs the Euclidean algorithm starting with the pair $(21, 17)$, and also with (a, b), doing to the symbols exactly what is done to the numbers.

The first few steps look like this:

$$
\begin{array}{ll}
(21, 17) & (a, b) \\
(17, 21 - 17) & (b, a - b) \\
(17, 4) & (b, a - b) \\
(17 - 4, 4) & (b - (a - b), a - b) \\
(13, 4) & (-a + 2b, a - b)
\end{array}
$$

So far, this gives $13 = -21 + 2 \times 17$ and $4 = 21 - 17$ in the form $21m + 17n$.

5.3.1 Complete the running of the Euclidean algorithm on $(21, 17)$, and hence find integers m and n such that $1 = 21m + 17n$.

5.3.2 Hence find integers x, y such that $21x + 17y = 3$.

5.4 Pell's Equation in Brahmagupta

Where linear Diophantine equations are concerned, Indian mathematics and Chinese mathematics are very similar. In fact, the resemblance is even greater than has been suggested so far, because Chinese remainder problems were also studied in India. This suggests possible contact and sharing of ideas. On the other hand, the two mathematical cultures diverge in other respects. The Chinese developed algebra and approximation methods for high-degree equations, but *not* integer solutions for nonlinear equations (except for the Pythagorean equation). The Indians made less progress in algebra, but had striking success finding integer solutions of the Pell equation—the first major advance in number theory since Diophantus.

The author of this advance was Brahmagupta, whose *Brâhma-sphuṭa-siddhânta* of 628 CE can be read in the English translation of Colebrooke (1817). Brahmagupta's treatment of the Pell equation

$$x^2 - Ny^2 = 1, \quad \text{where } N \text{ is a nonsquare integer,}$$

is based on his discovery (see Colebrooke (1817), p. 363) that

$$(x_1^2 - Ny_1^2)(x_2^2 - Ny_2^2) = (x_1 x_2 + Ny_1 y_2)^2 - N(x_1 y_2 + x_2 y_1)^2.$$

This identity generalizes the identity discovered by Diophantus

$$(x_1^2 + y_1^2)(x_2^2 + y_2^2) = (x_1 x_2 - y_1 y_2)^2 + (x_1 y_2 + x_2 y_1)^2,$$

to which we shall return later in connection with complex numbers. Like Diophantus's identity, Brahmagupta's is easily checked by multiplying out both sides, though not easily discovered in the first place.

Brahmagupta used his identity to find solutions of

$$x^2 - Ny^2 = 1$$

via a sequence of equations of the form

$$x^2 - Ny^2 = k_i.$$

His identity shows that if

$$x = x_1, \quad y = y_1 \quad \text{is a solution of} \quad x^2 - Ny^2 = k_1,$$

and

$$x = x_2, \quad y = y_2 \quad \text{is a solution of} \quad x^2 - Ny^2 = k_2,$$

then

$$x = x_1 x_2 + Ny_1 y_2, \quad y = x_1 y_2 + x_2 y_1 \quad \text{is a solution of} \quad x^2 - Ny^2 = k_1 k_2.$$

This is called *composition* of the triples (x_1, y_1, k_1) and (x_2, y_2, k_2) to form the triple $(x_1 x_2 + Ny_1 y_2, x_1 y_2 + x_2 y_1, k_1 k_2)$.

If $k_1 = 1$ and $k_2 = 1$, composition is a way to generate infinitely many solutions of $x^2 - Ny^2 = 1$ when one is known (if only one of k_1, k_2 is 1, compose the corresponding triple with itself). More surprisingly, it is often possible to find a solution of $x^2 - Ny^2 = 1$ from solutions of

$$x^2 - Ny^2 = k_1 \quad \text{and} \quad x^2 - Ny^2 = k_2 \quad \text{for integers } k_1, k_2 > 1.$$

The reason is that composing (x_1, y_2, k_1) with itself gives a solution of $x^2 - Ny^2 = k_1^2$, say $x = X$, $y = Y$, and hence a *rational* solution $x = X/k_1$, $y = Y/k_1$ of $x^2 - Ny^2 = 1$. With a bit of luck, this solution will be integral, or else it will yield an integral solution when composed further.

Example: $x^2 - 92y^2 = 1$. (This is Brahmagupta's first example; he says that "a person solving this problem within a year is a mathematician." See Colebrooke (1817), p. 364.)

Solution. Since $10^2 - 92 \times 1^2 = 8$, we have the triple $(10, 1, 8)$. Composing this with itself gives the triple

$$(10 \times 10 + 92 \times 1 \times 1, 10 \times 1 + 1 \times 10, 8 \times 8) = (192, 20, 64),$$

which means

$$192^2 - 92 \times 20^2 = 8^2.$$

Dividing both sides by 8^2 gives

$$24^2 - 92 \times (5/2)^2 = 1,$$

hence the new "nearly integer" triple $(24, 5/2, 1)$. Composing $(24, 5/2, 1)$ with itself finally gives the integer triple

$$(24^2 + 92 \times (5/2)^2, 24 \times (5/2) + (5/2) \times 24, 1) = (576 + 575, 120, 1)$$
$$= (1151, 120, 1).$$

Thus $x = 1151$, $y = 120$ is an integer solution of $x^2 - 92y^2 = 1$.   □

EXERCISES

5.4.1 Explain the solutions $x_{n+1} = x_n + 2y_n$, $y_{n+1} = x_n + y_n$ of $x^2 - 2y^2 = (-1)^n$ (the "side and diagonal numbers" of Section 3.4) in terms of composition.

5.4.2 Derive Brahmagupta's identity using the factorization

$$(x_1^2 - Ny_1^2)(x_2^2 - Ny_2^2) = (x_1 - \sqrt{N}y_1)(x_1 + \sqrt{N}y_1)(x_2 - \sqrt{N}y_2)(x_2 + \sqrt{N}y_2),$$

and combining the first factor with the third, and the second with the fourth.

5.4.3 Show that $\sqrt{N}$ is irrational when N is a nonsquare integer. Deduce that if $a_1 - \sqrt{N}b_1 = a_2 - \sqrt{N}b_2$ for integers a_1, b_1, a_2, b_2, then $a_1 = a_2$ and $b_1 = b_2$.

5.4.4 If $(x_3, y_3, 1)$ is the composite of $(x_1, y_1, 1)$ and $(x_2, y_2, 1)$, use Exercise 5.4.3 to show that x_3, y_3 may also be defined as the integers such that

$$(x_1 - \sqrt{N}y_1)(x_2 - \sqrt{N}y_2) = x_3 - \sqrt{N}y_3.$$

Now we free x and y from the restriction to integer or rational values, and define the Brahmagupta composite of *any* triples $(x_1, y_1, 1)$ and $(x_2, y_2, 1)$ to be $(x_1x_2 + Ny_1y_2, x_1y_2 + x_2y_1, 1)$.

5.4.5 (For readers familiar with hyperbolic functions.) Show that the functions $x = \cosh u$, $y = \frac{1}{\sqrt{N}}\sinh u$ define a one-to-one correspondence between the real numbers u and the points (x, y) on the branch of the hyperbola $x^2 - Ny^2 = 1$ where $x > 1$. Show also that the Brahmagupta composite of $(\cosh u_1, \frac{1}{\sqrt{N}}\sinh u_1, 1)$ and $(\cosh u_2, \frac{1}{\sqrt{N}}\sinh u_2, 1)$ is $(\cosh(u_1+u_2), \frac{1}{\sqrt{N}}\sinh(u_1+u_2), 1)$; hence Brahmagupta's composition corresponds to addition of real numbers u.

5.4.6 Use the functions $x = \cos\theta$ and $y = \sin\theta$ parameterizing the unit circle to show similarly that "Diophantus's composition" of (x_1, y_1) and (x_2, y_2) to form $(x_1x_2 - y_1y_2, x_1y_2 + x_2y_1)$ corresponds to addition of angles θ.

5.5   Pell's Equation in Bhâskara II

Brahmagupta found integer solutions of many Pell equations $x^2 - Ny^2 = 1$ by his composition method, but he was not able to apply it uniformly for all values of N. The best he could do was show that if $x^2 - Ny^2 = k$ has an integer solution for $k = \pm 1, \pm 2$, or ± 4 then $x^2 - Ny^2 = 1$ also has an integer solution. His proofs that composition succeeds in these cases may be found in Srinivasiengar (1967), p. 111.

The first general method for solving the Pell equation was given by Bhâskara II in his *Bîjaganita* of 1150 CE. He completed Brahmagupta's program by giving a method, called the *cakravâla* or *cyclic process*, which always succeeds in finding integers x, y, k with $x^2 - Ny^2 = k$ and $k = \pm 1, \pm 2$, or ± 4. Admittedly, Bhâskara II did not give a proof that the cyclic process always works—this was first done by Lagrange (1768)—but in fact it does. A proof using only concepts accessible to Bhâskara II may be found in Weil (1984), p. 22. We shall merely describe the cyclic process, and one of its most spectacular successes—the solution of $x^2 - 61y^2 = 1$.

Given relatively prime a and b such that $a^2 - Nb^2 = k$, we compose the triple (a, b, k) with the triple $(m, 1, m^2 - N)$ obtained from the trivial equation

$$m^2 - N \times 1^2 = m^2 - N.$$

The result is the triple $(am + Nb, a + bm, k(m^2 - N))$, which can be scaled down to the (possibly nonintegral) triple

$$\left(\frac{am + Nb}{k}, \frac{a + bm}{k}, \frac{m^2 - N}{k} \right).$$

We now choose m so that $(a + bm)/k = b_1$ is an integer, and it turns out that $(am + Nb)/k = a_1$ and $(m^2 - N)/k = k_1$ are integers, too. If we also choose m so that $m^2 - N$ is as small as possible, we are well on the road to a triple (a_i, b_i, k_i) with $k_i = \pm 1, \pm 2,$ or ± 4.

Example. $x^2 - 61y^2 = 1$. (This is Bhâskara's example. See Colebrooke (1817), pp. 176–178.)

Solution. The equation $8^2 - 61 \times 1^2 = 3$ gives us the triple $(a, b, k) = (8, 1, 3)$. We compose $(8, 1, 3)$ with $(m, 1, m^2 - 61)$, obtaining the triple $(8m + 61, 8 + m, 3(m^2 - 61))$ and hence

$$\left(\frac{8m + 61}{3}, \frac{8 + m}{3}, \frac{m^2 - 61}{3} \right).$$

Choosing $m = 7$ (because 7^2 is the nearest square to 61 for which 3 divides $8 + m$), we get the triple $(39, 5, -4)$, so already $k = -4$. We scale down further to the triple $(39/2, 5/2, -1)$. Composing $(39/2, 5/2, -1)$ with itself gives $(1523/2, 195/2, 1)$, and composing this again with $(39/2, 5/2, -1)$ gives the integer triple $(29718, 3805, -1)$. Finally, composing the latter with itself gives the triple $(1766319049, 226153980, 1)$.

Thus the equation $x^2 - 61y^2 = 1$ has integer solution $x = 1766319049$, $y = 226153980$. ☐

This amazing example was rediscovered by Fermat (1657), who posed the equation $x^2 - 61y^2 = 1$ as a challenge to his colleague Frenicle. The solution $x = 1766319049$, $y = 226153980$ is in fact the *minimal* nonzero solution of $x^2 - 61y^2 = 1$, which suggests that the Pell equation has a lot of hidden complexity—one does not expect such a short question to have such a long answer. Presumably Bhâskara II and Fermat knew that the Pell equation is particularly hard for $N = 61$. Among the Pell equations for $N \leq 100$, this has one of the largest minimal solutions, and it is much larger than any for $N < 61$.

The cyclic process is a little *too* successful on $N = 61$, because it terminates before anything "cyclic" becomes apparent. In fact, the cyclic process detects the same periodicity we previously observed in the continued fraction for $\sqrt{N}$ (Section 3.3), and the size of the minimal solution is related to the length of the period. These facts became clear only with the work of Lagrange (1768), which is based on a study of continued fractions.

For a solution of the Pell equation that avoids continued fractions, see Section 25.2.

EXERCISES

The surprising step in Bhaskara's process, where choosing the integer m so that $(a + bm)/k$ is an integer also produces integers $(am + Nb)/k$ and $(m^2 - N)/k$, deserves some explanation. It depends on choosing the initial a and b so that $\gcd(a, b) = 1$—as one normally would to make $a^2 - Nb^2 = k$ small—because there are counterexamples when $\gcd(a, b) > 1$.

5.5.1 Suppose we choose $a = 4, b = 2$ in $a^2 - 2b^2$, so $k = 8$. Find an m for which $(a + bm)/k$ is an integer but $(am + Nb)/k$ is not.

Supposing $\gcd(a, b) = 1$, however, we can prove that if $(a+bm)/k$ is an integer then so is $(am + Nb)/k$. It follows that $(m^2 - N)/k$ is too, thanks to the equation

$$\left(\frac{am + Nb}{k}\right)^2 - N\left(\frac{a + bm}{k}\right)^2 = \frac{m^2 - N}{k} \qquad (*)$$

to which this triple corresponds. The proof that $(am + Nb)/k$ is an integer goes as follows. At the end it involves the "method of finding 1."

5.5.2 Assuming $a + bm = kl$, substitute $kl - mb$ for one copy of a in the equation $a^2 - Nb^2 = k$, and hence show that k divides $b(am + Nb)$.

5.5.3 Substituting $kl - a$ for both copies of bm in the equation $a^2m^2 - Nb^2m^2 = km^2$, show that k divides $a^2(m^2 - N)$.

5.5.4 Deduce from Exercise 5.5.3, and the other form of equation $(*)$,

$$(am + Nb)^2 - N(a + bm)^2 = k(m^2 - N),$$

that k^2 divides $a^2(am + Nb)^2$, so that k divides $a(am + Nb)$.

5.5.5 Deduce from Exercises 5.5.2 and 5.5.4 that k divides $(ar + bs)(am + Nb)$ for any integers r and s, and hence that k divides $am + Nb$.

5.6   Rational Triangles

After the discovery of rational right-angled triangles, and their complete description by Euclid (Section 1.2), a question one might expect to arise is, what about rational triangles in general? Of course, any three rational numbers can serve as the sides of a triangle, provided the sum of any two of them is greater than the third. Thus a "rational triangle" should be one that is rational not only in its side lengths, but also in some other quantity, such as altitude or area. Since area $= \frac{1}{2}$base $\times$ altitude, a triangle with rational sides has rational area if and only all its altitudes are rational, so it is reasonable to define a *rational triangle* to be one with rational sides and rational area.

Many questions can be raised about rational triangles, but they rarely occur in Greek mathematics. As far as we know, the first to treat them thoroughly was Brahmagupta, in his *Brâhma-sphuṭa-siddhânta* of 628 CE. In particular, he found the following complete description of rational triangles.

Parameterization of rational triangles. *A triangle with rational sides a, b, c and rational area is of the form*

$$a = \frac{u^2}{v} + v, \quad b = \frac{u^2}{w} + w, \quad c = \frac{u^2}{v} - v + \frac{u^2}{w} - w$$

for some rational numbers u, v, and w.

Brahmagupta (see Colebrooke (1817), p. 306) actually has a factor $1/2$ in each of a, b, and c, but this is superfluous because, for example,

$$\frac{1}{2}\left(\frac{u^2}{v} + v\right) = \frac{(u/2)^2}{v/2} + v/2 = \frac{u_1{}^2}{v_1} + v_1,$$

where $u_1 = u/2$ and $v_1 = v/2$ are likewise rational. The formula is stated without proof, but it becomes easy to see if one rewrites a, b, c and makes the following stronger claim.

Any triangle with rational sides and rational area is of the form

$$a = \frac{u^2 + v^2}{v}, \quad b = \frac{u^2 + w^2}{w}, \quad c = \frac{u^2 - v^2}{v} + \frac{u^2 - w^2}{w}$$

for some rationals u, v, and w, with altitude h $= 2u$ splitting side c into segments $c_1 = (u^2 - v^2)/v$ and $c_2 = (u^2 - w^2)/w$.

The stronger claim says in particular that any rational triangle splits into two rational right-angled triangles. It follows from the parameterization of rational right-angled triangles, which was known to Brahmagupta.

Proof. For *any* triangle with rational sides a, b, c, the altitude h splits c into rational segments c_1 and c_2 (Figure 5.1). This follows from the

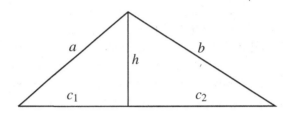

Figure 5.1: Splitting a rational triangle

Pythagorean theorem in the two right-angled triangles with sides c_1, h, a and c_2, h, b, respectively. Namely,

$$a^2 = c_1^2 + h^2,$$
$$b^2 = c_2^2 + h^2.$$

Hence, by subtraction,

$$a^2 - b^2 = c_1^2 - c_2^2 = (c_1 - c_2)(c_1 + c_2) = (c_1 - c_2)c,$$

so

$$c_1 - c_2 = \frac{a^2 - b^2}{c}, \quad \text{which is rational.}$$

But also

$$c_1 + c_2 = c, \quad \text{which is rational;}$$

hence

$$c_1 = \frac{1}{2}\left(\frac{a^2 - b^2}{c} + c\right), \quad c_2 = \frac{1}{2}\left(c - \frac{a^2 - b^2}{c}\right)$$

are both rational.

Thus if the area, and hence the altitude h, is also rational, the triangle splits into two rational right-angled triangles with sides c_1, h, a and c_2, h, b.

We know from Diophantus's method (Section 1.3) that any rational right-angled triangle with hypotenuse 1 has sides of the form

$$\frac{1-t^2}{1+t^2}, \quad \frac{2t}{1+t^2}, \quad 1 \qquad \text{for some rational } t,$$

or, writing $t = v/u$,

$$\frac{u^2 - v^2}{u^2 + v^2}, \quad \frac{2uv}{u^2 + v^2}, \quad 1 \qquad \text{for some rational } u, v.$$

Thus the arbitrary rational right-angled triangle with hypotenuse 1 is a multiple (by $v/(u^2 + v^2)$) of the triangle with sides

$$\frac{u^2 - v^2}{v}, \quad 2u, \quad \frac{u^2 + v^2}{v}.$$

The latter therefore represents all rational right-angled triangles with altitude $2u$, as the rational v varies. And it follows that any *two* rational

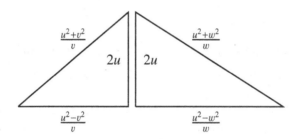

Figure 5.2: Assembling an arbitrary rational triangle

right-angled triangles with altitude $2u$ have sides

$$\frac{u^2 - v^2}{v}, \quad 2u, \quad \frac{u^2 + v^2}{v} \qquad \text{and} \qquad \frac{u^2 - w^2}{w}, \quad 2u, \quad \frac{u^2 + w^2}{w}$$

for some rational v and w. Putting the two together (Figure 5.2) gives an arbitrary rational triangle, and its sides and altitude are of the required form.

$\square$

EXERCISES

5.6.1 (Brahmagupta) Show that the triangle with sides 13, 14, 15 splits into two integer right-angled triangles.

5.6.2 Show that for *any* triangle with sides a, b, c and altitude h on side c there are *real* numbers u, v, w such that

$$a = \frac{u^2 + v^2}{v}, \quad b = \frac{u^2 + w^2}{w}, \quad c = \frac{u^2 - v^2}{v} + \frac{u^2 - w^2}{w},$$

with the side c split into parts $(u^2 - v^2)/v$ and $(u^2 - w^2)/w$ by the altitude $h = 2u$.

5.6.3 Define the *semiperimeter* of the triangle with sides a, b, and c to be $s = (a + b + c)/2$. Then, with the notation of Exercise 5.6.2, show that

$$s(s - a)(s - b)(s - c) = u^2(v + w)^2 \left(\frac{u^2}{vw} - 1 \right)^2.$$

5.6.4 Deduce from Exercise 5.6.3 that

$$\sqrt{s(s - a)(s - b)(s - c)} = u \left(\frac{u^2 - v^2}{v} + \frac{u^2 - w^2}{w} \right)$$

is the area of the triangle with sides a, b, and c. (This formula for area in terms of a, b, and c is named after the Greek mathematician Hero, or Heron, who lived in the first century CE.)

5.7   Biographical Notes: Brahmagupta and Bhâskara

Brahmagupta was born in 598 CE, the son of Jiṣṇagupta, and lived until at least 665. His book the *Brâhma-sphuṭa-siddhânta* describes him as the teacher from Bhillamâla, which is a town now known as Bhinmal in the Indian state of Gujurat. Very little is known for certain about his life except that he was prominent in astronomy as well as mathematics.

Apart from the contributions described above, Brahmagupta is known for introducing a general solution formula for the quadratic equation (see Section 6.3), and a remarkable formula for the area of a cyclic quadrilateral. The latter states that if a quadrilateral has sides a, b, c, and d, semiperimeter s, and all vertices on a circle, then its area is $\sqrt{(s - a)(s - b)(s - c)(s - d)}$. Notice that this generalizes the Heron formula mentioned in Exercise 5.6.4.

Brahmagupta's parameterization of rational triangles leads naturally to other rationality questions, about triangles and other figures. The most famous of these is probably: is there a *rational box*? That is, is there a solid with rational rectangular faces whose body diagonal and the three face diagonals are all rational? According to Dickson (1920), p. 497, a mathematician named Paul Halcke found a box with rational sides and face diagonals in 1719. Its sides are 44, 240, and 117. However, the body diagonal of this box is irrational, and it is still not known whether a rational box exists, despite the efforts of many mathematicians, including Euler and Mordell.

Bhâskara II was born in 1114 or 1115 and died around 1185. He was the son of Maheśvara, from the city of Bījāpur. A great admirer of Brahmagupta, Bhâskara became the greatest mathematician and astronomer in 12th-century India, serving as head of the observatory at Ujjain. His most famous work, the *Līlāvatī*, is said to be named after his daughter, to console her for an astrological forecast that went wrong.

The story goes that Bhâskara used his astronomical knowledge (which in those days included astrological "knowledge") to choose the most propitious date and time for his daughter's wedding. As the time approached, one of her pearls fell into the water clock as she leaned over it, stopping the outflow of water. Before anyone noticed, however, the crucial time passed, and the wedding had to be called off. The hapless Līlāvatī never married, and now she is remembered only through the book that bears her name.

6

Polynomial Equations

PREVIEW

The first phase in the history of algebra was the search for solutions of polynomial equations. The "degree of difficulty" of an equation corresponds rather neatly to the degree of the corresponding polynomial.

Linear equations are easily solved, and 2000 years ago the Chinese were even able to solve n equations in n unknowns by the method we now call "Gaussian elimination."

Quadratic equations are harder to solve, because they generally require the square root operation. But the solution—essentially the same as that taught in high schools today—was discovered independently in many cultures more than 1000 years ago.

The first really hard case is the cubic equation, whose solution requires both square roots and cube roots. Its discovery by Italian mathematicians in the early 16th century was a decisive breakthrough, and equations quickly became the language of virtually all mathematics. (See, for example, analytic geometry in Chapter 7 and calculus in Chapter 9.)

Despite this breakthrough, the problem of polynomial equations remained incompletely solved. The obstacle was the *quintic* equation—the general equation of degree 5. In the 1820s it finally became clear that the quintic equation is *not* solvable in the sense that equations of lower degree are solvable. But explaining why this is so requires a new, and more abstract, concept of algebra (see Chapter 19).

J. Stillwell, *Mathematics and Its History*, Undergraduate Texts in Mathematics,
DOI 10.1007/978-1-4419-6053-5_6, © Springer Science+Business Media, LLC 2010

6.1   Algebra

The word "algebra" comes from the Arabic word *al-jabr* meaning "restoring." It passed into mathematics through the book *Al-jabr w'al mûqabala* (Science of restoring and opposition) of al-Khwārizmī in 830 CE, a work on the solution of equations. In this context, "restoring" meant adding equal terms to both sides and "opposition" meant setting the two sides equal. For centuries, al-jabr more commonly meant the resetting of broken bones, and the surgical meaning accompanied the mathematical one when "al-jabr" became "algebra" in Spanish, Italian, and English. Even today the surgical meaning is included in the *Oxford English Dictionary*. Al-Khwārizmī's own name has given us the word "algorithm," so his work has had a lasting impact on mathematics, even though its content was quite elementary.

His algebra went no further than the solution of quadratic equations, which had already been understood by the Babylonians, presented from the geometric viewpoint by Euclid, and reduced to a formula by Brahmagupta (628) (see Section 6.3). Brahmagupta's work, the high point of Indian mathematics to that time, was more advanced than al-Khwārizmī's in several respects—notation, admission of negative numbers, and the treatment of Diophantine equations—even though it predated al-Khwārizmī and was very likely known to him. Indian mathematics had spread to the Arab world with the general promotion of culture by the eighth-century caliphs of Baghdad, and Arab mathematicians acknowledged the Indian origin of certain ideas, for instance, decimal numerals. Why then did al-Khwārizmī's work rather than Brahmagupta's become the definitive "algebra"?

Perhaps this is a case (like "Pell's equation," to mention another pertinent example) where a mathematical term caught on for accidental reasons. However, it may be that the time was ripe for the idea of algebra to be cultivated, and the simple algebra of al-Khwārizmī served this purpose better than those of his more sophisticated predecessors. In Indian mathematics, algebra was inseparable from number theory and elementary arithmetic. In Greek mathematics, algebra was hidden by geometry. Other possible sources of algebra, Babylonia and China, were lost or cut off from the West until it was too late for them to be influential. Arabic mathematics developed at the right time and place to absorb both the geometry of the West and the algebra of the East and to recognize algebra as a separate field with its own methods. The concept of algebra that emerged—the theory of polynomial equations—proved its worth by holding firm for 1000 years.

Only in the 19th century did algebra grow beyond the bounds of the theory of equations, and this was a time when most fields of mathematics were outgrowing their established habitats.

The early algebraic methods seemed only superficially different from geometric methods, as we shall see in the case of quadratic equations in Section 6.3. Algebraic methods for solving equations became distinct from, and superior to, the geometric only with the advent of new manipulative techniques and efficient notation in the 16th century (Section 6.5). Algebra did not break away from geometry, however, but actually gave geometry a new lease on life, thanks to the development of analytic geometry by Fermat and Descartes around 1630. This recombination of algebra and geometry at a higher level is discussed in Chapter 7. It led to the modern field of algebraic geometry.

The story of algebraic geometry unfolds along with the story of polynomial equations, becoming entwined with many other mathematical threads in the process. We shall study several of the decisive early events in this story. One we have already seen is Diophantus's chord and tangent method for finding rational solutions of equations (Section 3.5). Another relevant event, though not in fact historically connected with Western mathematics, was the method of elimination developed by Chinese mathematicians between the early Christian era and the Middle Ages. Since this method predates any comparable method in the West, and concerns equations of the lowest degree, it is logical to discuss it first.

6.2   Linear Equations and Elimination

The Chinese discovered a method for solving linear equations in any number of unknowns during the Han dynasty (206 BCE–220 CE). It appears in the famous book *Jiuzhang suanshu* (Nine Chapters of Mathematical Art; see Shen et al. (1999)), which was written during this period, and survives today in a third-century version with a commentary by Liu Hui. The method was essentially what we call "Gaussian elimination," systematically eliminating terms in a system

$$a_{11}x_1 + a_{12}x_2 + \cdots + a_{1n}x_n = b_1$$

$$\vdots$$

$$a_{n1}x_1 + a_{n2}x_2 + \cdots + a_{nn}x_n = b_n$$

by subtracting a suitable multiple of each equation from the one below it until a triangular system is obtained:

$$a'_{11}x_1 + a'_{12}x_2 + \cdots + a'_{1n}x_n = b'_1$$
$$a'_{22}x_2 + \cdots + a'_{2n}x_n = b'_2$$
$$\ddots \qquad \vdots$$
$$a'_{nn}x_n = b'_n$$

then solving for $x_n, x_{n-1}, \ldots, x_1$ in turn by successive substitutions. This type of calculation was particularly suited to a Chinese device called the counting board, which held the array of coefficients and facilitated manipulations similar to those we perform with matrices. For further details, see Li and Du (1987).

Around the 12th century, Chinese mathematicians discovered that elimination could be adapted to simultaneous polynomial equations in two or more variables. For example, one can eliminate y between the pair of equations

$$a_0(x)y^m + a_1(x)y^{m-1} + \cdots + a_m(x) = 0, \tag{1}$$
$$b_0(x)y^m + b_1(x)y^{m-1} + \cdots + b_m(x) = 0, \tag{2}$$

where the $a_i(x), b_j(x)$ are polynomials in x. The y^m term can be eliminated by forming the equation $b_0(x) \times (1) - a_0(x) \times (2)$, say,

$$c_0(x)y^{m-1} + c_1(x)y^{m-2} + \cdots + c_{m-1}(x) = 0. \tag{3}$$

We can form a second equation of degree $m - 1$ in y by multiplying (3) by y, then again eliminating y^m between (3) $\times y$ and (1), giving, say,

$$d_0(x)y^{m-1} + d_1(x)y^{m-2} + \cdots + d_{m-1}(x) = 0. \tag{4}$$

The problem is now reduced to eliminating y between the equations (3) and (4), which are of lower degree in y than (1) and (2). Thus one can continue inductively until an equation in x alone is obtained. This method was extended to four variables in the work of Zhū Shijié (1303) entitled *Siyuan yujian* (Jade Mirror of Four Unknowns).

As we shall see in Chapter 7, the two-variable polynomial problem arose in the West in the 17th century, in the context of finding intersections of curves. This led first to a rediscovery of the method of elimination for polynomials; only later was this method based on an understanding of linear equations. The well-known Cramer's rule for the solution of linear equations was named after its appearance in a book on algebraic curves, Cramer (1750).

EXERCISES

The first interesting case of elimination between two-variable polynomials occurs when the polynomials have degree 2. Geometrically, this amounts to finding the intersections of two conic sections.

6.2.1 Derive an equation that is linear in y from the two equations

$$x^2 + xy + y^2 = 1,$$
$$4x^2 + 3xy + 2y^2 = 3,$$

and hence show that $y = (1 - 2x^2)/x$.

6.2.2 Deduce that the intersections of the two curves in Exercise 6.2.1 occur where x satisfies $3x^4 - 4x^2 + 1 = 0$.

This example, where the two equations of degree 2 yield a single equation of degree 4 (= 2×2), illustrates a general phenomenon where degrees are multiplied. We shall observe other instances, and study it more deeply, as the book progresses.

The present example is not a typical equation of degree 4, since it is quadratic in $x^2 = z$. However, this makes it a lot easier to solve.

6.2.3 Solve $3z^2 - 4z + 1 = 0$ for $z = x^2$ by factorizing the left-hand side, and hence find four solutions for x.

Give geometric reasons why you would expect two curves of degree 2 to have up to four intersections. Could they have more than four?

The *Jade Mirror of Four Unknowns* does not go beyond four equations in four unknowns (hence the name). The idea is quite general, but it becomes hard to implement on the counting board when there are more than four unknowns. An amusing problem in three unknowns from the *Jade Mirror*, which does not require the full strength of the elimination method, is given in the exercises below.

6.2.4 Problem 2 in the *Jade Mirror* (see Hoe (1977), p. 135) is to find the side a of a right-angled triangle (a, b, c) such that

$$a^2 - (b + c - a) = ab,$$
$$b^2 + (a + c - b) = bc.$$

The *Jade Mirror* suggests choosing the unknowns $x = a$ and $y = b + c$. Using $a^2 = c^2 - b^2$, show that this implies

$$b = (y - x^2/y)/2,$$
$$c = (y + x^2/y)/2.$$

6.2.5 Deduce that the first two equations in Exercise 6.2.4 are equivalent, respectively, to

$$(-2 - x)y^2 + (2x + 2x^2)y + x^3 = 0,$$
$$(2 - x)y^2 + 2xy + x^3 = 0.$$

6.2.6 By subtracting one equation in Exercise 6.2.5 from the other, deduce that $y = x^2/2$. Substitute this back to obtain a quadratic equation for x, with solution $x = a = 4$. What are the values of b and c?

6.3   Quadratic Equations

As early as 2000 BCE, the Babylonians could solve a pair of simultaneous equations of the form

$$x + y = p,$$
$$xy = q,$$

which are equivalent to the quadratic equation

$$x^2 + q = px.$$

The original pair was solved by a method that gave the two roots of the quadratic,

$$x, y = \frac{p}{2} \pm \sqrt{\left(\frac{p}{2}\right)^2 - q},$$

when both were positive (the Babylonians did not admit negative numbers). The steps in the method were as follows:

(i)   Form $\frac{x+y}{2}$.

(ii) Form $\left(\frac{x+y}{2}\right)^2$.

(iii) Form $\left(\frac{x+y}{2}\right)^2 - xy$.

(iv) Form $\sqrt{\left(\frac{x+y}{2}\right)^2 - xy} = \frac{x-y}{2}$.

(v)   Find x, y by inspection of the values in (i), (iv).

(See Boyer (1968), p. 34, for an actual example.) Of course, these steps were not expressed in symbols but only applied to specific numbers. Nevertheless, a general method is implicit in the many specific cases solved.

An explicit general method, expressed as a formula in words, was given by Brahmagupta (628):

> To the absolute number multiplied by four times the [coefficient of the] square, add the square of the [coefficient of the] middle term; the square root of the same, less the [coefficient of the] middle term, being divided by twice the [coefficient of the] square is the value.

Colebrooke (1817), p. 346

This is the solution

$$x = \frac{\sqrt{4ac + b^2} - b}{2a}$$

of the equation

$$ax^2 + bx = c,$$

yet one wonders whether Brahmagupta understood it quite this way when, a few lines later, he gives another rule that is trivially equivalent to the first when expressed in our notation:

$$x = \frac{\sqrt{ac + (b/2)^2} - (b/2)}{a}.$$

The methods of the Babylonians and Brahmagupta clearly give correct solutions, but their basis is not clear. The meaning of square roots, for example, was not questioned as it was by Greeks. A rigorous basis for the solution of quadratic equations can be found in Euclid's *Elements*, Book VI. Proposition 28 can be interpreted as a solution of the general quadratic equation in the case where there is a positive root, as Heath (1925), Vol. 2, p. 263 explains. However, the algebraic interpretation is far from obvious even when one specializes the proposition, which is about parallelograms, to one about rectangles. It seems unlikely that Euclid was aware of the algebra, or he would have expressed it by much simpler geometry.

The transition from geometry to algebra can be seen in al-Khwārizmī's solution of a quadratic equation (Figure 6.1). The solution is still expressed in geometric language, but now the geometry is a direct embodiment of the algebra. It is really the standard algebraic solution, but with "squares" and "products" understood literally as geometric squares and rectangles. To solve $x^2 + 10x = 39$, represent x^2 by a square of side x, and $10x$ by two $5 \times x$ rectangles as in Figure 6.1. The extra square of area 25 "completes the square" of side $x + 5$ to one of area $25 + 39$, since 39 is the given value of $x^2 + 10x$. Thus the big square has area 64, hence its side $x + 5$ equals 8. This gives the solution $x = 3$.

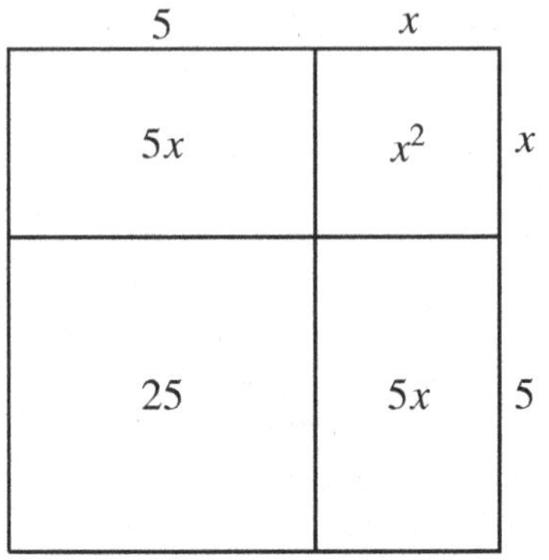

Figure 6.1: Solving a quadratic equation

Euclid and al-Khwārizmī did not admit negative lengths, so the solution $x = -13$ to $x^2 + 10x = 39$ does not appear. This is quite natural, since geometry admits only one square with area 64. Avoidance of negative coefficients, however, causes some unnatural algebraic complications. There is not one general quadratic equation, but three, corresponding to the different ways of distributing positive terms between the two sides: $x^2 + ax = b$, $x^2 = ax + b$, $x^2 + b = ax$.

EXERCISES

Quadratic equations arise frequently in geometry because distance is governed by a quadratic equation (ultimately, by the Pythagorean theorem). In fact, the points created from rational points by any ruler and compass construction can be found by solving a series of linear or quadratic equations, which is why they can be expressed by rational operations and square roots. This result, which was claimed in Section 2.3, can be proved as follows.

6.3.1 Show that the line through two rational points has an equation with rational coefficients.

6.3.2 Show that a circle whose center is a rational point and whose radius is rational has an equation with rational coefficients.

Your proof should show, more generally, that a line or circle constructed from *any* points has an equation with coefficients obtainable from the coordinates of the given points by rational operations. It then suffices to show that intersections of lines and circles can be obtained from the coefficients of their equations by rational operations and square roots.

6.3.3 Show that the intersection of two lines can be computed by rational operations.

6.3.4 Show that the intersection of a line and a circle can be computed by rational operations and a square root (because it depends on solving a quadratic equation).

The last, and hardest, case is finding the intersection of two circles. Fortunately, it is easy to reduce these two quadratic equations to the case just handled in Exercise 6.3.4.

6.3.5 The equations of any two circles can be written in the form

$$(x - a)^2 + (y - b)^2 = r^2,$$
$$(x - c)^2 + (y - d)^2 = s^2.$$

Explain why. Now subtract one of these equations from the other, and hence show that their common solutions can be found by rational operations and square roots.

When a sequence of quadratic equations is solved, the solution may involve *nested* square roots, such as $\sqrt{(5 + \sqrt{5})/2}$. This very number, in fact, occurs in the icosahedron, as one sees from Pacioli's construction in Section 2.2.

6.3.6 Show that the diagonal of a golden rectangle (which is also the diameter of an icosahedron of edge length 1) is $\sqrt{(5 + \sqrt{5})/2}$.

6.4 Quadratic Irrationals

The roots of quadratic equations with rational coefficients are numbers of the form $a + \sqrt{b}$, where a, b are rational. Euclid took the theory of irrationals further in Book X of the *Elements* with a very detailed study of numbers of the form $\sqrt{\sqrt{a} \pm \sqrt{b}}$, where a, b are rational. Book X is the longest book in the *Elements* and it is not clear why Euclid devoted so much space to this topic: perhaps because some of it is needed for the study of regular polyhedra in Book XIII (see Section 2.2 and Exercise 6.3.6), perhaps simply because it was Euclid's favorite topic, or perhaps it was one in which he had some original contributions to show off. It is said that Apollonius took the theory of irrationals further, but unfortunately his work on the subject is lost.

After this, there seems to have been no progress in the theory of irrationals until the Renaissance, except for a remarkable isolated result by Fibonacci (1225). Fibonacci showed that the roots of $x^3 + 2x^2 + 10x = 20$ are not any of Euclid's irrationals. This is *not* a proof, as some historians have thought, that the roots cannot be constructed by ruler and compass. Fibonacci did not rule out *all* expressions built from rationals and square roots; nevertheless, it was the first step into the world of irrationals beyond Euclid.

At this point it is worth asking how difficult it is to show that a specific number, say, $\sqrt[3]{2}$, cannot be constructed from rational numbers by square roots. The answer will depend on how well the reader manages the following exercises. The manipulation required would certainly not have been beyond the 16th-century algebraists. The subtle part is finding a suitable classification of expressions according to complexity—extending Euclid's classification to expressions in which radical signs are nested to arbitrary depth—and using induction on the level of complexity. This type of thinking did not emerge until the 1820s, hence the relatively late proof that $\sqrt[3]{2}$ is not constructible by ruler and compass, by Wantzel (1837).

EXERCISES

An elementary proof that $\sqrt[3]{2}$ is not constructible was found by the number theorist Edmund Landau (1877–1938) when he was still a student. It is broken down to easy steps below. But first we should check that $\sqrt[3]{2}$ is actually irrational.

6.4.1 Show that the assumption $\sqrt[3]{2} = m/n$, where m and n are integers, leads to a contradiction.

Landau's proof now organizes all numbers involved in a construction into sets $F_0, F_1, F_2, \ldots$, according to the "depth of nesting" of square roots.

6.4.2 Let

$$F_0 = \{\text{rationals}\}, \quad F_{k+1} = \{a + b\sqrt{c_k} : a, b, c_k \in F_k\} \quad \text{for some } c_k \in F_k.$$

Show that each F_k is a *field*, that is, if x, y are in F_k, then so are $x + y$, $x - y$, xy, and x/y (for $y \neq 0$).

We know from Exercise 6.4.1 that $\sqrt[3]{2}$ is not in F_0, but if it is constructible it will occur in some F_{k+1}. A contradiction now ensues by considering (hypothetically) the first such F_{k+1}.

6.4.3 Show that if $a, b, c \in F_k$ but $\sqrt{c} \notin F_k$, then $a + b\sqrt{c} = 0 \Leftrightarrow a = b = 0$. (For $k = 0$ this is in the *Elements*, Book X, Prop. 79.)

6.4.4 Suppose $\sqrt[3]{2} = a + b\sqrt{c}$, where $a, b, c \in F_k$, but that $\sqrt[3]{2} \notin F_k$. (We know that $\sqrt[3]{2} \notin F_0$ from Exercise 6.4.1.) Cube both sides and deduce from Exercise 6.4.3 that

$$2 = a^3 + 3ab^2c \quad \text{and} \quad 0 = 3a^2b + b^3c.$$

6.4.5 Deduce from Exercise 6.4.4 that $\sqrt[3]{2} = a - b\sqrt{c}$ also, and explain why this is a contradiction.

6.5   The Solution of the Cubic

> In our own days Scipione del Ferro of Bologna has solved the case of the cube and first power equal to a constant, a very elegant and admirable accomplishment. Since this art surpasses all human subtlety and the perspicuity of mortal talent and is a truly celestial gift and a very clear test of the capacity of men's minds, whoever applies himself to it will believe that there is nothing that he cannot understand. In emulation of him, my friend Niccolò Tartaglia of Brescia, wanting not to be outdone, solved the same case when he got into a contest with his [Scipione's] pupil, Antonio Maria Fior, and, moved by my many entreaties, gave it to me ... having received Tartaglia's solution and seeking a proof of it, I came to understand that there were a great many other things that could also be had. Pursuing this thought and with increased confidence, I discovered these others, partly by myself and partly through Lodovico Ferrari, formerly my pupil.
>
> Cardano (1545), p. 8

The solution of cubic equations in the early 16th century was the first clear advance in mathematics since the time of the Greeks. It revealed the power of algebra that the Greeks had not been able to harness, power that was soon to clear a new path to geometry, which was virtually a royal road (analytic geometry and calculus). Cardano's elation at the discovery is completely understandable. Even in the 20th century, personally discovering the solution of the cubic equation has been the inspiration for at least one distinguished mathematical career—see Kac (1984).

As far as the history of the original discovery goes, we do not know much more than Cardano tells us. Scipione del Ferro died in 1526, so the first solution was known before then. Tartaglia discovered his solution on February 12, 1535, probably independently, because he solved all problems in the contest with del Ferro's pupil Fior, whereas Fior did not. Cardano has been accused by almost everyone, from Tartaglia on, of stealing Tartaglia's solution, but his own account seems to distribute credit quite fairly. For more background, see the introduction and preface to Cardano (1545) and Crossley (1987).

Cardano presents algebra in the geometric style of al-Khwārizmī (whom he describes as the originator of algebra at the beginning of the book), with the case distinctions that result from avoidance of negative coefficients. By ignoring these complications, his solution can be described as follows. The cubic equation $x^3 + ax^2 + bx + c = 0$ is first transformed into one with no quadratic term by a linear change of variable, namely, $x = y - a/3$. One then has, say,

$$y^3 = py + q.$$

By setting $y = u + v$, the left-hand side becomes

$$(u^3 + v^3) + 3uv(u + v) = 3uvy + (u^3 + v^3),$$

which will equal the previous right-hand side if

$$3uv = p,$$
$$u^3 + v^3 = q.$$

Eliminating v gives a quadratic in u^3,

$$u^3 + \left(\frac{p}{3u}\right)^3 = q,$$

with roots

$$\frac{q}{2} \pm \sqrt{\left(\frac{q}{2}\right)^2 - \left(\frac{p}{3}\right)^3}.$$

By symmetry, we obtain the same values for v^3. And since $u^3 + v^3 = q$, if one of the roots is taken to be u^3, the other is v^3. Without loss of generality we can take

$$u^3 = \frac{q}{2} + \sqrt{\left(\frac{q}{2}\right)^2 - \left(\frac{p}{3}\right)^3},$$

$$v^3 = \frac{q}{2} - \sqrt{\left(\frac{q}{2}\right)^2 - \left(\frac{p}{3}\right)^3},$$

and hence

$$y = u + v = \sqrt[3]{\frac{q}{2} + \sqrt{\left(\frac{q}{2}\right)^2 - \left(\frac{p}{3}\right)^3}} + \sqrt[3]{\frac{q}{2} - \sqrt{\left(\frac{q}{2}\right)^2 - \left(\frac{p}{3}\right)^3}}.$$

EXERCISES

The two equations $3uv = p$, $u^3 + v^3 = q$ provide another instance of the phenomenon noted in Exercise 6.2.2: when a variable is eliminated between two equations, the degrees of the equations are multiplied.

6.5.1 The equation $3uv = p$ is of degree 2 in u and v, and $u^3 + v^3 = q$ is of degree 3. What about the equation obtained by eliminating v?

The Cardano formula produces some surprising results, which we look at again in Section 14.3. But first let us test it on a really simple cubic equation.

6.5.2 Use Cardano's formula to solve $y^3 = 2$. Do you get the obvious solution?

Now try one where the solution is less obvious.

6.5.3 Use Cardano's formula to solve $y^3 = 6y + 6$, and check your answer by substitution.

6.6   Angle Division

Another important contributor to algebra in the 16th century was Viète (1540–1603). He helped emancipate algebra from the geometric style of proof by introducing letters for unknowns and using plus and minus signs to facilitate manipulation. Yet at the same time he strengthened its ties with geometry at a higher level by relating algebra to trigonometry. A case in point is his solution of the cubic by circular functions (Viète (1591), Ch. VI, Theorem 3), which shows that solving the cubic is equivalent to trisecting an arbitrary angle.

Namely, if we take the cubic in the form

$$x^3 + ax + b = 0,$$

we can reduce it to an equation

$$4y^3 - 3y = c$$

with just one parameter, by setting $x = ky$ and choosing k so that

$$\frac{k^3}{ak} = \frac{-4}{3}, \qquad \text{or} \qquad k = \sqrt{\frac{-4a}{3}}.$$

The point of the expression $4y^3 - 3y$ is that

$$4\cos^3\theta - 3\cos\theta = \cos 3\theta;$$

hence by setting $y = \cos\theta$ we obtain

$$\cos 3\theta = c.$$

If we are given c, then we can construct a triangle with angle $\cos^{-1} c = 3\theta$. Trisection of this angle gives us the solution $y = \cos\theta$ of the equation. Conversely, the problem of trisecting an angle with cosine c is equivalent to solving the cubic equation $4y^3 - 3y = c$.

Of course, there is a problem with trigonometric interpretation when $|c| > 1$, which requires complex numbers for its resolution. Complex numbers are also involved in Cardano's formula, since the expression under the square root sign, $(q/2)^2 - (p/3)^3$, can be negative. It so happens that Viète's method requires complex numbers only when Cardano's does not, so between the two of them, complex numbers are avoided. Nevertheless, cubic equations are the birthplace of complex numbers, as we shall see, when we study complex numbers in more detail later.

Astonishingly, the problem of dividing an angle into any odd number of equal parts turns out to have an algebraic solution analogous to the algebraic solution of the cubic. Viète (1579) himself took the problem as far as finding expressions for $\cos n\theta$ and $\sin n\theta$ as polynomials in $\cos\theta$ and $\sin\theta$, at least for certain values of n. Newton read Viète in 1663–4 and found the equation

$$y = nx - \frac{n(n^2-1)}{3!}x^3 + \frac{n(n^2-1)(n^2-3^2)}{5!}x^5 + \cdots$$

relating $y = \sin n\theta$ and $x = \sin\theta$ (see Newton (1676a) in Turnbull (1960)). He asserted this result for arbitrary n, but we are interested in the case of odd integral n, when it reduces to a polynomial equation. The surprise is that Newton's equation then has a solution by nth roots analogous to the Cardano formula for cubics,

$$x = \frac{1}{2}\sqrt[n]{y + \sqrt{y^2-1}} + \frac{1}{2}\sqrt[n]{y - \sqrt{y^2-1}}, \tag{1}$$

although only for n of the form $4m+1$. This formula appears out of the blue in de Moivre (1707). (It also appears in the unpublished Leibniz (1675),

though without the restriction on n. See Schneider (1968), pp. 224–228.)
He does not explain how he found it, but it is comprehensible to us as

$$\sin \theta = \frac{1}{2} \sqrt[n]{\sin n\theta + i \cos n\theta} + \frac{1}{2} \sqrt[n]{\sin n\theta - i \cos n\theta}, \tag{2}$$

a consequence of *our* version of de Moivre's formula

$$(\cos \theta + i \sin \theta)^n = \cos n\theta + i \sin n\theta \tag{3}$$

when $n = 4m + 1$. (See Exercises 6.6.1 and 6.6.2.)

Viète himself came remarkably close to (3) in a posthumously pub-
lished work, Viète (1615). He observed that the products of $\sin \theta$, $\cos \theta$
that occur in $\cos n\theta$, $\sin n\theta$ are the alternate terms in the expansion of
$(\cos \theta + \sin \theta)^n$, except for certain minus signs. He failed only to notice
that the signs could be explained by giving $\sin \theta$ the coefficient i. In any
case, such an explanation would not have seemed natural to his contemporaries, who were far more comfortable with Cardano's formula than they
were with i. In Section 14.5 we shall see how the perception of de Moivre's
formula changed with the development of complex numbers.

EXERCISES

The reasons why (1) and (2) hold only for certain integer values, while (3)
holds for all, can be understood by actually working out $(\sin \theta + i \cos \theta)^n$.

6.6.1 Use (3) and $\sin \alpha = \cos(\pi/2 - \alpha)$, $\cos \alpha = \sin(\pi/2 - \alpha)$ to show that

$$(\sin \theta + i \cos \theta)^n = \begin{cases} \sin n\theta + i \cos n\theta & \text{when } n = 4m + 1 \\ -\sin n\theta - i \cos n\theta & \text{when } n = 4m + 3. \end{cases}$$

6.6.2 Deduce from Exercise 6.6.1 that (2) is correct for $n = 4m + 1$ and false for
$n = 4m + 3$, and hence that (1) is a correct relation between $y = \sin n\theta$ and
$x = \sin \theta$ only when $n = 4m + 1$.

6.6.3 Show that (1) is a correct relation between $y = \cos n\theta$ and $x = \cos \theta$ for *all*
n (de Moivre (1730)).

6.7   Higher-Degree Equations

The general fourth–degree, or *quartic*, equation

$$x^4 + ax^3 + bx^2 + cx + d = 0$$

was solved by Cardano's student Ferrari, and the solution was published
in Cardano (1545), p. 237. A linear transformation reduces the equation to
the form

$$x^4 + px^2 + qx + r = 0,$$

or

$$(x^2 + p)^2 = px^2 - qx + p^2 - r.$$

Then for any y,

$$(x^2 + p + y)^2 = (px^2 - qx + p^2 - r) + 2y(x^2 + p) + y^2$$
$$= (p + 2y)x^2 - qx + (p^2 - r + 2py + y^2).$$

The quadratic $Ax^2 + Bx + C$ on the right-hand side will be a square if
$B^2 - 4AC = 0$, which is a cubic equation for y. We can therefore solve for
y and take the square root of both sides of the equation for x, which then
becomes quadratic and hence also solvable. The final result is a formula for
x using just square and cube roots of rational functions of the coefficients.

   This impressive bonus to the solution of cubic equations raised hopes
that higher-degree equations could also be solved by formulas built from
the coefficients by rational operations and roots, and *solution by radicals*,
as it was called, became a major goal of algebra for the next 250 years.
However, all such efforts to solve the general equation of fifth degree (quintic) failed. The most that could be done was to reduce it to the form

$$x^5 - x - A = 0$$

with only one parameter. This was done by Bring (1786), and a sketch of
his method may be seen in Pierpont (1895). Bring's result appeared in a
very obscure publication and went unnoticed for 50 years, or it might have
rekindled hopes for the solution of the quintic by radicals. As it happened,
Ruffini (1799) offered the first proof that this is impossible. Ruffini's proof
was not completely convincing; however, he was vindicated when a satisfactory proof was given by Abel (1826), and again with the beautiful
general theory of equations of Galois (1831b).

   A positive outcome of Bring's result was the nonalgebraic solution of
the quintic by Hermite (1858). The reduction to an equation with one parameter opened the way to a solution by transcendental functions, analogous to Viète's solution of the cubic by circular functions. The appropriate
functions, the elliptic modular functions, had been discovered by Gauss,

Abel, and Jacobi, and Galois (1831a) had hinted at their relation to quintic equations. This extraordinary convergence of mathematical ideas was the subject of Klein (1884).

In view of the difficulties with the quintic, there was naturally very little progress with the general equation of degree n. However, two simple but important contributions were made by Descartes (1637). The first was the superscript notation for powers we now use: x^3, x^4, x^5, and so on. (Though not x^2, oddly enough. The square of x continued to be written xx until well into the next century.) The second was the theorem of Descartes (1637), p. 159, that a polynomial $p(x)$ with value 0 when $x = a$ has a factor $(x - a)$. Since division of a polynomial $p(x)$ of degree n by $(x - a)$ leaves a polynomial of degree $n - 1$, Descartes's theorem raised the hope of factorizing each nth-degree polynomial into n linear factors. As Chapter 14 shows, this hope was fulfilled with the development of complex numbers.

EXERCISES

The main steps in the proof of Descartes's theorem go as follows. If the first step does not seem sufficiently easy, begin with $a = 1$.

6.7.1 Show that $x^n - a^n$ has a factor $x - a$. What is the quotient $(x^n - a^n)/(x - a)$? (And what does this have to do with geometric series?)

6.7.2 If $p(x) = a_k x^k + a_{k-1} x^{k-1} + \cdots + a_1 x + a_0$, use Exercise 6.7.1 to show that $p(x) - p(a)$ has a factor $x - a$.

6.7.3 Deduce Descartes's theorem from Exercise 6.7.2.

6.8   Biographical Notes: Tartaglia, Cardano, and Viète

Little is known about Scipione del Ferro, the discoverer of the first solution to cubic equations, other than his dates (1465–1526) and the fact that he was a professor of arithmetic and geometry at Bologna from 1496. This has possibly resulted in Tartaglia and Cardano receiving more mathematical credit than they deserve. On the other hand, there is no denying that Tartaglia's and Cardano's personalities, their contrasting lives, and their quarrel make a story that is fascinating in its own right.

Niccolò Tartaglia (Figure 6.2) was born in Brescia in 1499 or 1500 and died in Venice in 1557. The name "Tartaglia" (meaning "stutterer") was actually a nickname; his real name is believed to have been Fontana.

Figure 6.2: Niccolò Tartaglia

Tartaglia's childhood was scarred by poverty, following the death of his father, a mail courier, around 1506, and injuries suffered when Brescia was sacked by the French in 1512. Despite taking refuge in a cathedral, Tartaglia received five serious head wounds, including one to the mouth, which left him with his stutter. His life was saved only by the devoted nursing of his mother, who literally licked his wounds. Around the age of 14 he went to a teacher to learn the alphabet, but he ran out of money for his lessons by the letter K. This much is in Tartaglia's own sketch of his life, Tartaglia (1546), p. 69. After that, the story goes, he stole a copybook and taught himself to read and write, sometimes using tombstones as slates for want of paper.

By 1534 he had a family and, still short of money, he moved to Venice. There he gave public mathematics lessons in the church of San Zanipolo and published scientific works. The famous disclosure of his method for solving cubic equations occurred on a visit to Cardano's house in Milan on March 25, 1539. When Cardano published it in 1545, Tartaglia angrily accused him of dishonesty. Tartaglia (1546), p. 120, claimed that Cardano had solemnly sworn never to publish the solution and to write it down only in cipher. Ferrari, who had been an 18-year-old servant of Cardano at the time, came to Cardano's defense, declaring that he had been present and there had been no promise of secrecy. In a series of 12 printed pamphlets, known as the *Cartelli* (reprinted by Masotti (1960)), Ferrari and Tartaglia traded insults and mathematical challenges; the two finally squared off in a public contest in the church of Santa Maria del Giardino, Milan, in 1548. It seems that Ferrari got the better of the exchange, as there was little subsequent improvement in Tartaglia's fortunes. He died alone, still impoverished, nine years later.

Apart from his solution of the cubic, Tartaglia is remembered for other contributions to science. It was he who discovered that a projectile should be fired at 45° to achieve maximum range (Tartaglia (1546), p. 6). His conclusion was based on incorrect theory, however, as is clear from Tartaglia's diagrams of trajectories—for example, Figure 6.3; Tartaglia (1546), p. 16.

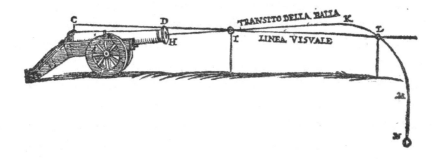

Figure 6.3: Tartaglia's trajectory of a cannonball

Tartaglia's Italian translation of the *Elements* was the first printed translation of Euclid in a modern language, and he also published an Italian translation of some of Archimedes' works. For information on these, and Tartaglia's mechanics, see Rose (1976), pp. 151–154.

Girolamo Cardano (Figure 6.4), often described in older English books by the anglicized name Jerome Cardan, was born in Pavia in 1501 and died in Rome in 1576. His father, Fazio, was a lawyer and physician who encouraged Girolamo's studies but otherwise seems to have treated him rather harshly, as did his mother, Chiara Micheri, whom Cardano described as "easily provoked, quick of memory and wit, and a fat, devout little woman." Cardano entered the University of Pavia in 1520 and completed a doctorate of medicine at Padua in 1526.

He married in 1531 and, after struggling until 1539 for acceptance, became a successful physician in Milan—so successful, in fact, that his fame

Figure 6.4: Girolamo Cardano

spread all over Europe. He evidently had a remarkable skill in diagnosis, though his contributions to medical knowledge were slight in comparison with those of his contemporaries Andreas Vesalius and Ambroise Paré. Mathematics was one of his many interests outside his profession. Cardano also secured a niche in the history of cryptography for an encoding device known as the Cardano grille (see Kahn (1967), pp. 143–145) and in the history of probability, where he was the first to make calculations, though not always correctly (see David (1962), pp. 40–60, and Ore (1953), which contains a translation of Cardano's book on games of chance).

The violence and intrigue of Renaissance Italy soured Cardano's life just as much as Tartaglia's, though in a different way. An uncle died of poisoning, attempts were made to poison both Cardano and his father (so Cardano said), and in 1560 Cardano's oldest son was beheaded for the crime of poisoning his wife. Cardano, who thought his son's only fault was to marry the girl in the first place, never got over this calamity. He could no longer bear to live in Milan and moved to Bologna. There he suffered another blow when his protégé Ferrari died in 1565—supposedly poisoned by his sister. In 1570 the Inquisition imprisoned Cardano for heresy. After a few months he recanted, was released, and moved to Rome.

In the year before he died, Cardano wrote *The Book of My Life* (Cardano (1575)), which is not so much autobiography as self-advertisement. It contains a few scenes from his childhood and returns again and again to the tragedy of his oldest son, but most of the book is devoted to boasting. There is a chapter of testimonials from patients, a chapter on important people who sought his services, a list of authors who cited his works, a list of his sayings he considered quotable, and a collection of tall stories that would have done Baron von Münchhausen proud. Admittedly, there is also a (very short) chapter called "Things in Which I Have Failed" and frequent warnings about the vanity of earthly things, but Cardano invariably tramples all such outbreaks of humility in his rush to admire other facets of his excellent self.

On the quarrel with Tartaglia, *The Book of My Life* is almost silent. Among the authors who have cited him, Cardano lumps Tartaglia with those of whom he "cannot understand by what impertinence they have managed to get themselves into the ranks of the learned." Only at the end of the book does Cardano concede that "in mathematics I received a few suggestions, but very few, from brother Niccolò." Thus we are forced back to the *Cartelli* and Tartaglia's writings. The most accessible analysis of these works, with translations of relevant passages, is in Ore (1953), Chapter 4.

François Viète (Figure 6.5) was born in 1540 in Fontenay-le-Comte, a town in what is now the Vendée department of France. His father, Etienne, was a lawyer, and his mother, Marguerite Dupont, was well connected to ruling circles in France. Viète was educated by the Franciscans in Fontenay and at the University of Poitiers. He received his bachelor's degree in law in 1560 and then returned to Fontenay to commence practice.

For the rest of his life he was engaged mainly in law or related judicial and court services, doing mathematics only in periods of leisure. His clients are said to have included Queen Mary of England and Eleanor of Austria, and from 1574 to 1584 he acted as an advisor and negotiator for King Henry III of France. At that stage he was banished through the efforts of political rivals, but he returned to court in 1589 when Henry III moved his seat of government from Paris to Tours. Following the assassination of Henry III in 1589, he served Henry IV until 1602. Viète died in 1603.

The most famous exploit of Viète's professional career was his deciphering of Spanish dispatches for Henry IV during the war against Spain. King Philip II of Spain, unable to believe that this was humanly possible,

Figure 6.5: François Viète

protested to the pope that the French were using black magic. The pope may well have been impressed, but not enough to believe that magic was involved, as the Vatican's own experts had broken one of Philip's codes 30 years earlier (see Kahn (1967), pp. 116–118).

An equally famous mathematical feat of Viète's, and equally magical to his contemporaries, was his solution of a 45th-degree equation posed by Adriaen van Roomen in 1593:

$$45x - 3795x^3 + 95634x^5 - \cdots + 945x^{41} - 45x^{43} + x^{45} = N.$$

Viète saw immediately that this equation resulted from the expansion of $\sin 45\theta$ in powers of $\sin \theta$, and he was able to give 23 solutions (he did not recognize negative solutions). This was one contest, incidentally, that did not generate any bitterness—it led to a firm friendship between the two mathematicians.

7

Analytic Geometry

PREVIEW

The first field of mathematics to benefit from the new language of equations was geometry. Around 1630, both Fermat and Descartes realized that geometric problems could be translated into algebra by means of *coordinates*, and that many problems could then be routinely solved by algebraic manipulation.

The language of equations also provides a simple but natural classification of curves by *degree*. The curves of degree 1 are the straight lines; the curves of degree 2 are the conic sections; so the first "new" curves are those of degree 3, the *cubic curves*.

Cubic curves exhibit new geometric features—cusps, inflections, and self-intersections—so they are considerably more complicated than the conic sections. Nevertheless, Newton attempted to classify them, and in doing so he discovered that cubic curves, when properly viewed, are not as complicated as they seem.

We will find our way to the "right" viewpoint in Chapters 8 and 15. In the meantime we discuss another theorem that depends on the "right" viewpoint: *Bézout's theorem*, according to which a curve of degree m always meets a curve of degree n in mn points.

J. Stillwell, *Mathematics and Its History*, Undergraduate Texts in Mathematics,
DOI 10.1007/978-1-4419-6053-5_7, © Springer Science+Business Media, LLC 2010

7.1   Steps Toward Analytic Geometry

The basic idea of analytic geometry is the representation of curves by equations, but this is not the whole idea. If it were, then the Greeks would be considered the first analytic geometers. Menaechmus was perhaps the first to discover equations of curves, along with his discovery of the conic sections, and we have seen how he used equations to obtain $\sqrt[3]{2}$ as the intersection of a parabola and a hyperbola (Section 2.4). Apollonius' study of conics used equations obtained as by-products of geometric arguments.

What was lacking in Greek mathematics was both the inclination and the technique to manipulate equations to obtain information about curves. The Greeks used curves to study algebra rather than the other way around. Menaechmus's construction of $\sqrt[3]{2}$ is a fine example of this: extraction of roots was not a given operation but one to be secured by geometric construction. Similarly, an equation was not an entity in its own right but a property of a curve that could be discovered after the curve had been constructed geometrically. This was a natural state of affairs as long as equations were written out in words. When, as in Apollonius, an equation takes half a page to write out, it is difficult to form a general concept of equation, function, or curve. Hence the lack of a general concept of curve in Greek mathematics—it was just too complicated to handle in their language.

In the Middle Ages the idea of coordinates emerged in a different way in the work of Oresme (around 1323–1382). Coordinates had been used in astronomy and geography since Hipparchus (around 150 BCE); in fact, Oresme called his coordinates "longitude" and "latitude," but he seems to have been the first to use them to represent functions such as velocity as a function of time. Setting up the coordinate system *before* determining the curve was Oresme's step beyond the Greeks, but he too lacked the algebra to go further.

The step that finally made analytic geometry feasible was the solution of equations and the improvement of notation in the 16th century, which we discussed in the previous chapter. This step made it possible to consider equations, and hence curves, in some generality and to have confidence in one's ability to manipulate them. As we shall see in the next section, the two founders of analytic geometry, Fermat and Descartes, were both strongly influenced by these developments.

For more details on the development of analytic geometry, the reader is referred to an excellent book by Boyer (1956).

7.1.1 Generalize the idea of Menaechmus to show that any cubic equation

$$ax^3 + bx^2 + cx + d = 0 \quad \text{with} \quad d \neq 0$$

may be solved by intersecting the hyperbola $xy = 1$ with a parabola.

7.2   Fermat and Descartes

There have been several occasions in the history of mathematics when an important discovery was made independently and almost simultaneously by two individuals: non-Euclidean geometry by Bolyai and Lobachevsky, elliptic functions by Abel and Jacobi, the calculus by Newton and Leibniz, for example. To the extent that we can rationally explain these remarkable events, it must be on the basis of ideas already "in the air," of conditions becoming favorable for their crystallization. As I tried to show in the previous section, conditions were favorable for analytic geometry at the beginning of the 17th century. Thus it is not completely surprising that the subject was independently discovered by Fermat (1629) and Descartes (1637). (Descartes's work *La Géométrie* may in fact have been started in the 1620s. In any case it is independent of Fermat, whose work was not published until 1679.)

It is a surprise to learn, however, that both Fermat and Descartes began with an analytic solution of the same classical geometric problem, the four-line problem of Apollonius, and that the main discovery of each was that second-degree equations correspond to conic sections. Up to this point Fermat was more systematic than Descartes, but that was as far as he went. He was content to leave his work in a "simple and crude" state, confident that it would grow in stature when nourished by new inventions.

Descartes, on the other hand, treated many higher-degree curves and clearly understood the power of algebraic methods in geometry. He wanted to withhold this power from his contemporaries, however, particularly the rival mathematician Roberval, as he admitted in a letter to Mersenne (see Boyer (1956), p. 104). *La Géométrie* was written to boast about his discoveries, not to explain them. There is little systematic development, and proofs are frequently omitted with a sarcastic remark such as, "I shall not stop to explain this in more detail, because I should deprive you of the pleasure of mastering it yourself" (p. 10). Descartes's conceit is so great that it is a pleasure to see him come a cropper occasionally, as on p. 91: "The ratios between straight and curved lines are not known, and I believe cannot be discovered by human minds." He was referring to the then-unsolved problem of determining the length of curves, but he spoke too soon, for in 1657 Neil and van Heuraet found the length of an arc of the semicubical parabola $y^2 = x^3$, and the calculus soon made such problems routine. (A full and interesting account of the story of arc length may be found in Hofmann (1974), Ch. 8.)

EXERCISES

As we now know, all conic sections may be given by the following standard form equations (from Section 2.4):

$$\frac{x^2}{a^2} + \frac{y^2}{b^2} = 1 \text{ (ellipse)}, \qquad y = ax^2 \text{ (parabola)}, \qquad \frac{x^2}{a^2} - \frac{y^2}{b^2} = 1 \text{ (hyperbola)}.$$

The reduction of an arbitrary quadratic equation in x and y to one of these forms depends on suitable choice of origin and axes, as Fermat and Descartes discovered. The main steps are outlined in the following exercises.

7.2.1 Show that a *quadratic form $ax^2 + bxy + cy^2$* may be converted to a form $a'x'^2 + b'y'^2$ by suitable choice of θ in the substitution

$$x = x' \cos\theta - y' \sin\theta,$$
$$y = x' \sin\theta + y' \cos\theta,$$

by checking that the coefficient of $x'y'$ is $(c - a)\sin 2\theta + b\cos 2\theta$.

7.2.2 Deduce from Exercise 7.2.1 that, by suitable rotation of axes, any quadratic curve may be expressed in the form $a'x'^2 + b'y'^2 + c'x' + d'y' + e' = 0$.

7.2.3 If $b' = 0$, but $a' \neq 0$, show that the substitution $x' = x'' + f$ gives either a standard-form parabola, or the "double line" $x''^2 = 0$.

(Why is this called a "double line," and is it a section of a cone?)

7.2.4 If both a' and b' are nonzero, show that a shift of origin gives the standard form for either an ellipse or a hyperbola, or else a pair of lines.

7.3   Algebraic Curves

I could give here several other ways of tracing and conceiving a series of curved lines, each curve more complex than any preceding one, but I think the best way to group together all such curves and then classify them in order is by recognizing the fact that all points of those curves which we may call "geometric," that is, those which admit of precise and exact measurement, must bear a definite relation to all points of a straight line, and that this relation must be expressed by means of a single equation.

Descartes (1637), p. 48

In this passage Descartes defines what we now call *algebraic curves*. The fact that he calls them "geometric" shows his attachment to the Greek idea that curves are the product of geometric constructions. He is using the notation of equations not to define curves directly but to restrict the notion of geometric construction more severely than the Greeks did, thereby restricting the concept of curve. As we saw in Section 2.5, the Greeks considered some constructions, such as rolling one circle on another, that are capable of producing transcendental curves. Descartes called such curves "mechanical" and found a way to exclude them by his restriction to curves "expressed by means of a single equation." It becomes clear in the lines following the preceding quotation that he means polynomial equations, since he gives a classification of equations by degree.

Descartes's rejection of transcendental curves was short-sighted, since the calculus soon provided techniques to handle them, but nevertheless it was fruitful to concentrate on algebraic curves. The notion of degree, in particular, was a useful measure of complexity. First-degree curves are the simplest possible, namely, straight lines. Those of second degree are the next simplest, conic sections. With third-degree curves one sees the new phenomena of inflections, double points, and cusps. Inflection and cusp are familiar from $y = x^3$ and $y^2 = x^3$, respectively; we also saw a cusp on the cissoid (Section 2.5). A classical example of a cubic with a double point is the *folium* (leaf) *of Descartes* (1638),

$$x^3 + y^3 = 3axy.$$

The "leaf" is the closed portion to the right of the double point; Descartes misunderstood the rest of the curve through neglect of negative coordinates. The true shape of the folium was first given by Huygens (1692). Figure 7.1 is Huygens's drawing, which also shows the asymptote to the curve.

An excellent account of the early history of curves can be found in Brieskorn and Knörrer (1981), Chapter 1. Many individual curves, with

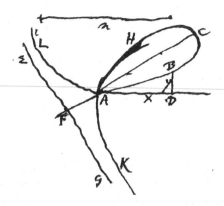

Figure 7.1: Huygens's drawing of the folium

diagrams, equations, and historical notes, can be found in Gomes Teixeira (1995a,b,c). The development of Descartes's concept of curve has been studied by Bos (1981).

EXERCISES

The folium is a cubic curve to which Diophantus's chord method (Section 3.5) applies. One takes the line $y = tx$ through the "obvious" rational point $(0, 0)$ on the curve, and finds its other point of intersection. This construction also enables us to express an arbitrary point (x, y) on the curve in terms of the parameter t.

7.3.1 Show that the folium of Descartes has parametric equations

$$x = \frac{3at}{1 + t^3}, \qquad y = \frac{3at^2}{1 + t^3}$$

and use these equations to show that it is tangential to the axes at 0.

7.3.2 Show that the equation $x^3 + y^3 = 3axy$ of the folium may be written in the form

$$x + y = \frac{3a}{\frac{x}{y} + \frac{y}{x} - 1}.$$

7.3.3 Show that x/y and y/x tend to -1 as $x \to \pm\infty$ on the folium, and hence deduce the equation of its asymptote from Exercise 7.3.2.

A whole family of "multileaved" curves was studied by Grandi (1723).

7.3.4 The *roses of Grandi* are given by the polar equations

$$r = a \cos n\theta$$

for integer values of n. Figure 7.2 shows some of these curves, as given by Grandi (1723). Show that the roses of Grandi are algebraic.

Figure 7.2: Roses of Grandi

7.3.5 Show that the "rose" for $n = 1$ is a circle and that the "rose" for $n = 2$ has cartesian equation
$$(x^2 + y^2)^3 = a^2(x^2 - y^2)^2.$$

7.4 Newton's Classification of Cubics

Since first- and second-degree curves are straight lines and conics, they were well understood before the advent of analytic geometry. Up to the end of the 18th century most mathematicians considered them not amenable to further clarification, and hence an unsuitable subject for the new methods. A famous example is the Greek-style treatment of planetary orbits in Newton's *Principia* (1687). The classical attitude to low-degree curves was summed up by d'Alembert in his article on geometry in the great French *Encyclopédie* (p. 637 of volume 7, 1757):

> Algebraic calculation is not to be applied to the propositions of elementary geometry because it is not necessary to use this calculus to facilitate demonstrations, and it appears that there are no demonstrations which can really be facilitated by this calculus except for the solution of problems of second degree by the line and circle.

Thus the first new problem opened up by analytic geometry, and also the first considered properly to belong to the subject, was the investigation of cubic curves. These curves were classified, more or less completely, by Newton (1695) (see Ball (1890) for a commentary).

Newton (1667) began this work with the general cubic in x and y,

$$ay^3 + bxy^2 + cx^2y + dx^3 + ey^2 + fxy + gx^2 + hy + kx + l = 0,$$

making a general transformation of axes, leading to an equation with 84 terms, then showing that the latter equation could be reduced to one of the forms

$$Axy^2 + By = Cx^3 + Dx^2 + Ex + F,$$
$$xy = Ax^3 + Bx^2 + Cx + D,$$
$$y^2 = Ax^3 + Bx^2 + Cx + D,$$
$$y = Ax^3 + Bx^2 + Cx + D.$$

Newton then divided the curves into species according to the roots of the right-hand side, obtaining 72 species (and overlooking 6). His paper does not contain detailed proofs; these were supplied by Stirling (1717), along with four of the species Newton had missed. Newton's classification was criticized by some later mathematicians, such as Euler, for lacking a general principle. A unifying principle was certainly desirable, to reduce the complexity of the classification. And such a principle was already implicit in one of Newton's passing remarks, Section 29, "On the Genesis of Curves by Shadows." This principle, which will be explained in the next chapter, reduces cubics to the five types seen in Figure 7.3 (taken from an English translation of Newton's paper published in 1710; see Whiteside (1964)).

The reader may wonder where the most familiar cubic, $y = x^3$, appears among these five. The answer is that it is equivalent to the one with a cusp, in Newton's Figure 75. This is explained in the next chapter.

EXERCISES

The cubic curves that Newton called "cuspidate" and "nodated" are algebraically simpler than the others. In particular, they can be parameterized by rational functions.

7.4.1 Find a parameterization $x = p(t)$, $y = q(t)$ of the semicubical parabola $y^2 = x^3$ by polynomials p and q, (i) by inspection, (ii) by finding the second intersection point of the line $y = tx$ through the cusp $(0,0)$.

7.4.2 Find rational functions $x = r(t)$, $y = s(t)$ that parameterize $y^2 = x^2(x + 1)$, by finding the second intersection of the line $y = tx$ through the double point of the curve.

Fig. 71.

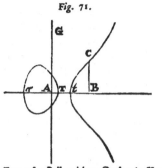

of the Form of a Bell, with an Oval at its Vertex. And this makes a *Sixty seventh Species.*

If two of the Roots are equal, a Parabola will be formed, either *Nodated* by touching an Oval,

Fig. 72.

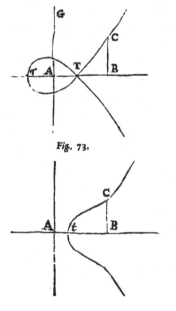

Fig. 73.

or *Punctate,* by having the Oval infinitely small. Which two *Species* are the *Sixty eighth* and *Sixty ninth.*

If three of the Roots are equal, the Parabola will be *Cuspidate* at the Vertex. And this is the

Fig. 75.

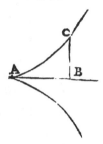

Neilian Parabola, commonly called Semi-cubical. Which makes the *Seventieth Species.*

If two of the Roots are impoffible, there will (See *Fig. 73.*)

Fig. 73.

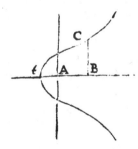

be a *Pure* Parabola of a Bell-like Form. And this makes the *Seventy first Species.*

Figure 7.3: Newton's classification of cubic curves

7.5   Construction of Equations, Bézout's Theorem

In Sections 7.1, 7.2, and 7.3 the development of analytic geometry is outlined from the first observations of equations as properties of curves to the full realization that equations *defined* curves and that the concept of (polynomial) equation was the key to the concept of (algebraic) curve. With hindsight, we can say that Descartes's *La Géométrie* (1637) was a major step in the maturation of the subject, but the book does not conclusively establish what analytic geometry is. In fact, it is largely devoted to two transitional topics in the development of the subject: the 16th-century theory of equations and the now almost forgotten discipline called "construction of equations."

The paradigm for construction of an equation was Menaechmus's construction of $\sqrt[3]{2}$ by intersecting a parabola and hyperbola. From a geometric point of view, one is using familiar curves (parabola and hyperbola) to construct a less familiar length ($\sqrt[3]{2}$). This becomes sharper when expressed algebraically: curves of degree 2 are being used to solve an equation of degree 3, $x^3 = 2$. In the 1620s Descartes discovered something more general: a method of solving any third- or fourth-degree equation by intersecting curves of degree 2, a parabola and a circle. His friend Beeckman (1628) reported in a note that "M. Descartes made so much of this invention that he confessed never to have found anything superior himself and even that nobody else had ever found anything better" (translation by Bos (1981), p. 330). Descartes was not as superior as he thought, since Fermat (1629) independently made the same discovery in an unpublished work, strengthening the already extraordinary coincidence between his work and that of Descartes. However, Fermat apparently did not pursue the idea further, and Descartes did.

In *La Géométrie* Descartes found a particular cubic curve, the so-called cartesian parabola, whose intersections with a suitable circle yield the solution of any given fifth- or sixth-degree equation. Descartes concludes the book with this result, blithely telling the reader that

> it is only necessary to follow the same method to construct all problems, more and more complex, ad infinitum; for in the case of a mathematical progression, whenever the first two or three terms are given, it is easy to find the rest.

> Descartes (1637), p. 240

In reality it was not easy, and efforts to find a satisfactory general construction for nth-degree equations petered out around 1750. The story of the rise and fall of this field of mathematics has been told by Bos (1981, 1984).

In their search for a general construction, mathematicians had casually assumed that a curve of degree m meets a curve of degree n in mn points. The first statement of this principle, which became known as Bézout's theorem, seems to have been made by Newton on May 30, 1665:

> For ye number of points in wch two lines may intersect can never bee greater y^n y^e rectangle of y^e numbers of their dimensions. And they always intersect in soe many points, excepting those wch are imaginarie onely.
>
> Newton (1665b), p. 498

Bézout's theorem leads one to expect that solutions of an equation $r(x) = 0$ of degree $k = m \cdot n$ might be obtainable from the intersections of a suitable degree m curve with a suitable degree n curve. In algebraic terms, one seeks equations

$$p(x, y) = 0, \tag{1}$$
$$q(x, y) = 0 \tag{2}$$

of degrees m, n respectively, from which elimination of y yields the given equation

$$r(x) = 0 \tag{3}$$

as "resultant." This is how mathematicians in the West first encountered the problem of elimination, which the Chinese had solved some centuries earlier (Section 6.2).

However, apart from the fact that construction of equations was inverse to elimination, and much harder, Western mathematicians needed two additional facts about elimination itself: first, that elimination between equations of degrees m and n gave a resultant of degree mn; second, that an equation of degree mn has mn roots. The second statement, as mentioned in Section 6.7, becomes a fact only when complex numbers are admitted. The first becomes a fact only when "points at infinity" are admitted. If, for example, (1) and (2) are equations of parallel lines, then (3) is of "degree 0" and has *no* solutions. However, one can consider parallel lines to meet "at infinity," and the geometric framework for this idea, projective geometry, developed at about the same time as analytic geometry. Unfortunately, it was not realized until the 19th century that projective geometry and analytic geometry needed each other. Until then, projective geometry developed without coordinates, and all attempts to prove Bézout's theorem—notably by Maclaurin (1720), Euler (1748b), Cramer (1750), and Bézout (1779)—foundered for want of a proper method for counting points at infinity. As a result, Bézout's theorem, which turned out to be the main achievement of the theory of construction of equations, was not properly proved until long after the theory itself had been abandoned.

The origins of projective geometry, and the fruits of its merger with analytic geometry, are discussed in Chapter 8.

EXERCISES

We know from Section 6.7 that an arbitrary quartic equation is equivalent to one of the form

$$x^4 + px^2 + qx + r = 0.$$

7.5.1 Show that any such equation may be solved by finding the intersection of the parabola $y = x^2$ with another quadratic curve (hence with a conic section).

7.5.2 Find two parabolas whose intersections give the solutions of $x^4 = x + 1$, and hence show that this quartic equation has two real roots.

7.6   The Arithmetization of Geometry

We have stressed that early analytic geometers—Descartes in particular—did not accept that geometry could be *based* on numbers or algebra. Perhaps the first to take the idea of arithmetizing geometry seriously was Wallis (1616–1703). Wallis (1657), Chs. XXIII and XXV, gave the first arithmetic treatment of Euclid's Books II and V, and Wallis (1655b) had earlier given the first purely algebraic treatment of conic sections. He initially derived equations from the classical definitions by sections of the cone but then proceeded conversely to derive their properties from the equations, "without the embranglings of the cone," as he put it.

Wallis was ahead of this time. Thomas Hobbes, introduced at the beginning of Chapter 2, described Wallis's treatise on conics as a "scab of symbols" and denounced "the whole herd of them who apply their algebra of geometry" (Hobbes (1656), p. 316, and Hobbes (1672), p. 447). The example and authority of Newton probably reinforced the opinion that algebra was inappropriate in the geometry of lines or conic sections; we saw in Section 7.4 how this remained the accepted view until at least 1750.

Algebra did not catch on in elementary geometry until it was taken up by Lagrange (1773b) and supported by influential textbooks of Monge and Lacroix around 1800. But by the time elementary geometry had been brought into the theory of equations, higher geometry had broken out, depending more and more on calculus and the emerging theories of complex functions, abstract algebra, and topology, which bloomed in the 19th century. Higher geometry broke away to form differential geometry and algebraic geometry, leaving the elementary residue we call "analytic geometry" today.

Despite its lowly status, analytic geometry was given an important foundational role by Hilbert (1899). Hilbert took Wallis's arithmetization to its logical conclusion by assuming only the real numbers and sets as given and constructing *Euclidean geometry* from them.

Thus from the set $\mathbb{R}$ of reals, one constructs the *Euclidean plane* as the set of ordered pairs (x, y) ("points") where $x, y \in \mathbb{R}$. A *straight line* is a set of points (x, y) in the plane such that $ax + by + c = 0$ for some constants a, b, c. Lines are *parallel* if their x and y coefficients are proportional. The *distance* between points (x_1, y_1) and (x_2, y_2) is defined to be $\sqrt{(x_2 - x_1)^2 + (y_2 - y_1)^2}$. As explained in Section 1.6, this definition is motivated by the Pythagorean theorem, which is the keystone in the bridge from arithmetic to geometry.

With these definitions, all axioms and propositions of Euclid's geometry become provable propositions about equations. For example, the axiom that nonparallel lines have a point in common corresponds to the theorem that linear equations

$$a_1 x + b_1 y + c_1 = 0,$$
$$a_2 x + b_2 y + c_2 = 0$$

have a solution when $a_1 b_2 - b_1 a_2 \neq 0$.

Hilbert did not believe, any more than Newton did, that numbers were the true subject matter of geometry. He strongly supported geometric intuition as a method of discovery, as the book Hilbert and Cohn-Vossen (1932) makes clear. The purpose of his arithmetization was to give a secure logical foundation to geometry after the 19th-century developments that discredited geometry and installed arithmetic as the ultimate authority in mathematics. This foundation is no longer quite as secure as it seemed in 1900, as we shall see in Chapter 24; nevertheless, it is still the most secure foundation we know.

7.7   Biographical Notes: Descartes

René Descartes (Figure 7.4) was born in La Haye (now called La Haye-Descartes) in the French province of Touraine in 1596 and died in Stockholm in 1650. His father, Joachim, was a councilor in the high court of Rennes in Brittany; his mother, Jeanne, was the daughter of a lieutenant general from Poitiers and the owner of property that was eventually to assure Descartes of financial independence. His mother died in 1597, and Descartes was raised by his maternal grandmother and a nurse. He does not seem to have been close to his father, brother, or sister, seldom mentioning them to others and writing to them only on matters of business.

Figure 7.4: René Descartes

Joachim Descartes was away from home for half the year because of his court duties, but he saw enough of René to observe his exceptional curiosity, calling him his "little philosopher." In 1606 he enrolled him in the Jesuit College of La Flèche, which had recently been founded by Henry IV in Anjou. The young Descartes was given special privileges at school, in recognition of his intellectual promise and delicate health. He was one of the few boys to have his own room, was permitted books forbidden to other students, and was allowed to stay in bed until late in the morning. Spending several morning hours in bed thinking and writing became his lifelong habit, and when he finally had to break it in the Swedish winter, the consequences were fatal.

The most dramatic event of his schooldays was the assassination of Henry IV in 1610. Since Henry IV was not only the founder of the school but also the most popular king in French history, his death was a profound shock. La Flèche became the venue for an elaborate funeral ceremony, the climax of which was the burial of the king's heart. Descartes was one of 24 students chosen to participate in the ceremony.

He left La Flèche in 1614 and, after legal studies at Poitiers, which seem to have left no impression on him, went to Holland as an unpaid volunteer in the army of Prince Maurice of Nassau in 1618. This was not an unusual decision for a young Frenchman of means at the time, since the Dutch were fighting France's enemy, Spain, and Descartes seems to have joined the army to see the world, not because of any taste for barracks life or combat. As it happened, there was then a lull in the war, and Descartes had two years of virtual leisure to reflect on science and philosophy.

When in Breda, on November 10, 1618, he saw a mathematical problem posted on a wall. Since his Dutch was not yet fluent, he asked a bystander to translate it for him. This was how Descartes met Isaac Beeckman, who became his first instructor in mathematics and a lifelong friend. The following November 10, Descartes was in Bavaria. He spent a day of intense thought in a heated room ("stove" he called it) and that night had a dream he later considered to be a revelation of the path he should follow in developing his philosophy. Whether the dream also revealed the path to analytic geometry, as some have conjectured, will probably never be known. Descartes's own description of the dream has been lost, and we have only a summary by his first biographer, Baillet (1691), p. 85, which is not helpful. In any case, it seems a little ludicrous to award Descartes priority over Fermat on the basis of a dream. Could a counterclaim of priority be lodged if the dream of a teenaged Fermat came to light?

In 1628 Descartes moved to Holland, where he spent most of the rest of his life. He lived a simple but leisurely life and finally settled down to working out the ideas conceived nine years earlier. The relative isolation suited him, as he was hostile to other scientific giants of his time such as Galileo, Fermat, and Pascal and preferred to communicate with scholars who could understand him without challenging his superiority. One such was Marin Mersenne, who had been a senior student at La Flèche in Descartes's time and was his main scientific contact in France. Others, with whom Descartes had extensive correspondences, were Princess Elizabeth of Bohemia and Queen Christina of Sweden.

A positive side to Descartes's intolerance of intellectual rivals was an apparently genuine interest in the affairs of his neighbors in Holland. He encouraged local youths who showed talent in mathematics, and he was known in the region as someone to turn to in times of trouble (see Vrooman (1970), pp. 194–196). The one serious love of his life was a servant girl named Helen, who bore him a daughter, Francine, in 1635. Admittedly, his interest in this case did not extend to marrying Helen, but the death of Francine from scarlet fever in 1640 was the greatest sorrow of his life.

In 1649 Descartes agreed to journey to Stockholm to become tutor to Queen Christina. This was the culmination of his correspondence with her and of negotiations through Descartes's friend Chanut, the French ambassador. The queen, who was noted for her physical as well as mental vigor, slept no more than five hours a night and rose at 4 a.m. Descartes had to arrive at 5 a.m. to give her lessons in philosophy. The program commenced on January 14, 1650, during the coldest winter in over 60 years. One can imagine the shock to Descartes's system of such early rising followed by a journey from the ambassador's residence to the palace. However, it was actually Chanut who succumbed to the cold first. On January 18 he came down with pneumonia, and Descartes apparently caught it from him. Chanut recovered but Descartes did not, and he died on February 11, 1650.

Descartes is, of course, as well known for his philosophy as his analytic geometry. The *Geometry* was originally an appendix to his main philosophical work, the *Discourse on Method*. The other appendices were the *Dioptrics*, a treatise on optics, and the *Meteorics*, the first attempt to give a scientific theory of the weather.

In the *Dioptrics*, Descartes did not inform his readers that Ptolemy, al-Haytham, Kepler, and Snell had already discovered the main principles of optics; nevertheless, he presented the subject with greater clarity and thoroughness than before, undoubtedly advancing both the theory and practice of optical instrumentation. As for the *Meteorics*, we now know how premature it was to attempt a theory of the weather in 1637, so it is understandable that this treatise has more misses than hits. His big hit was a correct explanation of rainbows (except for the colors, whose explanation was completed by Newton), which Descartes was able to give on the basis of his optics. More typical, unfortunately, was his explanation of thunder: it was caused by clouds bumping together, and not related to lightning. An excellent survey of Descartes's scientific work and philosophy, with a particularly detailed analysis of the *Geometry*, is given by Scott (1952).

8

Projective Geometry

PREVIEW

At about the same time as the algebraic revolution in classical geometry, a new *kind* of geometry also came to light: *projective* geometry. Based on the idea of projecting a figure from one plane to another, projective geometry was initially the concern of artists. In the 17th century, only a handful of mathematicians were interested in it, and their discoveries were not seen to be important until the 19th century.

The fundamental quantities of classical geometry, such as length and angle, are not preserved by projection, so they have no meaning in projective geometry. Projective geometry can discuss only things that *are* preserved by projection, such a points and lines.

Surprisingly, there are nontrivial theorems about points and lines. One of them was discovered by the Greek geometer Pappus around 300 CE, and another by the French mathematician Desargues around 1640.

Even more surprisingly, there is a *numerical* quantity preserved by projection. It is a "ratio of ratios" of lengths called the *cross-ratio*. In projective geometry, the cross-ratio plays a role similar to that played by length in classical geometry.

One of the virtues of projective geometry is that it simplifies the classification of curves. All conic sections, for example, are "projectively the same," and there are only five types of cubic curve.

The projective viewpoint also removes some apparent exceptions to the theorem of Bézout. For example, a line (curve of degree 1) always meets another line in exactly one point, because in projective geometry even parallel lines meet.

J. Stillwell, *Mathematics and Its History*, Undergraduate Texts in Mathematics,
DOI 10.1007/978-1-4419-6053-5_8, © Springer Science+Business Media, LLC 2010

8.1 Perspective

Perspective may be simply described as the realistic representation of spatial scenes on a plane. This of course has been a concern of painters since ancient times, and some Roman artists seem to have achieved correct perspective by the first century BCE; an impressive example is shown in Wright (1983), p. 38. However, this may have been a stroke of individual genius rather than the success of a theory, because the vast majority of ancient paintings show incorrect perspective. If indeed there was a classical theory of perspective, it was well and truly lost during the Dark Ages. Medieval artists made some charming attempts at perspective but always got it wrong, and errors persisted well into the 15th century. (Errors still survive in 20th-century mathematics texts. Figure 8.1 shows a 15th-century artistic example from Wright (1983), p. 41, alongside a 20th-century mathematical example from the exposé of Grünbaum (1985).)

 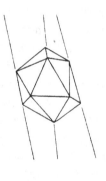

Figure 8.1: Errors in perspective

The discovery of a method for correct perspective is usually attributed to the Florentine painter–architect Brunelleschi (1377–1446), around 1420. The first published method appears in the treatise *On Painting* by Alberti (1436). The latter method, which became known as *Alberti's veil*, used a piece of transparent cloth, stretched on a frame, and set in front of the scene to be painted. Then, viewing the scene with one eye, in a fixed position, one could trace the scene directly onto the veil. Figure 8.2 shows

Figure 8.2: Dürer's depiction of Alberti's veil

this method, with a peephole to maintain a fixed eye position, as depicted by Dürer (1525).

Alberti's veil was fine for painting actual scenes, but to paint an imaginary scene in perspective some theory was required. The basic principles Renaissance artists used were the following:

(i) A straight line in perspective remains straight.

(ii) Parallel lines either remain parallel or converge to a single point (their *vanishing point*).

These principles suffice to solve a problem artists frequently encountered: the perspective depiction of a square-tiled floor. Alberti (1436) solved the special case of this problem in which one set of floor lines is horizontal, that is, parallel to the horizon. Alberti's method is shown in simplified form in

Figure 8.3. The nonhorizontal floor lines are determined by spacing them equally along the base line (imagined to touch the floor) and letting them converge to a vanishing point on the horizon. The horizontal floor lines are then determined by choosing one of them arbitrarily, thus determining one tile in the floor, and then producing the diagonal of this tile to the horizon. The intersections of this diagonal with the nonhorizontal lines are the points through which the horizontal lines pass. This is certainly true on the actual floor (Figure 8.4); hence it remains true in the perspective view.

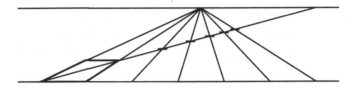

Figure 8.3: Alberti's method

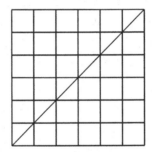

Figure 8.4: The actual floor

EXERCISES

In almost all paintings of tiled floors, one set of lines is parallel to the horizon. However, the principles (i) and (ii) suffice to generate a perspective view of a tiled floor given an arbitrarily situated tile, and they show that no measurement is needed to achieve equal spacing along the base line in Alberti's method.

8.1.1 Use the lines shown in Figure 8.5 to determine all lines in a pavement generated by the given tile.

8.1.2 By using diagonals as in Exercise 8.1.1, show how to generate the lines in the tiling when the baseline is parallel to the horizon, without making any measurements.

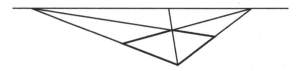

Figure 8.5: Tiled floor with arbitrary orientation

8.2 Anamorphosis

It is clear from the Alberti veil construction that a perspective view will not look absolutely correct except when seen from the viewpoint used by the artist. Experience shows, however, that distortion is not noticeable except from extreme viewing positions. Following the mastery of perspective by the Italian artists, an interesting variation developed, in which the picture looks right from only one, extreme, viewpoint. The first known example of this style, known as *anamorphosis*, is an undated drawing by Leonardo da Vinci from the *Codex Atlanticus* (compiled between 1483 and 1518). Figure 8.6 shows part of this drawing, a child's face which looks correct when viewed with the eye near the right-hand edge of the page.

Figure 8.6: Leonardo's drawing of a face

The idea was taken up by German artists around 1530. The most famous example occurs in Holbein's painting *The Two Ambassadors* (1533). A mysterious streak across the bottom of the picture becomes a skull when viewed from near the picture's edge. For an excellent view of this picture and a history of anamorphosis, see Baltrušaitis (1977) and Wright (1983), pp. 146–156. The art of anamorphosis reached its technically most advanced form in France in the early 17th century. It seems no coincidence that this was also the time and place of the birth of projective geometry. In fact, key figures in the two fields, Niceron and Desargues, were well aware of each other's work.

Niceron (1613–1646) was a student of Mersenne and, like him, a monk in the order of Minims. He executed some extraordinary anamorphic wall paintings, up to 55 meters long, and also explained the theory in *La perspective curieuse* (1638). Figure 8.7 is his illustration of anamorphosis of a chair (from Baltrušaitis (1977), p. 44). The anamorphosis, viewed normally, shows a chair like none ever seen, yet from a suitably extreme point one sees an ordinary chair in perspective.

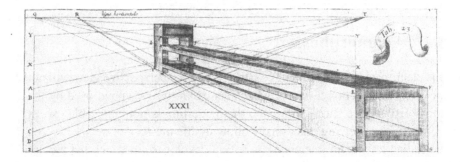

Figure 8.7: Niceron's chair

This example exposes an important mathematical fact: *a perspective view of a perspective view is not in general a perspective view*. Iteration of perspective views gives what we now call a *projective* view, and Niceron's chair shows that projectivity is a broader concept than perspectivity. As a consequence, *projective geometry*, which studies the properties that are invariant under projection, is broader than the theory of perspective. Perspective itself became a mathematical theory, called *descriptive geometry*, only at the end of the 18th century.

8.3   Desargues's Projective Geometry

The mathematical setting in which one can understand Alberti's veil is the family of lines ("light rays") through a point (the "eye"), together with a plane V (the "veil") (Figure 8.8). In this setting, the problems of perspective and anamorphosis were not very difficult, but the *concepts* were interesting and a challenge to traditional geometric thought. Contrary to Euclid, one had the following:

(i) Points at infinity ("vanishing points") where parallels met.

(ii) Transformations that changed lengths and angles (projections).

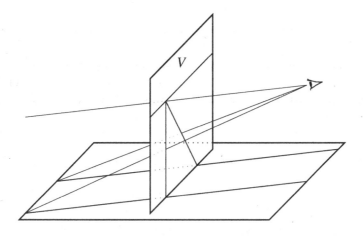

Figure 8.8: Seeing through Alberti's veil

The first to construct a mathematical theory incorporating these ideas was Desargues (1591–1661), although the idea of points at infinity had already been used by Kepler (1604), p. 93. The book of Desargues (1639), *Brouillon projet d'une atteinte aux événemens des rencontres du cône avec un plan* (Schematic Sketch of What Happens When a Cone Meets a Plane), suffered an extreme case of delayed recognition, being completely lost for 200 years. Fortunately, his two most important theorems, the so-called Desargues's theorem and the invariance of the cross-ratio, were published in a book on perspective, Bosse (1648). The text of Desargues (1639) and a portion of Bosse (1648) containing Desargues's theorem may be found in Taton (1951). An English translation, with an extensive historical and mathematical analysis, is in Field and Gray (1987).

Kepler and Desargues both postulated one point at infinity on each line, closing the line to a "circle of infinite radius." All the lines in a family of parallels share the same point at infinity. Nonparallel lines, having a finite point in common, do not have the same point at infinity. Thus any two distinct lines have exactly one point in common—a simpler axiom than Euclid's. Strangely enough, the line at infinity was only introduced into the theory by Poncelet (1822), even though it is the most obvious line in perspective drawing, the horizon. Desargues made extensive use of projections in the *Brouillon projet*; he was the first to use them to prove theorems about conic sections.

Desargues's theorem is a property of triangles in perspective illustrated by Figure 8.9. The theorem states that the points X, Y, Z at the intersections of corresponding sides lie on a line. This is obvious if the triangles are in space, since the line is the intersection of the planes containing them. The theorem in the plane is subtly but fundamentally different and requires a separate proof, as Desargues realized. In fact, Desargues's theorem was shown to play a key role in the foundations of projective geometry by Hilbert (1899) (see Section 20.7).

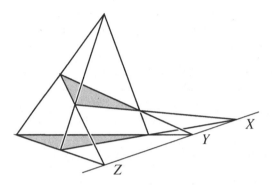

Figure 8.9: Desargues's theorem

The invariance of the cross-ratio answers a natural question first raised by Alberti: since length and angle are not preserved by projection, what is? No property of three points on a line can be invariant because it is possible to project any three points on a line to any three others (Exercise 8.3.1). At least four points are therefore needed, and the cross-ratio is in fact a projective invariant of four points. The cross-ratio $(ABCD)$ of points A, B, C, D on a line (in that order) is $\frac{CA}{CB} / \frac{DA}{DB}$. Its invariance is most simply seen by reexpressing it in terms of angles using Figure 8.10. Let O be any point outside the line and consider the areas of the triangles OCA, OCB, ODA, and ODB. First compute them from bases on AB and height h, then recompute using OA and OB as bases and heights expressed in terms of the sines of angles at O:

$$\frac{1}{2}h \cdot CA = \text{area } OCA = \frac{1}{2}OA \cdot OC \sin \angle COA,$$

$$\frac{1}{2}h \cdot CB = \text{area } OCB = \frac{1}{2}OB \cdot OC \sin \angle COB,$$

$$\frac{1}{2}h \cdot DA = \text{area } ODA = \frac{1}{2}OA \cdot OD \sin \angle DOA,$$

$$\frac{1}{2}h \cdot DB = \text{area } ODB = \frac{1}{2}OB \cdot OD \sin \angle DOB.$$

Substituting the values of CA, CB, DA, and DB from these equations we find, following Möbius (1827), the cross-ratio in terms of angles at O:

$$\frac{CA}{CB}\bigg/\frac{DA}{DB} = \frac{\sin \angle COA}{\sin \angle COB}\bigg/\frac{\sin \angle DOA}{\sin \angle DOB}.$$

Any four points A', B', C', D' in perspective with A, B, C, D from a point O have the same angles (Figure 8.10); hence they will have the same cross-ratio. But then so will any four points A'', B'', C'', D'' projectively related to A, B, C, D, since a projectivity is by definition the product of a sequence of perspectivities.

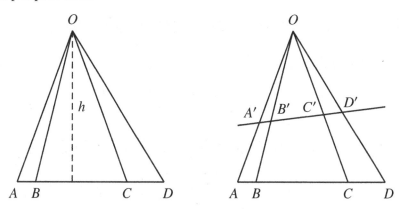

Figure 8.10: Evaluating the cross-ratio

As mentioned above, we cannot hope for an invariant that is simpler than the cross-ratio, because any three points in a line are projectively related to any other.

8.3.1 Show that any three points on a line can be sent to any other three points on a line by projection. (If you need a hint, see Figure 23.1.)

The case of Desargues's theorem in which the two triangles lie in the same plane is proved by viewing the plane in space. The setup for the proof is shown in Figure 8.11. The triangles $A_1B_1C_1$ and $A_2B_2C_2$ are in perspective from O in a plane Π, P is a point in space outside Π, and the line OD_1D_2 meets Π only at O.

8.3.2 Show that the triangles $A_1C_1D_1$ and $A_2C_2D_2$ are in different planes, and in perspective from O.

Thus it follows from the nonplanar version of Desargues's theorem that the intersections of the side pairs (A_1D_1, A_2D_2), (A_1C_1, A_2C_2), and (C_1D_1, C_2D_2) lie on a line.

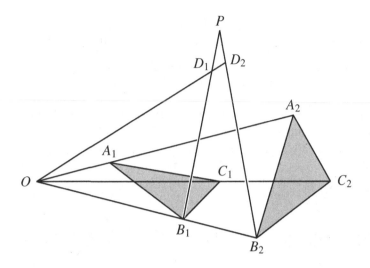

Figure 8.11: The planar Desargues's theorem

8.3.3 Show that these intersections are projected from P to the intersections of the side pairs (A_1B_1, A_2B_2), (A_1C_1, A_2C_2), and (C_1B_1, C_2B_2), and hence deduce the planar Desargues's theorem.

8.3.4 Does this proof capture your intuitive idea of looking at the planar Desargues configuration (Figure 8.9) and interpreting it three-dimensionally? If so, what does the point P represent?

8.4   The Projective View of Curves

The problems of perspective drawing mainly involved the geometry of straight lines. There were, it is true, problems such as drawing ellipses to look like perspective views of circles, but artists were generally content to solve such problems by interpolating smooth-looking curves in a suitable straight-line framework. An example is the drawing of a chalice by Uccello (1397–1475) in Figure 8.12.

A mathematical theory of perspective for curves became possible with the advent of analytic geometry. When a curve is specified by an equation $f(x, y) = 0$, the equation of any perspective view is obtainable by suitably transforming x and y. However, this transformational viewpoint, even though quite simple algebraically, emerged only with Möbius (1827). The first works in projective geometry, by Desargues (1639) and Pascal (1640),

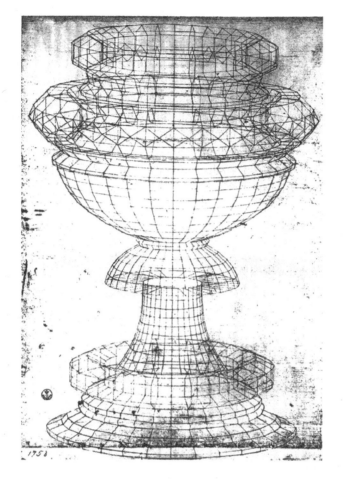

Figure 8.12: Drawing of a chalice by Uccello (Uffizi, Florence)

used the language of classical geometry, even though the language of equations was available from Descartes (1637). This was understandable, not only because the analytic method was so obscure in Descartes, but also because the advantages of the projective method could be more clearly seen when it was used in a classical setting. Desargues and Pascal confined themselves to straight lines and conic sections, showing how projective geometry could easily reach and surpass the results obtained by the Greeks. Moreover, the projective viewpoint gave something else that would have been incomprehensible to the Greeks: a clear account of the behavior of curves at infinity.

For example, Desargues (1639) (in Taton (1951), p. 137) distinguished
the ellipse, parabola, and hyperbola by their numbers of points at infinity:
0, 1, and 2, respectively. The points at infinity on the parabola and hyperbola can be seen quite plainly by tilting the ordinary views of them into
perspective views (Figures 8.13 and 8.14). The parabola has just one point
at infinity because it crosses each ray through 0, except the y-axis, at just
one finite point. As for the hyperbola, its two points at infinity are where it
touches its asymptotes, as seen in Figure 8.14. The continuation of the hyperbola above the horizon results from projecting the lower branch through
the same center of projection (Figure 8.15).

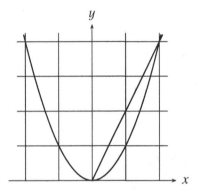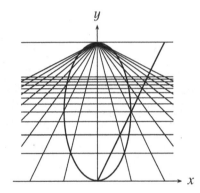

Figure 8.13: The parabola

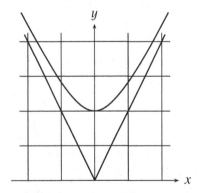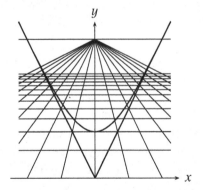

Figure 8.14: The hyperbola

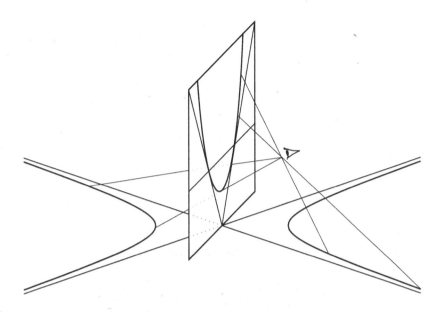

Figure 8.15: Branches of the hyperbola

Projective geometry goes beyond describing the behavior of curves at infinity. The line at infinity is no different from any other line and can be deprived of its special status. Thus, all projective views of a curve are equally valid, and one can say, for example, that all conic sections are ellipses when suitably viewed. This is no surprise if one remembers conic sections not as second-degree curves but as sections of the cone. Of course they all look the same from the vertex of the cone.

More surprisingly, a great simplification of cubic curves also occurs when they are viewed projectively. As mentioned in Section 7.4, Newton (1695) classified cubic curves into 72 types (and missed 6). However, in his Section 29, "On the Genesis of Curves by Shadows," Newton claimed that each cubic curve can be projected onto one of just five types. As mentioned in Section 7.4, this includes the result that $y = x^3$ can be projected onto $y^2 = x^3$. The proof of this is an easy calculation when coordinates are introduced (see Exercise 8.7.2), but one already gets an inkling of it from the perspective view of $y = x^3$. See Figure 8.16. The lower half of the cusp is the view of $y = x^3$ below the horizon; the upper half comes from projecting the view behind one's head through P to the picture plane in front.

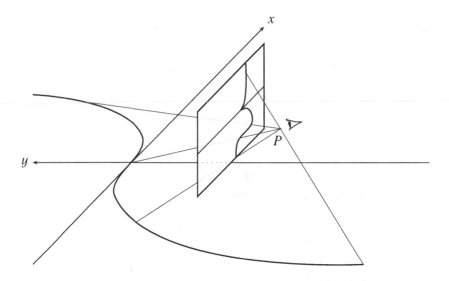

Figure 8.16: Perspective view of a cubic curve

Conversely, $y^2 = x^3$ has an inflection at infinity. Newton's projective classification comes about by studying the behavior at infinity of all cubics and observing that each has characteristics already possessed, not necessarily at infinity, by curves of the form

$$y^2 = Ax^3 + Bx^2 + Cx + D.$$

Newton had already divided these into five types in his analytic classification. They are the five shown in Figure 7.3. Newton's result was improved only in the 19th century, when projective classification over the complex numbers reduced the number of types of cubics to just three. We discuss this later in connection with the development of complex numbers (Section 16.5).

EXERCISES

As suggested above, the points at infinity of a curve may be counted by considering intersections of the curve with lines through the origin, and observing where they tend to infinity.

8.4.1 Use this method to explain why

- the hyperbola $xy = 1$ has two points at infinity,
- the curve $y = x^3$ has one point at infinity.

Figures 8.13 and 8.14 were made by taking Alberti's veil to be the (x, z)-plane in (x, y, z)-space, with the "eye" at $(0, -4, 4)$ viewing the (x, y)-plane.

8.4.2 Find the parametric equations of the line from $(0, -4, 4)$ to $(x', y', 0)$, and hence show that this line meets the veil where

$$x = \frac{4x'}{y' + 4}, \quad z = \frac{4y'}{y' + 4}.$$

8.4.3 Renaming the coordinates x, z in the veil as X, Y respectively, show that

$$x' = \frac{4X}{4 - Y}, \quad y' = \frac{4Y}{4 - Y}.$$

8.4.4 Deduce from Exercise 8.4.3 that the points (x', y') on the parabola $y = x^2$ have image on the veil

$$X^2 + \frac{(Y - 2)^2}{4} = 1,$$

and check that this is the ellipse shown in Figure 8.13.

8.5   The Projective Plane

The way in which projective geometry allows infinity to be put on the same footing as the finite points of the plane is intuitively clear when one thinks of the horizon in a picture, which is a line like any other. But what, mathematically speaking, is this line we see? We can model the situation mathematically by taking the plane we look at to be the plane $z = -1$ in the three-dimensional space with coordinates (x, y, z), and placing our eye at the origin $O = (0, 0, 0)$, as in Figure 8.17.

Points $P_1, P_2, P_3, \ldots$ in the plane lie on "lines of sight" $\mathcal{L}_1, \mathcal{L}_2, \mathcal{L}_3, \ldots$ through O, and as the point P_n tends to infinity its line of sight $\mathcal{L}_n$ tends to horizontal. Therefore, it is natural to interpret each horizontal line through O, which does not correspond to an actual point of the plane, as the line of sight to a "point at infinity" of the plane. More boldly, we can define the lines through O to *be* the points of a *projective plane*, called the *real projective plane* $\mathbb{RP}^2$, and the planes through O to be the lines of $\mathbb{RP}^2$—the so-called *projective lines*.

Modeling the points of the plane $z = -1$ by the non-horizontal lines through O enables us to *complete* this ordinary plane to a projective plane by using the remaining lines through O (which are not called "horizontal" for nothing!) to model the points on its horizon. Moreover, the horizontal

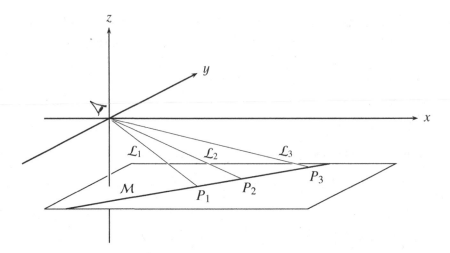

Figure 8.17: Vision of the plane

plane through O models the horizon line, reinforcing our intuition that the horizon is a line like any other.

This model of the projective plane is geometrically as natural as one could wish, and it answers certain questions that are confusing for vision alone. For example, we can see why it is proper for a line $\mathcal{M}$ in the ordinary plane to have only one point at infinity: because there is only *one* line through O to which the lines through $P_1, P_2, P_3, \ldots$ tend as P_n tends to infinity, namely, the parallel to $\mathcal{M}$ through O. Thus, Kepler and Desargues were not far wrong in thinking of a projective line as a circle. The two "ends" of the line are joined by its single point at infinity.

While a projective line is essentially a circle, a projective plane is *not* essentially a sphere, but something more peculiar, as was noticed by Klein (1874). $\mathbb{RP}^2$ is essentially a *sphere with antipodal points identified*, where antipodal points P, P' are pairs such as those shown in Figure 8.18: the diametrically opposite points at which a line through O meets the unit sphere with center O. "Identifying" the points P, P' means treating the pair (P, P') as a single point. This is appropriate since the pair corresponds to a single line through O, that is, to a single point of $\mathbb{RP}^2$.

The surface $\mathbb{RP}^2$ modeled by the pairs (P, P') is strikingly different from the sphere of individual points P. For example, on a sphere, any simple closed curve separates the surface into two parts. A "small" closed curve in $\mathbb{RP}^2$—that is, one strictly contained in a hemisphere of the model—also

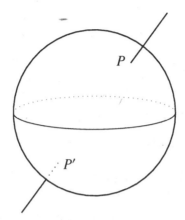

Figure 8.18: The projective plane of antipodal point pairs

separates it, but a "large" one may not. The equator, for instance, does not separate the upper hemisphere from the lower, because the hemispheres are the *same place* under antipodal point identification! A less paradoxical view of this is seen by going back to the model of $\mathbb{RP}^2$ whose elements are lines through O. The lines through the equator do not separate the lines through the upper hemisphere from the lines through the lower hemisphere, because these are the same lines.

EXERCISES

The model of the projective plane whose points are lines through O and whose lines are planes through O also helps in visualizing other basic properties of projective lines.

8.5.1 Use this interpretation of projective lines to show that all lines in a family of parallels have the same point at infinity.

8.5.2 Likewise, show that any two projective lines meet in exactly one point.

Now let us return to the interpretation of the projective plane as a surface, the sphere with antipodal points identified. The following result shows another way in which the projective plane differs from a sphere.

8.5.3 Show that a strip of the projective plane surrounding a projective line is a Möbius band (Figure 8.19).

8.5.4 Why is the Möbius band not a part of the sphere?

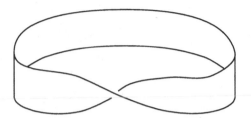

Figure 8.19: A Möbius band

8.6   The Projective Line

As we have seen, projective geometry arose from efforts to understand the relationship between two and three dimensions. But the idea arising from these efforts—that of *projection* or *projective transformations*—is interesting even in one dimension. In this section we make a more detailed study of projection from a line to a line, and use it to present a more sophisticated concept of projective line. In the process, we will meet the concept of *linear fractional transformation*, which plays a key role in many later developments. In particular, we will show how linear fractional transformations give a new insight into the invariance of the cross-ratio.

We start by viewing the line as the number line $\mathbb{R}$, and study how the numerical values of points are related when we project one line onto another. The simplest kind of projection is *parallel projection* (or projection from infinity) of a line onto a parallel line, as shown in Figure 8.20.

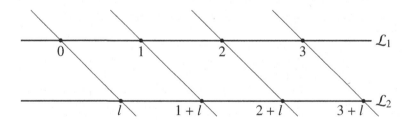

Figure 8.20: Projection from infinity

Clearly, when we make the natural choice of coordinates on the two lines, parallel projection sends x on $\mathcal{L}_1$ to $x + l$ on $\mathcal{L}_2$, for some constant l. We abbreviate this mapping of coordinates by $x \mapsto x + l$.

If we project from a point P at a finite distance, then it is likewise clear from Figure 8.21 (where we align the zero point on each line with P) that x on $\mathcal{L}_1$ is sent to kx on $\mathcal{L}_2$ for some nonzero constant k. We abbreviate this mapping of coordinates by $x \mapsto kx$ $(k \neq 0)$.

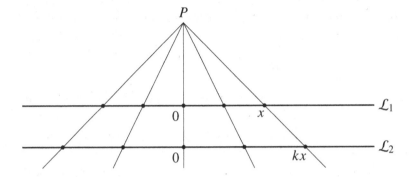

Figure 8.21: Projection from a finite point

A more remarkable case is shown in Figure 8.22, where we project a line $\mathcal{L}_1$ onto a perpendicular line $\mathcal{L}_2$ from a point not on either line, but equidistant from both. Then with suitable choice of coordinates, x on $\mathcal{L}_1$ is sent to $1/x$ on $\mathcal{L}_2$.

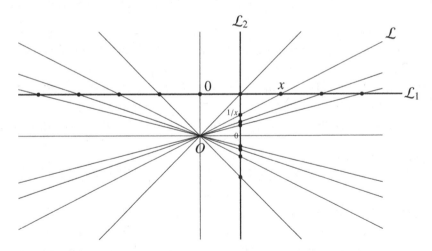

Figure 8.22: Projection of a line onto a perpendicular line

This makes $\mathcal{L}_2$ a highly distorted image of $\mathcal{L}_1$, with the equally spaced points $1, 2, 3, 4, \ldots$ on $\mathcal{L}_2$ going to the points $1, 1/2, 1/3, 1/4, \ldots$ on $\mathcal{L}_2$.

These image points tend to the point 0 on $\mathcal{L}_2$, which is *not* the projection of any point on $\mathcal{L}_1$. However, if we extend $\mathcal{L}_1$ by an extra point ∞—its point at infinity—then it seems right to view 0 on $\mathcal{L}_2$ as the projection of ∞ on the extended line $\mathcal{L}_1 \cup \{\infty\}$. It likewise seems right to extend $\mathcal{L}_2$ by its point ∞ at infinity, and to view this point as the projection of 0 on $\mathcal{L}_1$.

If we still claim that this map sends x to $1/x$, then we must admit that

$$1/0 = \infty \quad \text{and} \quad 1/\infty = 0.$$

We have legalized division by zero! Is this valid? In this limited setting, yes, because we are merely labeling each line $\mathcal{L}$ through O by two symbols: x and $1/x$. If $\mathcal{L}$ is neither vertical nor horizontal, then x and $1/x$ are the intersections of $\mathcal{L}$ with $\mathcal{L}_1$ and $\mathcal{L}_2$ respectively; if $\mathcal{L}$ is vertical, then $x = 0$ is its real intersection with $\mathcal{L}_1$ and $1/0 = \infty$ is its "intersection at infinity" with its parallel $\mathcal{L}_2$; if $\mathcal{L}$ is horizontal, then $1/x = 1/\infty = 0$ is its real intersection with $\mathcal{L}_2$ and ∞ is its "intersection at infinity" with its parallel $\mathcal{L}_1$.

Actually, division by zero is valid in the more general and interesting setting of linear fractional transformations:

$$f(x) = \frac{ax + b}{cx + d}, \quad \text{where} \quad ad - bc \neq 0.$$

These are precisely the functions obtainable as combinations of the functions $x \mapsto x + l$, $x \mapsto kx$ for $k \neq 0$, and $x \mapsto 1/x$, and they correspond to arbitrary projections of one projective line onto another. To be precise, each linear fractional function gives a well-defined and one-to-one map of $\mathbb{R} \cup \{\infty\}$ to itself, and these maps realize all projections of the projective line. These relationships are verified in the exercises below. Because of this, we call $\mathbb{R} \cup \{\infty\}$, together with its linear fractional functions, the *real projective line* $\mathbb{RP}^1$.

The linear fractional functions give $\mathbb{RP}^1$ its "projective" nature. Unlike the real line $\mathbb{R}$, $\mathbb{RP}^1$ has no concept of length, because length is not preserved by linear fractional functions. Not even the ratio of lengths is preserved, as one can see with the function $x \mapsto 1/x$. However, *the cross-ratio is preserved by linear fractional functions*, and hence by projections.

To see why, consider four points A, B, C, D on a line. If we view these points as numbers, then their cross-ratio (defined in Section 8.3) becomes

$$\frac{CA \cdot DB}{CB \cdot DA} = \frac{(C - A)(D - B)}{(C - B)(D - A)}.$$

The function $x \mapsto x + l$, which adds l to each of A, B, C, D, obviously does not change the cross-ratio. Neither does the function $x \mapsto kx$ for $k \neq 0$, which multiplies each of A, B, C, D by k. It is not so obvious that the cross-ratio is preserved by the function $x \mapsto 1/x$, which sends each of A, B, C, D to its reciprocal, but this is confirmed by a simple calculation. Thus the cross-ratio is preserved by all combinations of $x \mapsto x + l$, $x \mapsto kx$ for $k \neq 0$, and hence by all linear fractional functions.

EXERCISES

We can see why each linear fractional function is a combination of functions of the forms $x \mapsto x + l$, $x \mapsto kx$ for $k \neq 0$, and $x \mapsto 1/x$ by a suitable breakdown of the fraction $\frac{ax+b}{cx+d}$.

8.6.1 Show that $\frac{ax+b}{cx+d} = \frac{a}{c} + \frac{bc-ad}{c(cx+d)}$ if $c \neq 0$.

8.6.2 Deduce from Exercise 8.6.1 that the function $x \mapsto \frac{ax+b}{cx+d}$ is a combination of functions $x \mapsto x + l$, $x \mapsto kx$, and $x \mapsto 1/x$ when $c \neq 0$. What if $c = 0$?

8.6.3 What property of $\frac{ax+b}{cx+d}$ is controlled by the condition $ad - bc \neq 0$?

8.6.4 Verify that the cross-ratio $\frac{(C-A)(D-B)}{(C-B)(D-A)}$ remains unchanged when each of the points A, B, C, D is replaced by its reciprocal.

It follows that the cross-ratio is preserved by any linear fractional function. It remains to show that projections are realized by linear fractional functions. We have already done this for projection of a line onto a parallel line. Hence it remains to study projection of a line, say the x-axis, onto a line that intersects it, say $y = cx$.

8.6.5 Show that projection from the point (a, b) sends the point $x = t$ on the x-axis to the point on the line $y = cx$ for which

$$x = \frac{bt}{ct + b - ca},$$

which is a linear fractional function of t.

8.7 Homogeneous Coordinates

Representing the points of the projective plane $\mathbb{RP}^2$ by lines through O gives *coordinates* to $\mathbb{RP}^2$ via the coordinates (x, y, z) of three-dimensional space. Such coordinates were invented by Möbius (1827) and Plücker (1830), and they are called *homogeneous* because each algebraic curve in $\mathbb{RP}^2$ is expressed by a homogeneous polynomial equation $p(x, y, z) = 0$. The simplest case is that of a projective line, which, as we saw in Section

8.5, is represented by a plane through O. Its equation therefore has the form

$$ax + by + cz = 0, \quad \text{for some constants } a, b, c, \text{ not all zero.}$$

Such an equation is called *homogeneous of degree* 1, because each nonzero term is of degree 1 in the variables x, y, z.

The *homogeneous coordinates of a point P* in $\mathbb{RP}^2$ are simply the coordinates of *all* points on the line through O that represents P. It follows that if (x, y, z) is one coordinate triple for P, so is (tx, ty, tz) for any real number t. And if $p(x, y, z) = 0$ is the equation of a curve in $\mathbb{RP}^2$, the polynomial p must be such that

$$p(tx, ty, tz) = 0 \quad \text{for all real numbers } t.$$

It follows that $p(tx, ty, tz) = t^n p(x, y, z)$ for some n, called the *degree* of p.

A typical example is the equation

$$x^2 - yz = 0,$$

which is homogeneous of degree 2. To see what this curve looks like in an ordinary plane, such as $z = 1$, we substitute for the appropriate variable. With $z = 1$ we obtain

$$y = x^2,$$

which is the equation of a parabola in the plane $z = 1$. Thus $x^2 - yz = 0$ is the *projective completion* of a parabola, with a point at infinity added (namely, the y-axis).

But $x^2 - yz = 0$ is also the projective completion of a hyperbola. We see this by intersecting the projective curve with the plane $x = 1$, obtaining the hyperbola $yz = 1$. Surprising as this seems at first, it reflects a fact we already know from Section 8.4—that all conic sections are projectively the same.

Homogeneous coordinates also make it easy to show that certain cubic curves have the same projective completion (see Exercise 8.7.2).

Bézout's Theorem Revisited

As we saw in Section 7.5, to obtain Bézout's theorem that a curve of degree m meets a curve of degree n in mn points we need a precise account of points at infinity. Homogeneous coordinates simplify this problem by changing it to one about homogeneous polynomials. If C_m is a curve with homogeneous equation of degree m,

$$p_m(x, y, z) = 0, \tag{1}$$

and if C_n is a curve with homogeneous equation of degree n,

$$p_n(x, y, z) = 0, \tag{2}$$

one wishes to show that the equation

$$r_{mn}(x, y) = 0, \tag{3}$$

which results from eliminating z between (1) and (2), is homogeneous of degree mn. This is not hard to do (see exercises), but it seems that a homogeneous formulation of Bézout's theorem, with a rigorous proof that the resultant r_{mn} has degree mn, was not given until the late 1800s. According to Kline (1972), p. 553, the "proper count of multiplicities" was first made by Halphen in 1873.

An obvious condition must be included in the hypothesis of Bézout's theorem: that the curves C_m and C_n have no common component. The algebraic equivalent of this condition is that the polynomials p_m, p_n have no nonconstant common factor. Then the form of Bézout's theorem that can be proved with the help of homogeneous coordinates is *curves C_m, C_n with homogeneous equations $p_m(x, y, z) = 0$, $p_n(x, y, z) = 0$ of degrees m, n and no common component have intersections given by the solutions of a homogeneous equation $r_{mn}(x, y) = 0$ of degree mn.*

A useful consequence of Bézout's theorem is that curves C_m, C_n of degrees m, n with *more* than mn intersections have a common component.

EXERCISES

8.7.1 We know that the hyperbola $yz = 1$ has two points at infinity. To which lines through O do they correspond in the projective completion $x^2 - yz = 0$?

8.7.2 By considering the homogeneous polynomial equation $x^3 - y^2z = 0$, show that the cubic curves $y = x^3$ and $y^2 = x^3$ have the same projective completion.

As the Chinese discovered (see Section 6.2), the problem of elimination belongs to linear algebra. In the case of Bézout's theorem, this includes the determinant criterion for a set of homogeneous equations to have a nonzero solution, and it leads to an expression for the resultant r_{mn} as a determinant.

8.7.3 Suppose that

$$p_m(x, y, z) = a_0 z^m + a_1 z^{m-1} + \cdots + a_m,$$
$$p_n(x, y, z) = b_0 z^n + b_1 z^{n-1} + \cdots + b_n$$

are homogeneous polynomials of degrees m, n. Thus $a_i(x, y)$ is homogeneous of degree i and $b_j(x, y)$ is homogeneous of degree j. By multiplying p_m and p_n by suitable powers of z, show that the equations

$$p_m = 0 \qquad \text{and} \qquad p_n = 0$$

are equivalent to a system of $m + n$ homogeneous linear equations in the variables $z^{m+n-1}, \ldots, z^2, z^1, z^0$, which in turn is equivalent to

$$r_{mn}(x, y) \equiv \begin{vmatrix} a_0 & a_1 & \cdots & a_m & 0 & \cdots & 0 \\ 0 & a_0 & a_1 & \cdots & a_m & 0 & \cdots & 0 \\ \vdots & & \ddots & & & \ddots & \ddots & \\ & & & & & & & 0 \\ 0 & \cdots & 0 & a_0 & & \cdots & & a_m \\ b_0 & b_1 & \cdots & & b_n & 0 & \cdots & 0 \\ 0 & b_0 & b_1 & \cdots & & b_n & & \vdots \\ \vdots & & \ddots & & & & \ddots & 0 \\ 0 & \cdots & 0 & b_0 & & \cdots & & b_n \end{vmatrix} = 0.$$

8.7.4 Show that a polynomial $p(x, y)$ is homogeneous of degree $k \Leftrightarrow p(tx, ty) = t^k p(x, y)$.

8.7.5 Show that $r_{mn}(tx, ty) = t^{mn} r_{mn}(x, y)$. *Hint*: Multiply the rows of $r_{mn}(tx, ty)$ by suitable powers of t to arrange that each element in any column contains the same power of t. Then remove these factors from the columns so that $r_{mn}(x, y)$ remains.

8.8   Pascal's Theorem

Pascal's *Essay on Conics* (1640) was written in late 1639, when Pascal was 16. He probably had heard about projective geometry from his father, who was a friend of Desargues. The *Essay* contained the first statement of a famous result that became known as Pascal's theorem or the *mystic hexagram*. The theorem states that the pairs of opposite sides of a hexagon inscribed in a conic section meet in three collinear points. (The vertices of the hexagon can occur in any order on the curve. In Figure 8.23 the

order was chosen to enable the three intersections to lie inside the curve.)
Pascal's proof is not known, but he probably established the theorem for
the circle first, then trivially extended it to arbitrary conics by projection.

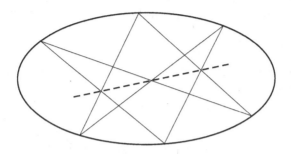

Figure 8.23: Pascal's theorem

Plücker (1847) threw new light on Pascal's theorem by showing it to
be an easy consequence of Bézout's theorem. Plücker used an auxiliary
theorem about cubics which can be bypassed, giving the following direct
deduction from Bézout's theorem.

Let $L_1, L_2, \ldots, L_6$ be the successive sides of the hexagon. The unions
of alternate sides, $L_1 \cup L_3 \cup L_5$ and $L_2 \cup L_4 \cup L_6$, can be regarded as cubic
curves

$$l_{135}(x, y, z) = 0, \qquad l_{246}(x, y, z) = 0,$$

where each l is a product of three linear factors. These two curves meet in
nine points: the six vertices of the hexagon and the three intersections of
opposite sides. Let

$$c(x, y, z) = 0 \tag{1}$$

be the equation of the conic that contains the six vertices.

We can choose constants α, β so that the cubic curve

$$\alpha l_{135}(x, y, z) + \beta l_{246}(x, y, z) = 0 \tag{2}$$

passes through any given point P. Let P be a point on the conic, distinct
from the six vertices. Then the curves (1), (2) of degrees 2, 3, have $7 > 2 \times 3$
points in common, and hence a common component by Bézout's theorem.
Since c has no nonconstant factor, by hypothesis, this common component
must be c itself. Hence

$$\alpha l_{135} + \beta l_{246} = cp \tag{3}$$

for some polynomial p, which must be linear since the left-hand side of (3) has degree 3 and c has degree 2. Since the curve $\alpha l_{135} + \beta l_{246} = 0$ passes through the nine points common to $l_{135} = 0$ and $l_{246} = 0$, while $c = 0$ passes through only six of them, the remaining three (the intersections of opposite sides) must be on the line $p = 0$.

EXERCISES

8.8.1 Generalize the preceding argument to show that if two curves of degree n meet in n^2 points, nm of which lie on a curve of degree m, then the remaining $n(n - m)$ points lie on a curve of degree $n - m$.

An important special case of Pascal's theorem was discovered around 300 CE by Pappus, and it is called the *theorem of Pappus*. In this theorem, the conic is a "degenerate" conic section, consisting of two straight lines.

The usual statement of Pappus's theorem, like that of Pascal's theorem, says that the intersections of opposite sides of the hexagon are in a straight line. However, if we avail ourselves of the freedom to take this line to be at infinity, then Pappus's theorem takes a form that is easier to visualize and prove.

8.8.2 Interpret Figure 8.24 as an illustration of Pappus's theorem.

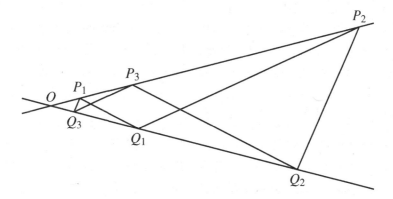

Figure 8.24: Illustration of Pappus's theorem

8.8.3 Write down a statement of the theorem corresponding to Figure 8.24, the conclusion of which is that P_1Q_3 and P_2Q_2 are parallel. (Equivalently, $OP_1/OP_2 = OQ_3/OQ_2$.)

8.8.4 Deduce the required equation from two other equations that express parallelism in Figure 8.24.

8.8.5 Also draw the figure and prove the theorem in the case where the two lines P_1P_2 and Q_1Q_2 do *not* meet at O, that is, when they too are parallel.

8.9   Biographical Notes: Desargues and Pascal

Girard Desargues was born in Lyons in 1591 and died in 1661. He was one
of nine children of Girard Desargues, a tithe collector, and Jeanne Croppet.
He was evidently brought up in Lyons, but information about his early life
is lacking. By 1626 he was working as an engineer in Paris and may have
used his expertise in the famous siege of La Rochelle in 1628, during which
a dike was built across the harbor to prevent English ships from relieving
the city.

In the 1630s he joined the circle of Marin Mersenne, which met regularly in Paris to discuss scientific topics, and in 1636 contributed a chapter
to a book of Mersenne on music theory. In the same year he published
a 12-page booklet on perspective, the first hint of his ideas in projective
geometry. His *Brouillon projet* (1639) was published in an edition of only
50 copies and won very little support. In fact, its reception was generally
hostile, and Desargues was engaged in a pamphleteering battle for years
with his detractors (see Taton (1951), pp. 36–45). At first his only supporters were Pascal, most of whose work on projective geometry is also lost,
and the engraver Abraham Bosse, who expounded Desargues's perspective
method (Bosse (1648)). Desargues became discouraged by the attacks on
his work and left the dissemination of his ideas up to Bosse, who was not
really mathematically equipped for the task. Projective geometry secured
a place in mathematics only with the publication of a book by Phillipe de
la Hire (1673), whose father, Laurent, had been a student of Desargues. It
seems quite likely that la Hire's book influenced Newton. For this and more
on Desargues's mathematical legacy, see Field and Gray (1987), Ch. 3.

Around 1645 Desargues turned his talents to architecture, perhaps to
demonstrate to his critics the practicality of his graphical methods. He was
responsible for various houses and public buildings in Paris and Lyons, ex-
celling in complex structures such as staircases. His best-known achievement in engineering, a system for raising water at the château of Beaulieu,
near Paris, is also interesting from the geometrical viewpoint. It makes
the first use of epicyclic curves (Section 2.5) in cogwheels, as was noted
by Huygens (1671). Huygens visited the château at the time when it was
owned by Charles Perrault, the author of *Cinderella* and *Puss in Boots*.

Desargues apparently returned to scientific circles in Paris toward the
end of his life—Huygens heard him give a talk on the existence of geometric points on November 9, 1660—but information about this period is

scanty. His will was read in Lyons on October 8, 1661, but the date and place of his death are unknown.

Blaise Pascal (Figure 8.25) was born in Clermont-Ferrand in 1623 and died in Paris in 1662. His mother, Antoinette Bagon, died when he was three, and Blaise was brought up by his father, Etienne. Etienne Pascal was a lawyer with an interest in mathematics who belonged to Mersenne's circle and, as mentioned earlier, was a friend of Desargues. He has a curve named after him, the *limaçon of Pascal*. In 1631 Etienne took Blaise and his two sisters to Paris and gave up all official duties to devote himself to their education. Thus Blaise Pascal never went to school or university, but by the age of 16 he was learned in Latin, Greek, mathematics, and science. And of course he had written his *Essay on Conics* and discovered Pascal's theorem.

Figure 8.25: Blaise Pascal

The *Essay on Conics* (1640) is a short pamphlet containing an outline of the great treatise on conics he had begun to prepare, and which is now lost. It includes a statement of Pascal's theorem for the circle. Pascal worked on his treatise until 1654, when it was nearly complete, but he never mentioned it thereafter. Leibniz saw the manuscript when he was in Paris in 1676, but no further sightings are known.

In 1640 Pascal and his sisters joined their father in Rouen, where he had become a tax official. Pascal got the idea of constructing a calculating machine to help his father in his work. He found a theoretical solution around the end of 1642, based on toothed wheels, but difficulties in the production of accurate parts delayed the appearance of the machine until 1645. This was the first working computer. The gear mechanism for addition seems rather obvious to us now, but in Pascal's day it already raised questions of the "Can a machine think?" kind. Pascal himself was sufficiently amazed by the mechanism to say that "the arithmetical machine produces effects which approach nearer to thought than all the actions of the animal. But it does nothing which would enable us to attribute will to it, as to the animals" (Pascal, *Pensées*, 340). The machine greatly impressed the French chancellor, and Pascal was granted exclusive rights to manufacture and sell it. Whether it was a commercial success is not known, but for a time, at least, Pascal was diverted by the opportunity to cash in on his ideas.

The direction of Pascal's life began to shift away from such worldly concerns in 1646, when his father was treated for a leg injury by two local bonesetters. The bonesetters were Jansenists, then a fast-growing sect within the Catholic church. Their influence resulted in the conversion of the whole family to Jansenism, and Pascal began to devote more time to religious thought. For some years, though, he continued with scientific work. In 1647 he investigated the variation of barometric pressure with altitude, resulting in his *New Experiments Concerning the Vacuum*, published the same year; in 1651 he did pioneering work in hydrostatics, resulting in his *Great Experiment Concerning the Equilibrium of Fluids*, published in 1663; and in 1654 he investigated the so-called Pascal's triangle, making fundamental contributions to number theory, combinatorics, and probability theory. For more on this, see Chapter 11. In 1654 Pascal experienced a "second conversion," which led to his almost complete withdrawal from the world and science and his increasing commitment to the Jansenist cause. Only in 1658 and 1659 did he concentrate at times on mathematics (on one occasion, so the story goes, to take his mind off the pain of a toothache). His favorite topic at this stage was the cycloid, the curve generated by a point on the circumference of a circle that rolls on a straight line. Later in the 17th century the cycloid became important in the development of mechanics and differential geometry (see Chapters 13 and 17).

Mathematicians are of course very sorry about Pascal's withdrawal from mathematics at an early age; however, it was not just religion that gained from Pascal's conversion. The *Provincial Letters*, which he wrote to promote Jansenist ideas, and his *Pensées*, which were edited by the Jansenists after his death, became classics of French literature. Undoubtedly Pascal is the only great mathematician whose standing is equally great among writers. Moreover, his devotion to the Jansenist ideal of serving the needy had one enduring practical consequence: his idea of a public transport system. Shortly before his death in 1662, Pascal saw the inauguration of the world's first omnibus service. Coaches could be taken from the Porte Sainte-Antoine to the Luxembourg in Paris for 5 sous, with profits being directed to the relief of the poor.

9

Calculus

PREVIEW

The shift towards algebraic thinking was not only a revolution in geometry. It was decisive in the second and greatest mathematical revolution of the 17th century: the invention of calculus. It is true that some results we now obtain by calculus were known to the ancients; for example, the area of the parabolic segment was found by Archimedes. But the *systematic computation* of areas, volumes, and tangents became possible only when symbolic computation—that is, algebra—became available.

The dependence of calculus on algebra is particularly clear in the work of Newton, whose calculus is essentially the algebra of infinite polynomials (power series). Moreover, Newton's starting point was a basic theorem about the polynomial $(1 + x)^n$, the *binomial theorem*, which he extended to fractional values of n.

The calculus of Leibniz was likewise based on algebra—in his case the algebra of *infinitesimals*. Despite doubts about the meaning and existence of infinitesimals, Leibniz and his followers obtained correct results by computing with them.

Results that we now obtain through a combination of algebra and limit processes were obtained by Leibniz through the algebra of infinitesimals. Our *derivative dy/dx* was, for Leibniz, literally the quotient of the infinitesimal dx by the infinitesimal dy. And our *integral $\int f(x)\,dx$* was, for Leibniz, literally the sum of the infinitesimals $f(x)\,dx$ (hence the symbol $\int$, which is an elongated S for "sum").

J. Stillwell, *Mathematics and Its History*, Undergraduate Texts in Mathematics, DOI 10.1007/978-1-4419-6053-5_9, © Springer Science+Business Media, LLC 2010

9.1   What Is Calculus?

Calculus emerged in the 17th century as a system of shortcuts to results obtained by the method of exhaustion and as a method for discovering such results. The types of problem for which calculus proved suitable were finding lengths, areas, and volumes of curved figures and determining local properties such as tangents, normals, and curvature—in short, what we now recognize as problems of integration and differentiation. Equivalent problems of course arise in mechanics, where one of the dimensions is time instead of distance; hence it was calculus that made mathematical physics possible—a development we shall consider in Chapter 13. In addition, calculus was intimately connected with the theory of infinite series, initiating developments that became fundamental in number theory, combinatorics, and probability theory.

The extraordinary success of calculus was possible, in the first instance, because it replaced long and subtle exhaustion arguments by short routine calculations. As the name suggests, calculus consists of *rules for calculating* results, not their logical justification. Mathematicians of the 17th century were familiar with the method of exhaustion and assumed they could always fall back on it if their results were challenged, but the flood of new results became so great that there was seldom time to do so. As Huygens (1659a), p. 337, wrote,

> Mathematicians will never have enough time to read all the
> discoveries in Geometry (a quantity which is increasing from
> day to day and seems likely in this scientific age to develop
> to enormous proportions) if they continue to be presented in a
> rigorous form according to the manner of the ancients.

The progress in geometry when Huygens wrote was indeed impressive, considering the very simple system of calculus then available. Virtually all that was known was the differentiation and integration of powers of x (possibly fractional) and implicit differentiation of polynomials in x, y. However, when allied with algebra and analytic geometry, this was sufficient to find tangents, maxima, and minima for all algebraic curves. And when allied with Newton's calculus of infinite series, discovered in the 1660s, the rules for powers of x formed a complete system for differentiation and integration of all functions expressible in power series.

The subsequent development of calculus is a puzzling exception to the normal process of simplification in mathematics. Nowadays we have a much less elegant system, which downplays the use of infinite series and complicates the system of rules for differentiation and integration. The rules for differentiation are still complete, given a sensible set of operations for constructing functions, but the rules for integration are pathetically incomplete. They do not suffice to integrate simple algebraic functions like $\sqrt{1 + x^3}$, or even rational functions with undetermined constants like $1/(x^5 - x - A)$. Moreover, it is only in recent decades that we have been able to tell *which* algebraic functions are integrable by our rules. (This little-known result is expounded by Davenport (1981).)

The conclusion seems to be that, apart from streamlining the language slightly, we cannot make calculus any simpler than it was in the 17th century! It is certainly easier to present the history of the subject if we refrain from imposing modern ideas. This approach also has the advantage of emphasizing the highly combinatorial nature of calculus—it is about *calculation*, after all. In view of the current controversy over the relative merits of calculus and combinatorics, it may be useful to remember that most classical combinatorics was part of the algebra of series, and hence a part of calculus. We develop this theme at greater length in the chapter on infinite series that follows.

Much has been written on the history of calculus, and some particularly useful books are Boyer (1959), Baron (1969), and Edwards (1979). However, historians tend to harp on the question of logical justification and to spend a disproportionate amount of time on the way it was handled in the 19th century. This not only obscures the boldness and vigor of early calculus, but it is overly dogmatic about the way in which calculus should be justified. Apart from the justification already available in the 17th century (the method of exhaustion), there is also a 20th-century justification (the theory of infinitesimals of Robinson (1966)), and the sheer diversity of foundations for calculus suggests that we have not yet got to the bottom of it.

9.2 Early Results on Areas and Volumes

The idea of integration is often introduced by approximating the area under curves $y = x^k$ by rectangles (Figure 9.1), say, from 0 to 1. If the base of the region is divided into n equal parts, then the heights of the rectangles are $(1/n)^k, (2/n)^k, \ldots, (n/n)^k$, and finding the area occupied by the rectangles depends on summing the series $1^k + 2^k + \cdots + n^k$. If the curve is revolved around the x-axis, then the rectangles sweep out cylinders of cross-sectional area πr^2, where $r = (1/n)^k, (2/n)^k, \ldots, (n/n)^k$, which necessitates summing the series $1^{2k} + 2^{2k} + \cdots + n^{2k}$.

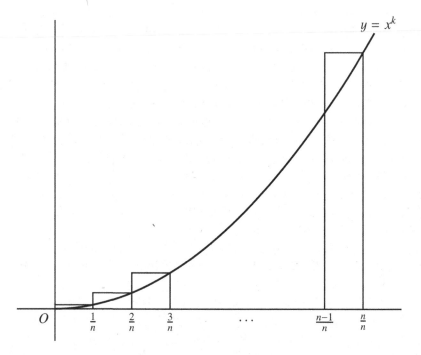

Figure 9.1: Approximating an area by rectangles

After the time of Archimedes, the first new results on areas and volumes were in fact based on summing these series. The Arab mathematician al-Haytham (around 965–1039) summed the series $1^k + 2^k + \cdots + n^k$ for $k = 1, 2, 3, 4$, and used the result to find the volume of the solid obtained by rotating the parabola about its base. See Baron (1969), p. 70, or Edwards (1979), p. 84, for al-Haytham's method of summing the series, and the exercises for another method.

Cavalieri (1635) extended these results up to $k = 9$, using them to obtain the equivalent of

$$\int_0^a x^k \, dx = \frac{a^{k+1}}{k+1}$$

and conjecturing this formula for all positive integers k. This result was proved in the 1630s by Fermat, Descartes, and Roberval. Fermat even obtained the result for fractional k (see Baron (1969), pp. 129, 185, and Edwards (1979), p. 116). Cavalieri is best known for his "method of indivisibles," an early method of discovery which considered areas divided into infinitely thin strips and volumes divided into infinitely thin slices. Archimedes' *Method* used similar ideas but, as mentioned in Section 4.1, this was not known until the 20th century. Remarkably, Cavalieri's contemporary Torricelli (the inventor of the barometer) speculated that such a method may have been used by the Greeks. Torricelli himself obtained many results using indivisibles, one being almost identical with an area determination for the parabola given by Archimedes in the *Method* (Torricelli (1644)). Another of his discoveries, which caused astonishment at the time, was that the infinite solid obtained by revolving $y = 1/x$ about the x axis from 1 to ∞ has finite volume (Torricelli (1643) and Exercise 9.2.3). The philosopher Hobbes (1672) wrote of Torricelli's result that "to understand this for sense, it is not required that a man should be 'a geometrician or logician, but that he should be mad."

EXERCISES

9.2.1 Find $1 + 2 + \cdots + n$ by summing the identity $(m + 1)^2 - m^2 = 2m + 1$ from $m = 1$ to n. Similarly find $1^2 + 2^2 + \cdots + n^2$ using the identity

$$(m + 1)^3 - m^3 = 3m^2 + 3m + 1$$

together with the previous result. Likewise, find $1^3 + 2^3 + \cdots + n^3$ using the identity

$$(m + 1)^4 - m^4 = 4m^3 + 6m^2 + 4m + 1$$

and so on.

9.2.2 Show that the approximation to the area under $y = x^2$ by rectangles in Figure 9.1 has value $(2n + 1)n(n + 1)/6n^3$, and deduce that the area under the curve is $1/3$.

9.2.3 Show that the volume of the solid obtained by rotating the portion of $y = 1/x$ from $x = 1$ to ∞ about the x-axis is finite. Show, on the other hand, that its surface area is infinite.

Cavalieri's most elegant application of his method of indivisibles was to prove Archimedes' formula for the volume of a sphere. His argument is simpler than that of Archimedes, and it goes as follows.

9.2.4 Show that the slice $z = c$ of the sphere $x^2 + y^2 + z^2 = 1$ has the same area as the slice $z = c$ of the cylinder $x^2 + y^2 = 1$ *outside* the cone $x^2 + y^2 = z^2$.

9.2.5 Deduce from Exercise 9.2.4, and the known volume of the cone, that the volume of the sphere is $2/3$ the volume of the circumscribing cylinder.

9.3   Maxima, Minima, and Tangents

The idea of differentiation is now considered to be simpler than integration, but historically it developed later. Apart from the construction of the tangent to the spiral $r = a\theta$ by Archimedes, no examples of the characteristic limiting process

$$\lim_{\Delta x \to 0} \frac{f(x + \Delta x) - f(x)}{\Delta x}$$

appeared until it was introduced by Fermat in 1629 for polynomials f and used to find maxima, minima, and tangents. Fermat's work, like his discovery of analytic geometry, was not published until 1679, but it became known to other mathematicians through correspondence after a more complicated tangent method was published by Descartes (1637).

Fermat's calculations involve a sleight of hand also used by Newton and others: introduction of a "small" or "infinitesimal" element E at the beginning, dividing by E to simplify, then omitting E at the end as if it were zero. For example, to find the slope of the tangent to $y = x^2$ at any value x, consider the chord between the points (x, x^2) and $(x + E, (x + E)^2)$ on it:

$$\text{slope} = \frac{(x + E)^2 - x^2}{E}$$
$$= \frac{2xE + E^2}{E} = 2x + E.$$

We now get the slope of the tangent by neglecting E. This procedure enraged the philosophers, who thought it was being claimed that $2x + E = 2x$ and at the same time $E \neq 0$. Of course, it is only necessary to claim that $\lim_{E \to 0}(2x + E) = 2x$, but 17th-century mathematicians did not know how to say this. In any case, they were too carried away with the power of the method to worry about such criticisms (and it was difficult to take philosophers seriously when they were as obstinate as Hobbes; see Section 9.2). Fermat's method applies to all polynomials $p(x)$, since the highest-degree term in $p(x+E)$ is always canceled by the highest-degree term in $p(x)$, leaving terms divisible by E. Fermat was also able to extend it to curves given by polynomial equations $p(x, y) = 0$. He did this in 1638 when Descartes, hoping to stump him, proposed finding the tangent to the folium.

The generality of Fermat's method entitles him to be regarded as one of the founders of calculus. He could certainly find tangents to all curves given by polynomial equations $y = p(x)$ and probably to all algebraic curves $p(x, y) = 0$. A completely explicit rule for the latter problem was found by Sluse about 1655 (but not published until Sluse (1673)) and by Hudde in 1657 (published in the 1659 edition of Descartes's *La Géométrie*, Schooten (1659)). In our notation, if

$$p(x, y) = \sum a_{ij} x^i y^j = 0,$$

then

$$\frac{dy}{dx} = -\frac{\sum i a_{ij} x^{i-1} y^j}{\sum j a_{ij} x^i y^{j-1}}.$$

Nowadays, this result is easily obtained by implicit differentiation (see the exercises below), but it can also be obtained by direct manipulation of polynomials.

EXERCISES

For evidence that tangents to algebraic curves may be found without calculus, it is enough to look more closely at what we called Diophantus's tangent method in Section 3.5. In his *Arithmetica*, Problem 18, Book VI (previously mentioned in Exercise 3.5.1), Diophantus finds the tangent $y = \frac{3x}{2} + 1$ to $y^2 = x^3 - 3x^2 + 3x + 1$ at the point $(0, 1)$, apparently by inspection. Without mentioning its geometric interpretation, he simply substitutes $\frac{3x}{2} + 1$ for y in $y^2 = x^3 - 3x^2 + 3x + 1$.

9.3.1 Check that this substitution gives the equation

$$x^3 - \frac{21}{4} x^2 = 0.$$

What is the geometric interpretation of the double root $x = 0$?

9.3.2 What would you substitute for y to find the tangent at $(0, 1)$ to the curve $y^2 = x^3 - 3x^2 + 5x + 1$?

These examples show how tangents can be found by looking for double roots, though it requires some foresight to make the right substitution. With calculus, the process is more mechanical.

9.3.3 Derive the formula of Hudde and Sluse by differentiating $\sum a_{ij} x^i y^j = 0$ with respect to x.

9.3.4 Use differentiation to find the tangent to the folium $x^3 + y^3 = 3axy$ at the point (b, c).

9.4   The *Arithmetica Infinitorum* of Wallis

Wallis's efforts to arithmetize geometry were noted in Section 7.6. In his *Arithmetica Infinitorum*, Wallis (1655a) made a similar attempt to arithmetize the theory of areas and volumes of curved figures. Some of his results were, understandably, equivalent to results already known. For example, he gave a proof that

$$\int_0^1 x^p \, dx = \frac{1}{p+1}$$

for positive integers p by showing that

$$\frac{0^p + 1^p + 2^p + \cdots + n^p}{n^p + n^p + n^p + \cdots + n^p} \to \frac{1}{p+1} \quad \text{as} \quad n \to \infty.$$

However, he made a new approach to fractional powers, finding $\int_0^1 x^{m/n} \, dx$ directly rather than by consideration of the curve $y^n = x^m$, as Fermat had done. He first found $\int_0^1 x^{1/2} \, dx$, $\int_0^1 x^{1/3} \, dx$, ... by considering the areas complementary to those under $y = x^2$, $y = x^3$, ... (Figure 9.2), then guessed the results for other fractional powers by analogy with those already obtained.

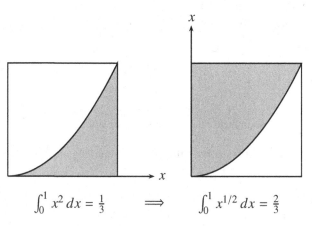

$$\int_0^1 x^2 \, dx = \tfrac{1}{3} \qquad \Longrightarrow \qquad \int_0^1 x^{1/2} \, dx = \tfrac{2}{3}$$

Figure 9.2: Areas used by Wallis

Like other early contributors to calculus, Wallis was ambivalent about quantities that tended to zero, treating them as nonzero one minute and zero the next. For this he received a ferocious blast from his arch-enemy Thomas Hobbes: "Your scurvy book of *Arithmetica infinitorum*; where your indivisibles have nothing to do, but as they are supposed to have quantity, that is to say, to be *divisibles*" (Hobbes (1656), p. 301). Quite apart from this fault, which is easily remedied by limit arguments, the reasoning of Wallis is extremely incomplete by today's standards. Observing a pattern in formulas for $p = 1, 2, 3$, for example, he will immediately claim a formula for all positive integers p "by induction" and for fractional p "by interpolation." His boldness reached new heights toward the end of the *Arithmetica infinitorum* in deriving his famous infinite product formula,

$$\frac{\pi}{4} = \frac{2}{3} \cdot \frac{4}{3} \cdot \frac{4}{5} \cdot \frac{6}{5} \cdot \frac{6}{7} \cdots$$

An exposition of his reasoning may be found in Edwards (1979), pp. 171–176, where it is described as "one of the more audacious investigations by analogy and intuition that has ever yielded a correct result."

   However, we must bear in mind that Wallis was offering primarily a method of discovery, and what a discovery he made! His infinite product for π was not the first ever given, since Viète (1593) had discovered

$$\frac{2}{\pi} = \cos\frac{\pi}{4} \cos\frac{\pi}{8} \cos\frac{\pi}{16} \cdots$$

$$= \sqrt{\frac{1}{2}} \cdot \sqrt{\frac{1}{2}\left(1 + \sqrt{\frac{1}{2}}\right)} \cdot \sqrt{\frac{1}{2}\left[1 + \sqrt{\frac{1}{2}\left(1 + \sqrt{\frac{1}{2}}\right)}\right]} \cdots$$

However, the formula of Viète is based on a clever but simple trick (see exercises), whereas that of Wallis is of deeper significance. By relating π to the integers through a sequence of rational operations, Wallis uncovered a sequence of fractions, obtained by terminating the product at the nth factor, that he called "hypergeometric." Similar sequences were later found to occur as coefficients in series expansions of many functions, which led to a broad class of functions being called "hypergeometric" by Gauss. Also, Wallis's product was closely related to two other beautiful formulas for π based on sequences of rational operations:

$$\frac{4}{\pi} = 1 + \cfrac{1^2}{2 + \cfrac{3^2}{2 + \cfrac{5^2}{2 + \cfrac{7^2}{2 + \cdots}}}}$$

and

$$\frac{\pi}{4} = 1 - \frac{1}{3} + \frac{1}{5} - \frac{1}{7} + \cdots.$$

The continued fraction was obtained by Brouncker from Wallis's product and also published in Wallis (1655b). The series is a special case of the series

$$\tan^{-1} x = x - \frac{x^3}{3} + \frac{x^5}{5} - \frac{x^7}{7} + \cdots$$

discovered by the Indian mathematician Mādhava in the 15th century (see Section 10.1) and rediscovered by Newton, Gregory, and Leibniz. Euler (1748a), p. 311, gave a direct transformation of the series for $\pi/4$ into Brouncker's continued fraction. Besides setting off this spectacular chain reaction, Wallis's method of "interpolation" had important consequences in the work of Newton, who used it to discover the general binomial theorem (Section 10.2).

EXERCISES

9.4.1 Use the identity $\sin x = 2\sin(x/2)\cos(x/2)$ to show that

$$\frac{\sin x}{2^n \sin(x/2^n)} = \cos \frac{x}{2} \cos \frac{x}{2^2} \cdots \cos \frac{x}{2^n},$$

whence

$$\frac{\sin x}{x} = \cos \frac{x}{2} \cos \frac{x}{2^2} \cos \frac{x}{2^3} \cdots.$$

9.4.2 Deduce Viète's product by substituting $x = \pi/2$.

The equation relating the series for $\pi/4$ to the continued fraction for $4/\pi$, namely

$$1 - \frac{1}{3} + \frac{1}{5} - \frac{1}{7} + \cdots = \cfrac{1}{1 + \cfrac{1^2}{2 + \cfrac{3^2}{2 + \cfrac{5^2}{2 + \cfrac{7^2}{2 + \cdots}}}}}$$

follows immediately from a more general equation

$$\frac{1}{A} - \frac{1}{B} + \frac{1}{C} - \frac{1}{D} + \cdots = \cfrac{1}{A + \cfrac{A^2}{B - A + \cfrac{B^2}{C - B + \cfrac{C^2}{D - C + \cdots}}}}$$

proved by Euler (1748a), p. 311. The following exercises give a proof of Euler's result.

9.4.3 Check that

$$\frac{1}{A} - \frac{1}{B} = \frac{1}{A + \dfrac{A^2}{B - A}} .$$

9.4.4 When $\frac{1}{B}$ on the left side in Exercise 9.4.3 is replaced by $\frac{1}{B} - \frac{1}{C}$, which equals $\dfrac{1}{B + \frac{B^2}{C-B}}$ by Exercise 9.4.3, show that B on the right should be replaced by $B + \frac{B^2}{C-B}$. Hence show that

$$\frac{1}{A} - \frac{1}{B} + \frac{1}{C} = \frac{1}{A + \dfrac{A^2}{B - A + \dfrac{B^2}{C - B}}} .$$

Thus when we modify the "tail end" of the series (replacing $\frac{1}{B}$ by $\frac{1}{B} - \frac{1}{C}$), only the "tail end" of the continued fraction is affected. This situation continues:

9.4.5 Generalize your argument in Exercise 9.4.4 to obtain a continued fraction for a series with n terms, and hence prove Euler's equation.

9.5   Newton's Calculus of Series

Newton made many of his most important discoveries in 1665/6, after studying the works of Descartes, Viète, and Wallis. In Schooten's edition of *La Géométrie* he encountered Hudde's rule for tangents to algebraic curves, which was virtually a complete differential calculus from Newton's viewpoint. Although Newton made contributions to differentiation that are useful to *us*—the chain rule, for example—differentiation was a minor part of *his* calculus, which depended mainly on the manipulation of infinite series. Thus it is misleading to describe Newton as a founder of calculus unless one understands calculus, as he did, as an algebra of infinite series. In this calculus, differentiation and integration are carried out term by term on powers of x and hence are comparatively trivial.

At the beginning of his main work on calculus, *A Treatise of the Methods of Series and Fluxions* (also known by its abbreviated Latin name of *De methodis*), Newton clearly states his view of the role of infinite series:

Since the operations of computing in numbers and with variables are closely similar ... I am amazed that it has occurred
to no one (if you except N. Mercator with his quadrature of
the hyperbola) to fit the doctrine recently established for decimal numbers in similar fashion to variables, especially since
the way is then open to more striking consequences. For since
this doctrine in species has the same relationship to Algebra
that the doctrine in decimal numbers has to common Arithmetic, its operations of Addition, Subtraction, Multiplication,
Division and Root extraction may be easily learnt from the
latter's.

Newton (1671), pp. 33–35

The quadrature (area determination) of the hyperbola mentioned by
Newton was the result that we would write as

$$\int_0^x \frac{dt}{1+t} = x - \frac{x^2}{2} + \frac{x^3}{3} - \frac{x^4}{4} + \cdots,$$

first published in Mercator (1668). Newton had discovered the same result
in 1665, and it was partly his dismay in losing priority that led him to
write *De methodis* and an earlier work *De analysi* (Newton (1669); the full
title in English is *On Analysis by Equations Unlimited in Their Number of
Terms*). Newton also independently discovered the series for $\tan^{-1} x$, $\sin x$,
and $\cos x$ in *De analysi*, without knowing that all three series had already
been discovered by Indian mathematicians. See Section 10.1.

The Mercator and Indian results were both obtained by the method of
expanding a geometric series and integrating term by term. In our notation,

$$\int_0^x \frac{dt}{1+t} = \int_0^x (1 - t + t^2 - t^3 + \cdots)\, dt$$

$$= x - \frac{x^2}{2} + \frac{x^3}{3} - \frac{x^4}{4} + \cdots$$

and

$$\tan^{-1} x = \int_0^x \frac{dt}{1+t^2}$$

$$= \int_0^x (1 - t^2 + t^4 - t^6 + \cdots)\, dt$$

$$= x - \frac{x^3}{3} + \frac{x^5}{5} - \frac{x^7}{7} + \cdots.$$

Newton routinely used these methods in *De analysi* and *De methodis*, but he greatly extended their scope by algebraic manipulation. Not only did he obtain sums, products, quotients, and roots, as foreshadowed in his introduction to *De methodis*, but his root extractions also extended to the general construction of *inverse functions* by the new idea of inverting infinite series. For example, after Newton (1671), p. 61, found the series $x - (x^2/2) + (x^3/3) - \cdots$, for $\int_0^x dt/(1 + t)$, which is $\log(1 + x)$, he set

$$y = x - \frac{x^2}{2} + \frac{x^3}{3} - \cdots \tag{1}$$

and solved (1) for x (which we recognize to be the exponential function e^y, minus 1). His method amounts to setting $x = a_0 + a_1 y + a_2 y^2 + \cdots$, substituting in the right-hand side of (1), and determining $a_0, a_1, a_2, \ldots$, successively by comparing with the coefficients on the left-hand side. Newton found the first few terms,

$$x = y + \frac{1}{2}y^2 + \frac{1}{6}y^3 + \frac{1}{24}y^4 + \frac{1}{120}y^5 + \cdots,$$

then confidently guessed that $a_n = 1/n!$ in the manner of Wallis. As he put it, "Now after the roots have been extracted to a suitable period, they may sometimes be extended at pleasure by observing the analogy of the series."

De Moivre (1698) gave a formula for inverting series that justifies such conclusions; Newton astonishes us by finding such an elegant result by such a forbidding method. His discovery of the sine series (Newton (1669), pp. 233, 237) is even more amazing. First he used the binomial series

$$(1 + a)^p = 1 + pa + \frac{p(p - 1)}{2!}a^2 + \frac{p(p - 1)(p - 2)}{3!}a^3 + \cdots$$

(though not with the natural choice $a = -x^2, p = -\frac{1}{2}$) to obtain

$$\sin^{-1} x = z = x + \frac{1}{2}\frac{x^3}{3} + \frac{1 \cdot 3}{2 \cdot 4}\frac{x^5}{5} + \frac{1 \cdot 3 \cdot 5}{2 \cdot 4 \cdot 6}\frac{x^7}{7} + \cdots$$

by term-by-term integration, and then casually stated "I extract the root, which will be

$$x = z - \frac{1}{6}z^3 + \frac{1}{120}z^5 - \frac{1}{5040}z^7 + \frac{1}{362880}z^9 - \cdots ,"$$

adding a few lines later that the coefficient of z^{2n+1} is $1/(2n + 1)!$.

EXERCISES

Newton inverted series by a tabular method like the following, which shows the coefficients of $1, y, y^2, y^3, \ldots$ in x and its powers.

	1	y	y^2	y^3	$\cdots$
x	a_0	a_1	a_2	a_3	$\cdots$
x^2	a_0^2	$2a_0 a_1$	$2a_0 a_2 + a_1^2$	$2a_0 a_3 + 2a_1 a_2$	$\cdots$

9.5.1 Use the rows shown to substitute series in powers of y for x and x^2 in $y = x - \frac{x^2}{2} + \cdots$, and hence show that $a_0 = 0$, $a_1 = 1$, and $a_2 = 1/2$ in turn, by comparing coefficients on the two sides of the equation.

9.5.2 Compute the first few entries in the third row of the table (the coefficients of x^3), and hence show that $a_3 = 1/6$.

This shows why the inverse function $x = e^y - 1$ has a power series that begins

$$y + \frac{1}{2}y^2 + \frac{1}{6}y^3 + \cdots .$$

9.5.3 Show that the binomial series gives

$$\frac{1}{\sqrt{1 - t^2}} = 1 + \frac{1}{2}t^2 + \frac{1 \cdot 3}{2 \cdot 4}t^4 + \frac{1 \cdot 3 \cdot 5}{2 \cdot 4 \cdot 6}t^6 + \cdots .$$

9.5.4 Use Exercise 9.5.3 and $\sin^{-1} x = \int_0^x dt / \sqrt{1 - t^2}$ to derive Newton's series for $\sin^{-1} x$.

9.6   The Calculus of Leibniz

Newton's epoch-making works (1669, 1671) were offered to the Royal Society and Cambridge University Press but, incredible as it now seems, were rejected for publication. One theory is that printers were short of paper, since large quantities had been destroyed by the great fire of London in 1666. At any rate, the first published paper on calculus was not by Newton but by Leibniz (1684). This led to Leibniz's initially receiving credit for the calculus and later to a bitter dispute with Newton and his followers over the question of priority for the discovery.

There is no doubt that Leibniz discovered calculus independently, that he had a better notation, and that his followers contributed more to the spread of calculus than did Newton's. Leibniz's work lacked the depth and virtuosity of Newton's, but then Leibniz was a librarian, a philosopher, and a diplomat with only a part-time interest in mathematics. His

Nova methodus (1684) is a relatively slight paper, though it does lay down some important fundamentals—the sum, product, and quotient rules for differentiation—and it introduces the dy/dx notation we now use. However, dy/dx was not just a symbol for Leibniz, as it is for us, but literally a quotient of *infinitesimals* dy and dx, which he viewed as differences (hence the symbol d) between neighboring values of y and x, respectively.

He also introduced the integral sign, $\int$, in his *De geometria* (1686) and proved the fundamental theorem of calculus, that integration is the inverse of differentiation. This result was known to Newton and even, in a geometric form, to Newton's teacher Barrow, but it became more transparent in Leibniz's formalism. For Leibniz, $\int$ meant "sum," and $\int f(x)\,dx$ was literally a sum of terms $f(x)dx$, representing infinitesimal areas of height $f(x)$ and width dx. The difference operator d yields the last term $f(x)\,dx$ in the sum, and dividing by the infinitesimal dx yields $f(x)$. So voila!

$$\frac{d}{dx}\int f(x)\,dx = f(x)$$

—the fundamental theorem of calculus.

Leibniz's strength lay in the identification of important concepts, rather than in their technical development. He introduced the word "function" and was the first to begin thinking in function terms. He made the distinction between algebraic and transcendental functions and, in contrast to Newton, preferred "closed-form" expressions to infinite series. Thus the evaluation of $\int f(x)\,dx$ for Leibniz was the problem of finding a known function whose derivative was $f(x)$, whereas for Newton it was the problem of expanding $f(x)$ in series, after which integration was trivial.

The search for closed forms was a wild goose chase but, like many efforts to solve intractable problems, it led to worthwhile results in other directions. Attempts to integrate rational functions raised the problem of factorization of polynomials and led ultimately to the fundamental theorem of algebra (see Chapter 14). Attempts to integrate $1/\sqrt{1-x^4}$ led to the theory of elliptic functions (Chapter 12).

As mentioned in Section 9.1, the problem of deciding which algebraic functions may be integrated in closed form has been solved only recently, though not in a form suitable for calculus textbooks, which continue to remain oblivious to most of the developments since Leibniz. (One thing that has changed: it is now much easier to publish a calculus book than it was for Newton!)

EXERCISES

Leibniz (1702) was stymied by the integral $\int \frac{dx}{x^4+1}$, because he did not spot the factorization of $x^4 + 1$ into real quadratic factors.

9.6.1 Writing $x^4 + 1 = x^4 + 2x^2 + 1 - 2x^2$ or otherwise, split $x^4 + 1$ into real quadratic factors.

9.6.2 Use the factors in Exercise 9.6.1 to express $\frac{1}{x^4+1}$ in the partial fraction form

$$\frac{x + \sqrt{2}}{q_1(x)} + \frac{x - \sqrt{2}}{q_2(x)},$$

where $q_1(x)$ and $q_2(x)$ are real quadratic polynomials.

9.6.3 Without working out all the details, explain how the partial fractions in Exercise 9.6.2 can be integrated in terms of rational functions and the $\tan^{-1}$ function.

9.7   Biographical Notes: Wallis, Newton, and Leibniz

John Wallis (Figure 9.3) was born in 1616 in Ashford, Kent, and died in Oxford in 1703. He was one of five children of John Wallis, the rector of Ashford, and Joanna Chapman. He had two older sisters and two younger brothers. Young John Wallis was recognized as the academic talent of the family and at 14 was sent to Felsted, Essex, to attend the school of Martin Holbech, a famous teacher of the time. At school he learned Latin, Greek, and Hebrew, but he did not meet mathematics until he was home on Christmas vacation in 1631. One of his brothers was learning arithmetic to prepare for a trade, and Wallis asked him to explain it. This turned out to be the only mathematical instruction Wallis ever received, even though he later studied at Emmanuel College in Cambridge.

As Wallis explained in his autobiography:

> Mathematicks were not, at that time, looked upon as Acca-
> demical Learning, but the business of Traders, Merchants, Sea-
> men, Carpenters, land-measurers, or the like; or perhaps some
> Almanak-makers in London. And of more than 200 at that
> time in our College, I do not know of any two that had more

Figure 9.3: John Wallis

of Mathematicks than myself, which was but very little; having never made it my serious studie (otherwise than as a pleasant diversion) till some little time before I was designed for a
Professor in it.

Wallis (1696), p. 27

At Emmanuel College, Wallis studied divinity from 1632 to 1640,
when he gained a master of arts degree. College life evidently agreed with
him, and he would have stayed on as a fellow, had there been a place available. He did become a fellow of Queens College, Cambridge, for a year
but, since fellows had to remain unmarried, relinquished the fellowship
when he married in 1645. Thus it was that Wallis spent most of the 1640s
in the ministry.

The 1640s were a decisive decade in English history, with the rise of
the parliamentary opposition to Charles I and the king's execution in 1649.
Partly by luck and partly by adaptation to the new political conditions,
Wallis changed the direction of his life toward mathematics. Early in the
conflict he found he had the very valuable ability to decipher coded messages. To quote the autobiography again:

> About the beginning of our Civil Wars, in the year 1642, a
> Chaplain of S^r. William Waller showed me an intercepted Let-
> ter written in Cipher. . . . He asked me (between jest and earnest)
> if I could make any thing of it. . . . I judged it could be no more
> than a new Alphabet and, before I went to bed, I found it out,
> which was my first attempt upon Deciphering.
>
> Wallis (1696), p. 37

This was the first in a series of successes Wallis had in codebreaking for
the Parliamentarians, which gained him not only political favor but also a
reputation for mathematical skill. (For more information on Wallis's cryptography see Kahn (1967), p. 166.) When the royalist Peter Turner was
expelled from the Savilian Chair of Geometry of Oxford in 1649, Wallis
was appointed in his place. At last his dormant mathematical ability had a
chance to develop, and from then on he was active in mathematics almost
continually until the end of his life.

Isaac Newton (Figure 9.4) was born on Christmas Day, 1642, at Woolsthorpe, Lincolnshire. His family background and early life did not augur
well for future greatness. Newton's father, also named Isaac, was fairly
well off but illiterate, and he died three months before Newton was born.
His mother, Hannah Ayscough, remarried when Newton was three, only
to abandon him on the insistence of his stepfather. The boy was left in
the care of the Ayscough family, a circumstance that helped his education
(Hannah's brother William had studied at Cambridge, and eventually he
directed Newton there) but did not compensate emotionally for the absence
of his father and mother. Newton became intensely neurotic, secretive, and
suspicious in later life; he never married and tended to make enemies rather
than friends.

The young Newton was more interested in building intricate machines,
such as model windmills, than academic studies, though once he set his
mind to it he became top of his school. In 1661 he entered Trinity College, Cambridge, as a sizar. Sizars had to earn their keep as servants to
the wealthier students, and it was indicative of his mother's meanness that
he had to become one, for she could afford to support him but chose not
to. Newton's early studies were in Aristotle, the standard curriculum of
the time. The first thinker to make an impression on him was Descartes,
whose works were then creating a stir in Cambridge. By 1664, in a series
of notes he called *Quaestiones quaedam philosophicae*, Newton was absorbed with questions of mechanics, optics, and the physiology of vision.

He was also struck by Descartes's geometry, preferring it to Euclid, which in his first encounter "he despised ... as a trifling book" (according to later reminiscences of de Moivre).

The years 1664 to 1666 were the most important in Newton's mathematical development and perhaps the most creative period in the life of any mathematician. In 1664 he devoured the mathematics of Descartes, Viète, and Wallis and began his own investigations. Late in 1664 he conceived the idea of curvature, from which much of differential geometry was to grow (see Chapter 17). The university was closed in 1665, which was the disastrous plague year in much of England. Newton returned to Woolsthorpe, where his mathematical reflections became an all-consuming passion. Fifty years later, Newton recalled the time as follows:

> In the beginning of the year 1665 I found the Method of approximating series & the Rule for reducing any dignity of any Binomial into such a series. The same year in May I found the method of Tangents of Gregory & Slusius, & in November had the direct method of fluxions & the next year in January had the Theory of Colours & in May following I had entrance into ye inverse method of fluxions. And in the same year I began to think of gravity extending to ye orb of the Moon & ... from Keplers rule of the periodical times of the Planets ... I deduced that the forces wch keep the Planets in the Orbs must [be] reciprocally as the squares of their distance from the centers. ... All this was in the two plague years of 1665–1666. For in those days I was in the prime of my age for invention & minded Mathematicks & Philosophy more then [*sic*] at any time since.
>
> Whiteside (1966), p. 32

In addition to the achievements mentioned, Newton's discoveries in this period included the series for $\log(1 + x)$ and, at least in preliminary form, the classification of cubic curves.

As we have seen, Newton's first attempts to publish his results were unsuccessful; nevertheless, there were some who read them and recognized his genius. In 1669 the Lucasian Professor of Mathematics at Trinity, Isaac Barrow, resigned to devote himself to theology, and Newton was appointed to the chair on Barrow's recommendation. Newton held the position until

1696, when he made the puzzling decision to accept the position of master of the Mint in London. The outstanding achievement of his Lucasian professorship was the classic *Principia* (1687), or, to give it its full title, *Philosophiae naturalis principia mathematica* (Mathematical Principles of Natural Philosophy).

The *Principia*, which developed the theory of gravitation based on Newton's inverse square law of 1665, owes its existence to a visit by Edmund Halley to Cambridge in 1684. The hypothesis of the inverse square law was in the air at this stage—Wren, Hooke, and Halley himself had thought of it—but a mathematical derivation of its consequences was lacking. Halley asked Newton what curve a planet would describe under this law and was delighted to learn that Newton had calculated it to be an ellipse. When asked to supply his demonstration, Newton had some trouble reconstructing it, eventually sending Halley a nine-page paper, *De motu corporum in gyrum* (On the Motion of Bodies in an Orbit), three months later. *De motu* was the *Principia* in embryonic form.

Figure 9.4: Isaac Newton

Realizing the importance of Newton's results, Halley communicated them to the Royal Society and urged Newton to expand them for publication. His prodding came at just the right time. The excitement over

Newton's early discoveries had died down, and for the preceding six or seven years he had been wasting his time on alchemical experiments. With his interest in mathematics rekindled, Newton devoted the next 18 months almost exclusively to *Principia* "so intent, so serious upon his studies, y^t he eat very sparingly, nay, ofttimes he has forget to eat at all," as a Cambridge contemporary noticed (see Westfall (1980), p. 406). When Book I was delivered to the Royal Society in April 1686, they were still reluctant to publish, and it took heroic efforts from Halley to bring them round. He not only risked his own money on the venture, but had to coax Newton to go through with it, as Newton flew into tantrums when Hooke raised his own claims of priority. Finally in 1687 the *Principia* was published and Newton's fame was secure, at least in Britain.

In the early 1690s Newton worked on revising the *Principia* and bringing some of his earlier investigations into order. As we have seen, the final form of his classification of cubic curves dates from this period. In 1693 he had a nervous breakdown, and this may have influenced him to leave Cambridge for the Mint in 1696. He did not completely abandon science, becoming president of the Royal Society in 1703, but his mathematical activity was mainly confined to the priority dispute with Leibniz over the invention of the calculus. Newton died in 1727 and was buried in Westminster Abbey. Westfall (1980) is an excellent recent biography.

Gottfried Wilhelm Leibniz (Figure 9.5) was born in Leipzig in 1646 and died in Hannover in 1716. His father, Friedrich, was professor of moral philosophy at Leipzig, and his mother, Katherina Schmuck, also came from an academic family. From the age of six Leibniz was given free access to his father's library, and he became a voracious reader. At 15 he entered the University of Leipzig and received a doctorate in law from Altdorf in 1666 (Leipzig refused him a doctorate because of his youth). During 1663, on a summer visit to the University of Jena, he learned a little of Euclid, but otherwise his studies were in law and philosophy, the subjects that were to be the basis of his subsequent career. The lack of early practice in mathematics left its mark on Leibniz's later mathematical style, in which good ideas are sometimes insufficiently developed through lack of technical skill. Often he seemed to lack not only the technique but also the patience to develop the ideas conceived by his wide-ranging imagination. It now appears that Leibniz was a pioneer in combinatorics, mathematical logic, and topology, but his ideas in these fields were too fragmentary to be of use to his contemporaries.

An interest in logic led Leibniz to his first mathematical venture, the essay *Dissertatio de arte combinatoria* (1666). His aim was "a general method in which all truths of reason would be reduced to a kind of calculation." Leibniz foresaw that permutations and combinations would be involved, but he did not make enough progress to interest 17th-century mathematicians in the project. The dream of a universal logical calculus was revived in the 19th century but finally shattered by the results of Gödel (1931) (see Chapter 24). Nevertheless, Leibniz benefited greatly from his work on combinatorics; it led him to his ideas in calculus.

Figure 9.5: Gottfried Wilhelm Leibniz

Following his doctorate in law, Leibniz had commenced a legal career in the service of the elector of Mainz. In 1672 his duties took him to Paris, where he met Huygens and for the first time gained a firm grasp of mathematics. The years 1672 to 1676 were crucial in Leibniz's mathematical life and have been covered in detail by Hofmann (1974). Beginning with "Pascal's triangle," which he had used in his *Dissertatio* (1666), Leibniz became interested in the differences between successive terms of series.

Using differences, he developed a method of interpolation for functions that, as we shall see in Section 10.2, was also the independent discovery of Newton and Gregory. Leibniz showed his discovery to Huygens, who encouraged him to use differences in the summation of infinite series by posing the problem of evaluating $\sum_{n=1}^{\infty} 1/n(n+1)$. Leibniz succeeded (after some time) and used the same method in other cases. This was his introduction to the infinite processes of the calculus and also, perhaps, the origin of his preference for "closed-form" solutions. In 1673 he advanced to a higher level, using term-by-term integration to discover

$$\frac{\pi}{4} = 1 - \frac{1}{3} + \frac{1}{5} - \frac{1}{7} + \cdots$$

and

$$\frac{1}{2}\log 2 = \frac{1}{2.4} + \frac{1}{6.8} + \frac{1}{10.12} + \cdots.$$

By 1676 he had virtually completed his formulation of the calculus, including the fundamental theorem, the dx notation, and the integral sign.

The first period of Leibniz's mathematical activity came to an end in 1676. He had failed to obtain an academic position in Paris or London and, seeking a better salary, he moved to Hannover to enter the service of the Duke of Brunswick-Lüneburg. His main duties were to act as adviser, librarian, and consultant on certain engineering works. When the duke died in 1679 his successor commissioned Leibniz to compile a genealogy of the House of Brunswick in order to bolster the family's dynastic claims. Leibniz threw himself into this project with a zeal that is hard to admire, given the purpose of the genealogy, though it did enable him to travel, visit libraries, and meet scholars throughout Europe. He helped found the journal *Acta Eruditorum* in 1682 and used it to publish his discoveries in calculus, as well as those of his brilliant successors, Jakob and Johann Bernoulli. This led to the rapid spread of Leibniz's notation and methods throughout the Continent.

With the succession of a new Duke of Brunswick in 1698, Leibniz fell somewhat from favor, though he retained his job and, with the support of other family members, founded the Berlin Academy in 1700 and became its first president. His final years were embittered by the priority dispute over the calculus and his employer's neglect. He was still doggedly trying to complete the history of the House of Brunswick when he died in 1716. His secretary was the only person to attend his funeral, and the history was not published until 1843.

10

Infinite Series

PREVIEW

As we saw in the previous chapter, many calculus problems have a solution that can be expressed as an infinite series. It is therefore useful to be able to recognize important individual series and to understand their general properties and capabilities. This is the aim of the present chapter.

Starting with the infinite geometric series, already known to Euclid, we discuss the handful of examples known before the invention of calculus. These include the *harmonic series* $1 + 1/2 + 1/3 + 1/4 + \cdots$, studied by Oresme around 1350, and the stunning series for the inverse tangent, sine, and cosine, discovered by Indian mathematicians in the 15th century.

The invention of calculus in the 17th century released a flood of new series, mostly of the form $a_0 + a_1 x + a_2 x^2 + \cdots$ (called *power series*), but also some variations, such as fractional power series.

The 18th century brought new applications. De Moivre (1730) used power series to find a formula for the nth term of the *Fibonacci sequence* $0, 1, 1, 2, 3, 5, 8, \ldots$. Euler (1748a) introduced a generalization of the harmonic series,

$$1 + 1/2^s + 1/3^s + 1/4^s + \cdots,$$

and showed that, for $s > 1$, it equals the *infinite product*

$$(1 - 1/2^s)^{-1}(1 - 1/3^s)^{-1}(1 - 1/5^s)^{-1} \cdots (1 - 1/p^s)^{-1} \cdots$$

over all the *prime* numbers p. This discovery of Euler's opened a new path to the secrets of the primes, exploration of which continues to this day.

J. Stillwell, *Mathematics and Its History*, Undergraduate Texts in Mathematics,   181
DOI 10.1007/978-1-4419-6053-5_10, © Springer Science+Business Media, LLC 2010

10.1 Early Results

Infinite series were present in Greek mathematics, though the Greeks tried to deal with them as finitely as possibly by working with arbitrary finite sums $a_1 + a_2 + \cdots + a_n$ instead of infinite sums $a_1 + a_2 + \cdots$. However, this is just the difference between potential and actual infinity. There is no question that Zeno's paradox of the dichotomy (Section 4.1), for example, concerns the decomposition of the number 1 into the infinite series

$$\frac{1}{2} + \frac{1}{2^2} + \frac{1}{2^3} + \frac{1}{2^4} + \cdots$$

and that Archimedes found the area of the parabolic segment (Section 4.4) essentially by summing the infinite series

$$1 + \frac{1}{4} + \frac{1}{4^2} + \frac{1}{4^3} + \cdots = \frac{4}{3}.$$

Both these examples are special cases of the result we express as summation of a geometric series

$$a + ar + ar^2 + ar^3 + \cdots = \frac{a}{1 - r} \quad \text{when} \quad |r| < 1.$$

The first examples of infinite series other than geometric series appeared in the Middle Ages. In a book from around 1350, called the *Liber calculationum*, Richard Suiseth (or Swineshead, known as the Calculator) used a very lengthy verbal argument to show that

$$\frac{1}{2} + \frac{2}{2^2} + \frac{3}{2^3} + \frac{4}{2^4} + \cdots = 2.$$

The argument is reproduced in Boyer (1959), p. 78. At about the same time, Oresme (1350b), pp. 413–421, summed this and similar series by geometric decomposition as in Figure 10.1, showing that

$$2 = \frac{1}{2} + \frac{2}{2^2} + \frac{3}{2^3} + \frac{4}{2^4} + \cdots.$$

Actually Oresme gives only the last picture in the figure, but it seems likely he arrived at it by cutting up an area of two square units as shown, judging from his opening remark: "A finite surface can be made as long as we wish, or as high, by varying the extension without increasing the size." The region constructed by Oresme, incidentally, is perhaps the first

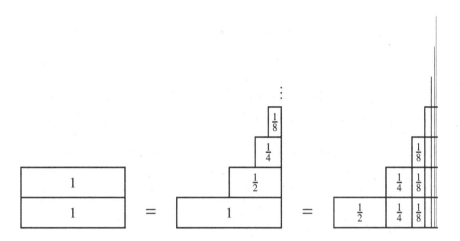

Figure 10.1: Oresme's summation

example of the phenomenon encountered by Torricelli (Section 9.2) in his
hyperbolic solid of revolution—infinite extent but finite content.

Another important discovery of Oresme (1350a) was the divergence of
the *harmonic series*

$$1 + \frac{1}{2} + \frac{1}{3} + \frac{1}{4} + \frac{1}{5} + \cdots .$$

His proof was by an elementary argument that is now standard:

$$1 + \left(\frac{1}{2}\right) + \left(\frac{1}{3} + \frac{1}{4}\right) + \left(\frac{1}{5} + \frac{1}{6} + \frac{1}{7} + \frac{1}{8}\right) + \cdots$$

$$> 1 + \left(\frac{1}{2}\right) + \left(\frac{1}{4} + \frac{1}{4}\right) + \left(\frac{1}{8} + \frac{1}{8} + \frac{1}{8} + \frac{1}{8}\right) + \cdots$$

$$= 1 + \frac{1}{2} + \frac{1}{2} + \frac{1}{2} + \cdots .$$

Thus by repeatedly doubling the number of terms collected in successive groups, we can indefinitely obtain groups of sum $> \frac{1}{2}$, enabling the
sum to grow beyond all bounds.

As mentioned in Section 9.4, the Indian mathematician Mādhava found
the series

$$\tan^{-1} x = x - \frac{x^3}{3} + \frac{x^5}{5} - \frac{x^7}{7} + \cdots$$

with its important special case

$$\frac{\pi}{4} = 1 - \frac{1}{3} + \frac{1}{5} - \frac{1}{7} + \cdots$$

in the 15th century. The series for π was the first satisfactory answer to
the classical problem of squaring the circle, for although the expression
is infinite (as it would have to be, in view of Lindemann's theorem on the
transcendence of π), the rule for generating successive terms is as finite and
transparent as it could possibly be. It is sad that the Indian series became
known in the West too late to have any influence or even to become well
known until recently. Rajagopal and Rangachari (1977, 1986) showed that
the series for $\tan^{-1} x$, $\sin x$, and $\cos x$ were known in the Kerala school of
Mādhava before 1540, and probably before 1500. For more recent information on the Kerala school, in the context of trigonometry and of Indian
mathematics in general, see Van Brummelen (2009) and Plofker (2009)
respectively.

EXERCISES

Oresme's proof by partitioning the harmonic series into

$$1 + \left(\frac{1}{2}\right) + \left(\frac{1}{3} + \frac{1}{4}\right) + \left(\frac{1}{5} + \frac{1}{6} + \frac{1}{7} + \frac{1}{8}\right) + \cdots$$

has the following geometric counterpart.

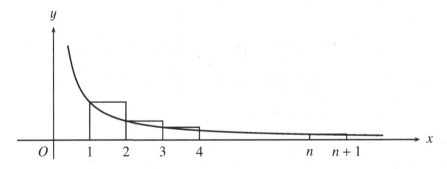

Figure 10.2: Comparing $1 + \frac{1}{2} + \frac{1}{3} + \cdots + \frac{1}{n}$ with an area

10.1.1 By referring to Figure 10.2, show that

$$1 + \frac{1}{2} + \frac{1}{3} + \cdots + \frac{1}{n} > \text{area under } y = \frac{1}{x} \text{ between } x = 1 \text{ and } x = n + 1.$$

10.1.2 Now partition this area under $y = 1/x$ into the pieces between $x = 1$ and $x = 2$, $x = 2$ and $x = 4$, $x = 4$ and $x = 8$, ..., and show that *all these pieces have the same area*. (This can even be done without using calculus, if you use the argument of Exercises 4.4.1 and 4.4.2.)

10.1.3 Deduce from Exercise 10.1.2 that the area from $x = 1$ to $x = n$, and hence the sum $1 + \frac{1}{2} + \frac{1}{3} + \cdots + \frac{1}{n}$, tends to infinity.

The area under $y = 1/x$ from $x = 1$ to $x = n + 1$ is of course $\log(n + 1)$, so Figure 10.2 shows that $1 + \frac{1}{2} + \frac{1}{3} + \cdots + \frac{1}{n} > \log(n + 1)$. As $n \to \infty$, these two functions of n remain about the same size.

10.1.4 By comparing the curved area with suitable rectangles beneath the curve, show that

$$\frac{1}{2} + \frac{1}{3} + \cdots + \frac{1}{n} < \log(n + 1),$$

and hence that $0 < 1 + \frac{1}{2} + \frac{1}{3} + \cdots + \frac{1}{n} - \log(n + 1) < 1$.

10.1.5 Also show, by a geometric argument, that $1 + \frac{1}{2} + \frac{1}{3} + \cdots + \frac{1}{n} - \log(n + 1)$ increases as n increases, so that it has a finite limit < 1.

The value of the limit is known as *Euler's constant* γ, and γ is approximately 0.577. However, little is known about the nature of γ—not even whether it is irrational.

10.2   Power Series

The Indian series for $\tan^{-1} x$ was the first example, apart from geometric series such as $1 + x + x^2 + x^3 + \cdots = 1/(1 - x)$, of a *power series*, that is, the expansion of a function $f(x)$ in powers of x. The idea of power series turned out to be fruitful not only in the representation of functions but even in the study of numerical series. Most of the interesting numerical series turned out to be instances of power series for particular values of x, for example, the series for $\pi/4$ is the $x = 1$ instance of the series for $\tan^{-1} x$.

The theory began with the series published by Mercator (1668):

$$\log(1 + x) = x - \frac{x^2}{2} + \frac{x^3}{3} - \frac{x^4}{4} + \cdots .$$

As we have seen, this was obtained by integrating the geometric series

$$\frac{1}{1 + x} = 1 - x + x^2 - x^3 + \cdots$$

term by term. Now the most important transcendental functions—logs, exponentials, and the related circular and hyperbolic functions—are obtained by integration and inversion from algebraic functions, and fairly simple algebraic functions at that. For example, e^y is the inverse function of $y = \log x$, and

$$\log(1 + x) = \int_0^x \frac{dt}{1+t},$$

$\sin y$ is the inverse function of $y = \sin^{-1} x$ and

$$\sin^{-1} x = \int_0^x \frac{dt}{\sqrt{1-t^2}}, \quad \tan^{-1} x = \int_0^x \frac{dt}{1+t^2},$$

and so on. Thus the key to finding power series is finding series expansions of simple algebraic functions. Once this is done, term-by-term integration and Newton's method of series inversion (Section 9.5) yield power series for all the common functions.

Rational functions, such as $1/(1 + t^2)$, can be expanded using geometric series; the crucial step was accomplished by Newton (1665a) when he discovered the general binomial theorem,

$$(1 + x)^p = 1 + px + \frac{p(p-1)}{2!}x^2 + \frac{p(p-1)(p-2)}{3!}x^3 + \cdots,$$

yielding the expansion of functions such as $1/\sqrt{1 - t^2} = (1 - t^2)^{-1/2}$. This theorem was also discovered independently by Gregory (1670). Both Newton and Gregory were inspired by the loose heuristic method of interpolation used by Wallis (1655a), but they refined it into a result now known as the *Gregory–Newton interpolation formula*:

$$f(a + h) = f(a) + \frac{h}{b}\Delta f(a) + \frac{(h/b)(h/b - 1)}{2!}\Delta^2 f(a) + \cdots, \quad (1)$$

where

$\Delta f(a) = f(a + b) - f(a),$

$\Delta^2 f(a) = \Delta f(a + b) - \Delta f(a) = f(a + 2b) - 2f(a + b) + f(a),$

$\Delta^3 f(a) = \Delta^2 f(a + b) - \Delta^2 f(a) = f(a + 3b) - 3f(a + 2b) + 3f(a + b) - f(a),$

$\vdots$

This wonderful formula finds the value of f at an arbitrary point $a + h$ from the values at an infinite arithmetic sequence of points $a, a + b, a + 2b, \ldots$.

The first n terms give an nth-degree polynomial in h taking the same values as f at $a, a+b, \ldots, a+nb$. Hence the formula is valid for any f that is the limit of its own approximating polynomials. This means all functions representable by power series, provided that the points $a, a+b, a+2b, \ldots$, are sensibly chosen. (The points $\pi, 2\pi, 3\pi, \ldots$, are a bad choice for $\sin x$, since the x-axis is a polynomial curve through all of them).

Newton discovered the formula (1) after his special investigations on interpolation that led to the binomial theorem. Gregory discovered the general formula first and then used it to derive the binomial theorem (see exercises below), all independently of Newton. It even appears that Gregory used the interpolation theorem to discover Taylor's theorem 44 years before Brook Taylor. There is strong evidence that Gregory used Taylor's series for other results (Gregory (1671)), and Taylor's series

$$f(a+h) = f(a) + hf'(a) + \frac{h^2}{2!}f''(a) + \cdots \qquad (2)$$

is just the limiting case of (1) as $b \to 0$. Indeed, this is how it was derived by Taylor (1715). The passage from (1) and (2) is simple if one assumes plausible limiting behavior for the infinite sum. Notice that

$$\frac{\Delta f(a)}{b} = \frac{f(a+b) - f(a)}{b} \to f'(a) \quad \text{as} \quad b \to 0$$

and similarly

$$\frac{\Delta^2 f(a)}{b^2} \to f''(a), \qquad \frac{\Delta^3 f(a)}{b^3} \to f'''(a),$$

and so on. We write (1) as

$$f(a+h) = f(a) + h\frac{\Delta f(a)}{b} + \frac{h(h-b)}{2!}\frac{\Delta^2 f(a)}{b^2} + \cdots$$

and observe that the nth term

$$\frac{h(h-b)(h-2b)\cdots(h-(n-1)b)}{n!}\frac{\Delta^n f(a)}{b^n} \to \frac{h^n}{n!}f^n(a) \quad \text{as} \quad b \to 0.$$

Assuming that the limit of the infinite sum is the sum of these limits, we then get Taylor's series (2) as the limit of (1) as $b \to 0$.

EXERCISES

Here is how to derive the general binomial series from the Gregory–Newton interpolation formula.

10.2.1 Show that

$$\Delta^n f(a) = \sum_{i=0}^{n} (-1)^{n-i} \binom{n}{i} f(a + ib),$$

where $\binom{n}{i}$ is the ordinary binomial coefficient.

10.2.2 If $a = 0$, $b = 1$, and $f(x) = (1 + k)^x$, show that $\Delta^n f(0) = k^n$ using the finite binomial series

$$(1 + h)^n = \sum_{i=0}^{n} \binom{n}{i} h^i.$$

10.2.3 Deduce the general binomial series

$$(1 + k)^x = 1 + xk + \frac{x(x - 1)}{2!} k^2 + \frac{x(x - 1)(x - 2)}{3!} k^3 + \cdots$$

using the Gregory–Newton interpolation formula.

10.3   An Interpolation on Interpolation

The importance of interpolation in the development of calculus seems to have been greatly underestimated. The topic rarely appears in calculus books today, and then only as a numerical method. Yet three of the most important founders of calculus, Newton, Gregory, and Leibniz, began their work with interpolation, and we have seen how this led to two of their most important results, the binomial theorem and Taylor's theorem. (For Leibniz's work, see Hofmann (1974).) With the relegation of interpolation to numerical methods, this connection has been lost. Of course, interpolation *is* a numerical method in practice, when one uses only a few terms of the Gregory–Newton series, but the full series is exact and hence of much greater interest. It was this interest in infinite expansions per se that set off Newton, Gregory, and Leibniz (as well as Wallis) from their predecessors in interpolation.

Interpolation goes back to ancient times as a method for estimating the values of functions between known values. But perhaps the first to glimpse the possibility of exact interpolation were Thomas Harriot (1560–1621) and Henry Briggs (1556–1630). A formula has been found in Harriot's papers that is equivalent to the first terms of the Gregory–Newton series (see Lohne (1965)). Lohne dates this work of Harriot at 1611. Briggs may have learned something about interpolation from Harriot when the two were at Oxford around 1620. Briggs's *Arithmetica logarithmica* (1624), which is concerned with the calculation of logarithms, uses series for interpolation, and in the process gives the first instance of the binomial theorem for a fractional exponent

$$(1 + x)^{1/2} = 1 + \frac{1}{2}x - \frac{1 \cdot 1}{2 \cdot 4}x^2 + \frac{1 \cdot 1 \cdot 3}{2 \cdot 4 \cdot 6}x^3 - \frac{1 \cdot 1 \cdot 3 \cdot 5}{2 \cdot 4 \cdot 6 \cdot 8}x^4 + \cdots.$$

Gregory knew of Briggs's work, and Newton certainly *could* have known of it, though no strong evidence that he did has yet been found. For more information on the history of interpolation, see Whiteside (1961) and Goldstine (1977).

10.4 Summation of Series

The results on infinite series that we have seen so far are mostly decompositions or expansions rather than summations. That is, one begins with a "known" quantity or function and decomposes it into an infinite series. Solutions of the converse problem, summation of a given series, were comparatively rare. Archimedes' summation of $1 + 1/4 + 1/4^2 + \cdots$ was one. Perhaps the next were of series such as $1/1 \cdot 2 + 1/2 \cdot 3 + \cdots + 1/n(n+1) + \cdots$, given by Mengoli (1650). The series $\sum 1/n(n+1)$ is easily summed because of the happy accident that

$$\frac{1}{n(n + 1)} = \frac{1}{n} - \frac{1}{n + 1},$$

whence

$$\frac{1}{1 \cdot 2} + \frac{1}{2 \cdot 3} + \cdots + \frac{1}{n(n + 1)} = \left(1 - \frac{1}{2}\right) + \left(\frac{1}{2} - \frac{1}{3}\right) + \cdots + \left(\frac{1}{n} - \frac{1}{n + 1}\right)$$

$$= 1 - \frac{1}{n + 1}.$$

By letting $n \to \infty$ we then obtain the sum 1 for the infinite series.

The first really tough summation problem was $1 + 1/2^2 + 1/3^2 + \cdots$. Mengoli tackled this without success, as did the brothers Jakob and Johann Bernoulli in a series of papers (1704). The Bernoulli brothers were able to sum similar series, rediscovering Mengoli's $\sum 1/n(n+1)$ and also summing $\sum 1/(n^2 - 1)$, but for $\sum 1/n^2$ itself they could obtain only trivial results such as

$$\frac{1}{2^2} + \frac{1}{4^2} + \frac{1}{6^2} + \cdots = \frac{1}{4}\left(1 + \frac{1}{2^2} + \frac{1}{3^2} + \cdots\right).$$

The solution was finally obtained by Euler (1734), long after the death of Jakob Bernoulli, and Johann Bernoulli exclaimed, "In this way my brother's most ardent wish is satisfied ... if only my brother were still alive!" (Johann Bernoulli, *Opera*, Vol. 4, p. 22). In fact, after hearing that the sum is $\pi^2/6$, Johann Bernoulli himself discovered a proof, which turned out to be the same as Euler's.

Euler (1707–1783) was the greatest virtuoso of series manipulation, and his first summation of $1 + 1/2^2 + 1/3^2 + \cdots$ was one of his most audacious. (Later he gave more rigorous proofs.) Consider the equation

$$\frac{\sin \sqrt{x}}{\sqrt{x}} = 1 - \frac{x}{3!} + \frac{x^2}{5!} - \frac{x^3}{7!} + \cdots = 0, \tag{1}$$

easily obtained from the sine series of Section 9.5. This equation has roots $x_1 = \pi^2$, $x_2 = (2\pi)^2$, $x_3 = (3\pi)^2, \ldots$, but *not* 0, because $\sin \sqrt{x}/\sqrt{x} \to 1$ as $x \to 0$. Now if a *polynomial* equation

$$1 + a_1 x + a_2 x^2 + \cdots + a_n x^n = 0$$

has roots $x = x_1, x_2, \ldots, x_n$, Descartes's factor theorem (Section 6.7) gives

$$1 + a_1 x + \cdots + a_n x^n = \left(1 - \frac{x}{x_1}\right)\left(1 - \frac{x}{x_2}\right)\cdots\left(1 - \frac{x}{x_n}\right). \tag{2}$$

Also

$$\frac{1}{x_1} + \frac{1}{x_2} + \cdots + \frac{1}{x_n} = -\text{coefficient of } x = -a_1,$$

since each x term in the expansion of the right-hand side of (2) comes from a term $-x/x_i$ in one factor multiplied by 1's in all the other factors. Assuming that this is also true of the "infinite polynomial" equation (1), we get

$$\frac{1}{x_1} + \frac{1}{x_2} + \frac{1}{x_3} + \cdots = -\text{coefficient of } x = -\left(-\frac{1}{3!}\right),$$

that is,

$$\frac{1}{\pi^2} + \frac{1}{(2\pi)^2} + \frac{1}{(3\pi)^2} + \cdots = \frac{1}{6}.$$

Hence

$$1 + \frac{1}{2^2} + \frac{1}{3^2} + \cdots = \frac{\pi^2}{6}. \qquad\qquad \text{Q.E.D.!}$$

EXERCISES

Euler's reasoning also leads to a correct infinite product formula for $\sin x$, which in turn gives the Wallis product for $\pi/4$ (Section 9.4).

10.4.1 Deduce an infinite product for $\frac{\sin \sqrt{x}}{\sqrt{x}}$ from Euler's reasoning, and hence show that

$$\sin x = x\left(1 - \frac{x^2}{\pi^2}\right)\left(1 - \frac{x^2}{2^2\pi^2}\right)\left(1 - \frac{x^2}{3^2\pi^2}\right)\cdots.$$

10.4.2 By substituting $x = \pi/2$ in the infinite product for $\sin x$, show that

$$\frac{2}{\pi} = \frac{1 \cdot 3}{2 \cdot 2} \cdot \frac{3 \cdot 5}{4 \cdot 4} \cdot \frac{5 \cdot 7}{6 \cdot 6} \cdots,$$

and hence obtain Wallis's product for $\pi/4$.

10.5   Fractional Power Series

The introduction of power series helped to make mathematicians conscious of the function concept (see also Section 13.6) by drawing attention to the generality of the expression $a_0 + a_1 x + a_2 x^2 + \cdots$. However, not every function $f(x)$ is expressible as a power series $a_0 + a_1 x + a_2 x^2 + \cdots$. This is obvious in the case of functions that tend to infinity as $x \to 0$, since the power series has value a_0 when $x \to 0$. For other functions, such as $f(x) = x^{1/2}$, the behavior at 0 disallows a power-series expansion for a more subtle reason. These functions have *branching behavior* at 0; they are *many-valued*, and hence they are not functions in the strict sense. The function $x^{1/2}$, for example, is two-valued because each number has two square roots, one the negative of the other.

Such behavior is not reflected in a power series $a_0 + a_1 x + a_2 x^2 + \cdots$, which can be assigned only one value for each value of x. All fractional powers of x are many-valued—$x^{1/3}$ is three-valued, $x^{1/4}$ is four-valued, and so on—and many-valued behavior is typical of algebraic functions in general. We say that y is an *algebraic function* of x if x and y satisfy a polynomial equation $p(x, y) = 0$. It follows from the impossibility of solving most polynomial equations by radicals (Section 6.7) that algebraic functions are not generally expressible by radicals, that is, by finite expressions built from $+, -, \times, \div$, and fractional powers.

Nevertheless, it was the remarkable discovery of Newton (1671) that any algebraic function y can be expressed as a *fractional power series* in x:

$$y = a_0 + a_1 x^{r_1} + a_2 x^{r_2} + a_3 x^{r_3} + \cdots,$$

where $r_1, r_2, r_3, \ldots$, are rational numbers. Furthermore, the series can be rewritten in the form

$$a_0 + b_1 x^{s_1}(c_{00} + c_{01}x + c_{02}x^2 + \cdots)$$
$$+ b_2 x^{s_2}(c_{10} + c_{11}x + c_{12}x^2 + \cdots)$$
$$\vdots$$
$$+ b_n x^{s_n}(c_{n0} + c_{n1}x + c_{n2}x^2 + \cdots)$$

that is, as a finite sum of ordinary power series with fractional powers of x as multipliers. This means that in the neighborhood of $x = 0$, the behavior of y is like that of a finite sum of fractional powers.

For example, if $y^2(1 + x)^2 = x$, we have

$$y = \frac{x^{1/2}}{1 + x}$$
$$= x^{1/2}(1 - x + x^2 - x^3 + \cdots),$$

and near the origin, y has behavior similar to $x^{1/2}$; in particular there are two values of y for each x. Newton's contribution was an ingenious algorithm for obtaining the successive powers of x. The fractional powers themselves were not properly understood until the variables x and y were taken to be complex. This was done in the 19th century, and on this basis a more rigorous derivation of Newton's series was given by Puiseux (1850). For this reason, the fractional power-series expansions of algebraic functions are now called *Puiseux expansions*.

EXERCISE

The impossibility of an ordinary power series for $x^{1/2}$ can be shown as follows.

10.5.1 Any ordinary power-series expansion of $x^{1/2}$ would have to be of the form

$$x^{1/2} = a_1 x + a_2 x^2 + a_3 x^3 + \cdots$$

because $x^{1/2} = 0$ when $x = 0$. Now square both sides and deduce a contradiction.

10.6   Generating Functions

Fibonacci (1202) introduced a famous sequence now known as the *Fibonacci sequence*

$$1, 2, 3, 5, 8, 13, 21, 34, 55, \ldots ,$$

in which each term (after the first two) is the sum of two preceding terms. Despite this simple law of formation, there is no obvious formula for the nth term of the sequence. Such a formula was not discovered for more than 500 years, by de Moivre (1730), and independently by Daniel Bernoulli (1728). In doing so, de Moivre introduced a powerful new application of infinite series, the method of *generating functions*. This method, which is of great importance in combinatorics, probability, and number theory, will be illustrated using the Fibonacci sequence itself.

It is technically convenient to begin with $F_0 = 0$ and $F_1 = 1$, then take subsequent terms as above (so $F_2 = 1$, $F_3 = 2$, $F_4 = 3$, ...) by defining

$$F_{n+2} = F_{n+1} + F_n \quad \text{for} \quad n \geq 0.$$

This is an example of *linear recurrence relation*, and it was to solve such relations in probability theory that de Moivre introduced generating functions. The generating function for the Fibonacci sequence is

$$f(x) = F_0 + F_1 x + F_2 x^2 + F_3 x^3 + \cdots .$$

We notice that

$$xf(x) = F_0 x + F_1 x^2 + F_2 x^3 + \cdots ,$$
$$x^2 f(x) = \qquad F_0 x^2 + F_1 x^3 + \cdots .$$

Hence

$$f(x) - xf(x) - x^2 f(x) = F_0 + F_1 x - F_0 x$$
$$+ (F_2 - F_1 - F_0)x^2$$
$$+ (F_3 - F_2 - F_1)x^3$$
$$+ \cdots ,$$

that is, $f(x)(1 - x - x^2) = F_0 + F_1 x - F_0 x = x$ because all the coefficients $F_{n+2} - F_{n+1} - F_n$ equal 0 by definition of the Fibonacci sequence. Thus

$$f(x) = \frac{x}{1 - x - x^2},$$

and using the roots $(-1 \pm \sqrt{5})/2 = 2/(1 \pm \sqrt{5})$ of $1 - x - x^2 = 0$ to factorize the denominator we get

$$f(x) = \frac{x}{[1 - ((1 + \sqrt{5})/2)x][1 - ((1 - \sqrt{5})/2)x]}.$$

Then splitting into partial fractions

$$f(x) = \frac{1}{\sqrt{5}} \left[\frac{1}{1 - ((1 + \sqrt{5})/2)x} - \frac{1}{1 - ((1 - \sqrt{5})/2)x} \right],$$

and using the geometric series expansions

$$\frac{1}{1 - ((1 + \sqrt{5})/2)x} = 1 + \frac{1 + \sqrt{5}}{2}x + \left(\frac{1 + \sqrt{5}}{2} \right)^2 x^2 + \cdots,$$

$$\frac{1}{1 - ((1 - \sqrt{5})/2)x} = 1 + \frac{1 - \sqrt{5}}{2}x + \left(\frac{1 - \sqrt{5}}{2} \right)^2 x^2 + \cdots,$$

we finally get

$$f(x) = \frac{1}{\sqrt{5}} \left[\frac{1 + \sqrt{5}}{2} - \frac{1 - \sqrt{5}}{2} \right] x + \cdots$$
$$+ \frac{1}{\sqrt{5}} \left[\left(\frac{1 + \sqrt{5}}{2} \right)^n - \left(\frac{1 - \sqrt{5}}{2} \right)^n \right] x^n + \cdots .$$

Equating this with the definition $f(x) = F_0 + F_1 x + F_2 x^2 + \cdots$ gives

$$F_n = \frac{1}{\sqrt{5}} \left[\left(\frac{1 + \sqrt{5}}{2} \right)^n - \left(\frac{1 - \sqrt{5}}{2} \right)^n \right]. \qquad (1)$$

No wonder a formula for F_n was hard to find! One would not have expected the irrational $\sqrt{5}$ to be involved in the integer-valued function F_n. The explanation is that the Fibonacci sequence actually defines $\sqrt{5}$, because $F_{n+1}/F_n \to (1 + \sqrt{5})/2$ (the golden ratio) as $n \to \infty$, so (1) in effect defines the individual terms of the Fibonacci sequence in terms of the sequence as a whole (or, if one prefers, in terms of the behavior of the sequence at infinity). The remarkable fact that the definition of F_n becomes explicit, rather than recursive, when expressed in terms of $(1 + \sqrt{5})/2$ is due to the simplicity of the generating function $f(x)$, which encodes the whole sequence.

The recursive property of Fibonacci numbers used in de Moivre's proof is that they satisfy a linear recurrence relation; that is, F_n is expressed as a fixed linear combination of earlier terms in the sequence. The proof is easily generalized to show that the generating function $\sum a_n x^n$ of any sequence $\{a_n\}$ defined by a linear recurrence relation is rational. Also, the proof can be reversed to show that the power series of any rational function has coefficients that satisfy a linear recurrence relation. Thus rational functions can be characterized in terms of their power series, a fact that was noticed by Kronecker (1881), Section IX.

EXERCISES

The formula $F_n = \frac{1}{\sqrt{5}}\left[\left(\frac{1+\sqrt{5}}{2}\right)^n - \left(\frac{1-\sqrt{5}}{2}\right)^n\right]$ gives several interesting limit and approximation properties of F_n. For example:

10.6.1 Show that $\frac{F_{n+1}}{F_n} \to \frac{1+\sqrt{5}}{2}$ as $n \to \infty$.

10.6.2 Show that F_n = nearest integer to $\frac{1}{\sqrt{5}}\left(\frac{1+\sqrt{5}}{2}\right)^n$.

10.6.3 Using $1/(1 + F_n/F_{n+1}) = F_{n+1}/F_{n+2}$, or otherwise, show that

$$\frac{1+\sqrt{5}}{2} = 1 + \cfrac{1}{1 + \cfrac{1}{1 + \cfrac{1}{1 + \cdots}}}.$$

10.7 The Zeta Function

The purpose of a generating function is to encode a complicated sequence by a function (of a real or complex variable) that is in some ways simpler. The method of encoding need not be as direct as taking the nth term of the sequence to be the coefficient of x^n. For example, a famous *product formula* of Euler (1748a), p. 288, encodes the sequence $2, 3, 5, 7, 11, \ldots$, of prime numbers as the following sum of powers of $1, 2, 3, 4, \ldots$

$$\zeta(s) = 1 + \frac{1}{2^s} + \frac{1}{3^s} + \frac{1}{4^s} + \cdots$$

(the *zeta function*). Euler's formula is

$$\frac{1}{(1 - 1/2^s)}\frac{1}{(1 - 1/3^s)}\frac{1}{(1 - 1/5^s)}\frac{1}{(1 - 1/7^s)}\frac{1}{(1 - 1/11^s)}\cdots$$

$$= 1 + \frac{1}{2^s} + \frac{1}{3^s} + \frac{1}{4^s} + \cdots.$$

The factors on the left-hand side are $(1-1/p_n^s)^{-1}$, where p_n is the nth prime. We expand each such factor as a geometric series

$$1 + \frac{1}{p_n^s} + \frac{1}{p_n^{2s}} + \frac{1}{p_n^{3s}} + \cdots.$$

Multiplying all these series together, we get the reciprocal of each possible product of primes, to the sth power, exactly once. That is, the left-hand side is the sum

$$1 + \sum \frac{1}{p_1^{m_1 s} p_2^{m_2 s} \cdots p_r^{m_r s}} = 1 + \sum \frac{1}{(p_1^{m_1} p_2^{m_2} \cdots p_r^{m_r})^s},$$

in which each product $p_1^{m_1} p_2^{m_2} \cdots p_r^{m_r}$ of primes occurs exactly once. But each natural number ≥ 2 is expressible in just one way as a product of primes (Section 3.3), hence the latter sum equals the right-hand side of Euler's formula

$$1 + \frac{1}{2^s} + \frac{1}{3^s} + \frac{1}{4^s} + \cdots.$$

Initially the exponent $s > 1$ was there only to ensure convergence. We saw in Section 10.1 that $\zeta(s)$ diverges when $s = 1$; it converges when $s > 1$. Riemann (1859) discovered that $\zeta(s)$ becomes much more powerful when s is taken to be a complex variable. In recognition of this, $\zeta(s)$ is often called the *Riemann* zeta function. Euler's result of Section 10.4 can be rephrased as $\zeta(2) = \pi^2/6$. The values of $\zeta(4), \zeta(6), \zeta(8), \ldots$ were also found by Euler and turn out to be rational multiples of $\pi^4, \pi^6, \pi^8, \ldots$, respectively. The values of $\zeta(3), \zeta(5), \ldots$ have no known relationship to π or other standard constants, though Apéry (1981) showed that $\zeta(3)$ is irrational. The most famous conjecture about $\zeta(s)$, and one of the most sought-after results in mathematics today, is the so-called *Riemann hypothesis*: $\zeta(s) = 0$ only when $\mathrm{Re}(s) = \frac{1}{2}$ (excluding the "trivial zeros" described below).

EXERCISES

Although $\zeta(s)$ is not defined for $s = 1$ (because this gives the divergent series $1 + \frac{1}{2} + \frac{1}{3} + \frac{1}{4} + \cdots$), this situation can be exploited to give a new proof that there are infinitely many primes. (Thus the Euler product formula encapsulates two apparently unrelated results—unique prime factorization, and the infinite number of primes.)

10.7.1 (Euler) Show that if there are only finitely many primes $p_1, \ldots, p_n$, then

$$\frac{1}{1 - 1/p_1} \cdot \frac{1}{1 - 1/p_2} \cdots \cdots \frac{1}{1 - 1/p_n} = 1 + \frac{1}{2} + \frac{1}{3} + \frac{1}{4} + \cdots.$$

Deduce that there are infinitely many primes.

The statement of the Riemann hypothesis needs some qualification, because $\zeta(s)$ can be defined for certain values of s for which the series $1 + \frac{1}{2^s} + \frac{1}{3^s} + \frac{1}{4^s} + \cdots$ is not meaningful. This follows from the formula

$$\zeta(1 - s) = 2(2\pi)^{-s} \cos\frac{s\pi}{2}\Gamma(s)\zeta(s)$$

discovered by Riemann and called the *functional equation* for the zeta function. The functional equation enables us to define $\zeta(1-s)$ when $\zeta(s)$ is known, and it also shows that there are certain "trivial zeros" of $\zeta(1-s)$, namely, where s satisfies $\cos\frac{s\pi}{2} = 0$.

10.7.2 Which s give a trivial zero of $\zeta(1-s)$?

The function Γ in the functional equation is the *gamma function*, introduced by Euler to generalize the factorial function: $\Gamma(n) = (n-1)!$ for integer values of n. An amusing consequence of the functional equation is that we can assign values to certain divergent series, such as $1 + 2 + 3 + 4 + \cdots$, by interpreting them as $\zeta(1-s)$, then reinterpreting $\zeta(1-s)$ by the functional equation.

10.7.3 By suitable reinterpretation, show that

$$1 + 2 + 3 + 4 + \cdots = -1/12.$$

Euler (1770a), p. 157, found another trick for the zeta function: giving a natural formula for the seemingly *un*natural Euler constant γ. Recall from Exercise 10.1.5 that γ is defined to be the limit of $1 + \frac{1}{2} + \frac{1}{3} + \cdots + \frac{1}{n} - \log(n+1)$ as $n \to \infty$.

10.7.4 Using the Mercator series for $\log(1 + \frac{1}{k})$, show that

$$\frac{1}{k} - \log(k+1) + \log(k) = \frac{1}{2k^2} - \frac{1}{3k^3} + \frac{1}{4k^4} - \cdots .$$

10.7.5 By adding the instances of the formula in Exercise 10.7.4 from $k = 1$ to $k = n$, show that

$$\left(1 + \frac{1}{2} + \frac{1}{3} + \cdots + \frac{1}{n}\right) - \log(n+1) =$$
$$\frac{1}{2}\left(\frac{1}{1^2} + \frac{1}{2^2} + \cdots + \frac{1}{n^2}\right) - \frac{1}{3}\left(\frac{1}{1^3} + \frac{1}{2^3} + \cdots + \frac{1}{n^3}\right) + \frac{1}{4}\left(\frac{1}{1^4} + \frac{1}{2^4} + \cdots + \frac{1}{n^4}\right) - \cdots .$$

10.7.6 Deduce from Exercise 10.7.5 that $\gamma = \frac{\zeta(2)}{2} - \frac{\zeta(3)}{3} + \frac{\zeta(4)}{4} - \frac{\zeta(5)}{5} + \cdots$.

10.8   Biographical Notes: Gregory and Euler

James Gregory was born in 1638 in Drumoak, near Aberdeen, the youngest of three sons of John Gregory, the town's minister. He received his early education from his mother, Janet Anderson, whose uncle Alexander had been secretary to Viète and editor of Viète's posthumously published works. The middle brother, David, also had mathematical ability and after their father's death in 1651, he encouraged James in his subsequent studies at grammar school and Marischal College in Aberdeen.

Figure 10.3: James Gregory

Gregory's first major achievement was the invention of the reflecting telescope, which he described in his book *Optica promota* of 1663. Unfortunately, he failed to get a satisfactory instrument constructed, and his design was overtaken by the simpler type invented by Newton. In the meantime, Gregory had decided to improve his scientific knowledge on the Continent, and he spent most of 1664 to 1668 studying mathematics in Italy. His teacher was Stefano degli Angeli (1623–1697) of Padua, from whom Gregory learned the methods of Cavalieri. The influence of the Italian school was evident in Gregory's geometric approach to integration problems in his first mathematical works, *Vera circuli et hyperbolae quadratura* (1667) and *Geometriae pars universalis* (1668), but so too was Gregory's originality. The books received glowing reviews in London and, when Gregory went there on his return from Italy, he was elected to the Royal Society.

The *Geometriae pars universalis* was mainly a systematization of the results in differentiation and integration then known, but it included the first published proof of the fundamental theorem of calculus. Important as this was, the theorem was not Gregory's alone, since Newton and Leibniz discovered it independently. What really set Gregory apart from other 17th-century mathematicians was the *Vera quadratura* (True Quadrature), an extraordinarily bold and imaginative attempt to prove that the numbers π and e are transcendental.

As mentioned in Section 2.3, transcendence of e and π was not proved until the 19th century, and certainly not by 17th-century methods, so it is understandable that Gregory's attempt fell short. Nevertheless, it is full of brilliant ideas: the unification of circular and hyperbolic functions (without the use of complex numbers), the concept of convergence, and the distinction between algebraic and transcendental functions. Gregory showed that areas cut off from both the circle and the hyperbola (giving π and various logarithms as special cases) could be obtained as the limit of alternate geometric and harmonic means:

$$i_{n+1} = \sqrt{i_n I_n},$$
$$\frac{1}{I_{n+1}} = \frac{1}{2}\left(\frac{1}{i_{n+1}} + \frac{1}{I_n}\right),$$
$$\lim_{n\to\infty} i_n = \lim_{n\to\infty} I_n = I.$$

If $i_0 = 2$ and $I_0 = 4$, then I (the *geometric–harmonic mean* of 2 and 4) is π. If, on the other hand, $i_0 = 99/20$ and $I_0 = 18/11$, then I is log 10. These examples given by Gregory illustrate the way his geometric–harmonic mean embraces both circular and hyperbolic functions. The alternating procedure used to define the mean had an interesting echo in the work of Gauss, who investigated the analogously defined *arithmetic–geometric mean* in the 1790s, with far-reaching results (Section 12.6).

In 1669 Gregory returned to Scotland to take up the chair of mathematics at St. Andrew's. He married a young widow, Mary Burnet, the daughter of artist George Jameson, who was also descended from the Anderson family. James and Mary had two daughters and a son, who became professor of medicine in Aberdeen. The rather impressive Gregory family tree may be found in Turnbull's short biography of Gregory (Turnbull (1939)).

Gregory stayed at St. Andrew's for five years, during which he obtained his important results on series. However, his contact with other scientists was restricted to letters from London, and on hearing of Newton's related results he assumed that he had been anticipated and did not publish. The lack of contact, and hostility to mathematics at St. Andrew's, led him to accept the offer of a chair at Edinburgh in 1674. Alas, he had been in Edinburgh barely a year when he collapsed, apparently from a stroke, while showing the moons of Jupiter to a group of students. He died a few days later, in October 1675, too soon for the world to have understood the importance of his work.

Leonhard Euler was born in Basel in 1707 and died in St. Petersburg in 1783. His father, Paul, studied theology at the University of Basel, where he also attended the mathematics lectures of Jakob Bernoulli. After graduation he became a Protestant minister and married a minister's daughter, Margarete Bruckner. Leonhard was the first of their six children. The family was quite poor and, soon after Euler's birth, moved to a village outside Basel where they lived in a two-room house. Euler received his first mathematical instruction at home from his father. He later moved back to Basel to attend secondary school, but mathematics was not taught there, so he took some private lessons from a university student.

At 13, Euler entered the University of Basel, which had become the mathematical center of Europe under Johann Bernoulli, the younger brother and successor of Jakob. Bernoulli advised Euler to study mathematics on his own and made himself available on Saturday afternoons to help with any difficulties. Euler's official studies were in philosophy and law. After receiving his master's degree in philosophy in 1723, he followed his father's wish by entering the department of theology. However, he was falling increasingly under the spell of mathematics and realized he would have to drop the idea of becoming a minister.

There were few opportunities for mathematicians in Switzerland, and in 1727 Euler left Basel for St. Petersburg. Johann Bernoulli's sons, Daniel and Nicholas, had been appointed to the new Academy of Sciences there, and they persuaded the authorities to find a place for Euler. Euler had already shown promise with a couple of papers in *Acta Eruditorum* and an honorable mention in the Paris Academy competition of 1727, but in St. Petersburg he surpassed all expectations, producing top-quality work at a rate that has astounded mathematicians ever since. The early years in St. Petersburg with the Bernoullis must have been a young mathematician's dream. Yet it is equally true that Euler's productivity was unaffected by later setbacks, including the loss of his sight. He filled half the pages published by the St. Petersburg Academy from 1729 until over 50 years after his death (!), and he also accounted for half the production of the Berlin Academy between 1746 and 1771.

The first major changes in Euler's life in St. Petersburg occurred in 1733, when Daniel Bernoulli returned to Basel. Euler then became professor of mathematics but also had to take over the Department of Geography. In the same year, he married a compatriot, Katharina Gsell, the daughter of an artist who taught in St. Petersburg. They were eventually to have

13 children, 5 of whom reached maturity. Euler's duties in geography included the preparation of a map of Russia, a task that strained his eyes and perhaps led to the fever that destroyed the sight of his right eye in 1738. Figure 10.4 is a portrait from his good side.

Figure 10.4: Leonhard Euler

By 1740 the political situation in St. Petersburg had become unsettled and Euler moved to Berlin, where Frederick the Great had just reorganized the Berlin Academy. Euler became director of the mathematical section and stayed in Berlin for 25 years. Some of his most famous works date from this period, in particular the *Introductio in analysin infinitorum* (1748a) and the *Letters à une princesse d'Allemagne sur divers sujets de physique et de philosophie*, one of the classics of popular science. However, Euler was not comfortable in Berlin. There were quarrels over the leadership of the Academy, and the cynical Frederick tended to sneer at the pious and unassuming Euler. In 1762 Catherine the Great came to the throne in Russia, and the St. Petersburg Academy, with which Euler had maintained contact throughout, began to look attractive again.

In 1766 he moved back to St. Petersburg with his family (as a bonus, his eldest son gained the chair of physics there). Soon after his arrival Euler suffered an illness that destroyed most of his remaining sight, and in 1771 he became completely blind. If anything, blindness concentrated Euler's mind more wonderfully. He had always had an extraordinary memory—knowing Virgil's *Aeneid* by heart, for example—and with assistance from two of his sons and other collaborators his flow of publications continued at a greater rate than ever. His *Algebra*, Euler (1770b), was dictated to his valet, yet it became the most successful mathematics textbook since Euclid's *Elements*.

One of Euler's most admirable qualities was a willingness to explain how his discoveries were made. Mathematicians of the 18th century were less secretive than their 16th- and 17th-century predecessors, but Euler was unique in revealing his preliminary guesses, experiments, and partial proofs. Some of the most interesting of these exposés are presented in the book Pólya (1954b) on plausible reasoning. Chapter 6 of the book, for example, includes a translation of the memoir in which Euler announced the pentagonal number theorem. It is impossible to summarize all of Euler's contributions to mathematics here, though several of the highlights are presented in the chapters that follow. The best summary available is in Yushkevich's article on Euler in the *Dictionary of Scientific Biography*.

11

The Number Theory Revival

PREVIEW

After the work of Diophantus, number theory in Europe languished for about 1000 years. In Asia there was significant progress, as we saw in Chapter 5, on topics such as Pell's equation. The first signs of reawakening in Europe came in the 14th century, when Levi ben Gershon found formulas for the numbers of permutations and combinations, using rudimentary induction proofs.

Interest in number theory gathered pace with the rediscovery of Diophantus by Bombelli, and the publication of a new edition by Bachet de Méziriac (1621). It was this book that inspired Fermat and launched number theory as a modern mathematical discipline.

Fermat mastered and extended the techniques of Diophantus, such as the chord and tangent method for finding rational points on cubic curves. He also shifted the emphasis from rational solutions to integer solutions. He proved "Fermat's little theorem" that $n^p - n$ is divisible by p for any prime p, and claimed "Fermat's last theorem" that $x^n + y^n = z^n$ has no positive integer solutions when $n > 2$.

We know that Fermat had a proof of his "last theorem" for $n = 4$, but he seems to have been mistaken in thinking that he could prove it for arbitrary n. The proof now known uses highly sophisticated ideas, not conceivable in the 17th century. Nevertheless, it is strangely appropriate that the modern proof reduces Fermat's last theorem to a problem about cubic curves.

J. Stillwell, *Mathematics and Its History*, Undergraduate Texts in Mathematics,   203
DOI 10.1007/978-1-4419-6053-5_11, © Springer Science+Business Media, LLC 2010

11.1   Between Diophantus and Fermat

Some important results in number theory were discovered in the Middle
Ages, though they failed to take root until they were rediscovered in the
17th century or later. Among these were the discovery of Pascal's triangle and the "Chinese remainder theorem" by Chinese mathematicians, and
formulas for permutations and combinations by Levi ben Gershon (1321).
The early development of the Chinese remainder theorem is discussed in
Chapter 5, and the theorem did not reemerge until after the period we are
about to discuss. A full account of its history may be found in Libbrecht
(1973), Chapter 5. Pascal's triangle, on the other hand, began to flourish in
the 17th century after a long dormancy, so it is of interest to see what was
known of it in medieval times and what Pascal did to revive it.

The Chinese used Pascal's triangle as a means of generating and tab-
ulating the binomial coefficients, that is, the coefficients occurring in the
formulas

$$(a + b)^1 = \qquad\qquad a + b$$
$$(a + b)^2 = \qquad\qquad a^2 + 2ab + b^2$$
$$(a + b)^3 = \qquad\qquad a^3 + 3a^2b + 3ab^2 + b^3$$
$$(a + b)^4 = \qquad\qquad a^4 + 4a^3b + 6a^2b^2 + 4ab^3 + b^4$$
$$(a + b)^5 = \qquad a^5 + 5a^4b + 10a^3b^2 + 10a^2b^3 + 5ab^4 + b^5$$
$$(a + b)^6 = \quad a^6 + 6a^5b + 15a^4b^2 + 20a^3b^3 + 15a^2b^4 + 6ab^5 + b^6$$
$$(a + b)^7 = \; a^7 + 7a^6b + 21a^5b^2 + 35a^4b^3 + 35a^3b^4 + 21a^2b^5 + 7ab^6 + b^7$$

and so on. When the binomial coefficients are tabulated as follows (with a
trivial row 1 added at the top, corresponding to the power 0 of $a + b$),

$$
\begin{array}{ccccccccccccc}
 & & & & & & 1 & & & & & & \\
 & & & & & 1 & & 1 & & & & & \\
 & & & & 1 & & 2 & & 1 & & & & \\
 & & & 1 & & 3 & & 3 & & 1 & & & \\
 & & 1 & & 4 & & 6 & & 4 & & 1 & & \\
 & 1 & & 5 & & 10 & & 10 & & 5 & & 1 & \\
1 & & 6 & & 15 & & 20 & & 15 & & 6 & & 1 \\
\end{array}
$$

$$1 \quad 7 \quad 21 \quad 35 \quad 35 \quad 21 \quad 7 \quad 1$$

and so on, the kth element $\binom{n}{k}$ of the nth row is the sum $\binom{n-1}{k-1} + \binom{n-1}{k}$ of
the two elements above it in the $(n-1)$th row, as follows from the formula

(Exercise 11.1.1)

$$(a + b)^n = (a + b)^{n-1}a + (a + b)^{n-1}b.$$

The triangle appears to a depth of six in Yáng Huí (1261) and to a depth of eight in Zhū Shijié (1303) (Figure 11.1). Yáng Huí attributes the triangle to Jia Xiàn, who lived in the 11th century.

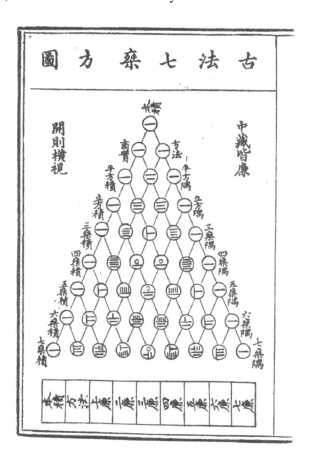

Figure 11.1: Chinese Pascal's triangle

The number $\binom{n}{k}$ appears in medieval Hebrew writings as the number of combinations of n things taken k at a time. Levi ben Gershon (1321) gives the formula

$$\binom{n}{k} = \frac{n!}{(n-k)!k!}$$

together with the fact that there are $n!$ permutations of n elements. In his treatment of permutations and combinations Levi ben Gershon comes very close to using mathematical induction, if not actually inventing it. As we now formulate this method of proof, a property $P(n)$ of natural numbers n is proved to hold for all n if one can prove $P(1)$ (the base step) and, for arbitrary n, one can prove $P(n) \Rightarrow P(n + 1)$ (the induction step). Rabinovitch (1970) offered an exposition of some of Levi ben Gershon's proofs that certainly seems to show a division into a base step and induction step, but the induction step needs some notational help to become a proof for truly arbitrary n. Levi ben Gershon does not say "Consider n elements $a, b, c, d, \ldots, e$," as we might, but only "Let the elements be a, b, c, d, e," since he does not have the device of ellipses.

In view of these excellent results, why do we call the table of binomial coefficients "Pascal's triangle"? It is of course not the only instance of a mathematical concept being named after a rediscoverer rather than a discoverer, but in any case Pascal deserves credit for more than just rediscovery. In his *Traité du triangle arithmétique*, Pascal (1654) united the algebraic and combinatorial theories by showing that the elements of the arithmetic triangle could be interpreted in two ways: as the coefficients of $a^{n-k}b^k$ in $(a + b)^n$ and as the number of combinations of n things taken k at a time. In effect, he showed that $(a + b)^n$ is a *generating function* for the numbers of combinations. As an application, he founded the mathematical theory of probability by solving the problem of division of stakes (Exercise 11.1.2), and as a method of proof he used mathematical induction for the first time in a really conscious and unequivocal way. Altogether, quite some progress!

In going to Pascal's work in 1654 we have overshot the end of the pre-Fermat period in number theory, since Fermat was already active in this field in the 1630s. However, it is convenient to have some background of binomial coefficients established, since Fermat's early work appears in this setting.

EXERCISES

The basic properties of the binomial coefficients, for example the fact that each is the sum of the two above it in Pascal's triangle, follow easily from their interpretation as the coefficients in the expansion of $(a + b)^n$.

11.1.1 Use the identity

$$(a + b)^n = (a + b)^{n-1}a + (a + b)^{n-1}b$$

to prove the sum property of binomial coefficients:

$$\binom{n}{k} = \binom{n-1}{k-1} + \binom{n-1}{k}.$$

This property gives an easy way to calculate Pascal's triangle to any depth, and hence compute a fair division of stakes in a game that has to be called off with n plays remaining. We suppose that players I and II have an equal chance of winning each play, and that I needs to win k of the remaining n plays to carry off the stakes.

11.1.2 Show that the ratio of I's winning the stakes to that of II's winning is

$$\binom{n}{n} + \binom{n}{n-1} + \cdots + \binom{n}{k} : \binom{n}{k-1} + \binom{n}{k-2} + \cdots + \binom{n}{0}.$$

The sum property of the binomial coefficients also explains the presence of some interesting numbers in Pascal's triangle.

11.1.3 Explain why the third diagonal from the left in the triangle, namely 1, 3, 6, 10, 15, 21, ..., consists of the triangular numbers.

11.1.4 The numbers on the next diagonal, namely 1, 4, 10, 20, 35..., can be called "tetrahedral numbers." Why is this an apt description?

11.2 Fermat's Little Theorem

The best-known theorem actually proved by Fermat (1640a), and known as his "little" or "lesser" theorem to distinguish it from his "last" or "great" theorem (next section), is the following.

If p is prime and n is relatively prime to p, then

$$n^{p-1} \equiv 1 \quad (\text{mod } p).$$

Equivalent statements of the conclusion, which avoid using the "congruent mod p" language unknown in Fermat's time, are

$$n^{p-1} - 1 \text{ is divisible by } p$$

or

$$n^p - n \text{ is divisible by } p.$$

The latter holds because $n^p - n = n(n^{p-1} - 1)$ is divisible by p only if $n^{p-1} - 1$ is, since p is prime and does not divide n.

Fermat's little theorem has recently become indispensable in areas of applied mathematics, such as cryptography, so it is thought-provoking to learn that it originated in one of the least applied problems in mathematics, the construction of perfect numbers. As we saw in Section 3.2, this depends on the construction of prime numbers of the form $2^m - 1$, and it was initially for this reason that Fermat became interested in conditions for $2^m - 1$ to have divisors. At the same time (mid-1630s) he was investigating the binomial coefficients, and the combination of these two interests very likely led to the discovery of his little theorem, for $n = 2$.

His actual proof is unknown, but various authors (for example, Weil (1984), p. 56) have pointed out that the theorem follows immediately from the fact that $\binom{p}{1}, \binom{p}{2}, \ldots, \binom{p}{p-1}$, for p prime, are divisible by p:

$$2^p = (1 + 1)^p = 1 + \binom{p}{1} + \binom{p}{2} + \cdots + \binom{p}{p-1} + 1,$$

hence

$$2^p - 2 = \binom{p}{1} + \binom{p}{2} + \cdots + \binom{p}{p-1}$$

is divisible by p, and therefore so is $2^{p-1} - 1$.

But how does one prove that $\binom{p}{1}, \binom{p}{2}, \ldots, \binom{p}{p-1}$ are divisible by p? This follows easily from the Levi ben Gershon formula

$$\binom{p}{k} = \frac{p!}{(p-k)!k!},$$

which shows that the prime p is a factor of the numerator but not of the denominator. The denominator nevertheless divides the numerator, since $\binom{p}{k}$ is an integer, so the factor must remain intact after the division has taken place. Fermat may not have had precisely this result, since he did not yet have Pascal's combinatorial interpretation of the binomial coefficients, but he did have the formula

$$n\binom{n+m-1}{m-1} = m\binom{n+m-1}{m},$$

which implies it and from which the divisibility property may be extracted (see Weil (1984), p. 47).

Thus far we have a proof of Fermat's little theorem for $n = 2$. Weil (1984) suggests two possible routes to the general theorem from this point.

One is by iteration of the binomial theorem, a method that was used in the first published proof of Fermat's theorem by Euler (1736). The other is by direct application of the *multinomial theorem*, the method of the earliest known proof, which is in an unpublished paper of Leibniz from the late 1670s (see Weil (1984), p. 56).

Just as

$$\text{coefficient of } a^{p-k}b^k \text{ in } (a+b)^p = p!/(p-k)!k!,$$

$$\text{coefficient of } a_1^{q_1} a_2^{q_2} \cdots a_n^{q_n} \text{ in } (a_1 + a_2 + \cdots + a_n)^p = p!/q_1!q_2!\cdots q_n!,$$

where $q_1 + q_2 + \cdots + q_n = p$ (Exercise 11.2.4). This *multinomial coefficient* is divisible by p, by the same argument as before, provided no $q_i = p$. Thus the coefficients of all but $a_1^p, a_2^p, \ldots, a_n^p$ in $(a_1 + a_2 + \cdots + a_n)^p$ are divisible by the prime p. It follows, by replacing each of the n terms $a_1, a_2, \ldots, a_n$ by 1, that

$$(1 + 1 + \cdots + 1)^p = 1^p + 1^p + \cdots + 1^p + \text{terms divisible by } p,$$

that is, $n^p - n$ is divisible by p. Then if n itself is relatively prime to p (hence not divisible by p), we have $n^{p-1} - 1$ divisible by p, or the general Fermat little theorem.

EXERCISES

The binomial theorem may be iterated to show that p divides $n^p - n$ as follows.

11.2.1 Use the result $2^p = (1 + 1)^p = 2 + \text{terms divisible by } p$, and its method of proof, to show that

$$3^p = (2 + 1)^p = 3 + \text{terms divisible by } p.$$

11.2.2 Build on the idea of Exercise 11.2.1 to show that $n^p - n$ is divisible by p for any positive integer n.

11.2.3 Observe the terms divisible by p in the first few rows of Pascal's triangle, computed in the previous section.

Like the binomial theorem, the multinomial theorem can be proved combinatorially by considering the number of ways a term $a_1^{q_1} a_2^{q_2} \cdots a_n^{q_n}$ can arise from the factors of $(a_1 + a_2 + \cdots + a_n)^p$.

11.2.4 Prove the formula for the multinomial coefficient given above by observing that the coefficient equals the number of ways of partitioning p things into disjoint subsets of sizes $q_1, q_2, \ldots, q_n$.

11.3   Fermat's Last Theorem

> On the other hand, it is impossible for a cube to be written as
> a sum of two cubes or a fourth power to be written as a sum
> of two fourth powers or, in general, for any number which
> is a power higher than second to be written as a sum of two
> like powers. I have a truly marvellous demonstration of this
> proposition which this margin is too small to contain.
>
> Fermat (1670), p. 241

This remark, written in the margin of his copy of Bachet's *Diophantus*
when he was studying that work in the late 1630s, is the second item in
Fermat's *Observations on Diophantus*, published posthumously in 1670.
Fermat was responding to Diophantus's treatment of the problem of expressing a square as a sum of two squares. As we saw in Chapter 1, this
is the problem of finding Pythagorean triples (a, b, c) or, equivalently, of
finding the rational points $(a/c, b/c)$ on the circle $x^2 + y^2 = 1$.

Fermat's last theorem, the claim that there are no triples (a, b, c) of
positive integers such that

$$a^n + b^n = c^n, \quad \text{where } n > 2 \text{ is an integer,}$$

became the most famous problem in mathematics. Many mathematicians
contributed solutions for particular values of n: Euler for $n = 3$, Fermat
himself for $n = 4$ (see next section), Legendre and Dirichlet for $n = 5$,
Lamé for $n = 7$, Kummer for all prime $n < 100$ except 37, 59, 67. A
thorough account of these early results may be found in Edwards (1977).
Of course it is sufficient to prove the theorem for prime exponents, since a
counterexample

$$a^n + b^n = c^n$$

for a nonprime exponent $n = mp$, where p is prime, would also be a counterexample

$$(a^m)^p + (b^m)^p = (c^m)^p$$

for the prime exponent p.

After Kummer, not much progress was made until the 1980s, when two
new approaches were opened up. Faltings (1983) showed that for each exponent n there were *at most finitely many counterexamples* to Fermat's last
theorem. This is a consequence of Faltings' much more general theorem,

settling a conjecture of Mordell (1922) that each curve of genus > 1 has at most finitely many rational points. The concept of genus is explained in Chapter 15. For the moment we mention only that the "Fermat curve"

$$x^n + y^n = 1$$

has genus 0 when $n = 2$, genus 1 when $n = 3$, and genus >1 otherwise. Thus Faltings' theorem showed that the Fermat curve could have at most finitely many rational points (and hence $a^n + b^n = c^n$ could have at most finitely many integer solutions) in the cases not already settled.

The second approach was initiated by Frey (1986), who made the astonishing suggestion that a counterexample $a^n + b^n = c^n$ to Fermat's last theorem might imply something impossible about the *cubic* curve

$$y^2 = x(x - a^n)(x + b^n).$$

At the time, the property in question—called *nonmodularity*—was only conjectured to be impossible, and it was also not known to be implied by a counterexample to Fermat's last theorem. However, Ribet (1990) proved that a counterexample implies nonmodularity, and in 1994 Andrew Wiles proved that nonmodularity is impossible for cubic curves of the above form. Thus no counterexample to Fermat's last theorem can exist.

There was a dramatic twist to this closing chapter in the story of Fermat's last theorem, because Wiles first announced his result in 1993 (after seven years working on it in seclusion), only to discover within months that there was a serious gap in his proof. However, with the help of Richard Taylor, the gap was filled in 1994, and the completed proof was published in Wiles (1995). The proof is highly sophisticated, but we can at least explain its general setting of cubic curves and elliptic functions; indeed these are important threads throughout the whole of this book.

11.4   Rational Right-Angled Triangles

The area of a right-angled triangle the sides of which are rational numbers cannot be a square number. This proposition, which is my own discovery, I have at length succeeded in proving, though not without much labour and hard thinking. I give the proof here, as this method will enable extraordinary developments to be made in the theory of numbers.

Fermat (1670), p. 271

This is number 45 of Fermat's *Observations on Diophantus*, responding to a problem posed by Bachet: to find a right-angled triangle whose area equals a given number. The observation is important not only for the theorem and the method announced, but also because it is followed by the only reasonably complete proof left by Fermat in number theory. As a bonus, the proof implicitly settles Fermat's last theorem for $n = 4$ (see exercises) and is an excellent illustration of his "method" of *infinite descent*, which did indeed lead to extraordinary developments in the theory of numbers. In what follows, the statements that make up Fermat's proof, appearing indented like the quote above, are expanded and expressed in modern notation following the reconstruction of Zeuthen (1903), p. 163. We use the translation of Fermat given by Heath (1910), p. 293, in his version of the reconstruction.

> If the area of a right-angled triangle were a square, there would exist two biquadrates the difference of which would be a square number. Consequently there would exist two square numbers the sum and difference of which would be squares.

By choosing a suitable unit of length, we can express the sides of a rational right triangle as a Pythagorean triple of relatively prime integers $p^2 - q^2$, $2pq$, $p^2 + q^2$, as noted in Section 1.2. Since their gcd is 1, $\gcd(p, q) = 1$ also. Therefore, since $2pq$ is even, $p^2 - q^2$ and its factors $p + q$, $p - q$ must be odd. Also, no two of p, q, $p + q$, $p - q$ have a common prime divisor, otherwise p, q would. Then if the area $pq(p + q)(p - q)$ is a square, its factors must all be squares:

$$p = r^2, \quad q = s^2, \quad p + q = r^2 + s^2 = t^2, \quad p - q = r^2 - s^2 = u^2. \quad (1)$$

Thus the sum and difference of the squares r^2, s^2 are also squares, so

$$r^4 - s^4 = (r^2 + s^2)(r^2 - s^2) = t^2 u^2 = v^2.$$

> Therefore we should have a square number which would be equal to the sum of a square and the double of another square, while the squares of which this sum is made up would themselves have a square number for their sum.

From (1) we have

$$t^2 - u^2 = 2s^2, \quad \text{that is,} \quad t^2 = u^2 + 2s^2. \quad (2)$$

And also from (1),

$$u^2 + s^2 = r^2.$$

But if a square is made up of a square and the double of another square, its side, as I can very easily prove, is also made up of a square and the double of another square.

Since $(t + u)(t - u) = t^2 - u^2 = 2s^2$ from (2), $(t + u)(t - u)$ is even. Then one of $t + u$, $t - u$ is even, and consequently so is the other. Put

$$t + u = 2w, \quad t - u = 2x. \tag{3}$$

Then

$$s^2 = (t + u)(t - u)/2 = 2wx.$$

Tracing back through (3), (2), (1) we see that any common divisor of w, x would also be common to t, u, to t^2, u^2, to r^2, s^2, and hence to p, q. Thus w, x are relatively prime and therefore, since wx is twice a square, we have either

$$w = y^2, \quad x = 2z^2 \qquad \text{or} \qquad w = 2z^2, \quad x = y^2.$$

In either case,

$$t = w + x = y^2 + 2z^2. \tag{4}$$

From this we conclude that the said side is the sum of the sides about the right angle in a right-angled triangle, and that the simple square contained in the sum is the base, and the double of the other square the perpendicular.

If we let y^2, $2z^2$ be the sides of a right triangle, then the hypotenuse h satisfies

$$h^2 = (y^2)^2 + (2z^2)^2 = \frac{1}{2}\left((y^2 + 2z^2)^2 + (y^2 - 2z^2)^2\right)$$

$$= \frac{1}{2}(t^2 + u^2) \qquad \text{by (3) and (4)}$$

$$= r^2. \qquad \text{by (1)}$$

Hence $h = r$ and the triangle is rational.

This right-angled triangle will thus be formed from two squares, the sum and difference of which will be squares. But both these squares can be shown to be smaller than the squares originally assumed to be such that both their sum and their difference are squares.

The original squares with sum and difference equal to squares were $p = r^2$, $q = s^2$, coming from the perpendicular sides $p^2 - q^2$ and $2pq$ of the rational right triangle whose area was assumed to be a square. We now have a rational (indeed integral) right triangle with perpendicular sides y^2, $2z^2$ whose area y^2z^2 is also a square. This triangle is smaller, since its hypotenuse r is less than side $2pq$ of the original triangle, so it gives a smaller pair of (integer) squares p', q', whose sum and difference are squares.

> Thus, if there exist two squares such that the sum and difference are both squares, there will also exist two other integer squares which have the same property but a smaller sum. By the same reasoning we find a sum still smaller than the last found, and we can go on *ad infinitum* finding integer square numbers smaller and smaller with the same property. This is, however, impossible because there cannot be an infinite series of numbers smaller than any given integer we please.

This contradiction means that the initial assumption of a rational right triangle with square area is false. The versions of Zeuthen and Heath proceed more directly to a contradiction than Fermat by observing that the descent from the hypothetical initial triangle to the one with area y^2z^2 can be iterated to give an infinite descending sequence of integer areas. Weil (1984), p. 77, shortens the proof even further.

The logical principle involved in Fermat's method of descent is of course the same as that on which mathematical induction is based: any set of natural numbers has a least member. However, the circumstances in which the two methods can be applied are quite different. With induction, one needs a suitable hypothesis on which to make the induction step; with descent, one needs a suitable quantity on which to descend. In practice, descent is a much more special method, being associated with geometric properties of certain curves: the curves of genus 1 we shall meet in Section 11.6 and later chapters (see also Weil (1984), p. 140). The general problem raised by Bachet—deciding which numbers n are the areas of rational right triangles—is intimately connected with the theory of genus 1 curves, and its 20th-century resurgence is beautifully covered by Koblitz (1985).

Exercises

Two of the propositions that arise in the descent from the hypothetical rational right triangle with square area are of independent interest and are also false because they imply the existence of such a triangle.

11.4.1 Show that the existence of squares r^2 and s^2 for which $r^2 + s^2$ and $r^2 - s^2$ are both squares implies the existence of a rational right triangle with square area.

11.4.2 Show that a nonzero integer solution of $r^4 - s^4 = v^2$ implies the existence of a rational right triangle with square area. (Hint: It's the same triangle as in Exercise 11.4.1.)

11.4.3 From Exercise 11.4.2, deduce Fermat's last theorem for $n = 4$.

The impossibility of a nonzero integer solution $r^4 - s^4 = v^2$ can also be shown by a more direct descent that avoids some of the steps used by Fermat. The main steps are as follows, assuming r, s, and hence v have no common prime divisor.

$$r^4 - s^4 = v^2 \implies r^2 = a^2 + b^2, \quad s^2 = 2ab, \quad v = a^2 - b^2$$

for some nonzero integers a, b

$$\implies a = c^2 - d^2, \quad b = 2cd$$

for some nonzero integers c, d

$$\implies c = e^2, d = f^2 \text{ and } c^2 - d^2 \text{ are squares}$$

because $s^2 = 4cd(c^2 - d^2)$

and c, d, $c^2 - d^2$ have no common prime divisor

$$\implies e^4 - f^4 = g^2$$

for an integer pair (e, f) smaller than (r, s).

11.4.4 Justify the steps in this argument.

11.5 Rational Points on Cubics of Genus 0

It may be doubted that Fermat had a correct proof of Fermat's last theorem because most of his work deals with curves of low degree (≤ 4), and it is highly unlikely that he could have foreseen Frey's reduction of the nth-degree Fermat problem to a question about cubic curves. Admittedly, we do not know for certain what Fermat's methods were, and he did not talk in terms of finding rational points on curves. Nevertheless, this is the most natural way to interpret his solutions of Diophantine equations and to link them with earlier and later results in the same vein by Diophantus and Euler, respectively. We have already described methods for finding rational points on curves of degree 2 (in Section 1.3) and 3 (in Section 3.5). Now we shall reexamine them from the point of view of genus, which becomes increasingly important as curves of higher degree are considered. In this section we confine attention to genus 0.

One of the properties of a curve C of degree 2 that we observed in Section 1.3 is that a rational line L through a rational point P on C meets C in a second rational point, provided the equation of C has rational coefficients. Also, one obtains all rational points Q on C in this way by rotating L about C. There is another important consequence of this construction, not depending on the rationality of C or L. It is that by expressing the x and y coordinates of Q in terms of the slope t of L we obtain a *parameterization of C by rational functions* (recall that a rational function need not have rational coefficients).

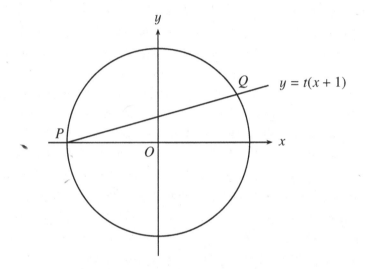

Figure 11.2: Parameterizing the circle

For example, this construction on the circle $x^2 + y^2 = 1$ in Section 1.3 gave the parameterization

$$x = \frac{1 - t^2}{1 + t^2}, \quad y = \frac{2t}{1 + t^2}$$

(Figure 11.2). Genus 0 curves can be defined as those that admit parameterization by rational functions. I shall now show that genus 0 includes some cubic curves by applying a similar construction to the folium of Descartes.

The folium was defined in Section 7.3 as the curve with equation

$$x^3 + y^3 = 3axy. \tag{1}$$

The origin O is an obvious rational point on the folium; moreover, O is a *double point* of the curve, as Figure 11.3 makes clear. The line $y = tx$ through O therefore meets the folium at one other point P, and varying t gives all other points P on the curve. By finding the coordinates of P as functions of t, we therefore obtain a parameterization.

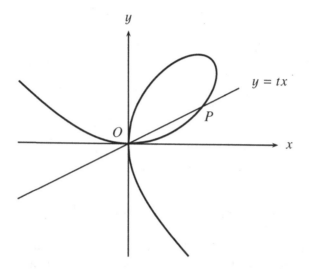

Figure 11.3: Parameterizing the folium

To find P we substitute $y = tx$ in (1), obtaining

$$x^3 + t^3 x^3 = 3axtx,$$

hence

$$x(1 + t^3) = 3at$$

and

$$x = \frac{3at}{1 + t^3}, \tag{2}$$

and therefore

$$y = \frac{3at^2}{1 + t^3}. \tag{3}$$

(These parametric equations were pulled out of the air in Exercise 7.3.1.) A similar construction applies to any cubic with a double point, or more generally to any curve of degree $n + 1$ with an n-tuple point; hence all such curves are of genus 0.

Exercises

It should be noted that a double point on a curve $p(x, y) = 0$ yields a *double root* of the equation $p(x, mx + c) = 0$ for the intersections of a line $y = mx + c$ through the double point.

11.5.1 Observe the double root of the equation obtained by substituting $y = tx$ in equation (1) above.

11.5.2 Explain, using the general double root property, why a line of rational slope through a rational double point on a cubic curve with rational coefficients necessarily meets the curve at another rational point.

We note also that, as in the construction for quadratic curves, *all* rational points on the folium are obtained by this method.

11.5.3 Show that if x and y are rational, then so is t in (2) and (3).

11.5.4 Deduce from Exercise 11.5.3 that the rational points on the folium are precisely those with rational t-values.

11.6   Rational Points on Cubics of Genus 1

We cannot yet give a precise definition of genus 1, but it so happens that this is the genus of all cubic curves that are not of genus 0. We know from Section 11.5 that cubics of genus 1 cannot have double points, and in fact they also cannot have cusps because both these cases lead to rational parameterizations. (For one case of a cusp, see Exercise 7.4.1.) What we have yet to exhibit are functions that do parameterize cubics of genus 1. Such functions, the *elliptic functions*, were not defined until the 19th century, and they were first used by Clebsch (1864) to parameterize cubics.

Many clues to the existence of elliptic functions were known before this, but at first they seemed to point in other directions. Initially, the mystery was how Diophantus and Fermat generated solutions of Diophantine equations. Newton's (1670s) interpretation of their results by the chord–tangent construction (Section 3.5) cleared up this first mystery—or would have if anyone had noticed it at the time. But before mathematicians really became conscious of the chord–tangent construction, they had to explain some puzzling relations between integrals of functions such as $1/\sqrt{ax^3 + bx^2 + cx + d}$, found by Fagnano (1718) and Euler (1768). Eventually Jacobi (1834) noticed that the chord–tangent construction explained this mystery too. Jacobi's explanation was cryptic, and, even though elliptic functions were then known in connection with integrals, they were not fully absorbed into number theory and the theory of curves until the appearance of Poincaré (1901).

The analytic origins of elliptic functions will be explained in the next chapter. In this section we shall prepare to link up with this theory by deriving the algebraic relation between collinear points on a cubic curve. A much deeper treatment of the whole story appears in Weil (1984).

We start with Newton's form of the equation for a cubic curve (Section 7.4):

$$y^2 = ax^3 + bx^2 + cx + d. \tag{1}$$

Figure 11.4 shows this curve when $y = 0$ for three distinct real values of x.

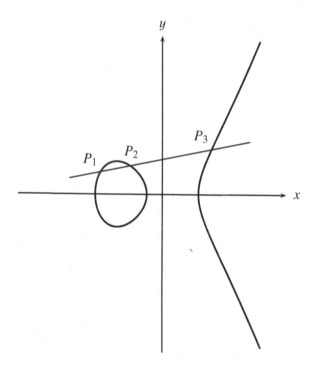

Figure 11.4: Collinear points on a cubic curve

In Section 3.5 we found that if a, b, c, d are rational, and if P_1, P_2 are rational points on the curve, then the straight line through P_1, P_2 meets the curve at a third rational point P_3. If the equation of this straight line is

$$y = tx + k, \tag{2}$$

then the result of substituting (2) in (1) is an equation

$$ax^3 + bx^2 + cx + d - (tx + k)^2 = 0 \qquad (3)$$

for the x coordinates x_1, x_2, x_3 of the three points P_1, P_2, P_3. But if the roots of (3) are x_1, x_2, x_3 its left-hand side must have the form

$$a(x - x_1)(x - x_2)(x - x_3).$$

In particular, the coefficient of x^2 must be

$$-a(x_1 + x_2 + x_3).$$

Comparing this with the actual coefficient of x^2 in (3), we find

$$b - t^2 = -a(x_1 + x_2 + x_3);$$

hence

$$x_3 = -(x_1 + x_2) - \frac{b - t^2}{a}. \qquad (4)$$

If $P_1 = (x_1, y_1)$, $P_2 = (x_2, y_2)$, then $t = (y_2 - y_1)/(x_2 - x_1)$, and substituting this in (4) we finally obtain

$$x_3 = -(x_1 + x_2) - \frac{b - [(y_2 - y_1)/(x_2 - x_1)]^2}{a}, \qquad (5)$$

giving x_3 as an explicit rational combination of the coordinates of P_1, P_2. If P_1, P_2 are rational points, then (5) shows that x_3 (and hence $y_3 = tx_3 + k$) is also rational, as we already knew.

What is unexpected is that (5) is also an *addition theorem* for elliptic functions. This has the consequence that the curve can be parameterized by elliptic functions $x = f(u)$, $y = g(u)$ such that (5) is precisely the equation expressing $x_3 = f(u_1 + u_2)$ in terms of $f(u_1) = x_1$, $f(u_2) = x_2$, $g(u_1) = y_1$, and $g(u_2) = y_2$. Thus the straight-line construction of x_3 from x_1 and x_2 can also be interpreted as *addition of parameter values*, u_1 and u_2 of x_1 and x_2. The first addition theorems were found by Fagnano (1718) and Euler (1768) by means of transformation of integrals. Euler realized that there was a connection between such transformations and number theory, but he could never quite put his finger on it. Even earlier, Leibniz had suspected such a connection when he wrote:

> I ... remember having suggested (what could seem strange to some) that the progress in our integral calculus depended in good part upon the development of that type of arithmetic which, so far as we know, Diophantus has been the first to treat systematically.

> Leibniz (1702), as translated by Weil (1984)

Jacobi (1834) apparently saw the connection for the first time after receiving a volume of Euler's works on the transformation of integrals, but considerable clarification of elliptic functions was needed before Jacobi's insight became generally available. We describe some of the main steps in this process of clarification in Chapters 12 and 16.

EXERCISES

A proof that genus 1 curves *cannot* be parameterized by rational functions can be modeled on Fermat's proof that $r^4 - s^4 = v^2$ is impossible in positive integers. The reason is that the behavior of rational functions is surprisingly similar to that of rational numbers, with polynomials playing the role of integers, and degree being the measure of size. The most convenient curve to illustrate the idea is $y^2 = 1 - x^4$, which happens to be of genus 1.

11.6.1 Show that a parameterization of $y^2 = 1 - x^4$ by rational functions of u implies that there are polynomials $r(u)$, $s(u)$, and $v(u)$ with

$$r(u)^4 - s(u)^4 = v(u)^2.$$

Now to imitate the rest of Fermat's proof (or the simplified version in Exercise 11.4.4) one needs a theory of divisibility for polynomials. Like the theory for natural numbers, this can be based on the Euclidean algorithm. It follows the same basic lines as in Section 3.3, so we shall omit it.

One also needs the formula for "Pythagorean triples" of rational functions. This can be found by the geometric method of Section 1.3, carried out in the "rational function plane" where each "point" is an ordered pair $(x(u), y(u))$ of rational functions.

11.6.2 Convince yourself that "lines" and "slope" make sense in the rational function plane, and hence show that each point $\neq (0, -1)$ on the "unit circle"

$$x(u)^2 + y(u)^2 = 1$$

is of the form

$$x(u) = \frac{1 - t(u)^2}{1 + t(u)^2}, \quad y(u) = \frac{2t(u)}{1 + t(u)^2}$$

for some rational function $t(u)$.

11.6.3 Deduce from Exercise 11.6.2 a formula for "Pythagorean triples" of polynomials, like Euclid's formula for ordinary Pythagorean triples.

It is now possible to imitate Fermat's proof, showing that $r(u)^4 - s(u)^4 = v(u)^2$ is impossible for polynomials, and hence that $y^2 = 1 - x^4$ has no parameterization by rational functions. It follows that the same is true of certain cubic curves.

11.6.4 Substitute $x = (X + 1)/X$ and $y = Y/X^2$ in $y^2 = 1 - x^4$, and hence show

$$Y^2 = \text{cubic polynomial in } X.$$

Deduce that if this cubic curve in X, Y has a rational parameterization, then so has $y^2 = 1 - x^4$.

11.7   Biographical Notes: Fermat

Pierre Fermat (Figure 11.5) was born in Beaumont, near Toulouse, in 1601 and died in Castres, also near Toulouse, in 1665. His life, like his mathematics, is not known in detail, but it seems to have been relatively uneventful. Fermat's father, Dominique, was a wealthy merchant and lawyer, his mother, Claire de Long, came from a prominent family, and they had two sons and two daughters. Pierre went to school in Beaumont, commenced university studies in Toulouse, and completed them with a law degree from Orléans in 1631. Thus Fermat's academic progress was far from meteoric, and not necessarily because he was distracted by mathematics. As far as we know, his earliest mathematical work was the analytical geometry of 1629 and, in the opinion of Weil (1984), his number theory did not mature until Fermat was in his late thirties.

On the evidence available, Fermat seems to defy the usual clichés about mathematical genius: he didn't start young, didn't work with passionate intensity, and was generally unwilling to publish his results (though he did sometimes boast about them). It is true that few mathematicians of Fermat's era actually did mathematics for a living, but Fermat was the purest of amateurs. It seems that mathematics never caused any interruption to his professional life.

In fact, after getting his law degree in 1631 he married a distant cousin on his mother's side, Louise de Long, collected a generous dowry, and settled into a comfortable legal career. His position entitled him to be addressed as Monsieur de Fermat, hence the name Pierre de Fermat by which he is now known. He and Louise had five children, the oldest of whom, Clement-Samuel, edited his father's mathematical works (Fermat (1670)).

Figure 11.5: Pierre de Fermat

Probably the most dramatic, and terrifying, experience of Fermat's life was his contracting the plague during an outbreak of the disease in Toulouse in 1652 or 1653. He was at first reported to be dead but was among the lucky few who recovered.

During the 1660s Fermat was in ill health. A meeting with Pascal in 1660 had to be called off because neither was well enough to travel. As a result, Fermat missed his only chance to meet a major mathematician. He never traveled far from Toulouse and all his work was done by correspondence, mostly with members of Mersenne's circle in Paris. After 1662 his letters cease to refer to scientific work, but he was signing legal documents until three days before his death. He died in Castres while on the court circuit, and was buried there. However, in 1675 his remains were transferred to the Fermat family vault in the Church of the Augustines in Toulouse.

Fermat's apparent refusal to put mathematics ahead of his profession makes the depth and range of his mathematical achievement all the more perplexing. We may never know enough about Fermat to understand his mathematical thought, but the attempts that have been made so far raise hopes that more can be done. Mahoney (1973) gives a survey of all of Fermat's mathematics but fails to do justice to the number theory. Weil (1984) has a brilliant analysis of Fermat's number theory, but other facets of Fermat's mathematics have yet to be analyzed with comparable insight.

12

Elliptic Functions

PREVIEW

Elliptic functions, like many innovations in mathematics, arose as a way
around an impasse. As we saw in Section 9.6, the search for closed-form
solutions in integral calculus foundered on integrands such as $1/\sqrt{1-x^4}$,
because no "known" function $f(x)$ has derivative $1/\sqrt{1-x^4}$.

Eventually, mathematicians accepted the fact that $\int_0^x \frac{dt}{\sqrt{1-t^4}}$ is a *new*
function. It is one of a family called the *elliptic integrals*, because one of
them is the integral that defines the arc length of the ellipse.

$\int_0^x \frac{dt}{\sqrt{1-t^4}}$ is the simplest elliptic integral to investigate, and many of
its properties were found by analogy with those of the arcsine integral
$\int_0^x \frac{dt}{\sqrt{1-t^2}}$. However, these were feats of virtuosity, like finding properties
of the arcsine integral without using the sine function.

The real innovation came around 1800, when Gauss realized that one
should *not* study the elliptic integral $u = \int_0^x \frac{dt}{\sqrt{1-t^4}}$, but rather its *inverse*
function x as a function of u (just as one should study the sine function
rather than the arcsine integral). He wrote $x = sl(u)$ and found that sl, like
the sine function, is *periodic*:

$$sl(u + 2\varpi) = sl(u), \quad \text{where } \varpi \text{ is a certain real number.}$$

More surprisingly, sl has second period $2i\varpi$, so sl is better viewed as a
function of *complex* numbers.

These results first became widely known when they were rediscovered
and published by Abel and Jacobi in the 1820s. Further insights into double
periodicity were obtained in the 1850s, as we will see in Chapter 16.

J. Stillwell, *Mathematics and Its History*, Undergraduate Texts in Mathematics,
DOI 10.1007/978-1-4419-6053-5_12, © Springer Science+Business Media, LLC 2010

12.1   Elliptic and Circular Functions

The story of elliptic functions is one of the most curious in the history of mathematics, beginning with a complicated analytic idea—integrals of the form $\int R(t, \sqrt{p(t)})\, dt$, where R is a rational function and p is a polynomial of degree 3 or 4—and reaching a climax with a simple geometric idea—the torus surface. Perhaps the best way to understand it is to compare it with a fictitious history of circular functions that begins with the integral $\int dt/\sqrt{1-t^2}$ and ends with the discovery of the circle. Unlikely as this fiction is, it was paralleled by the actual development of elliptic functions between the 1650s and the 1850s.

The late recognition of the geometric nature of elliptic functions was due to late recognition of the existence and geometric nature of complex numbers. In fact, the later history of elliptic functions unfolds alongside the development of complex numbers, which is the subject of Chapters 14 to 16. In the present chapter we are concerned mainly with the history up to 1800, before complex numbers entered in a really essential way. However, there are some subplots of the main story that do not require complex numbers for their understanding and nicely show the parallel with the fictitious history of circular functions. It is convenient to relate one of these now, because it illustrates the parallel in a simplified way and also ties up a loose end from Chapter 11—the parameterization of cubic curves.

12.2   Parameterization of Cubic Curves

To see how to construct parameterizing functions for a cubic curve, we first reconstruct the parameterizing functions

$$x = \sin u,$$

$$y = \cos u$$

for the circle $x^2 + y^2 = 1$, pretending that we do not know this curve geometrically but only as an algebraic relation between x and y.

The sine function can be defined as the inverse f of $f^{-1}(x) = \sin^{-1} x$, which in turn is definable as the integral

$$f^{-1}(x) = \int_0^x \frac{dt}{\sqrt{1-t^2}}.$$

Finally, the integral can be viewed as an outgrowth of the equation $y^2 = 1 - x^2$, because the integrand $1/\sqrt{1-x^2}$ is simply $1/y$. Why do we use this integrand rather than any other to define $u = f^{-1}(x)$ and hence obtain x as a function $f(u)$? The answer is that we then obtain y as $f'(u)$; hence x, y are both functions of the parameter u. This is confirmed by the following calculation:

$$f'(u) = \frac{dx}{du} = 1 \Big/ \frac{du}{dx}$$

and

$$\frac{du}{dx} = \frac{d}{dx} \int_0^x \frac{dt}{\sqrt{1-t^2}} = \frac{1}{\sqrt{1-x^2}} = \frac{1}{y};$$

hence $y = f'(u)$ (which of course is $\cos u$).

Exactly the same construction can be used to parameterize any relation of the form $y^2 = p(x)$. We put

$$u = g^{-1}(x) = \int_0^x \frac{dt}{\sqrt{p(t)}}$$

to get $x = g(u)$, and then find that $y = g'(u)$ by differentiation of u. Thus in a sense it is trivial to parameterize curves of the form $y^2 = p(x)$ (which we know from Section 8.4 to include all cubic curves, up to a projective transformation of x and y). As we shall see in the next section, the integrals $\int dt/\sqrt{p(t)}$ had been studied since the 1600s for p a polynomial of degree 3 or 4; however, no one thought to invert them until about 1800. Jacobi had a deep knowledge of both the integrals *and* inversion when he wrote his cryptic paper, Jacobi (1834), pointing out the relation between integrals and rational points on curves (cf. Sections 11.6 and 12.5). Thus it seems likely he understood the preceding parameterization, though such a parameterization was first given explicitly by Clebsch (1864).

EXERCISES

It may happen that the integral $\int_0^x dt/\sqrt{p(t)}$ does not converge because of the behavior of $1/\sqrt{p(t)}$ at $t = 0$. But in that case one can use the parameter $u = f^{-1}(x) = \int_a^x dt/\sqrt{p(t)}$ for some other value of a.

12.2.1 Check that $y = f'(u)$ remains true with this change of definition.

When the cubic curve is $y^2 = x^3$, which has a rational parameterization, the parameterizing functions constructed above indeed turn out to be rational.

12.2.2 Given $y = x^{3/2}$, find $x = f(u)$ and $y = f'(u)$, where $u = f^{-1}(x) = \int_a^x \frac{dt}{t^{3/2}}$.

12.3   Elliptic Integrals

Integrals of the form $\int R(t, \sqrt{p(t)})\, dt$, where R is a rational function and p is a polynomial of degree 3 or 4 without multiple factors, are called *elliptic integrals*, because the first example occurs in the formula for the arc length of the ellipse. (The functions obtained by inverting elliptic integrals are called *elliptic functions*, and the curves that require elliptic functions for their parameterization are called *elliptic curves*. This drift in the meaning of "elliptic" is rather unfortunate because the ellipse, being parameterizable by rational functions, is not an elliptic curve!)

Elliptic integrals arise in many important problems of geometry and mechanics, for example, finding arc lengths of the ellipse and hyperbola, period of the simple pendulum, and deflection of a thin elastic bar. See Chapter 13 and, for example, Melzak (1976), pp. 253–269. When these problems first arose in the late 17th century they posed the first obstacle to Leibniz's program of integration in "closed form" or "by elementary functions." As mentioned in Section 9.6, Leibniz considered the proper solution of an integration problem $\int f(x)\, dx$ to be a known function $g(x)$ with the property $g'(x) = f(x)$. The functions then "known," and now called "elementary," were those composed from algebraic, circular, and exponential functions and their inverses.

All efforts to express elliptic integrals in these terms failed, and as early as 1694 Jakob Bernoulli conjectured that the task was impossible. The conjecture was eventually confirmed by Liouville (1833), in the course of showing that a large class of integrals is nonelementary. In the meantime, mathematicians had discovered so many properties of elliptic integrals, and the elliptic functions obtained from them by inversion, that they could be considered known even if not elementary.

The key that unlocked many of the secrets of elliptic integrals was the curve known as the *lemniscate of Bernoulli* (Figure 12.1). This curve is mentioned briefly in Section 2.5 as one of the spiric sections of Perseus. It has cartesian equation

$$(x^2 + y^2)^2 = x^2 - y^2$$

and polar equation

$$r^2 = \cos 2\theta.$$

The first to consider it in its own right was Jakob Bernoulli (1694). He showed that its arc length is expressed by the elliptic integral $\int_0^x dt/\sqrt{1 - t^4}$,

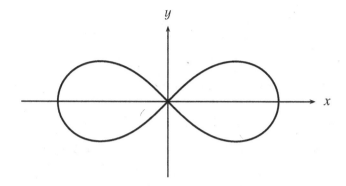

Figure 12.1: The lemniscate of Bernoulli

subsequently known as the *lemniscatic integral*, and thus he gave this formal expression a concrete geometric interpretation. Many later developments in the theory of elliptic integrals and functions grew out of interplay between the lemniscate and the lemniscatic integral. Being the simplest elliptic integral, or at any rate the most analogous to the arcsine in-
tegral $\int_0^x dt/ \sqrt{1 - t^2}$, the lemniscatic integral $\int_0^x dt/ \sqrt{1 - t^4}$ was the most
amenable to manipulation. It was often possible, after some property had
been extracted from the lemniscatic integral, to extend the argument to
more general elliptic integrals.

The most notable example of this methodology was in the discovery of
the addition theorems, which we discuss in the next section.

EXERCISES

The properties of the lemniscate mentioned above are easily proved by some
standard analytic geometry and calculus.

12.3.1 Deduce the cartesian equation of the lemniscate from its polar equation

$$r^2 = \cos 2\theta.$$

12.3.2 Use the polar equation of the lemniscate and the formula for the element
of arc in polar coordinates,

$$ds = \sqrt{(r\, d\theta)^2 + dr^2}$$

to deduce that arc length of the lemniscate is given by

$$s = \int \frac{d\theta}{r}.$$

12.3.3 Conclude, by changing the variable of integration to r, that the total length of the lemniscate is $4 \int_0^1 dr/\sqrt{1-r^4}$.

Unlike the arcsine integrand $1/\sqrt{1-t^2}$, which is rationalized by substituting $2v/(1+v^2)$ for t, the lemniscatic integrand $1/\sqrt{1-t^4}$ cannot be rationalized by replacing t by any rational function.

12.3.4 Explain how this follows from the exercises in Section 11.6.

It was this connection between the lemniscatic integral and Fermat's theorem on the impossibility of $r^4 - s^4 = v^2$ in positive integers that led Jakob Bernoulli to suspect the impossibility of evaluating the lemniscatic integral by known functions.

12.4   Doubling the Arc of the Lemniscate

An addition theorem is a formula expressing $f(u_1 + u_2)$ in terms of $f(u_1)$ and $f(u_2)$, and perhaps also $f'(u_1)$ and $f'(u_2)$. For example, the addition theorem for the sine function is

$$\sin(u_1 + u_2) = \sin u_1 \cos u_2 + \sin u_2 \cos u_1.$$

Since the derivative, $\cos u$, of $\sin u$ equals $\sqrt{1 - \sin^2 u}$, we can also write the addition theorem as

$$\sin(u_1 + u_2) = \sin u_1 \sqrt{1 - \sin^2 u_2} + \sin u_2 \sqrt{1 - \sin^2 u_1},$$

showing that $\sin(u_1 + u_2)$ is an algebraic function of $\sin u_1$ and $\sin u_2$.

To simplify the comparison with elliptic functions we consider the following special case of the sine addition theorem:

$$\sin 2u = 2 \sin u \sqrt{1 - \sin^2 u}. \tag{1}$$

If we let

$$u = \sin^{-1} x = \int_0^x \frac{dt}{\sqrt{1 - t^2}},$$

then

$$2u = 2 \int_0^x \frac{dt}{\sqrt{1 - t^2}}.$$

But from (1) we also have

$$2u = \sin^{-1}(2x \sqrt{1 - x^2}),$$

so

$$2 \int_0^x \frac{dt}{\sqrt{1-t^2}} = \int_0^{2x\sqrt{1-x^2}} \frac{dt}{\sqrt{1-t^2}}. \tag{2}$$

Bearing in mind that $\sin^{-1} x = \int_0^x dt/\sqrt{1-t^2}$ represents the angle u seen in Figure 12.2, equation (2) tells us that the angle (or arc length) u is doubled by going from x to $2x\sqrt{1-x^2}$. The latter number, since it is obtained from x by rational operations and square roots, is constructible from x by ruler and compass (confirming the geometrically obvious fact that an angle can be duplicated by ruler and compass).

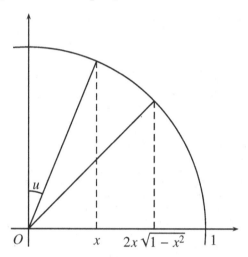

Figure 12.2: Doubling a circular arc

All this has a remarkable parallel in the properties of the lemniscate and its arc-length integral $\int_0^x dt/\sqrt{1-t^4}$. The discovery of a formula for doubling the arc of the lemniscate by Fagnano (1718) showed that geometric information could be extracted from the previously intractable elliptic integrals, and we can also view it as the first step toward the theory of elliptic functions. In our notation, Fagnano's formula is

$$2 \int_0^x \frac{dt}{\sqrt{1-t^4}} = \int_0^{2x\sqrt{1-x^4}/(1+x^4)} \frac{dt}{\sqrt{1-t^4}}. \tag{3}$$

Since $2x\sqrt{1-x^4}/(1+x^4)$ is obtained from x by rational operations and square roots, (3) shows, like (2), that the arc can be doubled by ruler and compass construction.

EXERCISES

Fagnano derived his formula by two substitutions that, as Siegel (1969), p. 3, points out, are analogous to a natural substitution for the arcsine integral. The following exercises compare the effect of the substitution $t = 2v/(1+v^2)$ in $dt/\sqrt{1-t^2}$ with analogous substitutions for t^2 in $dt/\sqrt{1-t^4}$.

12.4.1 Show that substituting $t = 2v/(1+v^2)$ gives $\sqrt{1-t^2} = (1-v^2)/(1+v^2)$ and hence that $dt/\sqrt{1-t^2} = 2dv/(1+v^2)$.

12.4.2 Show that $t^2 = 2v^2/(1+v^4)$ gives $\sqrt{1-t^4} = (1-v^4)/(1+v^4)$ and hence

$$\frac{dt}{\sqrt{1-t^4}} = \sqrt{2}\frac{dv}{\sqrt{1+v^4}}.$$

It follows that this change of variable corresponds to a certain relation between integrals, which turns out to be "half way" to the Fagnano formula.

12.4.3 Deduce from Exercise 12.4.2 that

$$\sqrt{2}\int_0^x \frac{dv}{\sqrt{1+v^4}} = \int_0^{\sqrt{2}x/\sqrt{1+x^4}} \frac{dt}{\sqrt{1-t^4}}.$$

To complete the journey to the Fagnano formula we make a second, similar, substitution that recreates the lemniscatic integral.

12.4.4 Similarly show that the substitution $v^2 = 2w^2/(1-w^4)$ gives

$$\frac{dv}{\sqrt{1+v^4}} = \sqrt{2}\frac{dw}{\sqrt{1-w^4}}.$$

12.4.5 Check that the result of the substitutions in Exercises 12.4.2 and 12.4.4 is

$$t = \frac{2w\sqrt{1-w^4}}{1+w^4}$$

and that the corresponding relation between integrals is the Fagnano duplication formula.

12.5   General Addition Theorems

Fagnano's duplication formula remained a little-known curiosity until Euler received a copy of Fagnano's works on December 23, 1751, a date later described by Jacobi as "the birth day of the theory of elliptic functions." Euler was the first to see that Fagnano's substitution trick was not just a curious fluke but a revelation into the behavior of elliptic integrals. With his superb manipulative skill Euler was quickly able to extend it to very general addition theorems; first to the addition theorem for the lemniscatic integral,

$$\int_0^x \frac{dt}{\sqrt{1 - t^4}} + \int_0^y \frac{dt}{\sqrt{1 - t^4}} = \int_0^{(x\sqrt{1-y^4}+y\sqrt{1-x^4})/(1+x^2y^2)} \frac{dt}{\sqrt{1 - t^4}} \, ;$$

then to $\int dt/ \sqrt{p(t)}$, where $p(t)$ is an arbitrary polynomial of degree 4. An ingenious reconstruction of Euler's train of thought, by analogy with the arcsine addition theorem

$$\int_0^x \frac{dt}{\sqrt{1 - t^2}} + \int_0^y \frac{dt}{\sqrt{1 - t^2}} = \int_0^{x\sqrt{1-y^2}+y\sqrt{1-x^2}} \frac{dt}{\sqrt{1 - t^2}} \, ,$$

has been given by Siegel (1969), pp. 1–10. Brilliant as his results were, Euler was dealing only with elliptic integrals, *not* with elliptic functions, their inverses, so one could still quibble with Jacobi's assessment. But one has to remember that Jacobi could see an elliptic function a mile off, probably more easily than we can see that the arcsine addition theorem is really a theorem about sines!

It should be mentioned that Euler's addition theorems do not cover all elliptic integrals. The general form $\int R(t, \sqrt{p(t)}) \, dt$ does, however, reduce to just three kinds, of which Euler's are the first and most important. The classical theory of elliptic integrals of the different kinds, with their various addition and transformation theorems, was systematized by Legendre (1825). Ironically, this was just before the appearance of elliptic functions, which made much of Legendre's work obsolete.

These early investigations exploited some of the formal similarities between $\int dt/ \sqrt{p(t)}$, where p is a polynomial of degree 4, and $\int dt/ \sqrt{q(t)}$, where q is a quadratic. There is no real difference if p is of degree 3, as an easy transformation shows (Exercise 12.5.1). This is why $\int dt/ \sqrt{p(t)}$ is also called an elliptic integral when p is of degree 3. In fact, it eventually turned out that the most convenient integral to use as a basis for the theory of elliptic functions is $\int dt/ \sqrt{4t^3 - g_2 t - g_3}$, whose inverse is known as the Weierstrass $\wp$-function.

The addition theorem for this integral is

$$\int_0^{x_1} \frac{dt}{\sqrt{4t^3 - g_2 t - g_3}} + \int_0^{x_2} \frac{dt}{\sqrt{4t^3 - g_2 t - g_3}} = \int_0^{x_3} \frac{dt}{\sqrt{4t^3 - g_2 t - g_3}} \, ,$$

where x_3 is none other than the x-coordinate of the third point on

$$y^2 = 4x^3 - g_2 x - g_3$$

of the straight line through (x_1, y_1) and (x_2, y_2) (see Section 11.6). Now that we know, from Section 12.2, that this curve is parameterized by $x = \wp(u)$, $y = \wp'(u)$, defined by inverting the integral, some connection between the geometry of the curve and the addition theorem is understandable. But the stunning simplicity of the relationship seems to demand a deeper explanation. This lies in the realm of complex numbers, which we shall enter briefly in the next section and more thoroughly in Sections 16.4 and 16.5.

EXERCISES

12.5.1 Show that the substitution $t = 1/u$ transforms

$$\frac{dt}{\sqrt{(t-a)(t-b)(t-c)}} \quad \text{into} \quad \frac{-du}{\sqrt{u(1-ua)(1-ub)(1-uc)}}.$$

Conversely, we can transform quartic polynomials under the square root sign to cubics, even in cases where the quartic is not of the form obtained in Exercise 12.5.1.

12.5.2 Transform

$$\frac{dt}{\sqrt{1-t^4}} \quad \text{into} \quad \frac{du}{\sqrt{\text{cubic polynomial in } u}}$$

by making a suitable substitution for t.

12.6   Elliptic Functions

The idea of inverting elliptic integrals to obtain elliptic functions is due to Gauss, Abel, and Jacobi. Gauss had the idea in the late 1790s but did not publish it; Abel had the idea in 1823 and published it in 1827, independently of Gauss. Jacobi's independence is not quite so clear. He seems to have been approaching the idea of inversion in 1827, but he was stung into action only by the appearance of Abel's paper. At any rate, his ideas subsequently developed at an explosive rate, and he published the first book on elliptic functions, the *Fundamenta nova theoriae functionum ellipticarum*, two years later (Jacobi (1829)).

    Gauss first considered inverting an elliptic integral in 1796, in the case of $\int dt / \sqrt{1-t^3}$. The following year he inverted the lemniscatic integral and made better progress. Defining the "lemniscatic sine function" $x = sl(u)$ by

$$u = \int_0^x \frac{dt}{\sqrt{1 - t^4}},$$

he found that this function is periodic, like the sine, with period

$$2\varpi = 4 \int_0^1 \frac{dt}{\sqrt{1 - t^4}}.$$

Gauss also noticed that $sl(u)$ invites complex arguments, since it follows from $i^2 = -1$ that

$$\frac{d(it)}{\sqrt{1 - (it)^4}} = i\frac{dt}{\sqrt{1 - t^4}};$$

hence $sl(iu) = isl(u)$ and the lemniscatic sine has a second period $2i\varpi$. Thus Gauss discovered *double periodicity*, one of the key properties of the elliptic functions, though at first he did not realize its universality. The scope and importance of elliptic functions hit him on May 30, 1799, when he discovered an extraordinary numerical coincidence. His diary entry of that day reads:

> We have established that the arithmetic-geometric mean between 1 and $\sqrt{2}$ is π/ϖ to 11 places; the demonstration of this fact will surely open up an entirely new field of analysis.

Gauss had been fascinated by the arithmetic–geometric mean since 1791, when he was 14. The arithmetic–geometric mean of two positive numbers a and b is the common limit, $agM(a, b)$, of the two sequences $\{a_n\}$ and $\{b_n\}$ defined by

$$a_0 = a, \qquad b_0 = b,$$
$$a_{n+1} = \frac{a_n + b_n}{2}, \qquad b_{n+1} = \sqrt{a_n b_n}.$$

For more information on the theory and history of the agM function, see Cox (1984).

It is indeed true that $agM(1, \sqrt{2}) = \pi/\varpi$, as Gauss soon proved, and the "entirely new field of analysis" he created from the merger of these ideas was extraordinarily rich. It encompassed elliptic functions in general, the theta functions later rediscovered by Jacobi, and the modular functions later rediscovered by Klein. The theory was not clearly improved until the

1850s, when Riemann showed that double periodicity becomes obvious when elliptic integrals are placed in a suitable geometric setting.

Unfortunately, Gauss released virtually none of his results on elliptic functions. Apart from publishing an expression for agM(a, b) as an elliptic integral (Gauss (1818)), he did nothing until Abel's results appeared in 1827—then promptly claimed them as his own. He wrote to Bessel (Gauss (1828)):

> I shall most likely not soon prepare my investigations on the transcendental functions which I have had for many years— since 1798. ...Herr Abel has now, as I see, anticipated me and relieved me of the burden in regard to one third of these matters.

It was disingenuous of Gauss to claim he had more results than Abel, because Abel also had results unknown to Gauss. True, Gauss had priority on the key ideas of inversion and double periodicity, but priority isn't everything, as Gauss himself perhaps knew. His own cherished discovery of the relation between agM and elliptic integrals had not only been found earlier, but even published by Lagrange (1785).

EXERCISES

The following exercises show how the lemniscatic sine and its derivative are quite analogous to the ordinary sine and its derivative, the cosine.

12.6.1 Show that $sl'(u) = \sqrt{1 - sl^4(u)}$.

12.6.2 Deduce from the Euler addition theorem (Section 12.4) that

$$sl(u + v) = \frac{sl(u)sl'(v) + sl(v)sl'(u)}{1 + sl^2(u)sl^2(v)}.$$

12.7   A Postscript on the Lemniscate

The duplication of the arc of the lemniscate had some interesting consequences for the lemniscate itself. Fagnano showed, by similar arguments, that a quadrant of the lemniscate can be divided into two, three, or five equal arcs by ruler and compass (see Ayoub (1984)). This raised a question: for which n can the lemniscate be divided into n equal parts by ruler and compass? Recall from Section 2.3 that the corresponding question for the circle had been answered by Gauss (1801), Art. 366. As mentioned in Section 2.3, the answer is $n = 2^m p_1 p_2 \cdots p_k$, where the p_i are distinct primes of the form $2^{2^h} + 1$. In the introduction to his theory (Art. 355), Gauss claims:

> The principles of the theory which we are going to explain actually extend much further than we will indicate. For they can be applied not only to circular functions but just as well to other transcendental functions, e.g. to those which depend on the integral $\int (1/\sqrt{1 - x^4})\,dx$.

However, his surviving papers do not include any result on the lemniscate as incisive as his result on the circle. There is only a diary entry of March 21, 1797, stating divisibility of the lemniscate into five equal parts.

The answer to the problem of dividing the lemniscate into n equal parts was found by Abel (1827), transforming Gauss's obscurity into crystal clarity: division by ruler and compass is possible for *precisely the same n* as for the circle. This wonderful result serves, perhaps better than any other, to underline the unifying role of elliptic functions in geometry, algebra, and number theory. A modern proof of it may be found in Rosen (1981).

12.8   Biographical Notes: Abel and Jacobi

Niels Henrik Abel was born in the small town of Finnøy, on the southwestern coast of Norway, in 1802 and died in Oslo in 1829. In his short life he managed to win the esteem of the best mathematicians in Europe, but he fell victim to official indifference, terrible family burdens, and tuberculosis. His heart-breaking story is not unlike that of his great contemporary in another field, the poet John Keats (1797–1823).

Like several mathematicians before him (Wallis, Gregory, Euler), Abel was the son of a Protestant minister. His father, Søren, distinguished himself in theology and philology at the University of Copenhagen and was a supporter of the new literary and social movements of his time. Søren's liberality, particularly toward the consumption of alcohol, was unfortunately not matched by good judgment, and his marriage to Anne Marie Simonsen in 1799 eventually led to disaster. The beautiful Anne Marie was a talented pianist and singer but completely irresponsible and later openly unfaithful to her husband. The family held together during Abel's early years, when he was educated by his father, but both parents were becoming frequently drunk and unstable by 1815, when Niels and his older brother, Hans Mathias, were sent to the Cathedral School in Oslo.

At first school was not much better than home. Some of its best teachers had gone to the recently opened Oslo University, and discipline had deteriorated to the point where fights between staff and students were common. The mathematics teacher, Bader, was particularly brutal, beating even good students like Abel, and injuring one boy severely enough to cause his death. This led to Bader's dismissal (without his being brought to court, however) and to the appointment of a new mathematics teacher, Bernt Michael Holmboe, in 1818. Although not a creative mathematician, Holmboe knew his subject and was an inspiring teacher. He introduced Abel to Euler's calculus texts, and Abel soon abandoned all other reading for the works of Newton, Lagrange, and Gauss, among others. By 1819 Holmboe was writing in his report book, "With the most excellent genius he combines an insatiable interest and desire for mathematics, so that if he lives he probably will become a great mathematician" (see Ore (1957), p. 33). Ore informs us that the last three words are a revision, probably of the phrase "the world's foremost mathematician," which Holmboe may have been asked to tone down by the school principal. Why Holmboe chose to balance the phrase with the ominous "if he lives" is a mystery, though uncomfortably close to correct prophecy.

During his last two years at the Cathedral School, around 1820, Abel believed he had discovered the solution of the quintic equation. The mathematicians in Oslo were skeptical but unable to fault Abel's argument, so it was sent to the Danish mathematician Ferdinand Degen. Degen, too, was unable to find an error, but he prudently asked Abel for more details and a numerical illustration. When Abel attempted to compute one he discovered his error. However, Degen also had another suggestion: Abel would do better to apply his energy to "the elliptic transcendentals."

Meanwhile, Abel's family was disintegrating. Hans Mathias, after a promising start at the Cathedral School, slipped to the bottom of the class and was sent home, eventually to become feeble-minded. His father drank himself to death in 1820, leaving the family penniless. Niels Henrik, now the oldest responsible member of the family, took steps that were to save his sister Elisabeth and younger brother Peder. He found another home for Elisabeth and took Peder with him when he entered the University of Oslo in 1821.

Before long, Abel had read most of the advanced mathematical works in the university library, and his own research began in earnest. By 1823 he had discovered the inversion that was the key to elliptic functions, proved the unsolvability of the quintic, and discovered a wonderful general theorem on integration, now known as Abel's theorem, which implicitly introduces the concept of genus. On a trip to Copenhagen in 1823 to tell Degen of these results, he met and fell in love with Christine ("Crelly") Kemp. Like Abel, she came from an educated but impoverished family; she was making a living for herself by tutoring. The remaining six years of Abel's life were consumed by the struggle for recognition of his mathematics and attempts to gain a position that paid enough to allow him to marry Crelly.

In 1824 he won a government grant to travel and meet other scientists, and he became engaged to Crelly at Christmas. She was now working in Oslo as a governess, a job that Abel had arranged for her. The grant was mainly intended to take him to Paris, but when he finally set off, late in 1825, he impulsively detoured to Berlin to visit friends. There he also met August Crelle, an engineer and amateur mathematician, who was about to found the first German mathematical journal. The meeting was fortuitous, because Crelle was able to give an international circulation to Abel's first important results, while Abel could supply papers of a quality that ensured the success of the new journal. In meeting influential mathematicians Abel was less lucky. He made no effort to visit Gauss while in Germany, being convinced that Gauss was "absolutely unapproachable," and failed to make an impression on Cauchy in Paris, though he presented him with a copy of the memoir on Abel's theorem. During his stay in Paris, Abel discovered his theorem on the lemniscate and sat for his only known portrait (Figure 12.3).

By the end of 1826 Abel was running out of money and eating only one meal a day. He feared he was losing touch with Crelly, since she had returned to Copenhagen and her letters were infrequent. He left Paris for Berlin on December 29, while he still had money to pay for the journey, and found a letter from Crelly waiting. Some good news at last! Crelly stood by him as ever, and their plans for the future were revived. Abel returned to Oslo in May 1827, via Copenhagen, and arranged another job in Norway for Crelly. Unfortunately, the university was still unwilling to give him more than a temporary appointment, which paid barely enough to meet his family's debts. In September 1827 Abel's first memoir on elliptic functions was published in Crelle's journal. In the same month, Jacobi appeared on the scene with the first announcement of his results. There were results

Figure 12.3: Niels Henrik Abel

that Abel knew how to prove, and when Jacobi's proofs appeared, some months later, Abel was shocked to see Jacobi using the method of inversion without acknowledging its previous appearance in Abel's paper. Abel was initially bitter over this blow and strove to "knock out" Jacobi with a second memoir. However, he ceased to bear a grudge after he learned how much Jacobi really admired his work. Jacobi in fact admitted that his first announcement had been based on guesswork and that he had realized inversion was the key to the proof only after reading Abel.

In May 1828 Abel finally received a decent job offer from Berlin, only to have it withdrawn two months later. Crelle had been working in support of Abel, but another candidate had pushed in ahead of him. Then a group of French mathematicians petitioned the king of Norway–Sweden to use his influence on Abel's behalf, but still the University of Oslo remained unmoved. By now, time was running out. Abel's health worsened and in January 1829 he began spitting blood. Crelle renewed his efforts in Berlin, but it was too late. Abel died on April 6, 1829, just two days before the arrival of a letter from Crelle informing him of his appointment as professor in Berlin.

For a recent biography of Abel, very thorough on his life and times, but less so on his mathematics, see Stubhaug (2000).

Carl Gustav Jacob Jacobi (Figure 12.4) was born in Potsdam in 1804 and died in Berlin in 1851. He was the second of three sons of Simon Jacobi, a banker. The oldest son, Moritz, became a physicist and inventor of a popular pseudoscience called "galvanoplastics," which made him more famous in his time than Carl. The youngest, Eduard, carried on the family business, and there was also a sister, Therese. Jacobi's mother's name has not come down to us, though her side of the family was also important, one of her brothers taking charge of Jacobi's education until he entered secondary school in 1816. He was promoted to the top class after only a few months, but he had to remain in school for four years, until he became old enough to enter university. During his school days Jacobi excelled in classics and history as well as mathematics. He studied Euler's *Introductio in analysin infinitorum* and attempted, like Abel, to solve the quintic equation.

Figure 12.4: Carl Gustav Jacob Jacobi

Entering the University of Berlin in 1821, Jacobi continued his broad classical education for two years, before private study of the works of Euler, Lagrange, Laplace, and Gauss convinced him that he had time only for mathematics. He gained his first degree in 1824 and began lecturing (on differential geometry) at the University of Berlin in 1825. Despite a reputation for bluntness and sarcasm, Jacobi made rapid progress in his career. He moved to Königsberg in 1826, becoming associate professor there in 1827 and full professor in 1832. Overriding Jacobi's sometimes abrasive manner were his exceptional energy and enthusiasm for both research and teaching. He managed to combine the two by lecturing up to 10 hours a week on elliptic functions, incorporating his latest discoveries. Such high-intensity instruction was unheard of then, as it is now, yet Jacobi built up a school of talented pupils.

In 1831 he married Marie Schwink, the daughter of a formerly wealthy man who had lost his fortune through speculation. Nine years later, with a growing family (eventually five sons and three daughters), Jacobi found himself in a similar predicament. His father's fortune had vanished and he had to support his widowed mother. In 1843 he suffered a breakdown from overwork, and diabetes was diagnosed. His friend Dirichlet managed to secure a grant for Jacobi to travel to Italy for the sake of his health. After eight months there, Jacobi was well enough to return. He was given permission to move to Berlin, because of its milder climate, and an increase in salary to meet the higher living costs in the capital. However, in 1849 the salary bonus was retracted. Jacobi had to move out of his house to an inn, and he sent the rest of his family to the small town of Gotha, where housing was cheaper. Early in 1851 he came down with influenza after visiting them. Before he had quite recovered, he was stricken with smallpox and died within a week.

Jacobi is remembered for his contributions to many fields of mathematics, including differential geometry, mechanics, and number theory as well as elliptic functions. He was a great admirer of Euler and planned the edition of Euler's works that eventually began to appear, on a reduced scale, in 1911. In fact, in many ways Jacobi was a second, if lesser, Euler. He saw elliptic functions not so much as things in themselves, as Abel did, but as a source of dazzling formulas with implications in number theory. An astounding collection of formulas may be found in his major work on elliptic functions, the *Fundamenta nova* (1829). At the same time, he was deeply impressed by Abel's ideas and selflessly campaigned to make them better known. He introduced the terms "Abelian integral" and "Abelian function" for the generalizations of elliptic integrals and functions considered by Abel as well as "Abelian theorem" for Abel's theorem, which he described as "the greatest mathematical discovery of our time."

13

Mechanics

PREVIEW

In Chapter 9 we introduced the concepts of derivative and integral geometrically, as tangents and areas respectively. Geometry was certainly an important source of calculus problems and concepts, but not the only one. From the beginning, mechanics was just as important.

Mechanics is conceptually important because the derivative and the integral are inherent in the concept of *motion*: velocity is the derivative of displacement (with respect to time), and displacement is the integral of velocity.

Also, mechanics was initially the only source of nonalgebraic curves; for example, the cycloid, which is generated by rolling a circle along a line. The "mechanical" curves spurred the development of calculus for the simple reason that they were not accessible to pure algebra.

An even greater spur was the development of *continuum mechanics*, which studies the behavior of such things as flexible and elastic strings, fluid motion, and heat flow. Continuum mechanics involves functions of several variables, and their various derivatives, hence *partial differential equations*.

Some of the most important partial differential equations, such as the *wave equation* and the *heat equation*, are clearly inseparable from their origins in continuum mechanics. Yet these very equations confronted mathematicians with basic questions in pure mathematics: for example, what is a function?

J. Stillwell, *Mathematics and Its History*, Undergraduate Texts in Mathematics,   243
DOI 10.1007/978-1-4419-6053-5_13, © Springer Science+Business Media, LLC 2010

13.1   Mechanics Before Calculus

The ambiguous title reflects the dual purpose of this section: to give a brief survey of the mechanics that came before calculus and to make the claim that mechanics was psychologically, if not logically, a prerequisite for calculus itself. The remainder of the chapter expands on this claim, showing how several important fields in calculus (and beyond) originated in the study of mechanical problems. Lack of space, not to mention lack of expertise, prevents my venturing far into the history of mechanical concepts. I assume some understanding of time, velocity, acceleration, force, and the like, and concentrate on the mathematics that emerged from reflection on these notions. These developments will be pursued as far as the 19th century. More details may be found in Dugas (1957, 1958) and Truesdell (1954, 1960). In the last 100 years, mathematics seems to have been the motivation for mechanics rather than the other way around. The outstanding mechanical concepts of the 20th century—relativity and quantum mechanics—would not have been conceivable without 19th-century advances in pure mathematics, some of which we discuss later.

It is mentioned in Section 4.5 that Archimedes made the only substantial contribution to mechanics in antiquity by introducing the basics of statics (balance of a lever requires equality of moments on the two sides) and hydrostatics (a body immersed in a fluid experiences an upward force equal to the weight of fluid displaced). Archimedes' famous results on areas and volumes were in fact discovered, as he revealed in his *Method*, by hypothetically balancing thin slices of different figures. Thus the earliest nontrivial results in calculus, if by calculus one means a method for discovering results about limits, relied on concepts from mechanics.

The medieval mathematician Oresme also was mentioned (Section 7.1) for his use of coordinates to give a geometric representation of functions. The relationship Oresme represented was in fact velocity v as a function of time t. He understood that displacement is then represented by the area under the curve, and hence in the case of constant acceleration (or "uniformly deformed velocity," as he called it) the displacement equals total time × velocity at the middle instant (Figure 13.1). This result is known as the "Merton acceleration theorem" (see, for example, Clagett (1959), p. 355) because it originated in the work of a group of mathematicians at Merton College, Oxford, in the 1330s. The first proofs were arithmetical and far less transparent than Oresme's figure.

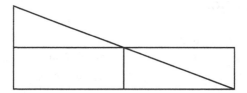

Figure 13.1: The Merton acceleration theorem

While constant acceleration was understood theoretically in the 1330s, it was not clear that it was actually a natural occurrence—namely, with falling bodies—until the time of Galileo (1564–1642). Galileo announced the equivalent result, that displacement of a body falling from rest at time $t = 0$ is proportional to t^2, in a letter (Galileo (1604)). At first he was uncertain whether this derived from a velocity proportional to time $v = kt$ (that is, constant acceleration) or one proportional to distance $v = ks$, but he resolved the question correctly in favor of $v = kt$ later (Galileo (1638)). By composing the uniformly increasing vertical velocity with constant horizontal velocity, Galileo derived for the first time the correct trajectory of a projectile: the parabola.

The motion of projectiles was of weighty importance in the Renaissance, and was presumably observed often, yet the trajectories suggested before Galileo were quite preposterous (see, for example, Figure 6.3). The belief, deriving from Aristotle, that motion could be sustained only by continued application of a force led mathematicians to ignore the evidence and to draw trajectories in which the horizontal velocity dwindled to zero. Galileo overthrew this mistaken belief by affirming the *principle of inertia*: a body not subject to external forces travels with constant velocity.

EXERCISES

Galileo's result that the path of a projectile is a parabola follows easily from the assumption that the only force is vertical, and hence the projectile has zero acceleration in the horizontal direction.

13.1.1 Show, by consideration of areas, that the distance covered in time t by a body moving with constant acceleration a is of the form $c + at^2/2$.

13.1.2 Assuming that a projectile has constant vertical acceleration a and zero horizontal acceleration, show that its position (x, y) at time t is given by the equations $x = bt, y = c + at^2/2$.

13.1.3 Deduce from Exercise 13.1.2 that the path of the projectile is a parabola.

13.2   The Fundamental Theorem of Motion

When Oresme assumed that the distance covered by any body equals the area under its velocity–time graph, he possibly thought of dividing the area under the graph into a number of thin vertical strips, corresponding to small time intervals (Figure 13.2). In each such interval the velocity is virtually constant; hence its product by the time interval—the area of the strip— virtually equals the distance covered in that time interval. By adding all the thin strips, one sees that total area equals total distance. Whether or not this was Oresme's argument, he had glimpsed an important relationship.

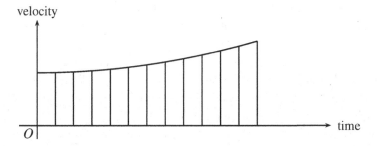

Figure 13.2: Distance represented by area

The *inverse* of this relationship, describing how velocity is derived from distance, was discovered in the 1630s. At this time, mathematicians became interested in the problem of finding tangents to curves, and they found it helpful to view the curve as the path of a moving point and the direction of the tangent as the instantaneous direction of its motion. In cases where the motion of the point could be viewed as the resultant of two velocities **u** and **v**, the direction of the tangent could be found as the *vector sum* **u** + **v** (Figure 13.3).

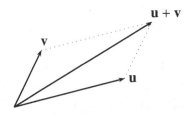

Figure 13.3: The vector sum of **u** and **v**

The idea of vector addition is due to Roberval, who used it to find the tangent to the *cycloid*, the curve generated by a point on a circle that rolls along a line. He found the velocity of the moving point as the vector sum of its velocities in the direction of the line (constant) and in the direction of the circle (constant in speed, but in the direction of the tangent to the circle at the current position of the point).

Now the *distance–time* graph of any point moving along a vertical line, with velocity v at time t, say, is generated by a point that moves with constant horizontal velocity 1 and vertical velocity v at time t (Figure 13.4).

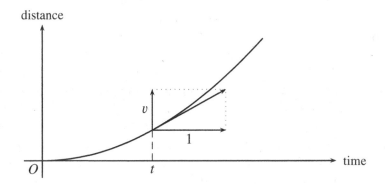

Figure 13.4: Distance–time graph of a vertically moving point

Thus the *slope of the tangent* at time t is simply $v/1 = v$. In other words, *velocity is the slope of the distance–time graph*. This relationship between velocity and distance was noticed around 1640 by Torricelli. Torricelli was also aware of the idea of distance as the area under the velocity–time graph (though he had not learned it from Oresme, but from his teacher Galileo), so it was probably he who first glimpsed the *inverse* relationship between velocity and distance:

> *Distance is the area of velocity (with respect to time).*
> *Velocity is the slope of distance (with respect to time).*

It seems reasonable to call this the *fundamental theorem of motion*.

It corresponds to a fundamental theorem about graphs, saying that "slope" is the *inverse operation* of "area." If we start with the graph of any quantity (not necessarily velocity), apply the "area" operation, and then the "slope" operation, we get back the original quantity. Such a theorem was first stated by Isaac Barrow in his *Lectiones Geometricae* of 1670. It was almost the fundamental theorem of *calculus*, but not quite, because the "calculus" was missing. There was not yet a general method for *calculating* slopes and areas of curves. However, the first step was to *notice* that there was an inverse relationship between slope and area, and the concept of motion brought this relationship to light.

The other crucial ingredient in mechanics is the relationship between force and motion. This was provided by Newton, whose starting point was Galileo's principle of inertia; indeed, it is often called Newton's first law. It is a special case of his second law, in Newton (1687), p. 13, that force is proportional to mass × acceleration. Under this law, the motion of a body is determined by composition of the forces acting on it. The correct law for the composition of forces, that forces add vectorially, had been discovered in the case of perpendicular forces by Stevin (1586) and in the general case by Roberval (published in Mersenne (1636)). The motion is thus determined by vector addition of the corresponding accelerations, the method Galileo used in the case of the projectile.

The determination of velocity and displacement from acceleration are problems of integration, so mechanics contributed a natural class of problems to calculus just as the subject was emerging. But more than this was true. The early practitioners of calculus believed that continuity was an essential attribute of functions, and ultimately the only way they were able to define continuity was by falling back on the dependence of a velocity or a displacement on time. From this viewpoint, *all* problems of integration and differentiation were problems of mechanics, and Newton described them as such when explaining how his calculus of infinite series could be applied:

> It now remains, in illustration of this analytical art, to deliver some typical problems and such especially as the nature of curves will present. But first of all I would observe that difficulties of this sort may all be reduced to these two problems alone, which I may be permitted to propose with regard to the space traversed by any local motion however accelerated or retarded:
>
> 1. Given the length of space continuously (that is, at every time), to find the speed of motion at any time proposed.
>
> 2. Given the speed of motion continuously, to find the length of space described at any time proposed.
>
> Newton (1671), p. 71

Of course we now know that the first problem requires differentiability rather than continuity for its solution, but the pioneers of calculus thought that differentiability was implied by continuity, and hence did not recognize it as a distinct notion. In fact it was a mechanical question—the problem of the vibrating string—whose investigation brought the distinction to light (see Section 13.6).

13.3   Kepler's Laws and the Inverse Square Law

Astronomy has been a powerful stimulus to mathematics since ancient times. The epicyclic theory of Apollonius and Ptolemy introduced an interesting family of algebraic and transcendental curves, as we saw in Section 2.5, and the theory itself ruled Western astronomy until the 17th century. Even Copernicus (1472–1543), when he overthrew Ptolemy's earth-centered system with a sun-centered system in his *De revolutionibus orbium coelestium* (1543), was unwilling to give up epicycles. Taking the sun as the center of the system simplifies the orbits of the planets but does not make them circular, so Copernicus, who accepted the Ptolemaic philosophy that orbits must be generated by circular motions, modeled them by epicycles. In fact he used more epicycles than Ptolemy.

Epicyles were finally overthrown by Kepler, who found that the simplest way to explain existing observations of Mars was to assume that its orbit was an ellipse, with the sun at one focus. This, and further study of the known planetary observations, led him to postulate three laws; the first two are in Kepler (1609), and the third in Kepler (1619). Kepler's laws are:

1. The orbit of each planet is an ellipse, with the sun at one focus.

2. The line from sun to planet sweeps out area at a constant rate.

3. The period of revolution of a planet is proportional to $R^{3/2}$, where R is half the major axis of the planet's orbit.

These laws, by including time as well as space (unlike Kepler's misguided attempt to explain the *size* of the planetary orbits in terms of the regular polyhedra) pointed the way to the next great advance in astronomy: Newton's explanation of Kepler's laws in terms of *gravitational force* in his *Principia* (1687). There he showed that the laws follow from the assumption of a gravitational force *between any two bodies, jointly proportional to their masses and inversely proportional to the square of the distance between them.*

We will not describe Newton's proofs here, though you are invited to try your hand at the second law in the exercises below. It is not very obvious why the inverse square law implies elliptic orbits. Instead, we will show why the inverse square law is *necessary* if Kepler's third law is to hold. Newton himself did this in his Corollary VI to Proposition IV in Book II of the *Principia*.

It suffices to consider the special case of a circular orbit, in which case symmetry tells us that the planet travels with constant speed in the tangential direction. We suppose that the radius of the orbit is r and the tangential speed is $v(r)$. Newton's second law says that the force exerted by the sun is proportional to the planet's acceleration towards the sun, so our first problem is to find the acceleration, towards the center of the circle, of a body traveling around a circle of radius r with constant speed $v(r)$.

Figure 13.5 shows the relevant information for calculating the acceleration: the velocity vectors $\mathbf{v}_1$ and $\mathbf{v}_2$ at two nearby points on the circle, and a diagram of these two vectors and the difference between them. The two

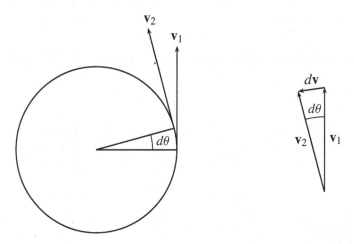

Figure 13.5: Velocity vectors of a planet in circular orbit

nearby points differ in position by a small angle $d\theta$, which is also the angle between the velocity vectors $\mathbf{v}_1$ and $\mathbf{v}_2$. Since both $\mathbf{v}_1$ and $\mathbf{v}_2$ have magnitude $v(r)$, their difference, $d\mathbf{v}$, has magnitude tending to $v(r)d\theta$ as $d\theta \to 0$, and direction tending to perpendicular to both $\mathbf{v}_1$ and $\mathbf{v}_2$; that is, towards the center of the circle.

It follows that the acceleration of the planet, $\frac{d\mathbf{v}}{dt}$, is directed towards the sun and has magnitude

$$\frac{v(r)d\theta}{dt}.$$

But also, because the planet covers arc length $r\,d\theta$ in time dt,

$$v(r) = \frac{d}{dt}r\,d\theta = r\frac{d\theta}{dt}.$$

Thus

$$\text{acceleration} = v(r)\frac{d\theta}{dt} = \frac{v(r)^2}{r}.$$

Since the circumference of the orbit is $2\pi r$, it follows that

$$\text{period of orbit} = \frac{2\pi r}{v(r)}.$$

On the other hand, according to Kepler's third law,

$$\text{period of orbit} = ar^{3/2}, \quad \text{for some constant } a.$$

Equating the two expressions of the period, we get

$$ar^{3/2} = \frac{2\pi r}{v(r)},$$

so

$$v(r) = \frac{b}{r^{1/2}}, \quad \text{for some constant } b.$$

Then, using the expression for acceleration above, we get

$$\text{acceleration} = \frac{v(r)^2}{r} = \frac{b}{r^2}, \quad \text{directed towards the sun.}$$

Finally, since force equals mass $\times$ acceleration by Newton's second law, we find

$$\text{force} = \frac{mb}{r^2}, \quad \text{directed towards the sun,}$$

where m is the mass of the planet.

EXERCISES

Kepler's second law does not depend on the inverse square law. Surprisingly, it holds for motion under *any* force directed towards a fixed "sun," as was shown by Newton in the *Principia*, Book I, Proposition II.

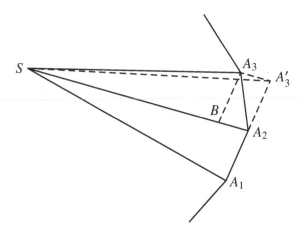

Figure 13.6: Motion under an impulsive central force

Newton approximates the smooth motion under continuous attraction towards the point S by motion under no force except for instantaneous "impulses" towards S at small, constant time intervals. Consequently, the planet travels in a polygonal path (denoted by ... $A_1A_2A_3$ ... in Figure 13.6) that changes direction at the points ..., $A_1, A_2, A_3,$ ... where the impulses towards S are received.

13.3.1 Explain why the lengths of the segments $A_1A_2, A_2A_3, \ldots$ are proportional to the speeds at which the planet travels along these segments.

Now suppose that the planet receives an impulse towards S at point A_2, changing its velocity by an amount proportional to length A_2B and in the direction of A_2B. Because of this impulse, the planet arrives at point A_3 at the end of the next time interval, instead of at the point A_3' that it would have reached had it continued with unchanged velocity.

13.3.2 Explain why $BA_2A_3'A_3$ is a parallelogram.

13.3.3 Deduce from Exercise 13.3.2 that the triangles SA_1A_2 and SA_2A_3' have equal area.

13.3.4 Deduce from Exercise 13.3.3 that the triangles SA_1A_2 and SA_2A_3 have equal area.

13.3.5 Now let the time interval between impulses tend to zero. Conclude that, when a planet P moves smoothly under a continuous force from S, the line SP sweeps out area at a constant rate.

13.4   Celestial Mechanics

Kepler was extremely lucky with his first law for two reasons. First, the system of one planet and the sun is exceptional in having a familiar curve (the ellipse) as the orbit. Second, if the observations at Kepler's disposal had been more accurate—so that the perturbing effects of other planets and moons were detectable—then he would have found that the orbits of the planets were not ellipses after all, but more complicated curves.

When Newton derived orbits from the inverse square law of gravitation in the *Principia*, Book I, Section III, he showed that there was a deeper level of explanation—the infinitesimal level—where simplicity could be attained even when it was not possible at the global level. The force on a given body B_1 is simply the vector sum of the forces due to the other bodies $B_2, \ldots, B_n$ in the system, determined by their masses and distances from B_1 by the inverse square law. By Newton's second law, this determines the acceleration of B_1. The accelerations of $B_2, \ldots, B_n$ are similarly determined, hence the behavior of the system is completely determined by the inverse square law, once initial positions and velocities are given. The inverse square law is an infinitesimal law in the sense that it describes the limiting behavior of a body—its acceleration—and not its global behavior such as the shape or period of its orbit.

As we now know, it is rarely possible to describe the global behavior of a dynamical system explicitly. So, by directing attention to infinitesimal behavior, Newton found the only viable basis for dynamics. Unfortunately, he communicated this insight poorly (for today's audience) by expressing it in geometrical terms, perhaps in the belief that calculus did not belong in a serious publication. By the 18th century this belief had been dispelled by Leibniz and his followers, and definitive formulations of dynamics in terms of calculus were given by Euler and Lagrange. The latter recognized that the infinitesimal behavior of a dynamical system was typically described by a system of *differential equations* and that the global behavior was derivable from these equations, in principle, by integration.

The question remained, however, whether the inverse square law did indeed account for the observed global behavior of the solar system. In a system with only two bodies, Newton (1687), p. 166, showed that each describes a conic section relative to the other—in normal cases an ellipse as stated by Kepler. With a three-body system, such as the earth–moon–sun, no simple global description was possible, and Newton could obtain only qualitative results through approximations. With the many bodies in the solar system, extremely complex behavior was possible, and for 100 years mathematicians were unable to account for some of the phenomena actually observed.

A famous example was the so-called secular variation of Jupiter and Saturn, which was detected by Halley in 1695 from the observations then available. For several centuries Jupiter had been speeding up (spiraling toward the sun) and Saturn had been slowing down (spiraling outward). The problem was to explain this behavior and to determine whether it would continue, with the eventual destruction of Jupiter and disappearance of Saturn. Euler and Lagrange worked on the problem without success; then, in the centenary year of *Principia*, Laplace (1787) succeeded in explaining the phenomenon. He showed that the secular variation was actually periodic, with Jupiter and Saturn returning to their initial positions every 929 years. Laplace viewed this as confirmation not only of the Newtonian theory but also of the stability of the solar system, though it seems that the latter is still an open question.

Laplace introduced the term "celestial mechanics" and left no doubt that the theory had arrived with his monumental *Mécanique céleste*, a work of five volumes that appeared between 1799 and 1825. In astronomy, the theory had its finest hour in 1846, with the discovery of Neptune, whose position had been computed by Adams and Leverrier from observed perturbations in the orbit of Uranus. The difficult question of stability was taken up again in the three-volume *Les méthodes nouvelles de la mécanique céleste* of Poincaré (1892, 1893, 1899). In this work Poincaré directed attention toward asymptotic behavior, in a sense complementing Newton's infinitesimal view with a view toward infinity, and his methods have become highly influential in 20th-century dynamics.

A surprising phenomenon brought to light by Poincaré is what we now call *chaos*, or "sensitive dependence on initial conditions." In many dynamical systems, such as a system of three or more bodies under their mutual gravitational attraction, a small change in initial conditions can produce a large change in outcome. Thus, even though the evolution of the system may be predictable in principle, it can be unpredictable in practice because the initial conditions need to be known with infinite accuracy.

An amazing example of "unpredictable" behavior occurs in the three-body system discovered by Sitnikov (1960). Sitnikov's system consists of two "suns" that revolve periodically about each other in ellipses, while an infinitesimal "planet" oscillates on a line through the center of mass of the two suns and perpendicular to their plane. Suppose we take the "year" for this system to be the period of revolution of the "suns." Also suppose that we have a record of the "years" in which the planet has crossed the plane of the "suns," and that this record extends infinitely far into the past. Thus the record of crossings is some nondecreasing sequence of integers, say

$$\ldots, \quad -1000000, \quad -997, \quad -300, \quad -14, \quad -13, \quad -3, \quad -2, \quad -1$$

The amazing property of Sitnikov's system is that *any nondecreasing sequence of integers can be realized in this way*. And therefore, *even if the years of all past crossings are known, the year of the next crossing cannot be predicted*.

Another astonishing kind of behavior that can occur in many-body systems is *escape to infinity in finite time*. That is, in a system of n bodies, all initially at finite distance and with finite velocity, one of the bodies can accelerate so fast that it reaches infinite distance (and infinite velocity) in finite time. This was first proved by Xia (1992), for $n = 5$.

For an entertaining account of the many kinds of pathological behavior recently discovered in celestial mechanics, see Diacu and Holmes (1996).

EXERCISES

Xia's example of escape to infinity seems implausible at first, because it seems impossible for a body to acquire the infinite energy required. However, infinite velocity (and hence infinite kinetic energy) can be acquired even in the system of two point bodies moving on a straight line. The catch is that the infinite velocity is acquired at the moment when the two bodies collide. We suppose that the two bodies lie on the x axis and that one of them is taken to be the origin.

13.4.1 Show that $\frac{d^2x}{dt^2} = \frac{d}{dx}\frac{1}{2}\left(\frac{dx}{dt}\right)^2$.

13.4.2 Deduce from Exercise 13.4.1 that, in the straight-line two-body problem, with one body at O, the other body reaches infinite speed when $x = 0$.

13.4.3 Also show (without finding a formula for distance) that the body reaches $x = 0$ in finite time.

13.5 Mechanical Curves

When Descartes gave his reasons for restricting *La Géométrie* to algebraic curves (which he called "geometric"; see Section 7.3), he explicitly excluded certain classical curves on the rather vague grounds that they

belong only to mechanics, and are not among those curves
that I think should be included here, since they must be conceived of as described by two separate movements whose relation does not admit of exact determination.

                                                Descartes (1637), p. 44

The curves that Descartes relegated "to mechanics" were those the
Greeks had defined by certain hypothetical mechanisms, for example, the
epicycles (described by rolling one circle on another) and the spiral of
Archimedes (described by a point moving at constant speed along a uniformly rotating line). He was probably aware that the spiral is transcendental by virtue of the fact that it meets a straight line in infinitely many points.
This is contrary to the behavior of an algebraic curve $p(x, y) = 0$, which
meets a straight line $y = mx + c$ in only finitely many points, corresponding
to the finitely many solutions of $p(x, mx + c) = 0$. This proof that there
are transcendental curves was given explicitly by Newton (1687), Lemma
XXVIII.

We do not know whether Descartes distinguished, say, the algebraic
epicycles from the transcendental ones; nevertheless, it is broadly true that
his "mechanical" curves were transcendental. This remained true with the
great expansion of mechanics and calculus in the 17th century, and indeed
most of the new transcendental curves originated in mechanics. In this
section we shall look at three of the most important of them: the catenary,
the cycloid, and the elastica.

The *catenary* is the shape of a hanging cord, assumed to be perfectly
flexible and with mass uniformly distributed along its length. In practice,
the flexibility and uniformity of mass are realized better by a hanging chain,
hence the name "catenary," which comes from the Latin *catena* for chain.
Hooke (1675) observed that the same curve occurs as the shape of an arch
of infinitesimal stones. The catenary looks very much like a parabola and
was at first conjectured to be one by Galileo. This was disproved by the
17-year-old Huygens (1646), though at the time Huygens was unable to
determine the correct curve. He did show, however, that the parabola was
the shape assumed by a flexible cord loaded by weights that are uniformly
distributed in the horizontal direction (as is approximately the case for the
cable of a suspension bridge).

The problem of the catenary was finally solved independently by Johann Bernoulli (1691), Huygens (1691), and Leibniz (1691), in response

to a challenge from Jakob Bernoulli in 1690. Johann Bernoulli showed that the curve satisfies the differential equation

$$\frac{dy}{dx} = \frac{s}{a},$$

where a is constant and s = arc length OP (Figure 13.7).

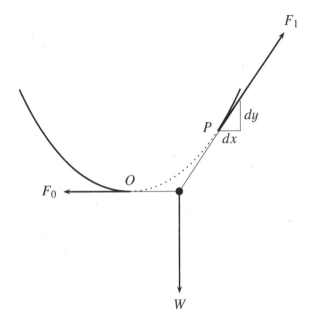

Figure 13.7: The catenary

He derived this equation by replacing the portion OP of the chain, which is held in equilibrium by the tangential force F_1 at P and the horizontal force F_0, which is independent of P, by a point mass W equal to the weight of OP (hence proportional to s) held in equilibrium by the same forces. Comparing the directions and magnitudes of the forces gives

$$\frac{dy}{dx} = \frac{W}{F_0} = \frac{s}{a}.$$

By ingenious transformations Bernoulli reduced the equation to

$$dx = \frac{a\,dy}{\sqrt{y^2 - a^2}},$$

in other words, to an integral. This solution was as simple as could be stated at the time, since x is a transcendental function of y and hence can be expressed, at best, as an integral. Today, of course, we recognize the function as one of the "standard" ones and abbreviate the solution as

$$y = a \cosh \frac{x}{a} - a.$$

The *cycloid* is the curve generated by a point on the circumference of a circle rolling on a straight line. Despite being a natural limiting case in the epicyclic family, the cycloid does not seem to have been investigated until the 17th century, when it became a favorite curve with mathematicians. It has many beautiful geometric properties, and even more remarkable mechanical properties. The first of these, discovered by Huygens (1659b), is that the cycloid is the *tautochrone* (equal-time curve). A particle constrained to slide along an inverted cycloid takes the same time to descend to the lowest point, regardless of its starting point.

Huygens (1673) made a classic application of this property to pendulum clocks, using a geometric property of the cycloid (Huygens, 1659c). If the pendulum, taken to be a weightless cord with a point mass at the end, is constrained to swing between two cycloidal "cheeks," as Huygens called them (Figure 13.8), then the point mass will travel along a cycloid.

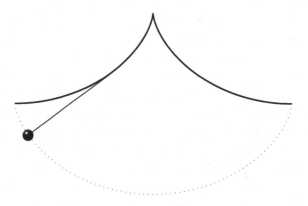

Figure 13.8: The cycloidal pendulum

Consequently, the period of the cycloidal pendulum is independent of amplitude. This makes it theoretically superior to the ordinary pendulum whose period, though approximately constant for small amplitudes, actually involves an elliptic function. In practice, problems such as friction make the cycloidal pendulum no more accurate than the ordinary pendulum, but its theoretical superiority shut the ordinary pendulum out of mechanics for some time. Newton's *Principia*, for example, often mentions the cycloidal pendulum but never the simple pendulum.

The second remarkable property of the cycloid is that it is the *brachistochrone*, the curve of shortest time. Johann Bernoulli (1696) posed the problem of finding the curve, between given points A and B, along which a point mass descends in the shortest time. He already knew that the solution was a cycloid, and solutions were found independently by Jakob Bernoulli (1697), l'Hôpital (1697), Leibniz (1697), and Newton (1697). The problem is deeper than that of the tautochrone, because the cycloid has to be singled out from *all possible* curves between A and B. Jakob Bernoulli's solution was the most profound because it recognized the "variable curve" aspect of the problem, and it is now considered to be the first major step in the development of the calculus of variations.

The *elastica* was another of Jakob Bernoulli's discoveries, and likewise important in the development of another field—the theory of elliptic functions. The elastica is the curve assumed by a thin elastic rod compressed at the ends. Jakob Bernoulli (1694) showed that the curve satisfies a differential equation that he reduced to the form

$$ds = \frac{dx}{\sqrt{1 - x^4}}.$$

To interpret this integral geometrically, he introduced the lemniscate and showed that its arc length is expressed by precisely the same integral. This was the beginning of the investigations of the lemniscatic integral, which included the important discoveries of Fagnano and Gauss mentioned in the last chapter. Euler's investigations of elliptic integrals were also stimulated by the elastica. Euler (1743) gave pictures of elasticas that show they have periodic forms (Figure 13.9). These drawings were the first to show the real period of elliptic functions, though of course periodicity was implicit in the first elliptic integral, the arc length of the ellipse (the real period being the circumference of the ellipse).

EXERCISES

The derivation of the cosh function from the catenary equation is helped by a tricky formula for $\frac{d^2y}{dx^2}$, based on Exercise 13.4.1, which you should verify first if it is not familiar to you.

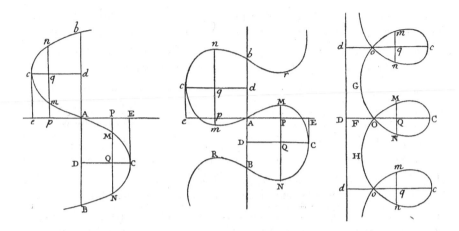

Figure 13.9: Forms of the elastica

13.5.1 Use

$$ds = \sqrt{dx^2 + dy^2} \quad \text{and} \quad \frac{d^2y}{dx^2} = \frac{d}{dy}\frac{1}{2}\left(\frac{dy}{dx}\right)^2$$

to transform the differential equation

$$\frac{dy}{dx} = \frac{s}{a}$$

to

$$\frac{dx}{dz} = \frac{a}{\sqrt{1 + z^2}}, \tag{1}$$

where $z = dy/dx$.

13.5.2 Solve (1) for x and hence show that the original equation has solution

$$y = a\cosh\frac{x}{a} + \text{const.}$$

It is considerably easier to solve the suspension bridge equation, which perhaps is why Huygens was able to do it at age 17, and before much calculus was known.

13.5.3 How should the formula $\frac{dy}{dx} = \frac{s}{a}$ be modified if the load is uniformly distributed in the horizontal direction (as in a suspension bridge)?

13.5.4 Solve the modified equation from Exercise 13.5.3, and hence show that the solution is a parabola.

Finally, we can verify that the catenary is indeed a transcendental curve.

13.5.5 Show that the functions sin and cos, and hence the functions sinh and cosh, are transcendental. *Hint:* You may need to use complex numbers.

13.6 The Vibrating String

The problem of the vibrating string is one of the most fertile in mathematics, being the source of such diverse fields as partial differential equations, Fourier series, and set theory. It is also remarkable in being perhaps the only setting in which the sense of hearing led to important mathematical discoveries. As we saw in Section 1.5, the Pythagoreans discovered the relationship between pitch and length by hearing the harmonious tones produced by two strings whose lengths were in a simple whole-number ratio. Thus in a sense it was possible to "hear the length of the string," and some later discoveries of mathematically significant properties of the strings—overtones, for example—were initially prompted by hearing. See Dostrovsky (1975).

Various authors in ancient times suggested that the physical basis of pitch was frequency of vibration, but it was not until the 17th century that the precise relationship between frequency and length was discovered, by Descartes's mentor Isaac Beeckman. In 1615 Beeckman gave a simple geometric argument to show that frequency is inversely proportional to length; hence the Pythagorean ratios of lengths can also be interpreted as (reciprocal) ratios of frequencies. The latter interpretation is more fundamental because frequency alone determines pitch, whereas length determines pitch only when the material, cross section, and tension of the string are fixed. The relation between frequency v, and tension T, cross-sectional area A, and length l was discovered experimentally by Mersenne (1625) to be

$$v \propto \frac{1}{l} \sqrt{\frac{T}{A}} .$$

The first derivation of Mersenne's law from mathematical assumptions was given by Taylor (1713), in a paper that marks the beginning of the modern theory of the vibrating string. In it he discovered the simplest possibility for the instantaneous shape of the string, the half sine wave

$$y = k \sin \frac{\pi x}{l} ,$$

and established generally that the force on an element is proportional to d^2y/dx^2.

The latter result was the starting point for a dramatic advance in the theory by d'Alembert (1747). Taking into account the dependence of y on time t as well as x, d'Alembert realized that acceleration should be expressed by $\partial^2 y / \partial t^2$ and the force found by Taylor by $\partial^2 y / \partial x^2$, so that *partial* derivatives are involved. Newton's second law then gives what is now called the *wave equation*,

$$\frac{\partial^2 y}{\partial x^2} = \frac{1}{c^2} \frac{\partial^2 y}{\partial t^2},$$

writing the constant of proportionality as $1/c^2$. Undeterred by the novelty of this partial differential equation, d'Alembert forged ahead to a general solution as follows. The equation may be simplified by a change of time scale $s = ct$ to

$$\frac{\partial^2 y}{\partial x^2} = \frac{\partial^2 y}{\partial s^2}. \tag{1}$$

The chain rule gives

$$d\left(\frac{\partial y}{\partial x} \pm \frac{\partial y}{\partial s}\right) = \frac{\partial^2 y}{\partial x^2} dx + \frac{\partial^2 y}{\partial x \partial s}(ds \pm dx) \pm \frac{\partial^2 y}{\partial s^2} ds$$

$$= \left(\frac{\partial^2 y}{\partial s^2} \pm \frac{\partial^2 y}{\partial x \partial s}\right)(ds \pm dx),$$

from which d'Alembert concluded that

$$\frac{\partial^2 y}{\partial s^2} + \frac{\partial^2 y}{\partial x \partial s}$$

is a function of $s + x$ and

$$\frac{\partial^2 y}{\partial s^2} - \frac{\partial^2 y}{\partial x \partial s}$$

is a function of $s - x$, whence, say,

$$\frac{\partial y}{\partial x} + \frac{\partial y}{\partial s} = \int \left(\frac{\partial^2 y}{\partial s^2} + \frac{\partial^2 y}{\partial x \partial s}\right) d(s + x) = f(s + x)$$

and similarly

$$\frac{\partial y}{\partial x} - \frac{\partial y}{\partial s} = g(s - x).$$

This gives

$$\frac{\partial y}{\partial x} = \frac{1}{2}\left(f(s + x) + g(s - x)\right), \qquad \frac{\partial y}{\partial s} = \frac{1}{2}\left(f(s + x) - g(s - x)\right),$$

and finally

$$y = \int \left(\frac{\partial y}{\partial x} dx + \frac{\partial y}{\partial s} ds \right)$$
$$= \int \frac{1}{2} \left(f(s+x)(ds+dx) - g(s-x)(ds-dx) \right)$$
$$= \Phi(s+x) + \Psi(s-x).$$

Reversing the argument, we see that the functions Φ and Ψ can be arbitrary, at least as long as they admit the various differentiations involved.

But how arbitrary *is* an arbitrary function? Is it as arbitrary as an arbitrarily shaped string? The vibrating-string problem caught 18th-century mathematicians unprepared to answer these questions. They had understood a function to be something expressed by a formula, possibly an infinite series, and this had been thought to guarantee differentiability. Yet the most natural shape of the vibrating string was one with a nondifferentiable point—the triangle of the plucked string as it is released—so nature seemed to demand an extension of the concept of function beyond the world of formulas.

The confusion was heightened when Daniel Bernoulli (1753) claimed, on physical grounds, that a general solution of the wave equation *could* be expressed by a formula, the infinite trigonometric series

$$y = a_1 \sin \frac{\pi x}{l} \cos \frac{\pi ct}{l} + a_2 \sin \frac{2\pi x}{l} \cos \frac{2\pi ct}{l} + \cdots .$$

This amounts to claiming that any mode of vibration results from the superposition of simple modes, a fact he considered to be intuitively evident. The nth term in the series represents the nth mode, generalizing Taylor's formula for the fundamental mode and building in the time dependence; but Daniel Bernoulli gave no method for calculating the coefficient a_n.

We now know that his intuition was correct and that the triangular wave form, among others, is representable by a trigonometric series. However, it was well into the 19th century before anything like a clear understanding of trigonometric series was obtained. The fact that the triangular wave could be represented by a series made it a bona fide function by classical standards; hence mathematicians were brought to the realization that a series representation does not guarantee differentiability. Later, continuity was also called into question, and infinitely subtle problems concerning the convergence of trigonometric series led Cantor to develop the theory of sets (see Chapter 24).

These remarkably remote consequences of what seemed at first to be a purely physical question were of course not the only fruits of the vibrating string investigations. Trigonometric series proved to be valuable all over mathematics, from the theory of heat, where Fourier applied them with such success that they became known as *Fourier series*, to the theory of numbers. Their most famous application to number theory is probably the Dirichlet (1837) proof that any arithmetic progression $a, a + b, a + 2b, \ldots$, where $\gcd(a, b) = 1$, contains infinitely many primes. Pythagoras would surely have approved!

EXERCISES

The simplest heat equation is the one-dimensional version,

$$\frac{\partial T}{\partial t} = \kappa \frac{\partial^2 T}{\partial x^2},$$

for the temperature T at time t and position x along an infinite straight wire. This equation may be derived from *Newton's law of cooling*, which asserts that the rate of heat flow between two points is proportional to their temperature difference.

Thus the approximate difference $\frac{\partial T}{\partial x} dx$ between T at x and $x + dx$ will induce heat to flow from $x + dx$ to x at a rate proportional to $\frac{\partial T}{\partial x} dx$. However, at the same time, heat will flow from $x - dx$ to x at approximately the same rate. To find the net flow toward x, and hence the rate $\frac{\partial T}{\partial t}$ of temperature increase, we need to take into account the rate of change of $\frac{\partial T}{\partial x}$, namely $\frac{\partial^2 T}{\partial x^2}$.

13.6.1 By pursuing this line of argument, give a plausible derivation of the heat equation

$$\frac{\partial T}{\partial t} = \kappa \frac{\partial^2 T}{\partial x^2}.$$

Sines and cosines arise from the heat equation when one solves it by the method of *separation of variables*.

13.6.2 Suppose the heat equation has a solution of the form $T(x, t) = X(x)Y(t)$, where X and Y are functions of the single variables x and t, respectively. Show that

$$\frac{1}{Y(t)} \frac{dY(t)}{dt} = \frac{\kappa}{X(x)} \frac{d^2 X(x)}{dx^2} = \text{constant}.$$

13.6.3 Now explain how sines and cosines are involved in solving for $X(x)$.

13.7   Hydrodynamics

The properties of fluid flow have been investigated since ancient times, initially in connection with practical questions such as water supply and water-powered machinery. However, nothing like a mathematical theory was obtained before the Renaissance, and until the advent of calculus it was possible to deal only with fairly coarse macroscopic quantities such as the average speed of emission from an opening in a container. Newton (1687), Book II, introduced infinitesimal methods into the study of fluids, but much of his reasoning is incomplete, based on inappropriate mathematical models, or simply wrong. As late as 1738, when the field of hydrodynamics finally got its name in the *Hydrodynamica* of Daniel Bernoulli, the basic infinitesimal laws of fluid motion had still not been discovered.

The first important law was discovered by Clairaut (1740), in a context that was essentially static. Clairaut was interested in a burning question of the time, the shape (or "figure") of the earth. Newton had argued that the earth must bulge somewhat at the equator as a result of its spin. Natural as this seems now (and indeed then, since the phenomenon was clearly observable in Jupiter and Saturn), it was opposed by the anti-Newtonian Cassini, who argued for a spindle-shaped earth, elongated toward the poles. Clairaut actually took part in an expedition to Lapland that confirmed Newton's conjecture by measurement, but he also attacked the problem theoretically by studying the conditions for the equilibrium of a fluid mass.

He considered the vector field of force acting on the fluid and observed that it must be what we now call a *conservative*, or *potential* field. That is, the integral of the force around any closed path must be zero; otherwise the fluid would circulate. Equivalently, the integral between any two points must be independent of the path. In the special two-dimensional case where there are components P, Q of force in the x and y directions, the quantity to be integrated is $P\,dx + Q\,dy$.

Clairaut argued that for the integral to be path-independent, this quantity must be a complete differential

$$df = \frac{\partial f}{\partial x}dx + \frac{\partial f}{\partial y}dy.$$

Consequently, $P = \partial f/\partial x$, $Q = \partial f/\partial y$ and P, Q satisfy the condition

$$\frac{\partial P}{\partial y} = \frac{\partial Q}{\partial x}. \tag{1}$$

This condition is indeed necessary, but the existence of the potential f involved more mathematical subtleties than could have been foreseen at the time. Clairaut derived the corresponding equations for the components P, Q, R in the physically more natural three-dimensional case and went as far as studying the equipotential surfaces f = constant. He also found a satisfying solution to the problem of the figure of the earth. When the force at a point is the resultant of gravity and the rotational force, then an ellipsoid of revolution is an equilibrium figure, with the axis of rotation being the shorter axis of the ellipse (Clairaut (1743), p. 194).

The two-dimensional equation (1), despite being physically special if not unnatural, turned out to have a deep mathematical significance. This was discovered in the dynamic situation, with P, Q taken to be components of velocity rather than force. In this case, (1) still holds when the flow is independent and irrotational as d'Alembert (1752) showed by an argument similar to Clairaut's. The crucial additional fact that now emerges is that P, Q satisfy a second relation,

$$\frac{\partial P}{\partial x} + \frac{\partial Q}{\partial y} = 0, \tag{2}$$

derived by d'Alembert as a consequence of the incompressibility of the fluid. He considered an infinitesimal rectangle of fluid with corners at the points (x, y), $(x+dx, y)$, $(x, y+dy)$, $(x+dx, y+dy)$, and the parallelogram into which it is carried in an infinitesimal time interval by the known velocities (P, Q), $(P + (\partial P/\partial x)dx, Q + (\partial Q/\partial x)dx)$, ... Equating areas of the two parallelograms leads to (2). In three dimensions one similarly gets

$$\frac{\partial P}{\partial x} + \frac{\partial Q}{\partial y} + \frac{\partial R}{\partial z} = 0,$$

but the significance of (1) and (2), as d'Alembert discovered, is that they can be combined into a single fact about the complex function $P + iQ$. This flash of inspiration became the basis for the theory of complex functions developed in the 19th century by Cauchy and Riemann (see Section 16.1).

EXERCISES

To understand the concept of irrotational flow more directly, it helps to consider a flow that is clearly *rotational*, for example a rigid rotation of the plane about the origin at constant angular velocity ω.

13.7.1 For this flow, show that the velocity components at the point (x, y) are

$$P = -\omega y, \quad Q = \omega x,$$

and deduce that $\frac{\partial P}{\partial y} - \frac{\partial Q}{\partial x} = -2\omega$.

Thus the quantity $\frac{\partial P}{\partial y} - \frac{\partial Q}{\partial x}$ is a measure of the amount of rotation of the flow. It is, in fact, sometimes called the "rotation" but it is more often called the *curl*, a term James Clerk Maxwell introduced in 1870.

The quantity $\frac{\partial P}{\partial x} + \frac{\partial Q}{\partial y}$ is called the *divergence* because it measures the amount of "expansion" of the fluid. As one would expect, the divergence is zero for the rigid flow above.

13.7.2 Check that the divergence is zero for the rigid rotation about the origin.

A more direct way to see that divergence is zero for any incompressible flow in the plane is to consider a *fixed* rectangle, with fluid flowing through it.

Consider the rectangle with corners fixed in the plane at (x, y), $(x + dx, y)$, $(x, y + dy)$, $(x + dx, y + dy)$, and consider the instantaneous flux of fluid through it. Fluid flows in the x end at speed P, so the influx is proportional to $P\, dy$, and it flows out the $x + dx$ end at speed $P + (\partial P/\partial x)\, dx$, etc.

13.7.3 Show that the net influx of fluid is

$$-\left(\frac{\partial P}{\partial x} + \frac{\partial Q}{\partial y} \right) dx\, dy,$$

and hence that the divergence is zero for incompressible flow.

13.7.4 Show similarly that

$$\frac{\partial P}{\partial x} + \frac{\partial Q}{\partial y} + \frac{\partial R}{\partial z} = 0$$

for an incompressible flow in three dimensions.

13.8   Biographical Notes: The Bernoullis

Undoubtedly the most outstanding family in the history of mathematics was the Bernoulli family of Basel, which included at least eight excellent mathematicians between 1650 and 1800. Three of these, the brothers Jakob (1654–1705) and Johann (1667–1754) and Johann's son Daniel (1700–1782), were among the great mathematicians of all time, as one may guess from their contributions already mentioned in this chapter. In fact, all the mathematicians Bernoulli were important in the history of mechanics. One can trace their influence in this field in Szabó (1977), which also contains portraits of most of them, and in Truesdell (1954, 1960). However, Jakob, Johann, and Daniel are of interest from a wider point of view, in mathematics, as well as in their personal lives. The Bernoulli family, with all its mathematical talent, also had more than its share of arrogance and jealousy, which turned brother against brother and father against son. In three successive generations, fathers tried to steer their sons into non-mathematical careers, only to see them gravitate back to mathematics. The fiercest conflict occurred among Jakob, Johann, and Daniel.

Figure 13.10: Portrait of Jakob Bernoulli by Nicholas Bernoulli

Jakob, the first mathematician in the family, was the oldest son of Nicholas Bernoulli, a successful pharmacist and civic leader in Basel, and Margaretha Schönauer, the daughter of another wealthy pharmacist. There were three other sons: Nicholas, who became an artist and in 1686 painted the portrait of Jakob seen here (Figure 13.10); Johann; and Hieronymus, who took over the family business. Their father's wish was that Jakob should study theology, which he initially did, obtaining his licentiate in 1676. However, Jakob also began to teach himself mathematics and astronomy, and he traveled to France in 1677 to study with the followers of Descartes. In 1681 his astronomy brought him into conflict with the theologians. Inspired by the appearance of a great comet in 1680, he published a pamphlet that proposed laws governing the behavior of comets and claiming that their appearances could be predicted. His theory was not actually correct (this was six years before *Principia*), but it certainly clashed with the theology of the time, which exploited the unexpectedness of comets in claiming they were signs of divine displeasure. Jakob decided that his future was in mathematics rather than theology, and he adopted the motto *Invito Patre, Sidera verso* (Against my father's will, I will turn to the stars). He made a second study tour, to the Netherlands and England, where he met Hooke and Boyle, and began to lecture on mechanics in Basel in 1683.

He married Judith Stepanus in 1684, and they eventually had a son and daughter, neither of whom became a mathematician. In a sense, the mathematical heir of Jakob was his nephew Nicholas (son of the painter), who carried on one of Jakob's most original lines of research, probability theory. He arranged for the posthumous publication of Jakob's book on the subject, the *Ars conjectandi* (1713), which contains the first proof of a law of large numbers. Jakob Bernoulli's law described the behavior of long sequences of trials for which a positive outcome has a fixed probability p (such trials are now called Bernoulli trials). In a precise sense, the proportion of successful trials will be "close" to p for "almost all" sequences.

In 1687 Jakob became professor of mathematics in Basel and, together with Johann (whom he had been secretly teaching mathematics), set about mastering the new methods of calculus that were then appearing in the papers of Leibniz. This proved to be difficult, perhaps more for Jakob than Johann, but by the 1690s the brothers equaled Leibniz himself in the brilliance of their discoveries. Jakob, the self-taught mathematician, was the slower but more penetrating of the two. He sought to get to the bottom of every problem, whereas Johann was content with any solution, the quicker the better.

Johann was the tenth child of the family, and his father intended him to have a business career. When his lack of aptitude for business became clear, he was allowed to enter the University of Basel in 1683 and became a master of arts in 1685. During this time he also attended his brother's lectures and, as mentioned earlier, learned mathematics from him privately. Their rivalry did not come to the surface until the catenary contest of 1690, but Jakob may have felt uneasy about his younger brother's talent as early as 1685. In that year he persuaded Johann to take up the study of medicine, making the highly optimistic forecast that it offered great opportunities for the application of mathematics. Johann went into medicine quite seriously, obtaining a licentiate in 1690 and a doctorate in 1695, but by that time he was more famous as a mathematician. With the help of Huygens he gained the chair of mathematics in Groningen, and thus became free to concentrate on his true calling.

The great applications of mathematics to medicine did not eventuate, though Johann Bernoulli did make an amusing application of geometric series that still circulates today as a piece of physiological trivia. In his *De nutritione*, Johann Bernoulli (1699) used the assumption that a fixed proportion of bodily substance, homogeneously distributed, is lost each day and replaced by nutrition to calculate that almost all the material in the body would be renewed in three years. This result provoked a serious theological dispute at the time, since it implied the impossibility of resurrecting the body from all its past substance.

Johann Bernoulli made several important contributions to calculus in the 1690s, outside mechanics. One was the first textbook in the subject, the *Analyse des infiniment petits*. This was published under the name of his student, the Marquis l'Hôpital (1696), apparently in return for generous financial compensation. Another contribution, made jointly with Leibniz, was the technique of partial differentiation. The two kept this discovery secret for 20 years in order to use it as a "secret weapon" in problems about families of curves (see Engelsman (1984)). Other discoveries still remain outside the territory usually explored in calculus, for example,

$$\int_0^1 x^x \, dx = 1 - \frac{1}{2^2} + \frac{1}{3^3} - \frac{1}{4^4} + \cdots .$$

This startling result of Johann Bernoulli (1697) can be proved using a suitable series expansion of x^x and integration by parts (see exercises).

The rivalry between Jakob and Johann turned to open hostility in 1697 over the *isoperimetric problem*, the problem of finding the curve of given length that encloses the greatest area. Jakob correctly recognized that this was a calculus of variations problem but withheld his solution, whereas Johann persisted in publicizing an incorrect solution and claiming that Jakob had no solution at all. Jakob presented his solution to the Paris Academy in 1701, but it somehow remained in a sealed envelope until after his death. Even when the solution was made public in 1706, Johann refused to admit his own error or the superiority of Jakob's analysis.

Johann was married to Dorothea Falkner, the daughter of a parliamentary deputy in Basel, and through his father-in-law's influence was awarded the chair of Greek in Basel in 1705. This enabled him to return to Basel from Groningen, but his real goal was the chair of mathematics, not Greek. Jakob was then in ill health, and his last days were embittered by the belief that Johann was plotting to take his place, using the Greek offer as a stepping stone. This is precisely what happened, for when Jakob died in 1705 Johann became the professor of mathematics.

With the death of Jakob and the virtual retirement of Leibniz and Newton, Johann enjoyed about 20 years as the leading mathematician in the world. He was particularly proud of his successful defense of Leibniz against the supporters of Newton:

> When in England war was declared against M. Leibniz for the honour of the first invention of the new calculus of the infinitely small, I was despite my wishes involved in it; I was pressed to take part. After the death of M. Leibniz the contest fell to me alone. A crowd of English antagonists fell upon my body. It was my lot to meet the attacks of Messrs Keil, Taylor, Pemberton, Robins and others. In short I alone like the famous Horatio Cocles kept at bay at the bridge the entire English army.
>
> Translation by Pearson (1978), p. 235

His portrait from this era shows the Bernoulli arrogance at its peak (Figure 13.11).

Johann Bernoulli finally met his match at the hands of his own pupil Euler in 1727. There was no open warfare, just a polite exchange of correspondence on the logarithms of negative numbers, but it revealed that Johann Bernoulli understood some of his own results less well than Euler did. Johann Bernoulli persisted in his stubborn misunderstanding for another 20 years, while Euler went on to develop his brilliant theory of complex logarithms and exponentials (see Section 16.1). Johann Bernoulli seems not to have minded his pupil's success at all; instead, he became consumed with jealousy over the success of his son Daniel.

Figure 13.11: Johann Bernoulli

Daniel Bernoulli (Figure 13.12) was the middle of Johann's three sons, all of whom became mathematicians. The oldest, Nicholas (called Nicholas II by historians to distinguish him from the first mathematician Nicholas), died of a fever in St. Petersburg in 1725 at the age of 30. The youngest, Johann II, was the least distinguished of the three, but he fathered the next generation of Bernoulli mathematicians, Jakob II and Johann III.

Daniel's path to mathematics was very similar to his father's. During his teens he was tutored by his older brother; his father wanted him to go into business, but when that career failed Daniel was permitted to study medicine.

He gained his doctorate in 1721 and made several attempts to win the chair of anatomy and botany in Basel, finally succeeding in 1733. By that

Figure 13.12: Daniel Bernoulli

time, however, he had drifted into mathematics, with such success that he had been called to the St. Petersburg Academy. During his years there (1725–1733) he conceived his ideas on modes of vibration and produced the first draft of his *Hydrodynamica*. Although he missed finding the basic partial differential equations of hydrodynamics, the *Hydrodynamica* made other important advances. One was the systematic use of a principle of conservation of energy; another was the kinetic theory of gases, including the derivation of Boyle's law that is now standard.

Unfortunately, publication of the *Hydrodynamica* was delayed until 1738. This left Daniel's priority open to attack, and the one to take advantage of him was his own father. The self-styled Horatius of the priority dispute between Leibniz and Newton attempted the most brazen priority theft in the history of mathematics by publishing a book on hydrodynamics in 1743 and dating it 1732. Daniel was devastated, and wrote to Euler:

> Of my entire *Hydrodynamics*, not one iota of which do in fact I owe to my father, I am all at once robbed completely and lose thus in one moment the fruits of the work of ten years. All propositions are taken from my *Hydrodynamics*, and then my father calls his writings *Hydraulics, now for the first time disclosed*, 1732, since my *Hydrodynamics* was printed only in 1738.
>
> Daniel Bernoulli (1743), in the Truesdell (1960) translation

The situation was not quite as clear-cut as Daniel claimed (a detailed assessment is in Truesdell (1960)), but at any rate Johann Bernoulli's move backfired. His reputation was so tarnished by the episode that he did not even receive credit for parts of his work that *were* original. Daniel went on to enjoy fame and a long career, becoming professor of physics in 1750 and lecturing to enthusiastic audiences until 1776.

EXERCISES

13.8.1 Use integration by parts to show that

$$\int_0^1 x^n (\log x)^n \, dx = \frac{(-1)^n n!}{(n+1)^{n+1}}.$$

13.8.2 Deduce that

$$\int_0^1 x^x \, dx = 1 - \frac{1}{2^2} + \frac{1}{3^3} - \frac{1}{4^4} + \cdots$$

using a series expansion of $x^x = e^{x \log x}$.

14

Complex Numbers in Algebra

PREVIEW

The next three chapters revisit the topics of algebra, curves, and functions, observing how they are simplified by the introduction of complex numbers. That's right: the so-called "complex" numbers actually make things simpler.

In the present chapter we see where complex numbers came from (not from quadratic equations, as you might expect, but from *cubic* equations) and observe how they simplify the study of polynomial equations. Equations become simpler because they always have solutions in the complex numbers, and it follows that they have the "right" number of solutions.

One of the reasons for the simplifying power of complex numbers is their two-dimensional nature. The extra dimension gives more room for solutions of equations to exist. For example, the equation $x^n = 1$, which has only one or two solutions in the real numbers, has n different solutions in the complex numbers, equally spaced around the unit circle.

More generally, complex numbers give a way to divide any angle into n equal parts. This comes about because *multiplication* of complex numbers involves *addition* of angles, and is related to the famous *de Moivre formula* in trigonometry.

The equation $x^n = 1$ is not the only one with the "right" number of solutions in the complex numbers. In fact, any equation of degree n has n complex solutions, when solutions are properly counted. This is the *fundamental theorem of algebra*, and it follows from intuitively simple properties of the plane and continuous functions.

J. Stillwell, *Mathematics and Its History*, Undergraduate Texts in Mathematics,
DOI 10.1007/978-1-4419-6053-5_14, © Springer Science+Business Media, LLC 2010

14.1 Impossible Numbers

Over the last few chapters it has often been claimed that certain mysteries—
de Moivre's formula for $\sin n\theta$ (Section 6.6), factorization of polynomials
(Section 6.7), classification of cubic curves (Section 8.4), branch points
(Section 10.5), genus (Section 11.3), and behavior of elliptic functions
(Sections 11.6 and 12.6)—are clarified by the introduction of complex
numbers. That complex numbers do all this and more is one of the miracles of mathematics. At the beginning of their history, complex numbers $a + b\sqrt{-1}$ were considered to be "impossible numbers," tolerated
only in a limited algebraic domain because they seemed useful in the solution of cubic equations. But their significance turned out to be geometric
and ultimately led to the unification of algebraic functions with conformal
mapping, potential theory, and another "impossible" field, non-Euclidean
geometry. This resolution of the paradox of $\sqrt{-1}$ was so powerful, unexpected, and beautiful that only the word "miracle" seems adequate to
describe it.

In the present chapter we shall see how complex numbers emerged
from the theory of equations and enabled its fundamental theorem to be
proved—at which point it became clear that complex numbers had meaning
far beyond algebra. Their impact on curves and function theory, which is
where conformal mapping and potential theory come in, is described in
Chapters 15 and 16. Non-Euclidean geometry had entirely different origins
but arrived at the same place as function theory in the 1880s, thanks to
complex numbers. This unexpected meeting is described in Chapter 18,
after some geometric preparations in Chapter 17.

14.2 Quadratic Equations

The usual way to introduce complex numbers in a mathematics course
is to point out that they are needed to solve certain quadratic equations,
such as the equation $x^2 + 1 = 0$. However, this did not happen when
quadratic equations first appeared, since at that time there was no *need* for
all quadratic equations to have solutions. Many quadratic equations are implicit in Greek geometry, as one would expect when circles, parabolas, and
the like are being investigated, but one does not demand that every geometric problem have a solution. If one asks whether a particular circle and
line intersect, say, then the answer can be yes or no. If yes, the quadratic

equation for the intersection has a solution; if no, it has no solution. An "imaginary solution" is uncalled for in this context.

Even when quadratic equations appeared in algebraic form, with Diophantus and the Arab mathematicians, there was initially no reason to admit complex solutions. One still wanted to know only whether there were real solutions, and if not the answer was simply—no solution. This is plainly the appropriate answer when quadratics are solved by geometrically completing the square (Section 6.3), as was still done up to the time of Cardano. A square of negative area did not exist in geometry. The story might have been different had mathematicians used symbols more and dared to consider the symbol $\sqrt{-1}$ as an object in its own right, but this did not happen until quadratics had been overtaken by cubics, at which stage complex numbers became unavoidable, as we shall now see.

14.3   Cubic Equations

The del Ferro–Tartaglia–Cardano solution of the cubic equation

$$y^3 = py + q$$

is

$$y = \sqrt[3]{\frac{q}{2} + \sqrt{\left(\frac{q}{2}\right)^2 - \left(\frac{p}{3}\right)^3}} + \sqrt[3]{\frac{q}{2} - \sqrt{\left(\frac{q}{2}\right)^2 - \left(\frac{p}{3}\right)^3}}$$

as we saw in Section 6.5. The formula involves complex numbers when $(q/2)^2 - (p/3)^3 < 0$. However, it is not possible to dismiss this as a case with no solution, *because a cubic always has at least one real root* (since $y^3 - py - q$ is positive for sufficiently large positive y and negative for sufficiently large negative y). Thus the Cardano formula raises the problem of reconciling a real value, found by inspection, say, with an expression of the form

$$\sqrt[3]{a + b\sqrt{-1}} + \sqrt[3]{a - b\sqrt{-1}}$$

Cardano did not face up to this problem in his *Ars magna* (1545). He did, it is true, once mention complex numbers, but in connection with a quadratic equation and accompanied by the comment that these numbers were "as subtle as they are useless" (Cardano (1545), Ch. 37, Rule II).

The first to take complex numbers seriously and use them to achieve the necessary reconciliation was Bombelli (1572). Bombelli worked out the formal algebra of complex numbers, with the particular aim of reducing expressions $\sqrt[3]{a + b\sqrt{-1}}$ to the form $c + d\sqrt{-1}$. His method enabled him to show the reality of some expressions resulting from Cardano's formula. For example, the solution of

$$x^3 = 15x + 4$$

is

$$x = \sqrt[3]{2 + 11\sqrt{-1}} + \sqrt[3]{2 - 11\sqrt{-1}}$$

according to the formula. On the other hand, inspection gives the solution $x = 4$. Bombelli had the hunch that the two parts of x in the Cardano formula were of the form $2 + n\sqrt{-1}$, $2 - n\sqrt{-1}$, and he found by cubing these expressions formally [using $(\sqrt{-1})^2 = -1$] that indeed

$$\sqrt[3]{2 + 11\sqrt{-1}} = 2 + \sqrt{-1},$$

$$\sqrt[3]{2 - 11\sqrt{-1}} = 2 - \sqrt{-1},$$

hence the Cardano formula also gives $x = 4$.

Figure 14.1 is a facsimile of the manuscript page on which Bombelli stated his result. It is not hard to pick out the preceding expressions when one allows for the notation and the fact that $11\sqrt{-1}$ is written as $\sqrt{0 - 121}$.

Much later, Hölder (1896) showed that any algebraic formula for the solution of the cubic must involve square roots of quantities that become negative for particular values of the coefficients. A proof of Hölder's result may be found in van der Waerden (1949), p. 180.

EXERCISES

14.3.1 Check that $(2 + \sqrt{-1})^3 = 2 + 11\sqrt{-1}$.

It is possible to work backwards and concoct a cubic equation with an "obvious" solution that can be reconciled with the hideous solution in the Cardano formula. Here is an example.

14.3.2 Check that $(3 + \sqrt{-1})^3 = 18 + 26\sqrt{-1}$.

14.3.3 Hence explain why

$$6 = (3 + \sqrt{-1}) + (3 - \sqrt{-1}) = \sqrt[3]{18 + 26\sqrt{-1}} + \sqrt[3]{18 - 26\sqrt{-1}}.$$

Figure 14.1: Bombelli's manuscript

14.3.4 Find p and q such that

$$18 = \frac{q}{2} \quad \text{and} \quad 26\sqrt{-1} = \sqrt{\left(\frac{q}{2}\right)^2 - \left(\frac{p}{3}\right)^3}.$$

14.3.5 Check that 6 is a solution of the equation $x^3 = px + q$ for the values of p and q found in Exercise 14.3.4.

14.4   Wallis's Attempt at Geometric Representation

Despite Bombelli's successful use of complex numbers, most mathematicians regarded them as impossible, and of course even today we call them "imaginary" and use the symbol i for the imaginary unit $\sqrt{-1}$. The first attempt to give complex numbers a concrete interpretation was made by

Wallis (1673). This attempt was unsatisfactory, as we shall see, but nevertheless an interesting "near miss." Wallis wanted to give a geometric interpretation to the roots of the quadratic equation, which we shall write as

$$x^2 + 2bx + c^2 = 0, \qquad b, c \geq 0.$$

The roots are

$$x = -b \pm \sqrt{b^2 - c^2}$$

and hence real when $b \geq c$. In this case the roots can be represented by points P_1, P_2 on the real number line that are determined by the geometric construction in Figure 14.2. When $b < c$, lines of length b attached to Q are too short to reach the number line, so the points P_1, P_2 "cannot be had in the line," and Wallis seeks them "out of that line ... (in the same Plain)." He is on the right track, but he arrives at unsuitable positions for P_1, P_2 by sticking too closely to his first construction.

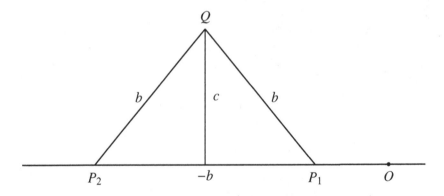

Figure 14.2: Wallis's construction of real roots

Figure 14.3 compares his representation of $P_1, P_2 = -b \pm i\sqrt{c^2 - b^2}$ when $b < c$ with the modern representation. Apparently Wallis thought $+$ and $-$ should continue to correspond to "right" and "left," though this has the unacceptable consequence that $i = -i$ (let $b \to 0$ in his representation). This was an understandable oversight, since in Wallis's time even negative numbers were still under suspicion, and there was confusion about the meaning of $(-1) \times (-1)$, for example. Confusion was compounded by the introduction of square roots, and as late as 1770 Euler gave a "proof" in his *Algebra* that $\sqrt{-2} \times \sqrt{-3} = \sqrt{6}$ (Euler (1770b), p. 43).

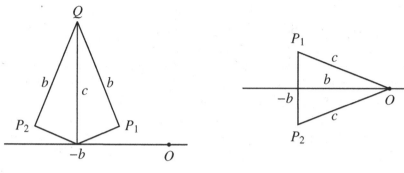

Wallis's representation   Modern representation

Figure 14.3: Construction of complex roots

EXERCISES

The claim that $\sqrt{-2} \times \sqrt{-3} = \sqrt{6}$ is wrong only if one uses the convention that $\sqrt{6}$ means the *positive* square root of 6, as we normally do today. It is not unreasonable to let $\sqrt{6}$ denote the *pair* $\pm \sqrt{6}$ of square roots of 6, in which case Euler's claim is correct.

14.4.1 Supposing that $\sqrt{-2}$ denotes the pair of square roots of -2, that $\sqrt{-3}$ denotes the pair of square roots of -3, and that $\sqrt{-2} \times \sqrt{-3}$ denotes all possible products, show that

$$\sqrt{-2} \times \sqrt{-3} = \sqrt{6}.$$

14.4.2 Is it also true (as in the usual interpretation) that

$$\sqrt{-2} \times \sqrt{-3} = -\sqrt{6}?$$

14.5   Angle Division

In Section 6.6 we saw how Viète related angle trisection to the solution of cubic equations, and how Leibniz (1675) and de Moivre (1707) solved the angle n-section equation by the Cardano-type formula

$$x = \frac{1}{2}\sqrt[n]{y + \sqrt{y^2 - 1}} + \frac{1}{2}\sqrt[n]{y - \sqrt{y^2 - 1}}. \tag{1}$$

We also saw how this and Viète's formulas for $\cos n\theta$ and $\sin n\theta$ could easily be explained by the formula

$$(\cos\theta + i\sin\theta)^n = \cos n\theta + i\sin n\theta \tag{2}$$

usually associated with de Moivre. Actually, de Moivre never stated (2) explicitly. The closest he came was to give a formula for $(\cos\theta + i\sin\theta)^{1/n}$ in de Moivre (1730). (See Smith (1959) for a series of extracts from the work of de Moivre on angle division). It seems that the clues in the algebra of circular functions were not strong enough to reveal (2) until a deeper reason for it had been brought to light by calculus.

Complex numbers made their entry into the theory of circular functions in a paper on integration by Johann Bernoulli (1702). Observing that $\sqrt{-1} = i$ makes possible the partial fraction decomposition

$$\frac{1}{1+z^2} = \frac{1/2}{1+zi} + \frac{1/2}{1-zi},$$

Bernoulli saw that integration would give an expression for $\tan^{-1}z$ as an imaginary logarithm, though he did not write down the expression in question and was evidently puzzled as to what it could mean. In Section 16.1 we shall see how Euler clarified Johann Bernoulli's discovery and developed it into the beautiful theory of complex logarithms and exponentials. What is relevant here is that Johann Bernoulli (1712) took up the idea again, and this time he carried out the integration to obtain an algebraic relation between $\tan n\theta$ and $\tan\theta$. His argument is as follows. Given

$$y = \tan n\theta, \qquad x = \tan\theta,$$

we have

$$n\theta = \tan^{-1}y = n\tan^{-1}x;$$

hence, taking differentials gives

$$\frac{dy}{1+y^2} = \frac{n\,dx}{1+x^2},$$

or

$$dy\left(\frac{1}{y+i} - \frac{1}{y-i}\right) = n\,dx\left(\frac{1}{x+i} - \frac{1}{x-i}\right).$$

Integration gives

$$\log(y+i) - \log(y-i) = n\log(x+i) - n\log(x-i),$$

that is,

$$\log\frac{y+i}{y-i} = \log\left(\frac{x+i}{x-i}\right)^n,$$

whence

$$(x - i)^n(y + i) = (x + i)^n(y - i). \tag{3}$$

This formula was the first of the de Moivre type actually to use i explicitly and the first example of a phenomenon later articulated by Hadamard: the shortest route between two truths in the real domain sometimes passes through the complex domain. Solving (3) for y as a function of x expresses $\tan n\theta$ as a rational function of $\tan \theta$, which is difficult to obtain using real formulas alone. In fact, it is easy to show from (3) that y is the quotient of the polynomials consisting of alternate terms in $(x + 1)^n$, provided with alternate $+$ and $-$ signs (see exercises).

During the 18th century, mathematicians were ambivalent about $\sqrt{-1}$. They were willing to use it en route to results about real numbers but doubted whether it had a concrete meaning of its own. Cotes (1714) even used $a + \sqrt{-1}b$ to represent the point (a, b) in the plane (as Euler did later), apparently without noticing that (a, b) was a valid *interpretation* of $a + \sqrt{-1}b$. Since results about $\sqrt{-1}$ were suspect, they were often left unstated when it was possible to state an equivalent result about reals. This may explain why de Moivre stated (1) but not (2). Another example of the avoidance of results about $\sqrt{-1}$ is the remarkable theorem on the regular n-gon discovered by Cotes in 1716 and published posthumously in Cotes (1722):

If $A_0, \ldots, A_{n-1}$ are equally spaced points on the unit circle with center O, and if P is a point on OA_0 such that $OP = x$, then (Figure 14.4)

$$PA_0 \cdot PA_1 \cdots PA_{n-1} = 1 - x^n.$$

This theorem not only relates the regular n-gon to the polynomial $x^n - 1$ but in fact geometrically realizes the *factorization of $x^n - 1$ into real linear and quadratic factors.* By symmetry one has $PA_1 = PA_{n-1}, \ldots,$ so

$$PA_0 \cdot PA_1 \cdots PA_{n-1} = \begin{cases} PA_0 \cdot PA_1^2 \cdot PA_2^2 \cdots PA_{(n-1)/2}^2 & n \text{ odd,} \\ PA_0 \cdot PA_1^2 \cdot PA_2^2 \cdots PA_{n/2-1}^2 PA_{n/2} & n \text{ even.} \end{cases}$$

$PA_0 = 1 - x$ is a real linear factor, as is $PA_{n/2}$ when n is even, and it follows from the cosine rule in triangle OPA_k that

$$PA_k^2 = 1 - 2x\cos\frac{2k\pi}{n} + x^2.$$

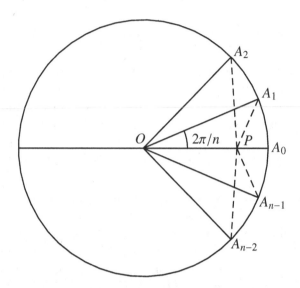

Figure 14.4: Cotes's theorem

The easiest route from here to the theorem is by splitting PA_k^2 into complex linear factors and using de Moivre's theorem, though we can only speculate that this was Cotes's method, since he stated the theorem without proof. There is a second half to Cotes's theorem, which similarly decomposes $1 + x^n$ into real linear and quadratic factors. These factorizations were needed to integrate $1/(1 \pm x^n)$ by resolution into partial fractions, which was in fact Cotes's main objective. Such problems were then high on the mathematical agenda, and they motivated subsequent research into the factorization of polynomials, in particular the first attempts to prove the fundamental theorem of algebra.

EXERCISES

Johann Bernoulli's formula relating $y = \tan n\theta$ to $x = \tan \theta$ is false for some values of n, because it neglects a possible constant of integration. The result of integration should be

$$\log(y + i) - \log(y - i) = n \log(x + i) - n \log(x - i) + C,$$

for some C, leading to

$$\frac{y + i}{y - i} = D\frac{(x + i)^n}{(x - i)^n}, \qquad (*)$$

for some constant D (equal to e^C). Sometimes $D = 1$ gives the correct formula, but sometimes we need $D = -1$.

14.5.1 Show that $D = 1$ gives the correct formula when $n = 1$.

14.5.2 Using formulas for $\sin 2\theta$ and $\cos 2\theta$, or otherwise, show that

$$\tan 2\theta = \frac{2 \tan \theta}{1 - \tan^2 \theta},$$

and check that this follows from (*) for $D = -1$, but not for $D = 1$.

14.5.3 Use the formula in Exercise 14.5.2 to express $\tan 4\theta$ in terms of $\tan 2\theta$, and hence in terms of $\tan \theta$.

14.5.4 Letting $y = \tan 4\theta$ and $x = \tan \theta$, express the result of Exercise 14.5.3 as

$$y = \frac{4x - 4x^3}{x^4 - 6x^2 + 1},$$

and check that this follows from (*) when $D = -1$.

14.6   The Fundamental Theorem of Algebra

The fundamental theorem of algebra is the statement that every polynomial equation $p(z) = 0$ has a solution in the complex numbers. As Descartes observed (Section 6.7), a solution $z = a$ implies that $p(z)$ has a factor $z - a$. The quotient $q(z) = p(z)/(z - a)$ is then a polynomial of lower degree; hence if every polynomial equation has a solution, we can also extract a factor from $q(z)$, and if $p(z)$ has degree n, we can go on to factorize $p(z)$ into n linear factors. The existence of such a factorization is of course another way to state the fundamental theorem.

Initially, interest was confined to polynomials $p(z)$ with real coefficients, and in this case d'Alembert (1746) observed that if $z = u + iv$ is a solution of $p(z) = 0$, then so is its conjugate $\bar{z} = u - iv$. Thus the imaginary linear factors of a real $p(z)$ can always be combined in pairs to form real quadratic factors:

$$(z - u - iv)(z - u + iv) = z^2 - 2uz + (u^2 + v^2).$$

This gave another equivalent of the fundamental theorem: each (real) polynomial $p(z)$ can be expressed as a product of real linear and quadratic factors. The theorem was usually stated in this way during the 18th century, when its main purpose was to make possible the integration of rational functions (see Section 14.5). This also avoided mention of $\sqrt{-1}$.

It has often been said that attempts to prove the fundamental theorem began with d'Alembert (1746), and that the first satisfactory proof was given by Gauss (1799).   This opinion should not be accepted without question, since the source of it is Gauss himself. Gauss (1799) gave a critique of proofs from d'Alembert on, showing that they all had serious weaknesses, then offered a proof of his own. His intention was to convince readers that the new proof was the first valid one, even though it used one unproved assumption (which is discussed further in the next section). The opinion as to which of two incomplete proofs is more convincing can of course change with time, and I believe that Gauss (1799) might be judged differently today. We can now fill the gaps in d'Alembert (1746) by appeal to standard methods and theorems, whereas there is still no easy way to fill the gap in Gauss (1799).

Both proofs depend on the geometric properties of the complex numbers and the concept of continuity for their completion. The basic geometrical insight—that the complex number $x + iy$ can be identified with the point (x, y) in the plane—mysteriously eluded all mathematicians until the end of the 18th century. This was one of the reasons that d'Alembert's proof was unclear, and the use of this insight by Argand (1806) was an important step in d'Alembert's reinstatement. Gauss seems to have had the same insight but concealed its role in his proof, perhaps believing that his contemporaries were not ready to view the complex numbers as a plane.

As for the concept of continuity, neither Gauss nor d'Alembert understood it very well. Gauss (1799) seriously understated the difficulties involved in the unproved step, claiming that "no one, to my knowledge, has ever doubted it. But if anybody desires it, then on another occasion I intend to give a demonstration which will leave no doubt" (translation from Struik (1969), p. 121). Perhaps to preempt criticism, he gave a second proof, Gauss (1816), in which the role of continuity was minimized. The second proof is purely algebraic except for the use of a special case of the intermediate value theorem. Gauss assumed that a polynomial function $p(x)$ of a real variable x takes all values between $p(a)$ and $p(b)$ as x runs from a to b. The first to appreciate the importance of continuity for the fundamental theorem of algebra was Bolzano (1817), who proved the continuity of polynomial functions and attempted a proof of the intermediate value theorem. The latter proof was unsatisfactory because Bolzano had no clear concept of real number on which to base it, but it did point in the right direction. When a definition of real numbers emerged in the 1870s (for example, with Dedekind cuts; Section 4.2), Weierstrass (1874) rigorously established the basic properties of continuous functions, such as the intermediate value theorem and extreme value theorem. This completed not only the second proof of Gauss but also the proof of d'Alembert, as we shall see in the next section.

EXERCISES

Complex roots of an equation with real coefficients occur in conjugate pairs because of the fundamental properties of conjugates.

14.6.1 Show directly from the definition $\overline{u + iv} = u - iv$ that

$$\overline{z_1 + z_2} = \overline{z_1} + \overline{z_2} \quad \text{and} \quad \overline{z_1 \cdot z_2} = \overline{z_1} \cdot \overline{z_2}$$

for any complex numbers z_1, z_2.

14.6.2 Deduce from Exercise 14.6.1 that $p(\overline{z}) = \overline{p(z)}$ for any polynomial $p(z)$ with real coefficients, and hence that the complex roots of $p(z) = 0$ occur in conjugate pairs.

14.7   The Proofs of d'Alembert and Gauss

The key to d'Alembert's proof is a proposition now known as *d'Alembert's lemma*: if $p(z)$ is a nonconstant polynomial function and $p(z_0) \neq 0$, then any neighborhood of z_0 contains a point z_1 such that $|p(z_1)| < |p(z_0)|$.

The proof of this lemma offered by d'Alembert depended on solving the equation $w = p(z)$ for z as a fractional power series in w. As mentioned in Section 9.5, such a solution was claimed by Newton (1671), but it was made clear and rigorous only by Puiseux (1850). Thus d'Alembert's argument did not stand on solid ground, and in any case it was unnecessarily complicated.

A simple elementary proof of d'Alembert's lemma was given by Argand (1806). Argand was one of the co-discoverers of the geometric representation of complex numbers (probably the first was Wessel (1797), but his work remained almost unknown for 100 years), and he offered the following proof as an illustration of the effectiveness of the representation.

The value of $p(z_0) = x_0 + iy_0$ is interpreted as the point (x_0, y_0) in the plane, so that $|p(z_0)|$ is the distance of (x_0, y_0) from the origin. We wish to find a Δz such that $p(z_0 + \Delta z)$ is nearer to the origin than $p(z_0)$. If

$$p(z) = a_0 z^n + a_1 z^{n-1} + \cdots + a_n,$$

then

$$p(z_0 + \Delta z) = a_0(z_0 + \Delta z)^n + a_1(z_0 + \Delta z)^{n-1} + \cdots + a_n$$
$$= a_0 z_0^n + a_1 z_0^{n-1} + \cdots + a_n + A_1 \Delta z + A_2(\Delta z)^2 + \cdots + A_n(\Delta z)^n$$

for some constants A_i depending on z_0, not all zero,

because p is not constant

$$= p(z_0) + A \Delta z + \varepsilon,$$

where $A = A_i(\Delta z)^i$ contains the first nonzero A_i and $|\varepsilon|$ is small compared with $|A\Delta z|$ when $|\Delta z|$ is small (because ε contains higher powers of Δz). It is then clear (Figure 14.5) that by choosing the direction of Δz so that $A\Delta z$ is opposite in direction to $p(z_0)$, we get $|p(z_0 + \Delta z)| < |p(z_0)|$. This completes the proof of d'Alembert's lemma.

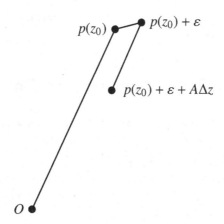

Figure 14.5: Construction for d'Alembert's lemma

To complete the proof of the fundamental theorem of algebra, take an arbitrary polynomial p and consider the continuous function $|p(z)|$. Since $p(z) \approx a_0 z^n$ for $|z|$ large, $|p(z)|$ increases with $|z|$ outside a sufficiently large circle $|z| = R$. We now get a z for which $|p(z)| = 0$ from the extreme value theorem of Weierstrass (1874); a continuous function on a closed bounded set assumes maximum and minimum values. By this theorem, $|p(z)|$ takes a minimum value for $|z| \leq R$. The minimum is ≥ 0 by definition, and if it is > 0 we get a contradiction by d'Alembert's lemma: either a point z with $|z| \leq R$ where $|p(z)|$ takes a value less than its minimum or a point z with $|z| > R$ where $|p(z)|$ is less than its values on $|z| = R$. Thus there is a point z where $|p(z)| = 0$ and hence $p(z) = 0$.

The proof of Gauss also used the fact that $p(z)$ behaves like its highest-degree term $a_0 z^n$ for $|z|$ large and likewise relied on a continuity argument to show that $p(z) = 0$ inside some circle $|z| = R$. Gauss considered the real and imaginary parts of $p(z)$, $\text{Re}[p(z)]$ and $\text{Im}[p(z)]$, and investigated the curves

$$\text{Re}[p(z)] = 0 \quad \text{and} \quad \text{Im}[p(z)] = 0.$$

(These are easily seen to be algebraic curves $p_1(x, y) = 0$ and $p_2(x, y) = 0$ by expanding the powers $z^k = (x + iy)^k$ and collecting real and imaginary terms.) His aim was to find a point where these curves meet, because at such a point

$$0 = \text{Re}[p(z)] = \text{Im}[p(z)] = p(z).$$

For $|z|$ large, the curves are close to the curves $\text{Re}(a_0 z^n)$ and $\text{Im}(a_0 z^n) = 0$, which are families of straight lines through the origin. Moreover, the lines where $\text{Re}(a_0 z^n) = 0$ *alternate* with those where $\text{Im}(a_0 z^n) = 0$ as one makes a circuit around the origin. For example, Figure 14.6 shows $\text{Re}(z^2) = 0$ and $\text{Im}(z^2) = 0$ as alternate solid and dashed lines. It follows that the curves $\text{Re}[p(z)] = 0$ and $\text{Im}[p(z)] = 0$ meet a sufficiently large circle $|z| = R$ *alternately*. Up to this point the argument is comparable to d'Alembert's lemma, and it can be made just as rigorous.

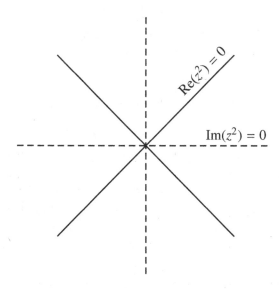

Figure 14.6: Lines for the Gauss proof

To complete this proof we have to show that the curves meet inside the circle, and this is the step Gauss thought nobody could doubt. He assumed that the separate pieces of the algebraic curve $\text{Re}[p(z)] = 0$ outside the circle $|z| = R$ would join inside the circle, as would the separate pieces of $\text{Im}[p(z)] = 0$. Then, since the pieces of $\text{Re}[p(z)] = 0$ alternate with those of $\text{Im}[p(z)]$ on $|z| = R$, it would be "patently absurd" for their connecting pieces inside the circle not to meet. One has only to visualize a situation like that seen in Figure 14.7 to feel sure that Gauss was right. However, the existence of the connecting pieces is extremely hard to prove (and proving that they meet is not trivial either, being at least as hard as the intermediate value theorem). The first proof was given by Ostrowski (1920).

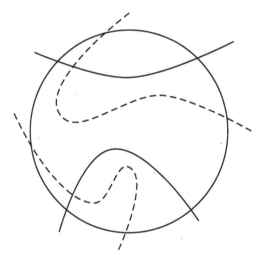

Figure 14.7: Curves for the Gauss proof

From our present perspective, d'Alembert's route to the fundamental theorem of algebra seems basically easy because it proceeds through general properties of continuous functions. The route of Gauss, although appearing equally easy from a distance, goes through the still-unfamiliar territory of real algebraic curves. The intersections of real algebraic curves are harder to understand than the intersections of complex algebraic curves, and in retrospect they are harder to understand than the fundamental theorem of algebra. Indeed, as we shall see in the next chapter, the fundamental theorem gives us Bézout's theorem, which in turn settles the problem of counting the intersections of complex algebraic curves.

The expression in d'Alembert's lemma for $p(z_0 + \Delta z)$ is an instance of *Taylor's series*, previously discussed in Section 10.2. When the function is a polynomial p, as here, its Taylor series is finite because p has only finitely many nonzero derivatives.

14.7.1 Show that $A_1 = na_0 z_0^{n-1} + (n-1)a_1 z_0^{n-2} + \cdots + a_{n-1}$ and that the latter expression is $p'(z_0)$.

14.7.2 Show that $A_2 = \frac{n(n-1)}{2} a_0 z_0^{n-2} + \frac{(n-1)(n-2)}{2} a_1 z_0^{n-3} + \cdots + a_{n-2}$ and that the latter expression is $p''(z_0)/2$.

14.7.3 Using the binomial theorem, show that $A_k = p^{(k)}(z_0)/k!$, and hence that

$$p(z_0 + \Delta z) = a_0 z_0^n + a_1 z_0^{n-1} + \cdots + a_n + A_1 \Delta z + A_2 (\Delta z)^2 + \cdots + A_n (\Delta z)^n$$

is an instance of the Taylor series formula.

14.8   Biographical Notes: d'Alembert

Jean le Rond d'Alembert (Figure 14.8) was born in Paris in 1717 and died there in 1783. He was the illegitimate son of the Chevalier Destouches-Canon, a cavalry officer, and salon hostess Madame de Tencin. His mother abandoned him at birth near the church of St. Jean-le-Rond in the cloisters of Notre Dame, and so he was christened Jean le Rond, following the custom for foundlings. He was subsequently located by his father, who found a home for him with a glazier named Rousseau and his wife. The name d'Alembert came later, for reasons that are unclear.

The Rousseaus must have been devoted foster parents, for d'Alembert lived with them until 1765. He received an annuity from his father, who also arranged for him to be educated at the Jansenist Collège de Quatre-Nations in Paris. There he received a good grounding in mathematics and developed a permanent distaste for theology. After brief studies in law and medicine he turned to mathematics in 1739.

In that year he began sending communications to the Académie des Sciences, and his ambition and talent rapidly carried him to fame. He became a member of the Académie in 1741 and published his best-known work, the *Traité de dynamique*, in 1743. Having struggled to the top from humble beginnings, d'Alembert did not want to lose his position. Once in the Académie, his struggle was to stay ahead of his rivals. Whether by accident or inborn competitiveness, d'Alembert always seemed to be working on the same problems as other top mathematicians—initially Clairaut,

Figure 14.8: Jean Baptiste le Rond d'Alembert

later Daniel Bernoulli and Euler. He was always fearful of losing priority and fell into a cycle of hasty publication followed by controversy over
the meaning and significance of his work. Despite the fact that he was an
excellent writer (elected to the Académie Française in 1754), his mathematics was almost always poorly presented. Many of his best ideas were
not understood until Euler overhauled them and gave them masterly expo-
sitions. Since Euler often did this without giving credit, d'Alembert was
understandably furious, but he squandered his energy in quarreling instead
of giving his own work the exposition it deserved.

Another reason for d'Alembert's lack of attention to his mathematics was his involvement in the broader intellectual life of his time. When

d'Alembert came on the scene in the 1740s, mathematics was enjoying great prestige in philosophical circles, largely because of Newton's success in explaining the motions of the planets. Mathematics was a model of rational inquiry that, it was hoped, would allow the proper organization of all knowledge and the proper conduct of all human affairs. The movement to reorganize knowledge and conduct along rational lines became known as the Enlightenment, and it was particularly strong in France, where philosophers saw it as a means to overthrow existing institutions, particularly the Church. Around 1745, d'Alembert became immersed in the ferment of the Enlightenment, then bubbling in the salons and cafés of Paris. He made friends with the leading lights—Diderot, Condillac, Rousseau—and was in demand at the most fashionable salons for his wit and gift for mimicry.

The Enlightenment was not all talk, however, and one of its most solid achievements was the 17-volume *Encyclopédie*, edited by Diderot between 1745 and 1772. D'Alembert wrote the introduction to the *Encyclopédie*, the *Discours préliminaire*, and in it summed up his views on the unity of all knowledge. It contributed greatly to the success of the project, and was the main reason for his election to the Académie Française. D'Alembert was also scientific editor and wrote many of the mathematics articles. Eventually a split developed among the encyclopedists between the extreme materialists, led by Diderot, and the more moderate faction of Voltaire. Diderot leaned toward biology, for which he conjectured an absurd pseudomathematical basis, while deploring the "impracticality" of ordinary mathematics. D'Alembert sided with Voltaire and cut his ties with the *Encyclopédie* in 1758.

Nevertheless, intellectual fashion was moving away from mathematics, and in the 1760s d'Alembert found himself with only one philosopher friend still interested in it, the probability theorist Condorcet. At about this time, d'Alembert met the one love of his life, Julie de Lespinasse. Julie was the cousin of Madame du Deffand, whose salon d'Alembert attended. After a quarrel over poaching the salon's members, Julie set up a salon of her own, with d'Alembert's help. When Julie became ill with smallpox, d'Alembert nursed her back to health; when he himself fell sick, she persuaded him to move in with her. This was in 1765, when he finally left his foster home. For the next ten years his life revolved around Julie's salon, and her death in 1776 came as a cruel blow. Humiliation was added to sorrow when he discovered from her letters that she had been passionately involved with other men throughout their time together.

D'Alembert spent his last seven years in a small apartment in the Louvre, to which he was entitled as permanent secretary of the Académie Française. He found himself unable to work in mathematics, although it was the only thing that still interested him, and he became gloomy about the future of mathematics itself. Despite his gloom, he did what he could to support and encourage young mathematicians. Perhaps the finest achievement of d'Alembert's later years was to launch the careers of Lagrange and Laplace, whose work in mechanics ultimately completed much of his own. It must have given him some satisfaction to anticipate the future successes of his gifted protégés, even though they effectively ended the theory of mechanics as d'Alembert knew it. What he could not have anticipated was that a minor element of his work, the use of complex numbers, would blossom in the next century (see Sections 16.1 and 16.2) and that mathematics would break out of the bounds set by 18th-century thinking.

15

Complex Numbers and Curves

PREVIEW

The fundamental theorem of algebra—that a polynomial of degree k has exactly k complex roots—enables us to get the "right" number of intersections between a curve of degree m and a curve of degree n. However, it is not enough to introduce complex coordinates: getting the right count of intersections also requires us to adjust our viewpoint in two other ways.

1. We must count intersections according to their *multiplicity*, which amounts to counting a root $x = r$ of a polynomial equation $p(x) = 0$ as many times as the factor $(x - r)$ occurs in $p(x)$.

2. We must view curves *projectively*, so that intersections at infinity are included.

For these reasons, and others, algebraic geometry moved to the setting of complex projective space in the 19th century. In this chapter we see how this viewpoint affects our picture of algebraic curves.

The simplest such curve is the *complex projective line*, which turns out to look like a sphere. Other algebraic curves also look like surfaces, but they can be more complicated than the sphere.

It was discovered by Riemann that *rational curves* (curves that can be parameterized by rational functions) are essentially the same as the sphere, but nonrational curves have "holes" and hence are essentially different. This discovery reveals the role of *topology* in the study of algebraic curves.

J. Stillwell, *Mathematics and Its History*, Undergraduate Texts in Mathematics,   295
DOI 10.1007/978-1-4419-6053-5_15, © Springer Science+Business Media, LLC 2010

15.1   Roots and Intersections

There is a close connection between intersections of algebraic curves and roots of polynomial equations, going back as far as Menaechmus's construction of $\sqrt[3]{2}$ (a root of the equation $x^3 = 2$) by intersecting a parabola and a hyperbola (Section 2.4). The most direct connection, of course, occurs in the case of a polynomial curve

$$y = p(x) \tag{1}$$

whose intersections with the axis $y = 0$ are just the real roots of the equation

$$p(x) = 0. \tag{2}$$

If (2) has k real roots, then the curve (1) has k intersections with the axis $y = 0$. Here we must count intersections the same way we count roots, according to *multiplicity*. A root r of (2) has multiplicity μ if the factor $(x - r)$ occurs μ times in $p(x)$, and the root r is then counted μ times.

This way of counting is also geometrically natural because if, for example, the curve $y = p(x)$ meets the axis $y = 0$ with multiplicity 2 at 0, then a line $y = \varepsilon x$ "close" to the axis meets the curve twice—once near the intersection with the axis and once precisely there. The intersection of $y = x^2$ with $y = 0$ (Figure 15.1) can therefore be considered as two *coincident points* to which the distinct intersections with $y = \varepsilon x$ tend as $\varepsilon \to 0$. Likewise, an intersection of multiplicity 3 can be explained as the limit of three distinct intersections, for example, of $y = \varepsilon x$ with $y = x^3$ (Figure 15.2)

Figure 15.1: Intersection of multiplicity 2

At first glance this idea seems to break down with multiplicity 4, since $y = \varepsilon x$ meets $y = x^4$ at only two points, $x = 0$ and $x = \sqrt[3]{\varepsilon}$. The explanation is that there are also two complex roots in this case ($\sqrt[3]{\varepsilon}$ times the two

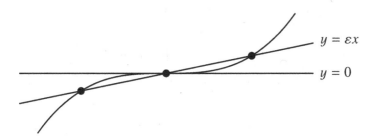

$$y = \varepsilon x$$

$$y = 0$$

Figure 15.2: Intersection of multiplicity 3

complex cube roots of 1), hence we cannot neglect complex roots if we want to get the geometrically "correct" number of intersections.

The fundamental theorem of algebra (Section 14.6) gives us n roots of an nth-degree equation (2) and consequently n intersections of the polynomial curve (1) with the axis $y = 0$. To get n roots, however, we have to admit complex values of x, hence we have to consider "curves" for which x and y are complex in order to obtain n intersections. This, and other tidy consequences of the fundamental theorem of algebra (for example, the "coincident point" interpretation of multiplicity; see Exercise 15.1.1), persuaded 18th-century mathematicians to admit complex numbers into the theory of curves before complex numbers themselves were understood— and even before the fundamental theorem of algebra was proved.

The most elegant consequence was Bézout's theorem that a curve C_m of degree m meets a curve C_n of degree n at mn points. As we saw in Section 8.6, if homogeneous coordinates are used to take account of points at infinity, then the intersections of C_m and C_n correspond to the solutions of an equation $r_{mn}(x, y) = 0$, which is homogeneous of degree mn. We can now use the fundamental theorem of algebra to show that $r_{mn}(x, y)$ is the product of mn linear factors as follows:

$$r_{mn}(x, y) = y^{mn} r_{mn}\left(\frac{x}{y}, 1\right)$$

$$= y^{mn} \prod_{i=1}^{p} \left(b_i \frac{x}{y} - a_i\right) \qquad \text{for some } p \leq mn$$

by the fundamental theorem, since $r_{mn}(x/y, 1)$ is a polynomial of degree

$p \leq mn$ in the single variable x/y. But then

$$r_{mn}(x, y) = y^{mn-p} \prod_{i=1}^{p} (b_i x - a_i y)$$

$$= \prod_{i=1}^{mn} (b_i x - a_i y)$$

since each factor y in front (if any) is trivially of the form $b_i x - a_i y$.

It follows that the equation $r_{mn}(x, y) = 0$ has mn solutions, and hence there are mn intersections of C_m and C_n, counting multiplicities.

EXERCISES

15.1.1 Show that $y = \varepsilon x$ meets $y = x^n$ in n distinct points when $\varepsilon \neq 0$, and list them (for example, with the help of de Moivre's theorem).

If a curve K has a double point at O, then a line $y = tx$ may have double contact with K at O even though nearby lines $y = (t + \varepsilon)x$ do not meet K at nearby points other than O. In this case the double contact may be explained as contact with the two branches of the curve at O.

15.1.2 Consider the lines $y = tx$ through the double point O of $y^2 = x^2(x + 1)$. Show that each such line has double contact with the curve at O, except when $t = \pm 1$. How do you account for the multiplicities when $t = \pm 1$?

15.1.3 Show that $y = tx$ also has double contact with $y^2 = x^3$ at its cusp point O. Try to explain this by viewing $y^2 = x^3$ as the result of "shrinking the loop" of $y^2 = x^2(x + \varepsilon)$ (letting $\varepsilon \to 0$).

15.1.4 Show that the line $y = tx$ has double contact at O with the lemniscate $(x^2 + y^2)^2 = x^2 - y^2$ except for two values of t, for which it has quadruple contact.

15.1.5 Explain the multiplicities found in Exercise 15.1.4 with the help of the known shape of the lemniscate (Figure 12.1).

15.2   The Complex Projective Line

We saw in Section 8.5 that adding a point at infinity to the real line $\mathbb{R}$ in $\mathbb{R} \times \mathbb{R}$ forms a closed curve that is qualitatively like a circle. Indeed, a real projective line in the sphere model of the real projective plane $\mathbb{RP}^2$ has much the same geometric properties as a great circle on a sphere, after one allows for the fact that antipodal points on the sphere are the same point on

$\mathbb{RP}^2$. The situation with the complex "line" $\mathbb{C}$ is similar but more difficult to visualize. $\mathbb{C}$ is already two-dimensional, as we saw in Gauss's proof of the fundamental theorem of algebra; hence the complex "plane" $\mathbb{C} \times \mathbb{C}$ is four-dimensional and virtually impossible to visualize.

To avoid an excursion into four-dimensional space, we first revise our approach to the real projective line. In Section 8.5 we considered ordinary lines L, in a horizontal plane not passing through the origin, and extended each to a projective line whose "points" are the lines through the origin O, in the plane through O and L. The nonhorizontal lines in this family correspond to points of L, and the horizontal line in the family to the point at infinity of L. We now use this construction again to demonstrate directly the qualitative, or more precisely *topological*, equivalence between a projective line and a circle (Figure 15.3).

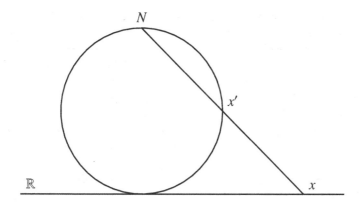

Figure 15.3: The real projective line

The origin N is taken to be the top point of a circle that, at its bottom point, touches our line $L = \mathbb{R}$. There is a continuous one-to-one correspondence between lines through N and points of the circle. Each nonhorizontal line corresponds to its intersection $x' \neq N$ with the circle, while the horizontal line corresponds to N itself. Thus the projective completion of $\mathbb{R}$, which we now call $\mathbb{RP}^1$, is *topologically the same* as the circle, in the sense that there is a continuous one-to-one correspondence between them. Moreover, we can understand projective completion of $\mathbb{R}$ topologically as a process of adding one "point" that is "approached" as one tends to infinity, in either direction, along $\mathbb{R}$, for as x tends to infinity in either direction, x' tends to the same point, N, on the circle.

We can now view projective completion of $\mathbb{C}$ in the same way using Figure 15.4, which shows the so-called *stereographic projection* of the plane $\mathbb{C}$ into a sphere. Each point $z \in \mathbb{C}$ is projected to a point z' on the tangential sphere S by the ray through z and the north pole N of S. This establishes a continuous one-to-one correspondence between points z of $\mathbb{C}$ and points $z' \neq N$ on S. Moreover, as z tends to infinity in any direction, z' tends to N; hence the projective completion of $\mathbb{CP}^1$ of $\mathbb{C}$ is topologically the same as the complete sphere S, with the point at ∞ of $\mathbb{C}$ corresponding to N.

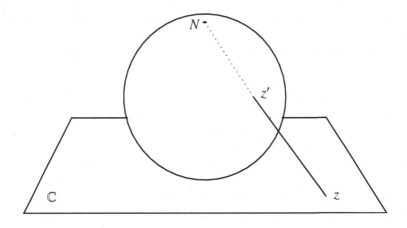

Figure 15.4: The complex projective line

Since one also wants to complete $\mathbb{C}$ by a point ∞ in this way for complex analysis, geometry and analysis are both served by passing from $\mathbb{C}$ to $\mathbb{CP}^1$. Gauss seems to have been the first to appreciate the advantages of $\mathbb{C} \cup \{\infty\}$ over $\mathbb{C}$; hence one often calls $\mathbb{CP}^1$ the *Gauss sphere* in analysis. (Unfortunately, only a few unpublished, undated fragments of Gauss's work on this topic seem to have survived; see Gauss (1819).) Algebraic geometers call $\mathbb{CP}^1$ the (complex) projective line, since it is the formal equivalent of a real line, even though it is topologically a surface. Similarly, complex curves are topologically surfaces, known to analysts as *Riemann surfaces*, though algebraic geometers prefer to call them "curves."

The "surface" viewpoint is helpful when studying intrinsic properties of complex curves. For example, *genus* (introduced in connection with parameterization in Sections 11.3 to 11.5) turns out to have a very simple meaning in the topology of surfaces (see Section 15.4). On the other hand, the "curve" viewpoint is helpful when studying intersections of curves and their embedding in $\mathbb{C} \times \mathbb{C}$ or its projective completion $\mathbb{CP}^2$. Instead of trying to imagine two planes meeting in a single point of $\mathbb{C} \times \mathbb{C}$, for example, it is better to imagine the intersection as analogous to that of real lines in a real plane—as the single solution of two linear equations. After all, we are working with $\mathbb{C}$ to remove anomalies that occur with $\mathbb{R}$, not for the sake of doing something different, and we expect that much of the behavior of real curves will recur with complex ones.

EXERCISES

Since addition and multiplication are continuous functions, it is quite easy to find one-to-one continuous maps between certain complex algebraic curves and the sphere.

15.2.1 Show that the projective completion of the curve $Y = X^2$ is topologically a sphere by considering its parameterization

$$X = t, \quad Y = t^2,$$

where t ranges over the sphere $\mathbb{C} \cup \{\infty\}$. Namely, show that the mapping $t \mapsto (t, t^2)$ is one-to-one and continuous.

15.2.2 Similarly show that the projective completion of $Y^2 = X^3$ is topologically a sphere by considering its parameterization

$$X = t^2, \quad Y = t^3$$

and the continuous mapping $t \mapsto (t^2, t^3)$.

15.2.3 Consider the mapping of the t sphere onto the projective completion of $Y^2 = X^2(X + 1)$ defined by $t \mapsto P(t)$, where $P(t)$ is the third intersection of the curve with the line $Y = tX$ through the double point (found in Exercise 7.4.2).

Show that this mapping is continuous and that it is one-to-one except at the points $t = \pm 1$, which are both mapped to the point O on the curve. Conclude that the curve is topologically the same as a sphere with two points identified (Figure 15.5).

15.3   Branch Points

The key to the topological form of a complex curve $p(x, y) = 0$ lies in its *branch points*, the points α where the Newton–Puiseux expansion of y begins with a fractional power of $(x - \alpha)$ (see Section 10.5). The nature of

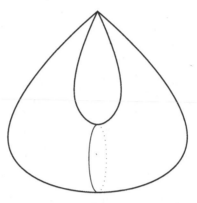

Figure 15.5: The singular sphere

branch points was first described by Riemann (1851) as part of a revolutionary new geometric theory of complex functions. Riemann's idea, one of the most illuminating in the history of mathematics, was to represent a relation $p(x, y) = 0$ between complex x and complex y by covering a plane (or sphere) representing the x variable by a surface representing the y variable, the point or points of the y surface over a given point $x = \alpha$ being those values of y that satisfy $p(\alpha, y) = 0$.

If the equation $p(\alpha, y) = 0$ is of degree n in y, there will in general be n distinct y values for a given α, consequently n "sheets" of the y surface lying over the x-plane in the neighborhood of $x = \alpha$. At finitely many exceptional values of x, sheets merge due to concidence of roots, and the Newton–Puiseux theory says that at such a point y behaves like a fractional power of x at 0. Our main problem, therefore, is to understand the behavior of the Riemann surface for $y = x^{m/n}$ in the neighborhood of 0.

The idea can be grasped sufficiently well from seeing the special case $y = x^{1/2}$. If we consider the unit disk in the y-plane and try to deform it so that the points $y = \pm \sqrt{x}$ lie above the point x in the unit disk of the x-plane, then the result is something like Figure 15.6.

The angles θ on the disk boundaries are the arguments of the corresponding points $e^{i\theta} = \cos \theta + i \sin \theta$, as we explain in Section 16.1. If

$$x = e^{i\theta} = e^{i(\theta + 2\pi)}$$

then

$$y = e^{i\theta/2}, \qquad e^{i(\theta/2 + \pi)},$$

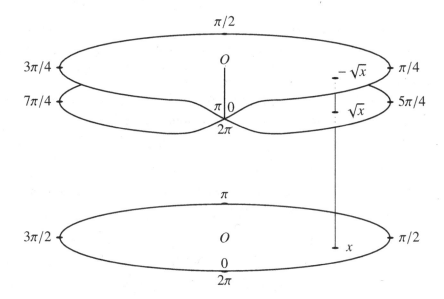

Figure 15.6: Branch point for the square root

giving the values shown. A more graphic depiction of the branch point
is seen in Figure 15.7, taken from an early textbook on Riemann's theory
(Neumann (1865), endpaper).

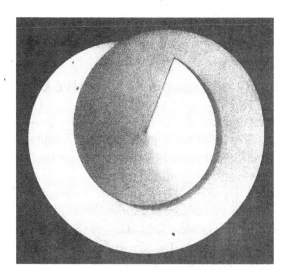

Figure 15.7: Neumann's picture of a branch point

It should be noted that the awkward appearance of the branch point, in particular the line of self-intersection, is a consequence of representing the relation $y^2 = x$ in fewer dimensions than the four it really requires. If we similarly attempt to represent the relation $y^2 = x$ between real x and y by laying the y-axis along the x-axis so that $y = \pm\sqrt{x}$ are on top of x, then the result is an awkward folded "branch point" at 0 (Figure 15.8). This is a consequence of trying to represent the relation in one dimension. In reality, as the second part of the figure shows, when viewed as a curve in the plane the relation is just as smooth at 0 as anywhere else. (Notice, incidentally, that the folded line in Figure 15.8, the real y-axis, corresponds to the self-intersection line in Figure 15.7.)

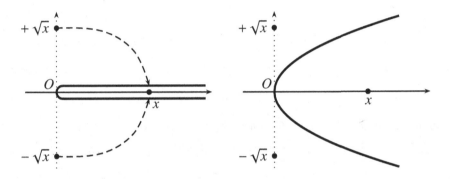

Figure 15.8: A one-dimensional branch point

15.4   Topology of Complex Projective Curves

To understand the complete structure of the complex projective curve defined by $y^2 = x$ we need to know its behavior at infinity. At ∞ there is another branch point like the one at 0 (just replace x by $1/u$ and y by $1/v$ and notice that we are looking at $v^2 = u$ near $y = 0$, $v = 0$—the same situation as before). The topological nature of the relation between x and y can then be captured by the model seen in Figure 15.9. A sphere (the x sphere) is covered by two spheres (like skins of an onion), slit along a line from 0 to ∞ and cross-joined. The slit from 0 to ∞ is arbitrary, but the cross-joining is necessary to produce the branch point structure at 0 and ∞.

The covering of the x sphere by this two-sheeted surface expresses the "covering projection map" $(x, y) \mapsto x$ from a general point on the curve

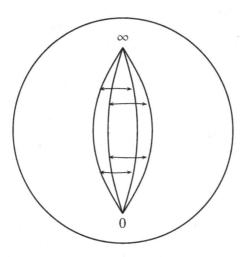

Figure 15.9: Covering the sphere

$y^2 = x$ to its x coordinate and shows that it is two-to-one except at the branch points 0, ∞. The two-sheeted surface itself captures the intrinsic topological structure of the curve, and this structure can be more readily seen by separating the two skins from the x sphere and each other, then joining the required edges (Figure 15.10). Edges to be joined are labeled by the same letters, and we see that the resulting surface is topologically a sphere.

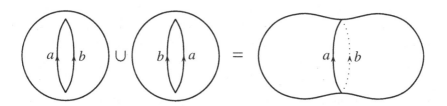

Figure 15.10: Joining the separated sheets

This result could have been obtained more directly by projecting each point (x, y) on the curve to y, since this is a one-to-one continuous map between the curve and the y-axis, which we know to be topologically a sphere (when ∞ is included). The curve here was modeled by cutting and joining sheets on the sphere because this method extends to all algebraic curves. The Newton–Puiseux theory implies that any algebraic relation $p(x, y) = 0$ can be modeled by a finite-sheeted covering of the sphere, with finitely many branch points. The most general branch point structure is given by a prescription for cross-joining (permuting) the sheets, and by slitting the sheets between branch points (or, if necessary, to an auxiliary point) they can be rejoined to produce the prescribed branching behavior.

The most interesting case of this method is the cubic curve

$$y^2 = x(x - \alpha)(x - \beta).$$

This relation defines a covering in the x sphere that is two-sheeted, since for each x there are $+$ and $-$ values for y, with branch points at 0, α, β, and ∞. (The branch point at ∞ is explained in the exercises below.) Thus if we slit the sheets from 0 to α and from β to ∞, the required joining is like that shown in Figure 15.11. We find, as Riemann did, that the surface is a torus, and hence *not* topologically the same as a sphere. This discovery proved to be a revelation for the understanding of cubic curves and elliptic functions, as we see in the next chapter.

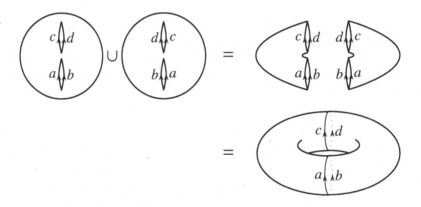

Figure 15.11: Joining the sheets of a cubic curve

One quickly sees that by considering relations of the form

$$y^2 = (x - \alpha_1)(x - \alpha_2) \cdots (x - \alpha_{2n})$$

it is possible to obtain Riemann surfaces of the form shown in Figure 15.12. These surfaces are distinguished topologically from each other by the number of "holes": 0 for the sphere, 1 for the torus, and so on. This simple topological invariant turns out to be the *genus*, which also determines the type of functions that can parameterize the corresponding complex curve. Other geometric and analytic properties of genus will unfold over the next few chapters. The topological importance of genus was established by Möbius (1863), when he showed that any closed surface in ordinary space is topologically equivalent to one of the form seen in Figure 15.12.

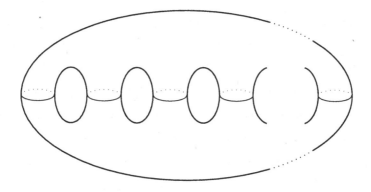

Figure 15.12: A general Riemann surface

EXERCISES

We can transfer the "one-dimensional branch point" (Figure 15.8) to infinity to see the topology of the real projective curve $y^2 = x$.

15.4.1 Explain why the real projective curve $y^2 = x$ has a branch point at infinity like the one at 0, and hence conclude that this curve is topologically a circle.

The explanation of the branch point at infinity of a cubic curve goes as follows.

15.4.2 Use the substitution $x = 1/u$, $y = 1/v$ to show that the curve

$$y^2 = x(x - \alpha)(x - \beta)$$

behaves at infinity as the curve

$$v^2 = u^3(1 - u\alpha)^{-1}(1 - u\beta)^{-1}$$

does at 0, which in turn is qualitatively like the behavior of

$$v = u^{3/2}.$$

15.4.3 Show, by considering the points lying above $u = e^{i\theta}$, that $v = u^{3/2}$ has a branch point at 0 like that of $v = u^{1/2}$.

15.5   Biographical Notes: Riemann

Bernhard Riemann (Figure 15.13) was born in the village of Breselenz, near Hannover, in 1826, and died at Selasca in Italy in 1866. He was the second of six children of Friedrich Riemann, a Protestant minister, and Charlotte Ebell. Up to the age of 13 he was taught by his father, with the help of the village schoolmaster, but he showed such a grasp of mathematics that sometimes they were unable to follow him. In 1840 Riemann went to live with his grandmother in Hannover in order to attend secondary school. After her death in 1842 he continued his studies at a school in Lüneburg, which was nearer to home, his father having moved to a new parish in the village of Quickborn. In Lüneburg it was his good fortune to have a headmaster who recognized his talent and gave him books by Euler and Legendre to read. The story goes that he mastered Legendre's 800-page *Théorie des Nombres* in six days.

Figure 15.13: Bernhard Riemann

The bright side of Riemann's life, described above, was not unlike Abel's. But, as with Abel, there was a darker side as well. Riemann's family was also poor and suffered from tuberculosis. His mother, three sisters, and Riemann himself eventually died from the disease. At least Riemann was spared the family discord and very early death that made Abel's life so tragic. At all times he maintained a close and loving relationship with his family, he lived long enough to marry and become a father, and he also had time to develop his major ideas to maturity and to gain a significant following. Riemann's published work—just a single volume—is in fact less copious than that of any important mathematician who lived to his fortieth year. But no other single volume has had such an impact on modern mathematics.

Riemann's career as a mathematician began soon after he entered the University of Göttingen in 1846. He intended to follow in his father's footsteps by studying theology but, like Euler and the Bernoullis before him, he found the call of mathematics too strong and obtained his father's permission to switch fields. The switch to mathematics was in recognition of where his greatest talent lay, not because of disdain for theology or philosophy. In fact, Riemann was deeply pious and well read in philosophy—so much so that readers ever since have lamented the influence of German philosophical writing on his style.

Göttingen in 1846 was not the mecca for mathematicians one would have expected it to be with the great Gauss in the chair of mathematics. Professors kept aloof from students and did not encourage original thinking or lecture on current research. Even Gauss himself taught only elementary courses. After a year, Riemann transferred to the University of Berlin, where the atmosphere was more democratic and where Jacobi, Dirichlet, Steiner, and Eisenstein shared their latest ideas. Riemann was too shy to immerse himself fully in this radically different environment, but he became friendly with Eisenstein, who was just three years his senior, and learned a great deal from Dirichlet. Riemann's later work made highly original use of some of Dirichlet's ideas, in particular a quasi-physical principle (actually first stated by Kelvin) that Riemann called *Dirichlet's principle*. Among the remarkable conclusions he drew from this principle was the theorem that curves of topological genus 0 are precisely those that can be parameterized by rational functions.

Dirichlet's forte was the use of analysis in pure mathematics, particularly in number theory, and Riemann too has been broadly classified as an analyst. However, he was not a specialist as analysts usually are today. His field was all of mathematics, seen from the analytic viewpoint. He saw where analysis could be used to illuminate mathematics from number theory to geometry, but he also saw where analysis itself was in need of illumination from outside. The concept of a Riemann surface, and the topological concept of genus in particular, made many previously hard-won results of analysis almost obvious. A vivid example of the illumination of analysis by topology is Riemann's explanation of the double periodicity of elliptic functions, which we shall see in Section 16.4.

Riemann surfaces were introduced in Riemann's doctoral thesis (Riemann (1851)). He had returned to Göttingen in 1849 and, after gaining his doctorate, began working to qualify for a Privatdozent (lecturing) position. One of the requirements was an essay, which he met with a memoir on Fourier series in which he introduced the "Riemann integral" concept. The Riemann integral is not really one of Riemann's best ideas—although it is the one best known to students today—since the integral later introduced by Lebesgue is far better suited to the subject (see Chapter 24). The other requirement was a lecture, for which he had to submit three titles to the university faculty. Gauss chose the third, which was the most difficult, on the foundations of geometry. However, Riemann rose brilliantly to the occasion, and his lecture *Über die Hypothesen, welche der Geometrie zu Grunde liegen* became one of the classics of mathematics (Riemann (1854b)). In it he introduced the main ideas of modern differential geometry: n-dimensional spaces, metrics and curvature, and the way in which curvature controls global geometric properties of a space. In the special case of two dimensions, these ideas had already been grasped by Gauss (see Chapter 17), so it was a joy and a revelation to Gauss, then in the last year of his life, to see how much further Riemann had carried them.

Riemann succeeded in becoming a lecturer and had the satisfaction of attracting an unexpectedly large class (eight students!). During the next few years he developed the material for perhaps his greatest work, Riemann (1857), which did for algebraic geometry what his earlier Riemann (1854b) did for differential geometry. One of his students at this time was Dedekind, who later recast Riemann's theory into the more algebraic form that is used today. Dedekind also coedited Riemann's collected works and wrote an essay on Riemann's life (Dedekind (1876)), which is the main biographical source for this section. The lecturer's position was very productive mathematically, but it brought in only voluntary fees from students, and Riemann was close to starvation. Other setbacks he suffered were the death of his father and sister Clara and a nervous breakdown brought on by overwork.

When Gauss died in 1855 and was succeeded by Dirichlet there was an unsuccessful move to appoint Riemann as associate professor. This move failed, but Riemann was granted a regular salary, and when Dirichlet died in 1859 Riemann succeeded him. In 1862 he married Elise Koch, a friend of his sisters, and their daughter, Ida, was born in Pisa in 1863. Riemann had begun traveling to Italy for the sake of his health in 1862, and he spent much time there during his remaining years. He loved Italy and its art treasures and also received a warm reception from Italian mathematicians. Two of his friends in Pisa, Enrico Betti and Eugenio Beltrami, were inspired by Riemann's ideas to make important contributions to topology and differential geometry. Beltrami saw how Riemann's concept of curved space could be used as a basis for non-Euclidean geometry, a revolutionary discovery that even Riemann may not have anticipated (see Chapter 18).

Riemann's sojourn in Italy was all too short. He died at Selasca on Lake Maggiore in the summer of 1866, with his wife beside him. Dedekind described his last days as follows (not in his usual style, but no doubt sensitive to the feelings of Riemann's widow):

> On the day before his death he lay beneath a fig tree, filled with joy at the glorious landscape, writing his last work, unfortunately left incomplete. His end came gently, without struggle or death agony; it seemed as though he followed with interest the parting of the soul from his body; his wife had to give him bread and wine, he asked her to convey his love to those at home, saying "Kiss our child." She said the Lord's prayer with him, he could no longer speak; at the words "Forgive our trespasses" he raised his eyes devoutly, she felt his hand in hers becoming colder, and after a few more breaths his pure, noble heart ceased to beat. The gentle mind implanted in him in his father's house stayed with him all his life, and he remained true to his God as his father had, though not in the same form.

Dedekind (1876)

It was said of Abel that he left enough to keep mathematicians busy for 500 years, and the same might be said of Riemann. Today, more than 130 years after Riemann's death, the major unsolved problem in pure mathematics is the so-called *Riemann hypothesis*, a conjecture made casually by Riemann (1859) in his paper on the distribution of prime numbers. Riemann considered Euler's function (discussed in Section 10.7),

$$\zeta(s) = 1 + \frac{1}{2^s} + \frac{1}{3^s} + \cdots,$$

introducing the zeta notation for it, and extended it to complex values of s. He observed that if $\zeta(s) = 0$, then $0 \leq \mathrm{Re}(s) \leq 1$, and added that it was quite likely that all zeros of $\zeta(s)$ had real part $1/2$. He did not pursue the matter further, since his initial observation was enough for his purpose, which was to derive an infinite series for $F(x)$, the number of primes less than a positive integer x. Mathematicians later realized that Riemann's hypothesis governs the distribution of prime numbers to an extraordinary extent, which is why its proof is so eagerly sought. Since all the efforts of the best mathematicians have failed so far, perhaps only another Riemann will succeed.

16

Complex Numbers and Functions

PREVIEW

The insight into algebraic curves afforded by complex coordinates—that a complex curve is topologically a surface—has important repercussions for functions defined as integrals of algebraic functions, such as the logarithm, exponential, and elliptic functions.

The complex logarithm turns out to be "many-valued," due to the different paths of integration in the complex plane between the same endpoints. It follows that its inverse function, the exponential function, is *periodic*. In fact, the complex exponential function is a fusion of the real exponential function with the sine and cosine: $e^{x+iy} = e^x(\cos y + i \sin y)$.

The double periodicity of elliptic functions also becomes clear from the complex viewpoint. The integrals that define them are taken over paths on a *torus* surface, on which there are two independent closed paths.

The two-dimensional nature of complex numbers imposes interesting and useful constraints on the nature of *differentiable* complex functions. Such functions define *conformal* (angle-preserving) maps between surfaces. Also, their real and imaginary parts satisfy equations, called the *Cauchy–Riemann* equations, that govern fluid flow. So complex functions can be used to study the motion of fluids.

Finally, the Cauchy–Riemann equations imply *Cauchy's theorem*. This fundamental theorem guarantees that differentiable complex functions have many good features, such as power series expansions.

J. Stillwell, *Mathematics and Its History*, Undergraduate Texts in Mathematics,   313
DOI 10.1007/978-1-4419-6053-5_16, © Springer Science+Business Media, LLC 2010

16.1   Complex Functions

When Bombelli (1572) introduced complex numbers, he implicitly introduced complex functions as well. The solution y of the cubic equation
$y^3 = py + q$,

$$y = \sqrt[3]{\frac{q}{2} + \sqrt{\left(\frac{q}{2}\right)^2 - \left(\frac{p}{3}\right)^3}} + \sqrt[3]{\frac{q}{2} - \sqrt{\left(\frac{q}{2}\right)^2 - \left(\frac{p}{3}\right)^3}},$$

involves the cube root of a complex argument when $(q/2)^2 < (p/3)^3$. It
could have been a revelation to see that complex numbers explain the coincidence of algebraic (Cardano) and geometric (Viète) solutions of the cubic
equation, and more generally the Leibniz–de Moivre theorem that

$$x = \frac{1}{2}\sqrt[n]{y + \sqrt{y^2 - 1}} + \frac{1}{2}\sqrt[n]{y - \sqrt{y^2 - 1}},$$

when $x = \sin\theta$ and $y = \sin n\theta$ (Section 6.6). In the case of the cubic,
this revelation can now be savored in Needham (1997), pp. 59–60. But
mathematicians were not concerned about the meaning of these complex
functions as long as they produced results that could be checked by algebra.

The need to *understand* complex functions became pressing only with
transcendental functions, particularly those defined by integration. A key
example is the logarithm function, which comes from integrating $dz/(1+z)$.
Once this function was understood, the reason for algebraic miracles like
the Leibniz–de Moivre theorem became much clearer.

Johann Bernoulli (1702) opened the story of the complex logarithm
when he noted that

$$\frac{dz}{1 + z^2} = \frac{dz}{2(1 + z\sqrt{-1})} + \frac{dz}{2(1 - z\sqrt{-1})}$$

and drew the conclusion that "imaginary logarithms express real circular
sectors." He did not actually perform the integration, but he may have
found

$$\tan^{-1} z = \frac{1}{2i} \log \frac{i - z}{i + z},$$

since Euler gives him credit for a similar formula when writing to him in
Euler (1728b). However, this may have been the young Euler's deference

to his former teacher, because Johann Bernoulli showed poor understanding of logarithms as the correspondence continued. He persistently claimed that $\log(-x) = \log(x)$ on the grounds that

$$\frac{d}{dx}\log(-x) = \frac{1}{x} = \frac{d}{dx}\log(x)$$

despite a reminder from Euler (1728b) that equality of derivatives does not imply equality of integrals. Euler went on to suggest that the complex logarithm had infinitely many values.

In the meantime, Cotes (1714) had also discovered a relation between complex logarithms and circular functions:

$$\log(\cos x + i \sin x) = ix.$$

Recognizing the importance of this result, he entitled his work *Harmonia mensurarum* (Harmony of measures). The "measures" in question were the logarithm and inverse tangent functions, which "measure" the hyperbola and the circle, respectively, via the integrals $\int dx/(1+x)$ and $\int dx/(1+x^2)$. A wide class of integrals had been reduced to these two types, but it was not understood why two apparently unrelated "measures" should be required. Cotes's result was the first (apart from the near-miss of Johann Bernoulli (1702)) to relate the two, showing that in the wider domain of complex functions the logarithm and inverse circular functions are essentially the same.

The most compact statement of their relationship was attained around 1740, when Euler shifted attention from the logarithm function to its inverse, the exponential function. The definitive formula

$$e^{ix} = \cos x + i \sin x$$

was first published by Euler (1748a), who derived it by comparing series expansions of both sides. Euler's formulation in terms of the single-valued function e^{ix} gave a simple explanation of the many values of the logarithm (which Cotes had missed) as a consequence of the periodicity of cos and sin. A direct explanation, based on the definition of log as an integral, was not possible until Gauss (1811) clarified the meaning of complex integrals and pointed out their dependence on the path of integration (see Section 16.3).

Euler's formula also shows that

$$(\cos x + i \sin x)^n = e^{inx} = \cos nx + i \sin nx$$

and hence gives a deeper explanation of the Leibniz–de Moivre formula. More generally, the addition theorems for cos and sin (Section 12.4) could be seen as consequences of the much simpler addition formula for the exponential function

$$e^{u+v} = e^u \cdot e^v.$$

The imaginary function e^{ix} was so much more coherent than its real constituents $\cos x$ and $\sin x$ that it was difficult to do without it, and Euler's formula gave mathematicians a strong push toward the eventual acceptance of complex numbers. A more detailed account of the role of the logarithm and exponential functions in the development of complex numbers may be found in Cajori (1913).

At almost the same time that Euler elucidated cos and sin, d'Alembert found many real functions occurring naturally in pairs as the real and imaginary parts of complex functions—in hydrodynamics. As mentioned in Section 13.7, d'Alembert (1752) discovered the equations

$$\frac{\partial P}{\partial y} - \frac{\partial Q}{\partial x} = 0, \tag{1}$$

$$\frac{\partial P}{\partial x} + \frac{\partial Q}{\partial y} = 0 \tag{2}$$

relating the velocity components P, Q in two-dimensional steady irrotational fluid flow. Equations (1) and (2) come from the requirements that $Q\,dx + P\,dy$ and $P\,dx - Q\,dy$ be complete differentials, in which case another complete differential is

$$Q\,dx + P\,dy + i(P\,dx - Q\,dy) = (Q + iP)\left(dx + \frac{dy}{i}\right) = (Q + iP)d\left(x + \frac{y}{i}\right).$$

D'Alembert concluded that this means $Q + iP$ is a function f of $x + y/i$, so that $Q = \mathrm{Re}(f)$ and $P = \mathrm{Im}(f)$.

To feel the force of this result, one has to forget the modern definition of function, under which $u(x, y) + iv(x, y)$ is a function of $x + iy$ for *any* functions u, v. In the 18th-century context, a "function" $f(x + iy)$ of $x + iy$ was calculable from $x + iy$ by elementary operations; at worst, $f(x + iy)$ was a power series in $x + iy$. This imposes a strong constraint on u, v, namely that

$$\frac{\partial u}{\partial x} = \frac{\partial v}{\partial y}, \quad \frac{\partial u}{\partial y} = -\frac{\partial v}{\partial x}.$$

These were just the equations d'Alembert found in his hydrodynamical investigations, but they came to be named the Cauchy–Riemann equations, because the latter mathematicians stressed their key role in the study of complex functions. The concept of complex function was solidified when Cauchy (1837) showed that a function $f(z)$, where $z = x + iy$, merely had to be differentiable in order to be expressible as a power series in z. Thus it suffices to define a complex function $f(z)$ to be one that is differentiable with respect to z in order to guarantee that f is defined with 18th-century strictness. It follows, in particular, that the first derivative of f entails derivatives of all orders and that the values of f in any neighborhood determine its values everywhere. This "rigidity" in the notion of complex function is enough of a constraint to enable nontrivial properties to be proved, but at the same time it leaves enough flexibility—one might say "fluidity"—to cover important general situations.

EXERCISES

Euler's derivation of $e^{ix} = \cos x + i \sin x$ is easy to explain using the power series

$$e^y = 1 + \frac{y}{1!} + \frac{y^2}{2!} + \frac{y^3}{3!} + \cdots$$

and

$$\sin x = x - \frac{x^3}{3!} + \frac{x^5}{5!} - \frac{x^7}{7!} + \cdots$$

found in Section 9.5.

16.1.1 Assuming that the series for e^y is also valid for $y = ix$, show that

$$e^{ix} = \left(1 - \frac{x^2}{2!} + \frac{x^4}{4!} - \frac{x^6}{6!} + \cdots\right) + i\left(x - \frac{x^3}{3!} + \frac{x^5}{5!} - \frac{x^7}{7!} + \cdots\right).$$

16.1.2 Assuming it is valid to differentiate the sine series term by term, show that

$$\cos x = 1 - \frac{x^2}{2!} + \frac{x^4}{4!} - \frac{x^6}{6!} + \cdots,$$

and hence that $e^{ix} = \cos x + i \sin x$.

Another consequence of $e^{ix} = \cos x + i \sin x$ is that $i = \cos \frac{\pi}{2} + i \sin \frac{\pi}{2} = e^{i\pi/2}$, which allows us to evaluate the outlandish number i^i.

16.1.3 Show that i^i has a real value (Euler (1746)). What is it?

16.1.4 Using the fact that $e^{2in\pi} = 1$ for any integer n, give a formula for all values of i^i (Euler (1746)).

16.2   Conformal Mapping

Another important general situation clarified by complex functions is the problem of conformal mapping. Mapping a sphere (the earth's surface) onto a plane is a practical problem that has attracted the attention of mathematicians since ancient times. Before the 18th century, the most notable mathematical contributions to mapping were stereographic projection (Section 15.2), due to Ptolemy around 150 CE, and the Mercator projection used by G. Mercator in 1569 (this Mercator was Gerard, not the Nicholas who discovered the series for $\log(1 + x)$). Both these projections were conformal, that is, angle-preserving, or what 18th-century mathematicians preferred to call "similar in the small." This means that the image $f(R)$ of any region R tends toward an exact scale map of R as the size of R tends to 0. Since "similarity in the large" is clearly impossible—for example, a great circle cannot be mapped to a closed curve that divides the plane into two equal parts—conformality is the best one can do to preserve the appearance of regions on the sphere. Preservation of angles was intentional in the Mercator projection, whose purpose was to assist navigation, and in the case of stereographic projection conformality was first noticed by Harriot around 1590 (see Lohne (1979)).

Advances in the theory of conformal mapping were made by Lambert (1772), Euler (1777) (sphere onto plane), and Lagrange (1779) (general surface of revolution onto plane). All these authors used complex numbers, but Lagrange's presentation is the clearest and most general. Using the method of d'Alembert (1752), he combined a pair of differential equations in two real variables into a single equation in one complex variable and arrived at the result that any two conformal maps of a surface of revolution onto the (x, y)-plane are related via a complex function $f(x + iy)$ mapping the plane onto itself. These results were crowned by the result of Gauss (1822) generalizing Lagrange's theorem to conformal maps of an arbitrary surface onto the plane.

Conversely, a complex function $f(z)$ defines a map of the z plane onto itself, and it is easy to see that this map is conformal. In fact, this is a consequence of the differentiability of f. To say that the limit

$$\lim_{\delta z \to 0} \frac{f(z_0 + \delta z) - f(z_0)}{\delta z}$$

exists is to say that the mapping of the disk $\{z : |z - z_0| < |\delta z|\}$ around z_0 to the region around $f(z_0)$ tends to a scale mapping as $|\delta z|$ tends to 0. If the derivative is expressed in polar form as

$$f'(z_0) = re^{i\alpha},$$

then r is the scale factor of this limit mapping and α is the angle of rotation. Riemann (1851) seems to have been the first to take the conformal mapping property as a basis for the theory of complex functions. His deepest result in this direction was the *Riemann mapping theorem*, which states that any region of the plane bounded by a simple closed curve can be mapped onto the unit disk conformally, and hence by a complex function. The proof of this theorem in Riemann (1851) depends on properties of potential functions, which Riemann justified partly by appeal to physical intuition—the so-called *Dirichlet's principle*. Such reasoning went against the growing tendency toward rigor in 19th-century analysis, and stricter proofs were given by Schwarz (1870) and Neumann (1870). However, Riemann's faith in the physical roots of complex function theory was eventually justified when Hilbert (1900b) put Dirichlet's principle on a sound basis.

EXERCISES

The claim that differentiability of $f(z)$ implies that f is a conformal mapping should be qualified by the condition $f'(z) \neq 0$, because if the scale factor tends to 0 then f cannot be said to be a scale mapping. At points where $f'(z) = 0$ one may find that angles are altered. Here is an example.

16.2.1 Show that $f(z) = z^2$ defines a conformal mapping except at $z = 0$, where it doubles angles.

This is no surprise if we view $z \mapsto z^2$ as a two-sheeted covering of the plane $\mathbb{C}$ (compare with Section 15.4).

16.2.2 Show that the map $z \mapsto z^2$ is two-to-one except at $z = 0$, and relate the angle doubling at $z = 0$ to the branch point of the covering.

16.2.3 Similarly describe the behavior of the map $z \mapsto z^3$ at $z = 0$.

16.3   Cauchy's Theorem

We have seen that interesting complex functions arise from integration. For example, the elliptic functions come from inversion of elliptic integrals (Section 12.3). However, it is not at first clear what the integral $\int_{z_0}^{z} f(t)\,dt$ means when z_0, z are complex numbers. It is natural, and not technically difficult, to define $\int_{z_0}^{z} f(t)\,dt$ as $\int_C f(t)\,dt$, the integral of f along a curve

C from z_0 to z; the problem is that $\int_C f(t)\,dt$ appears to depend on C and hence may not be anything like a function of z, as one would wish.

The first to recognize and resolve this problem seems to have been Gauss. In a letter to Bessel, Gauss (1811) raised the problem and claimed its resolution as follows:

> Now how is one to think of $\int \Phi(z)\,dz$ for $z = a + ib$? Evidently, if one wishes to start from clear concepts, one must assume that z changes by infinitely small increments (each of the form $\alpha + i\beta$) from that value for which the integral is to be 0 to $c = a + ib$, and then *sum* all the $\phi(z)\,dz$. ...But now ...continuous transition from one value of z to another $a + ib$ takes place along a curve and hence is possible in infinitely many ways. I now conjecture that the integral $\int_0^c \phi(z)\,dz$ will always have the same value after two different transitions if $\phi(z)$ never becomes infinite within the region enclosed by the two curves representing the transitions.

Translation of Gauss (1811) in Birkhoff (1973), p. 31

In the same letter, Gauss also observed that if $\phi(z)$ *does* become infinite in the region, then in general $\int_0^c \phi(z)\,dz$ *will* take different values when integrated along different curves. He saw in particular that the infinitely many values of $\log c$ corresponded to the different ways a path from 1 to c could wind around $z = 0$, the point where $\phi(z) = 1/z$ becomes infinite.

The theorem that $\int_{z_0}^z f(t)\,dt$ is independent of the path throughout a region where f is finite (and differentiable, which went without saying for Gauss) is now known as *Cauchy's theorem*, since Cauchy was the first to offer a proof and to develop the consequences of the theorem. An equivalent and more convenient statement is that $\int_C f(t)\,dt = 0$ for any closed curve C in a region where f is differentiable. Cauchy presented a proof to the Paris Academy in 1814 but first published it later (Cauchy (1825)). In Cauchy (1846) he presented a more transparent proof, based on the Cauchy–Riemann equations and the theorem of Green (1828) and Ostrogradsky (1828), which relates a line integral to a surface integral. The latter theorem, usually known as *Green's theorem*, is a generalization of the fundamental theorem of calculus to real functions $f(x, y)$ of two variables and can be stated as follows: if C is a simple closed curve bounding a region $\mathcal{R}$

and f is suitably smooth, then

$$\int_C f\,dx = \iint_{\mathcal{R}} \frac{\partial f}{\partial y}\,dx\,dy,$$

$$\int_C f\,dy = -\iint_{\mathcal{R}} \frac{\partial f}{\partial x}\,dx\,dy,$$

where $\iint_{\mathcal{R}}$ denotes the surface integral over $\mathcal{R}$ and $\int_C$ denotes the line integral around C in the counterclockwise sense. (The difference in sign in the two formulas reflects the different sense of C when x and y are interchanged.)

Cauchy's theorem follows from Green's theorem by an easy calculation. If

$$f(t) = u(t) + iv(t)$$

is the decomposition of f into real and imaginary parts, and if we write

$$dt = dx + i\,dy,$$

then

$$
\begin{aligned}
\int_C f(t)\,dt &= \int_C (u + iv)(dx + i\,dy) \\
&= \int_C (u\,dx - v\,dy) + i\int_C (v\,dx + u\,dy) \\
&= \iint_{\mathcal{R}} \left(\frac{\partial u}{\partial y} + \frac{\partial v}{\partial x}\right)dx\,dy + i\iint_{\mathcal{R}} \left(\frac{\partial v}{\partial y} - \frac{\partial u}{\partial x}\right)dx\,dy \\
&= 0,
\end{aligned}
$$

since

$$\frac{\partial u}{\partial y} + \frac{\partial v}{\partial x} = 0 \quad \text{and} \quad \frac{\partial v}{\partial y} - \frac{\partial u}{\partial x} = 0$$

by the Cauchy–Riemann equations. This proof requires f to have a continuous first derivative in order to be able to apply Green's theorem. The restriction of continuity of $f'(t)$ in the proof was removed by Goursat (1900). As it happens, if f' exists, it will have not only continuity but also derivatives of all orders. This follows from one of the remarkable consequences Cauchy (1837) drew from the assumption $\int_C f(t)\,dt = 0$, namely, that f has a power-series expansion. By Goursat (1900), then, differentiability of a complex function is enough to guarantee a power-series expansion.

A generalization of this result to f that become infinite at isolated points was made by Laurent (1843) (f then has an expansion including negative powers, called the *Laurent expansion*) and to "many-valued" f with branch points by Puiseux (1850) (f then has an expansion in fractional powers, the *Newton–Puiseux expansion*).

<small>EXERCISES</small>

The Cauchy–Riemann equations follow easily from the existence of $f'(z)$, that is, from the condition that

$$\lim_{\delta z \to 0} \frac{f(z + \delta z) - f(z)}{\delta z}$$

have the same value, regardless of the path along which $\delta z \to 0$.

16.3.1 Suppose $f(z) = u(x, y) + iv(x, y)$ and $\delta z = \delta x + i\delta y$. By letting $\delta z \to 0$ along the x-axis ($\delta y = 0$) and along the y-axis ($\delta x = 0$), and equating the resulting values of $f'(z)$, show that

$$\frac{\partial u}{\partial x} = \frac{\partial v}{\partial y}, \quad \frac{\partial u}{\partial y} = -\frac{\partial v}{\partial x}.$$

These equations give a convenient test for a function $u(x, y) + iv(x, y)$ to be a differentiable function of $z = x + iy$.

16.3.2 Check that $u(x, y) = x^2 - y^2$ and $v(x, y) = 2xy$ satisfy the Cauchy–Riemann equations.

16.3.3 Express $x^2 - y^2 + 2ixy$ as a function of $z = x + iy$.

16.4   Double Periodicity of Elliptic Functions

The view of complex integration provided by Cauchy's theorem is one step toward understanding elliptic integrals such as $\int_0^z dt/\sqrt{t(t - \alpha)(t - \beta)}$. The other important step is the idea of a Riemann surface (Section 15.4), which enables us to visualize the possible paths of integration from 0 to z. The "function" $1/\sqrt{t(t - \alpha)(t - \beta)}$ is of course two-valued and, by an argument like that in Section 15.4, is represented by a two-sheeted covering of the t sphere, with branch points at 0, α, β, ∞. Thus the paths of integration, correctly viewed, are curves on this surface, which is topologically a torus (again, as in Section 15.4).

Now a torus contains certain closed curves that do not bound a piece of the surface, such as the curves C_1 and C_2 shown in Figure 16.1. There is no region $\mathcal{R}$ bounded by C_1 or C_2; hence Green's theorem does not apply, and we in fact obtain nonzero values

$$\omega_1 = \int_{C_1} \frac{dt}{\sqrt{t(t-\alpha)(t-\beta)}},$$

$$\omega_2 = \int_{C_2} \frac{dt}{\sqrt{t(t-\alpha)(t-\beta)}}.$$

Consequently the integral

$$\Phi^{-1}(z) = \int_0^z \frac{dt}{\sqrt{t(t-\alpha)(t-\beta)}}$$

will be ambiguous: for each value $\Phi^{-1}(z) = w$ obtained for a certain path C from 0 to z we also obtain the values $w + m\omega_1 + n\omega_2$ by adding to C a detour that winds m times around C_1 and n times around C_2. (For topological reasons, this is essentially the most general path of integration.)

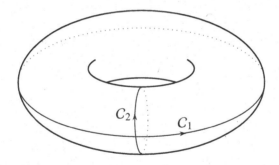

Figure 16.1: Nonbounding curves on the torus

It follows that the inverse relation $\Phi(w) = z$, the elliptic function corresponding to the integral, satisfies

$$\Phi(w) = \Phi(w + m\omega_1 + n\omega_2)$$

for any integers m, n. That is, Φ is doubly periodic, with periods ω_1, ω_2. This intuitive explanation of double periodicity is due to Riemann (1851), who later (Riemann (1858a)) developed the theory of elliptic functions from this standpoint.

Remarkable series expansions of elliptic functions, which exhibit the double periodicity analytically, were discovered by Eisenstein (1847). The precedents for Eisenstein's series, as Eisenstein himself pointed out, were partial fraction expansions of circular functions discovered by Euler, for example

$$\pi \cot \pi x = \sum_{n=-\infty}^{\infty} \frac{1}{x+n}$$

(Euler (1748a), p. 191). It is obvious (at least formally, though one has to be a little careful about the meaning of this summation to ensure convergence) that the sum is unchanged when x is replaced by $x + 1$; hence the period 1 of $\pi \cot \pi x$ is exhibited directly by its series expansion. Eisenstein showed that doubly periodic functions could be obtained by analogous expressions, such as

$$\sum_{m,n=-\infty}^{\infty} \frac{1}{(z + m\omega_1 + n\omega_2)^2},$$

which again (with suitable interpretation to ensure convergence) are obviously unchanged when z is replaced by $z + \omega_1$ or $z + \omega_2$. Hence we obtain a function with periods ω_1, ω_2. The function above is in fact identical (up to a constant) with the Weierstrass $\wp$-function, mentioned in Section 12.5 as the inverse to the integral $\int dt / \sqrt{4t^3 - g_2 t - g_3}$. Weierstrass (1863), p. 121, found the relations between g_2, g_3 and the periods ω_1, ω_2:

$$g_2 = 60 \sum \frac{1}{(m\omega_1 + n\omega_2)^4},$$
$$g_3 = 140 \sum \frac{1}{(m\omega_1 + n\omega_2)^6},$$

where the sums are over all pairs $(m, n) \neq (0, 0)$. Elegant modern accounts of the Eisenstein and Weierstrass theories may be found in Weil (1976) and Robert (1973).

EXERCISES

The precise definition of the Weierstrass $\wp$-function is

$$\wp(z) = \frac{1}{z^2} + \sum_{m,n \neq 0,0}^{\infty} \left(\frac{1}{(z + m\omega_1 + n\omega_2)^2} - \frac{1}{(m\omega_1 + n\omega_2)^2} \right).$$

This series has better convergence than the Eisenstein series given above, but its double periodicity is not quite so obvious. We can establish double periodicity by differentiating and integrating as follows (which is valid because of the convergence properties of the Weierstrass series).

16.4.1 By differentiating term by term, show that

$$\wp'(z) = -2 \sum_{m,n=-\infty}^{\infty} \frac{1}{(z + m\omega_1 + n\omega_2)^3},$$

and conclude that $\wp'(z + \omega_1) = \wp'(z)$ and $\wp'(z + \omega_2) = \wp'(z)$.

16.4.2 By integrating the equations just obtained, show that

$$\wp(z + \omega_1) - \wp(z) = c \quad \text{and} \quad \wp(z + \omega_2) - \wp(z) = d,$$

for some constants c and d.

16.4.3 Deduce from Exercise 16.4.2 that

$$\wp\left(\frac{\omega_1}{2}\right) - \wp\left(-\frac{\omega_1}{2}\right) = c \quad \text{and} \quad \wp\left(\frac{\omega_2}{2}\right) - \wp\left(-\frac{\omega_2}{2}\right) = d.$$

16.4.4 But $\wp(z) = \wp(-z)$ (why?); hence conclude that $\wp$ is doubly periodic.

16.5   Elliptic Curves

We have seen that nonsingular cubic curves of the form

$$y^2 = ax^3 + bx^2 + cx + d \tag{1}$$

are important not only among the cubic curves themselves (see Newton's classification, Sections 7.4 and 8.4), but also in number theory (Section 11.6) and the theory of elliptic functions (Section 12.2). One of the great achievements of 19th-century mathematics was the synthesis of a unified view of all these manifestations of cubic curves. The view was glimpsed by Jacobi (1834), and it came more clearly into focus with the development of complex analysis between Riemann (1851) and Poincaré (1901). The theory of elliptic curves, as the unified view has come to be known, continues to inspire researchers today, since it seems to encompass some of the most fascinating problems of number theory. We now know, for example, how to derive Fermat's last theorem from properties of elliptic curves (see Section 11.3).

Jacobi saw, at least implicitly, that the curve (1) could be parameterized as

$$x = f(z), \qquad y = f'(z), \tag{2}$$

where f and its derivative f' are elliptic functions. Knowing that f and f' are doubly periodic, with the same periods ω_1, ω_2, say, he would have seen that this gave a map of the z plane $\mathbb{C}$ onto the curve (1) for which the preimage of a given point on (1) is a set of points in $\mathbb{C}$ of the form

$$z + \Lambda = \{z + m\omega_1 + n\omega_2 : m, n \in \mathbb{Z}\},$$

where

$$\Lambda = \{m\omega_1 + n\omega_2 : m, n \in \mathbb{Z}\}.$$

Λ is called the *lattice of periods* of f. The numbers $z + m\omega_1 + n\omega_2$ in $z + \Lambda$ are also called "equivalent with respect to Λ." One such equivalence class is shown by asterisks in Figure 16.2.

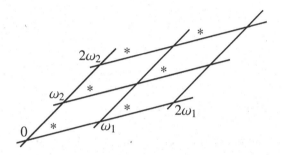

Figure 16.2: Lattice-equivalent points

The parameterization (2) means that there is a one-to-one correspondence between the points $(f(z), f'(z))$ of the curve and the equivalence classes $z + \Lambda$. Today we express this relation by saying that the curve is *isomorphic to the space* $\mathbb{C}/\Lambda$ of these equivalence classes. Jacobi might have seen, though it was probably not of interest to him, that $\mathbb{C}/\Lambda$ is a torus. One sees this by taking one parallelogram in $\mathbb{C}$, which includes a representative of each equivalence class, and identifying the equivalent points on its boundary (that is, pasting opposite sides together, as in Figure 16.3). Of course, the torus form of (1) eventually came to light through the Riemann surface construction given in Section 15.4.

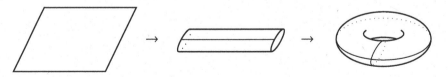

Figure 16.3: Construction of a torus by pasting

An elegant way of demonstrating both the double periodicity of elliptic functions and the parameterization of cubic curves was given by Weierstrass (1863). Beginning with the function

$$\sum_{m,n=-\infty}^{\infty} \frac{1}{(z + m\omega_1 + n\omega_2)^2},$$

which, as mentioned in Section 16.4, makes the double periodicity evident, Weierstrass defined the function

$$\wp(z) = \frac{1}{z^2} + \sum_{m,n\neq 0,0}^{\infty} \left(\frac{1}{(z + m\omega_1 + n\omega_2)^2} - \frac{1}{(m\omega_1 + n\omega_2)^2} \right),$$

which has better convergence properties and is also doubly periodic. He then showed by simple computations with series that

$$\wp'(z)^2 = 4\wp(z)^3 - g_2\wp(z) - g_3,$$

where g_2, g_3 are the constants depending on ω_1, ω_2, which were defined in Section 16.4. It follows that the point $(\wp(z), \wp'(z))$ lies on the curve

$$y^2 = 4x^3 - g_2 x - g_3, \tag{3}$$

and a little further checking shows that (3) is in fact isomorphic to $\mathbb{C}/\Lambda$, where Λ is the lattice of periods of $\wp$. The parameterization of all curves (1) by elliptic functions follows by making a linear transformation.

The reason for saying that the curve and $\mathbb{C}/\Lambda$ are "isomorphic" (which comes from the Greek for "same form") is not only because they both have the form of a torus. They also have the same *algebraic structure*, which comes to light when we consider their natural "addition" operation.

Once the curve (1) is parameterized as

$$x = f(z), \qquad y = f'(z),$$

one sees a natural "addition" of points on the curve induced by adding their parameter values. Because of the double periodicity of f and f', this "addition" is simply ordinary addition in $\mathbb{C}$, modulo Λ. In particular, it is immediate that "addition of points" has some properties of ordinary addition, such as commutativity and associativity. However, as mentioned in Section 11.6, addition of parameter values z is also reflected in the geometry of the curve. The most concise statement of the relationship, due to

Clebsch (1864), is that if z_1, z_2, z_3 are parameter values of three collinear points, then

$$z_1 + z_2 + z_3 = 0 \quad \text{mod} \ (\omega_1, \omega_2)$$

(or $z_1 + z_2 + z_3 \in \Lambda$). This means that "addition of points" also has an elementary geometric interpretation, for which, incidentally, the algebraic properties are far less obvious.

On the other hand, the straight-line interpretation of "addition" gives the simplest explanation of the addition theorems for elliptic functions. As we saw in Section 11.6, the value of $f(z_3)$ is easy to compute as a rational function of $f(z_1)$, $f'(z_1)$, $f(z_2)$, $f'(z_2)$ when z_1, z_2, z_3 are the parameter values of collinear points. Originally, of course, the formula was obtained by Euler, with great difficulty, by manipulating the integral inverse to f (see Section 12.5).

Another reason to accept $\mathbb{C}/\Lambda$ as the "right" view of the curve is that it gives an answer to the seemingly unrelated question of classification by projective equivalence. Recall from Section 8.4 that Newton had reduced cubics to the cusp type, the double-point type, and three nonsingular types using real projective transformations. All cubics with a cusp are, in fact, equivalent to $y^2 = x^3$, and all with a double point are equivalent to $y^2 = x^2(x + 1)$, while the distinction between the nonsingular types disappears over the complex numbers, where, as we now know, all are equivalent to tori $\mathbb{C}/\Lambda$. The problem that remains is to decide projective equivalence among the nonsingular cubics. Salmon (1851) showed that this was determined by a certain complex number τ, which can be computed from the equation of the curve. He defined τ geometrically, so that its projective invariance was obvious, with no thought of elliptic functions. But τ turned out to be nothing but ω_1/ω_2, which means that two nonsingular cubics are projectively equivalent if and only if their period lattices Λ have the same shape.

EXERCISES

Strictly speaking, the ratio $\tau = \omega_1/\omega_2$ determines only the shape of the *parallelogram* with vertices 0, ω_1, ω_2, and $\omega_1 + \omega_2$.

16.5.1 Explain how both the angle between adjacent sides of this parallelogram, and the ratio between their lengths, may be extracted from $\tau = \omega_1/\omega_2$.

The lattice of periods

$$\Lambda = \{m\omega_1 + n\omega_2 : m, n \in \mathbb{Z}\}$$

can be viewed as the set of vertices in a tiling of the plane by copies of this parallelogram, as in Figure 16.2. However, infinitely many *differently shaped parallelograms* give the same Λ. Thus the number τ alone should not be taken to characterize the shape of Λ.

16.5.2 Show that Λ may also be tiled by copies of a parallelogram with shape given by $\tau + 1$.

16.5.3 More generally, show that Λ may be generated by any two of its elements, $\omega_1' = a\omega_1 + b\omega_2$ and $\omega_2' = c\omega_1 + d\omega_2$ provided $ad - bc = \pm 1$. *Hint*: Write down a product of matrices transforming the column vector of (ω_1, ω_2) to (ω_1', ω_2') and back to (ω_1, ω_2), and take its determinant.

16.5.4 Deduce from Exercise 16.5.3 that the lattice $\Lambda = \{m\omega_1 + n\omega_2 : m, n \in \mathbb{Z}\}$ has shape characterized by the whole family of complex numbers

$$\frac{a\tau + b}{c\tau + d} \quad \text{where} \quad \tau = \frac{\omega_1}{\omega_2} \quad \text{and} \quad a, b, c, d \text{ are integers with } ad - bc = \pm 1.$$

There are functions of the complex variable τ that depend only on the lattice Λ, and hence take the same value for each number $(a\tau + b)/(c\tau + d)$ characterizing the lattice shape.

16.5.5 Consider g_2 and g_3 from Section 16.4, which are obviously functions $g_2(\Lambda)$ and $g_3(\Lambda)$ of the lattice Λ. Show that g_2^3/g_3^2 and $g_2^3/(g_2^3 - 27g_3^2)$ are both functions of τ.

The latter function is none other than the famous *modular function* mentioned in Section 6.7 in connection with the solution of the quintic equation. For more information on its amazing properties, see McKean and Moll (1997).

16.6   Uniformization

The characteristic of nonsingular cubics that allows their parameterization by elliptic functions is their topological form. The two periods correspond to the two essentially different circuits around the torus (Figure 16.1).

A representation of the x and y values on a curve by simultaneous functions of a single parameter z is sometimes called a *uniform* representation, and so the problem of parameterizing all algebraic curves in this way came to be known as the *uniformization* problem. Once the elliptic case was understood, it became clear that a solution of the uniformization problem for arbitrary algebraic curves would depend on a better understanding of surfaces: their topology, the periodicities associated with their closed curves, and the way these periodicities could be reflected in $\mathbb{C}$. These problems were first attacked by Poincaré and Klein in the 1880s, and their work led to the eventual positive solution of the uniformization problem by Poincaré (1907) and Koebe (1907).

Even more important than the solution of this single problem, however, was the amazing convergence of ideas in the preliminary work of Poincaré and Klein. They discovered that multiple periodicities are reflected in $\mathbb{C}$ by groups of transformations, and that the transformations in question are of the simple type $z \mapsto (az + b)/(cz + d)$, called *linear fractional*. Linear fractional transformations generalize the linear transformations $z \mapsto z + \omega_1$, $z \mapsto z + \omega_2$ naturally associated with the periods of elliptic functions. However, while the transformations $z \mapsto z + \omega_1$, $z \mapsto z + \omega_2$ are algebraically and geometrically transparent—they commute, and they generate the general transformations $z \mapsto z + m\omega_1 + n\omega_2$, which are simply translations of the plane—the more general linear fractional transformations are not as easily understood. Linear fractional transformations do not normally commute, and their mastery requires a simultaneous grasp of algebraic, geometric, and topological aspects.

The simultaneous view proved to be enormously fruitful in the development of group theory and topology, as we shall see in Chapters 19 and 22. Geometry was also given a new lease of life when Poincaré (1882) discovered that linear fractional transformations give a natural interpretation of non-Euclidean geometry, a field that until then had been a curiosity on the fringes of mathematics. In the next two chapters we look at the origins of non-Euclidean geometry and see how the subject was transformed by Poincaré's discovery.

EXERCISES

The first example, beyond the elliptic functions, of periodicity under linear fractional transformations is seen in the modular function derived in the previous exercise set. It turns out that the periodicity of the modular function can be generated by two transformations: $z \mapsto z + 1$ and $z \mapsto -1/z$.

16.6.1 Check that $z \mapsto z + 1$ and $z \mapsto -1/z$ are among the transformations

$$z \mapsto \frac{az + b}{cz + d}, \quad \text{where } a, b, c, d \text{ are integers with } ad - bc = \pm 1.$$

16.6.2 Show that the transformations $z \mapsto z + 1$ and $z \mapsto -1/z$ do not commute.

16.6.3 Show that both $z \mapsto z + 1$ and $z \mapsto -1/z$ map the half-plane $\{\text{Im } z > 0\}$ onto itself, and that $z \mapsto -1/z$ exchanges the inside and outside of the unit circle.

16.7   Biographical Notes: Lagrange and Cauchy

Joseph Louis Lagrange (Figure 16.4) was born in Turin in 1736 and died in Paris in 1813. He was the oldest of 11 children of Giuseppe Lagrangia, treasurer of the Office of Public Works in Turin, and Teresa Grosso, the daughter of a physician, and a member of the wealthy Conti family. Despite this background, Lagrange's family was not well off, since his father had made some unwise financial speculations. Lagrange eventually came to appreciate the loss of his chance to become a wealthy idler, saying, "If I had inherited a fortune I should probably not have cast my lot with mathematics."

Figure 16.4: Joseph Louis Lagrange

His prowess in mathematics developed with amazing speed after he first encountered calculus in 1753, at the age of 17. By 1754 he was writing to Euler about his discoveries, and in 1755 he was made professor at the Royal Artillery School in Turin. As early as 1756 he was offered a superior position in Prussia, but he was too shy, or too reluctant to leave home, to accept it. As his reputation grew, he also won the support of d'Alembert. When Euler left Berlin in 1766, d'Alembert arranged for Lagrange to take Euler's place. In 1767, perhaps missing the company of his family in Turin, Lagrange married his cousin Vittoria Conti. In a letter to d'Alembert in 1769 he said he had chosen a wife "who is one of my good cousins and who even lived for a long time with my family, is a very good housewife and has no pretensions at all," adding that they had no children and did not want any.

Notwithstanding this lackluster beginning, and the ill health of both Lagrange and his wife, the marriage strengthened over the years. Lagrange nursed Vittoria as her health worsened, and he was heartbroken when she died in 1783. He became deeply depressed about his work and the future of mathematics itself, writing to d'Alembert, "I cannot say that I shall still be doing mathematics 10 years from now. It also seems to me that the mine is already too deep, and that unless new veins are discovered it will have to be abandoned." Not long before this, Lagrange had completed one of his greatest works, the *Mécanique analytique*, but when a copy of the book reached him from the printer he left it unopened on his desk.

Frederick II died in 1786, and Lagrange's position in Berlin became less secure. After receiving several offers from Italy and France, he accepted a position at the Paris Academy in 1787. The change of scene did not appreciably revive his general spirits or enthusiasm for mathematics. Though always welcome at social and scientific gatherings, he was always politely detached, sympathetic but uninvolved. At least it can be said of his detachment that it enabled him to survive the 1789 revolution, which took the lives of his more committed friends Condorcet and Lavoisier. The revolution did in fact stir some activity in Lagrange. In 1790 he became a member of the commission on weights and measures, which introduced the metric system now used universally in science. An interesting glimpse of mathematics during the revolution, in the form of a "panel discussion" between Lagrange, Laplace, and members of a student audience, may be found in Dedron and Itard (1973), pp. 302–310.

In 1792 Lagrange married Renée-Françoise-Adelaïde Le Monnier, the teenaged daughter of an astronomer colleague. His interest in life and mathematics revived, and even in his seventies he made some brilliant contributions to celestial mechanics, which he incorporated in a second edition of the *Mécanique analytique*. When he died in 1813 he was buried in the Pantheon in Paris.

Lagrange is known for an uncompromisingly formal approach to analysis and mechanics. He viewed all functions as power series and attempted to reduce all mechanics to the analysis of such functions, without use of geometry. He was proud of the fact that the *Mécanique analytique* contained no diagrams. His fear that mathematics would have to be abandoned "if new veins were not discovered" was of course unfounded, but understandable as an admission of the limitations of his own approach. The great advances of 19th-century analysis were due, more than anything else, to a revival of geometry. In particular, Lagrange's own view of functions as power series became intelligible only in the domain of complex functions, when it emerged from the geometric theory of complex integration discovered by Gauss and Cauchy.

Augustin-Louis Cauchy (Figure 16.5) was born in Paris in 1789, only weeks after the storming of the Bastille, but he was anything but a child of the revolution. His father, Louis-François, was a lawyer and government official who fled to Paris with his wife, Marie-Madeleine Desestre, during the Terror. Augustin-Louis was the first of their six children. Throughout his life Cauchy was to hold extreme antirevolutionary and royalist views. The family settled in the village of Arcueil, and Cauchy received his early education from his father. He also had the benefit of contact with famous scientists who came to visit Laplace, who was a neighbor. Lagrange is said to have predicted that Cauchy would become a scientific genius but advised his father not to show him a mathematics book before he was 17.

Figure 16.5: Augustin-Louis Cauchy

When Napoleon took power at the end of the 18th century, Cauchy's father returned to government service, and the family moved back to Paris.

Cauchy concentrated on classics in secondary school, which he completed in 1804, but then gravitated toward a scientific career. He entered the École Polytechnique in 1805, transferred to the École des Ponts in 1807, and began working as an engineer around 1809. In 1810 he went to Cherbourg to help in the construction of Napoleon's naval base, carrying with him, so it was said, Laplace's *Mécanique céleste* and Lagrange's *Traité des fonctions analytiques*.

His first important mathematical work was the solution of a problem posed to him by Lagrange: to show that any convex polyhedron is rigid. (More precisely, to show that the dihedral angles of a convex polyhedron are uniquely determined by its faces.) An accessible proof of his result, which deserves to be better known, is in Lyusternik (1966). Cauchy's theorem partially settled a conjecture of Euler that any closed surface is rigid, and was in fact the best positive result obtainable, since Connelly (1977) has found a nonconvex polyhedron that is *not* rigid. Cauchy's second major discovery was his proof, in 1812, of Fermat's conjecture that every integer is the sum of at most n n-agonal numbers (see Section 3.2).

The Cauchy integral theorem, submitted to the French Academy in 1814, carried him into the mathematical mainstream. He also managed to catch the political tide, which was turning royalist again, and became a member of the Academy on the expulsion of some republican members in 1816. At the same time he became a professor at the École Polytechnique, where the 1820s saw the publication of his classic analysis texts and also one of his most important creations, the theory of elasticity. He also gained additional chairs at the Sorbonne and the Collège de France. He and Aloïse de Bure were married in 1818, and they had two daughters.

The mild revolution of 1830, which replaced the Bourbon King Charles X with the Orléans King Louis-Phillipe, was a catastrophe in Cauchy's view. From principles that were curious, though certainly firmly held, Cauchy refused to take the oath of allegiance to the new king. This meant he had to resign his chairs, but Cauchy went further than this—he left his family and followed the old king into exile. He did not return to Paris until 1838, and it was another 10 years before he regained one of his former chairs. Ironically, he had the revolution to thank for this, because it abolished the oath of allegiance. He returned to the Sorbonne and kept up a steady flow of mathematical papers until his death in 1857.

17

Differential Geometry

PREVIEW

As mentioned in Chapter 13, calculus made it possible to study nonalgebraic curves: the "mechanical" curves, or *transcendental* curves as we now call them. Calculus computes not only their basic features, such as tangents and area, but also more sophisticated properties such as *curvature*. Curvature turns out to be a fundamental concept of geometry, not only for curves, but also for higher-dimensional objects.

The concept of curvature is particularly interesting for surfaces, because it can be defined *intrinsically*. The intrinsic curvature, or *Gaussian* curvature as it is known, is unaltered by bending the surface, so it can be defined without reference to the surrounding space.

This opens the possibility of studying the *intrinsic* surface geometry. On any smooth surface one can define the distance between any two points (sufficiently close together), and hence "lines" (curves of shortest length), angles, areas, and so on.

The question then arises, to what extent does the intrinsic geometry of a curved surface resemble the classical geometry of the plane? For surfaces of *constant* curvature, the difference is reflected in two of Euclid's axioms: the axiom that straight lines are infinite, and the parallel axiom.

On surfaces of constant positive curvature, such as the sphere, all lines are finite and there are no parallels. On surfaces of zero curvature there may also be finite straight lines; but if all straight lines are infinite the parallel axiom holds. The most interesting case is constant negative curvature, because it leads to a realization of *non-Euclidean geometry*, as we will see in Chapter 18.

J. Stillwell, *Mathematics and Its History*, Undergraduate Texts in Mathematics,   335
DOI 10.1007/978-1-4419-6053-5_17, © Springer Science+Business Media, LLC 2010

17.1   Transcendental Curves

We saw in Chapter 9 that the development of calculus in the 17th century
was greatly stimulated by problems in the geometry of curves. Differentiation grew out of methods for the construction of tangents, and integration
grew out of attempts to find areas and arc lengths. Not only did calculus un-
lock the secrets of the classical curves and of the algebraic curves defined
by Descartes; it also extended the concept of curve itself. Once it became
possible to handle slopes, lengths, and areas with precision, it also became
possible to use these quantities to define new, nonalgebraic curves. These
were the curves called "mechanical" by Descartes (Sections 7.3 and 13.5)
and "transcendental" by Leibniz. In contrast to algebraic curves, which
could be studied in some depth by purely algebraic methods, transcendental curves were inseparable from the methods of calculus. Hence it is not
surprising that a new set of geometric ideas, the ideas of "infinitesimal" or
differential geometry, first emerged from the investigation of transcendental curves.

A more surprising by-product of the investigation of transcendental
curves was the first solution of the ancient problem of arc length. The
problem was first posed for an algebraic curve, the circle, by the Greeks
and in this case it is equivalent to an area problem ("squaring the circle"),
since both area and arc length of the circle depend on the evaluation of π.
As we now know, π is a transcendental number (Section 2.3), so the arc
length problem for the circle has no solution by the elementary means allowed by the Greeks. The first curve whose arc length could be found by
elementary means was discovered by Harriot around 1590. It is the curve
defined by the polar equation

$$r = e^{k\theta}$$

known as the *logarithmic* or *equiangular* spiral.

Harriot did not have the exponential function and knew the curve only
by its equiangular property, which is that the tangent makes a constant angle ϕ (depending on k) with the radius vector. The spiral turned up in his
researches on navigation and map projections (Section 16.2) as the projection of a *rhumb line* on the sphere (Figure 17.1). A rhumb line is a curve
that meets the meridians at a constant angle; in practical terms, it represents
the course of a ship sailing in a fixed compass direction.

Not having the tools of calculus, Harriot had to rely on ingenious geometry and a simple limit argument. His construction is illustrated in Figure

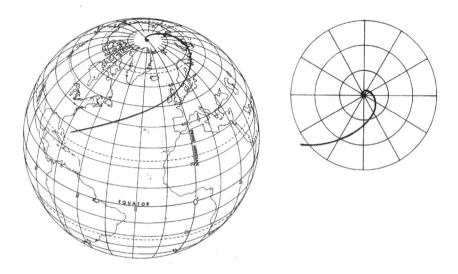

Figure 17.1: A rhumb line and its projection

17.2, from Lohne (1979) p. 273. The spiral of angle $55°$ is approximated by a polygon with sides $s_1, s_2, s_3, \ldots$, which yield triangles $T_1, T_2, T_3, \ldots$ when connected to the origin p. The triangles $T_1, T_2, T_3, \ldots$ can be re-assembled to form triangle ABT, whose area therefore equals that of the spiral (when the areas of overlapping turns are added together). Also

$$BT + TA = s_1 + s_2 + s_3 + \cdots = \text{length of the spiral.}$$

When the approximation is made with shorter sides $s_1', s_2', s_3', \ldots$, but otherwise in the same way, the *same triangle ABT* results: the isosceles triangle with base a and base angles $55°$. Hence we have also found length and area of the smooth curve.

Harriot's work was not published, and the arc length of the equiangular spiral was rediscovered by Torricelli (1645). Gradually the problem of arc length became understood more systematically as a problem of integration, though usually a rather intractable one. The first solution for an algebraic curve was for the "semicubical parabola" $y^2 = x^3$, by Neil and Heuraet in 1657. Soon after this Wren solved the problem for the cycloid, and his solution was given by Wallis (1659). A remarkable feature of Wren's result is that the length of one arch of the cycloid is a rational multiple (namely, 4) of the diameter of the generating circle.

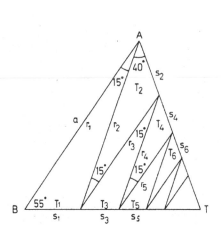
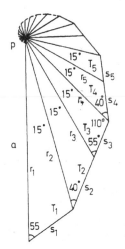

Figure 17.2: Constructing the area of a spiral

As mentioned in Section 13.5, other extraordinary properties of the cycloid are related to mechanics, and one of these will be reinterpreted geometrically in the next section. One transcendental curve that we did not discuss in connection with mechanics is the *tractrix* of Newton (1676b). Newton defined this curve by the property that the length of its tangent from point of contact to the *x*-axis is constant (Figure 17.3). It follows that the curve satisfies

$$\frac{dy}{ds} = \frac{y}{a},$$

where *s* denotes arc length. By using $ds = \sqrt{dx^2 + dy^2}$, this differential equation can be solved to give

$$x = a \log \frac{a + \sqrt{a^2 - y^2}}{y} - \sqrt{a^2 - y^2},$$

the equation for the curve given, in more geometric language, by Huygens (1693b). Huygens pointed out that the curve could be interpreted as the path of a stone pulled by a string of length *a* (hence the name "tractrix"). Thus the tractrix, too, has some mechanical significance. In fact it can be constructed from the famous mechanical curve, the catenary, by a method we shall see in the next section. However, its most important role was in the generation of the *pseudosphere*, a surface discussed in Section 17.4.

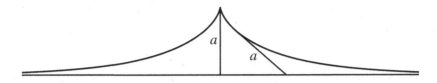

Figure 17.3: The tractrix

The arc length of $y^2 = x^3$ is today a fairly routine exercise with the arc length

integral $\int \sqrt{1 + \left(\dfrac{dy}{dx}\right)^2}\, dx.$

17.1.1 Show that the arc length of $y = x^{3/2}$ between O and $x = a$ is

$$\frac{8}{27}\left(\left(1 + \frac{9a}{4}\right)^{3/2} - 1\right).$$

Likewise, it is easy for us to derive properties of the logarithmic spiral from its polar equation and knowledge of the exponential function.

17.1.2 Show that the logarithmic spiral is *self-similar*. That is, magnifying $r = e^{k\theta}$ by a factor m to $r = me^{k\theta}$ gives a curve that is congruent to the original (in fact, it results from a rotation of the original).

Jakob Bernoulli was so impressed by this property of the logarithmic spiral that he arranged to have the spiral engraved on his tombstone, with a motto: *Eadem mutata resurgo* ("Though changed, I arise again the same"). (See Jakob Bernoulli (1692) p. 213.)

17.1.3 Deduce the equiangular property of the logarithmic spiral from its self-similarity.

The equation of the tractrix given above can be derived as follows.

17.1.4 Explain why the constant tangent property implies $\frac{dy}{ds} = \frac{y}{a}$, then multiply both sides of this equation by $\frac{ds}{dx} = \sqrt{1 + (\frac{dy}{dx})^2}$, and deduce that

$$\frac{dx}{dy} = \pm \frac{\sqrt{a^2 - y^2}}{y}.$$

17.1.5 Check by differentiation that $x = a \log \dfrac{a + \sqrt{a^2 - y^2}}{y} - \sqrt{a^2 - y^2}$ satisfies the differential equation found in Exercise 17.1.4, and also show that x has the appropriate value when $y = a$.

17.2   Curvature of Plane Curves

One of the most important ideas in differential geometry is that of *curvature*. The development of this idea from curves to surfaces and then to higher-dimensional spaces has had many important consequences for mathematics and physics, among them clarification of both the mathematical and physical meaning of "space," "space-time," and "gravitation." In this section we shall look at the beginnings of the theory of curvature in the 17th-century theory of curves. As discussed here, this theory concerns *plane* curves only; space curves involve an additional consideration of *torsion* (twisting), which will not concern us.

Just as the direction of a curve C at point P is determined by its straight-line approximation, that is, tangent, at P, the curvature of C at P is determined by an approximating circle. Newton (1665c) was the first to single out the circle that defines the curvature: the circle through P whose center R is the limiting position of the intersection of the normal through P and the normal through a nearby point Q on the curve (Figure 17.4). R is called the *center of curvature*, $RP = \rho$ the *radius of curvature*, and $1/\rho = \kappa$ the *curvature*. It follows that the circle of radius r has constant curvature $1/r$. The only other curve of constant curvature is the straight line, which has curvature 0. This is a consequence of the formula for curvature discovered by Newton (1671):

$$\rho = \frac{[1 + (dy/dx)^2]^{3/2}}{d^2y/dx^2} .$$

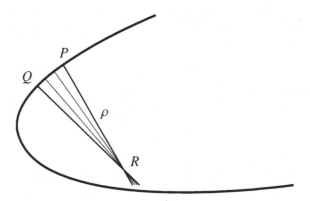

Figure 17.4: Normals through nearby points on a curve

There is an interesting relationship between a curve C and the locus C' of the center of curvature of C. C is the so-called *involute* of C', which, intuitively speaking, is the path of the end of a piece of string as it is unwound from C' (Figure 17.5). It is intuitively clear that Q, the end of the string, is instantaneously moving in a circle with center at P, the point where the string is tangential to C'.

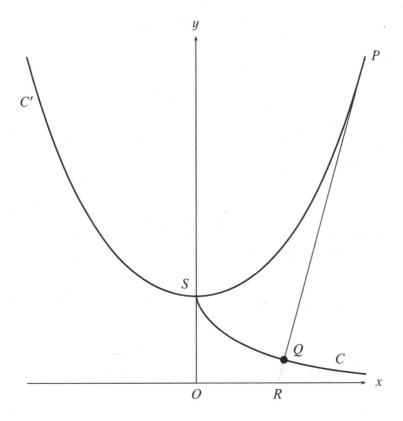

Figure 17.5: Construction of the involute

The geometric property of the cycloid that Huygens (1673) used to design the cycloidal pendulum (Section 13.5) can now be seen as simply this: the involute of a cycloid is another cycloid. Two other stunning results on involutes were obtained by the Bernoulli brothers. Jakob Bernoulli (1692) found that the involute of the logarithmic spiral is another logarithmic spiral, and Johann Bernoulli (1691) found that the tractrix is the involute of the catenary.

Another useful and intuitive definition of curvature, which turns out to be equivalent to the preceding one, was given by Kaestner (1761). He defined curvature as the rate at which the tangent turns, that is, $d\theta/ds = \lim_{\Delta x \to 0}(\Delta\theta/\Delta s)$, where $\Delta\theta$ is the angle between the tangents at points separated by an arc of the curve of length Δs. It follows from this definition that $\int_C \kappa\, ds = 2\pi$ for a simple closed curve C, since the tangent makes one complete turn on a circuit around C. We shall see in Section 17.6 that this result has a very interesting generalization for curves on nonplanar surfaces.

EXERCISES

Despite the complexity of the Newton curvature formula, it is easy enough to solve for y when the curvature κ is zero.

17.2.1 Use the formula to show that $\kappa = 0$ implies that y is a linear function of x.

17.2.2 Show that $d\theta/ds = 1/r$ for the circle of radius r, and deduce that $d\theta/ds = \kappa$ for any curve.

The description of the tractrix as the involute of the catenary is convenient for studying the pseudosphere. We therefore work out some steps in this approach in the following exercises. The curve C' in Figure 17.5 is now assumed to be the catenary $y = \cosh x$, which meets the y-axis at the point S where $y = 1$.

17.2.3 Using the arc-length integral on the catenary $y = \cosh x$ between $S = (0, 1)$ and $P = (\sigma, \cosh\sigma)$, show that

$$\text{arc length } PS = \sinh\sigma = PQ.$$

17.2.4 Also find the equation of the tangent at P, and use it to show that $R = (\sigma - \coth\sigma, 0)$. Then use the value of PQ to show that

$$QR = \frac{1}{\sinh\sigma} = \frac{1}{PQ}.$$

17.2.5 Finally, use the length of PQ again to show that $Q = (\sigma - \tanh\sigma, \operatorname{sech}\sigma)$, and show that the parametric equations of the tractrix C,

$$x = \sigma - \tanh\sigma, \quad y = \operatorname{sech}\sigma,$$

imply the cartesian equation of the tractrix (with $a = 1$),

$$x = \log\frac{1 + \sqrt{1 - y^2}}{y} - \sqrt{1 - y^2}.$$

17.3   Curvature of Surfaces

The first approach to defining curvature at a point P of a surface S in three-dimensional space was to express it in terms of the curvature of the plane curves, by considering sections of S by planes through the normal at P. Of course, different planes normal to the surface at P may cut the surface in quite different curves, with different curvatures, as the example of the cylinder shows (Figure 17.6).

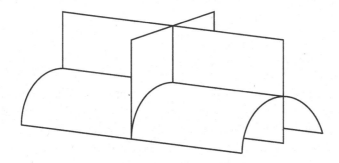

Figure 17.6: Sections of the cylinder

However, among these curves there will be one of maximum curvature and one of minimum curvature (which may be negative, since we give a sign to curvature according to the side on which the center of curvature lies). Euler (1760) showed that these two curvatures κ_1 and κ_2, called the *principal curvatures*, occur in perpendicular sections and that together they determine the curvature κ in a section at angle α to one of the principal sections by

$$\kappa = \kappa_1 \cos^2 \alpha + \kappa_2 \sin^2 \alpha.$$

This is as far as one can go as long as the curvature of surfaces is subordinated to the curvature of plane curves. A deeper idea occurred to Gauss in the course of his work in geodesy (surveying and mapmaking): curvature of a surface may be detectable *intrinsically*, that is, by measurements that take place entirely on the surface. The curvature of the earth, for example, was known on the basis of measurements made by explorers and surveyors, *not* (in the time of Gauss) by viewing it from space. Gauss (1827) made the extraordinary discovery that the quantity $\kappa_1 \kappa_2$ can be defined intrinsically and hence can serve as an intrinsic measure of curvature. He was so proud of this result that he called it the *theorema egregium* (excellent theorem).

It follows in particular that $\kappa_1\kappa_2$, which is called the *Gaussian curvature*, is unaffected by bending (without creasing or stretching).

The plane, for example, has $\kappa_1 = \kappa_2 = 0$ and thus zero Gaussian curvature. Hence so has any surface obtained by bending a plane, such as a cylinder. We can verify the theorema egregium in this case, because one of the principal curvatures of a cylinder is obviously zero.

Surfaces S_1, S_2 obtained from each other by bending are said to be *isometric*. More precisely, S_1 and S_2 are isometric if there is a one-to-one correspondence between points P_1 of S_1 and points P_2 of S_2 such that

distance between P_1 and P_1' in S_1 = distance between P_2 and P_2' in S_2,

where the distances are measured *within* the respective surfaces. A more precise statement of the theorema egregium then is: if S_1, S_2 are isometric, then S_1, S_2 have the same Gaussian curvature at corresponding points. The converse statement is not true: there are surfaces S_1, S_2 that are not isometric even though there is a one-to-one (and continuous) correspondence between them for which Gaussian curvature is the same at corresponding points. An example is given in Strubecker (1964, Vol. 3, p. 121), involving surfaces of nonconstant Gaussian curvature.

For surfaces of constant Gaussian curvature there is better agreement between isometry and curvature, as we shall see in the next section. From now on, unless otherwise qualified, "curvature" will mean Gaussian curvature.

17.4   Surfaces of Constant Curvature

The simplest surface of constant positive curvature is the sphere of radius r, which has curvature $1/r^2$ at all points. Other surfaces of curvature $1/r^2$ may be obtained by bending portions of the sphere; however, all such surfaces have either edges or points where they are not smooth, as was proved by Hilbert (1901). The plane, as we have observed, has zero curvature, and so have all surfaces obtained by bending the plane or portions of it.

It remains to investigate whether there are surfaces of constant *negative* curvature. In ordinary space, such a surface has principal curvatures of opposite sign at each point, giving it the appearance of a saddle (Figure 17.7). A number of surfaces of constant negative curvature were given by Minding (1839). The most famous of them is the *pseudosphere*, the surface of revolution obtained by rotating a tractrix about the x-axis (Figure 17.8).

This surface was investigated as early as 1693 by Huygens, who found its surface area, which is finite, and the volume and center of mass of the solid it encloses, which are also finite (Huygens (1693a)).

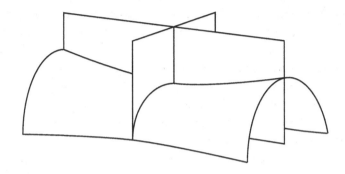

Figure 17.7: A saddle

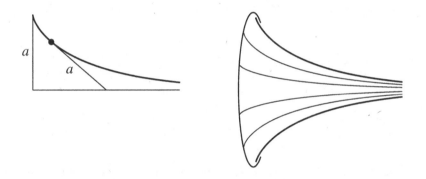

Figure 17.8: The tractrix and the pseudosphere

The pseudosphere is in some ways the negative-curvature counterpart of the cylinder, and hence one may wonder whether there is a surface of constant negative curvature that is more like a plane. Hilbert (1901) proved that there is no smooth unbounded surface of constant negative curvature in ordinary space, so this rules out planelike surfaces and also accounts for the "edge" on the pseudosphere. One can, however, obtain a "plane" of negative curvature by introducing a nonstandard notion of length into the Euclidean plane. This discovery of Beltrami (1868a) is discussed in the next chapter, along with other implications of negative curvature for non-Euclidean geometry.

These geometric implications can also be glimpsed if we return to the question whether surfaces S_1, S_2 of equal curvature are isometric. Even with constant curvature this is still not true, since a plane is not isometric to a cylinder. What *is* true, though, is that any sufficiently small portion of the plane can be mapped isometrically into any part of the cylinder. Minding (1839) showed that the analogous result is true for any two surfaces S_1, S_2 of the same constant curvature. Taking $S_1 = S_2$, this result can be interpreted as saying that *rigid motion* is possible within S_1; a body within S_1 can be moved, without any shrinking or stretching, to any part of S_1 large enough to contain it. The latter restriction is necessary, for example, for the pseudosphere since it becomes arbitrarily narrow as $x \to \infty$.

The possibility of rigid motion was fundamental to Euclid's geometry of the plane, and with the discovery of curved surfaces that support rigid motion, Euclid's geometry could be seen as a special case—the zero curvature case—of something broader. The broader notion of geometry on a surface begins to take shape once one has an appropriate notion of "straight line." This is developed in the next section.

Exercises

The construction of the tractrix as the involute of the catenary in Section 17.2 gives a remarkable insight into the two principal curvatures of the pseudosphere, enabling us to see why the pseudosphere has constant negative curvature.

17.4.1 Interpreting PQ in Figure 17.5 as the radius of curvature of the tractrix, and hence as the curvature of a section of the pseudosphere, suggest an interpretation of QR as a radius of curvature.

17.4.2 Assuming that PQ and QR are in fact principal radii of curvature, deduce from Exercise 17.2.4 that

Gaussian curvature of the pseudosphere at any point $= -1$.

17.5   Geodesics

A "straight line," or *geodesic* as it is called, can be defined equivalently by a shortest-distance property or a zero-curvature property. The shortest-distance definition was historically first, even though it is mathematically deeper and subject to the inconvenience that a geodesic segment is *not* necessarily the shortest path between two points. On a sphere, for example, there are two geodesic segments between two nearby points P_1, P_2: the short portion and the long portion of the great circle through P_1, P_2. We

can cover both by saying that a geodesic gives the shortest distance between any two of its points that are sufficiently close together. In talking
about shortest distance, even between nearby points P_i, P_j, one still has
the calculus of variations problem of finding which curve from P_i to P_j has
minimum length. Nevertheless, this is how geodesics were first defined, by
Jakob and Johann Bernoulli; and Euler (1728a) found a differential equation for geodesics from this approach.

A more elementary approach is to define the *geodesic curvature* κ_g at
P of a curve C on a surface S as the ordinary curvature of the orthogonal projection of C in the tangent plane to S at P. As one might expect,
geodesic curvature can also be defined intrinsically, and κ_g was introduced
in this way by Gauss (1825). A geodesic is then a curve of zero geodesic
curvature. This is the definition of Bonnet (1848).

The latter definition immediately shows that great circles on the sphere
are geodesics, since their projections onto tangent planes are straight lines.
Other examples are the horizontal lines, vertical circles, and helices on the
cylinder (Figure 17.9). These all come from straight lines on the plane that
is rolled up to form the cylinder. Geodesics on the pseudosphere, and other
surfaces of negative curvature, are not all so simple to describe. However,
the next chapter shows that they become simple when one maps the surface
of constant negative curvature suitably onto a plane.

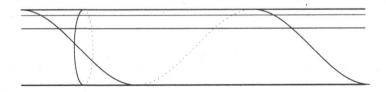

Figure 17.9: Geodesics on the cylinder

EXERCISES

17.5.1 Are the circles on the pseudosphere, in planes perpendicular to its axis,
geodesics? Give a qualitative argument to support your answer.

It may be easier to answer this question if one first considers the *cone*, a
surface also obtained by bending the plane. To avoid worrying about the apex,
where the cone is not smooth, we omit this point.

17.5.2 Show that the circles on the cone, in planes perpendicular to its axis, are *not* geodesics.

17.5.3 Show that there are nonsmooth geodesics on the cone, that is, curves of geodesic curvature zero except at certain points where they have no tangent.

17.6   The Gauss–Bonnet Theorem

In Section 17.2 we observed that

$$\int_C \kappa\, ds = 2\pi$$

for a simple closed curve C in the plane. This result has a profound generalization to curved surfaces known as the *Gauss–Bonnet* theorem. On a curved surface, κ must be replaced by the geodesic curvature κ_g, and the theorem states that

$$\int_C \kappa_g\, ds = 2\pi - \iint_{\mathcal{R}} \kappa_1 \kappa_2\, dA,$$

where A denotes area and $\mathcal{R}$ is the region enclosed by C (Bonnet (1848)). Gauss himself published only a special case, or rather the limit of a special case, in which C is a geodesic triangle. In this case, of course, $\kappa_g = 0$ along the sides of C, and κ_g becomes infinite at the corners. By rounding off the corners by small arcs ds, one sees (Figure 17.10) that

$$\int_{C^*} \kappa_g\, ds \cong \alpha' + \beta' + \gamma',$$

where α', β', γ' are the external angles of the triangle and C^* is the rounded approximation to the triangle C.

Then by letting the size of the round-offs tend to zero one gets

$$\int_{C^*} \kappa_g\, ds = \alpha' + \beta' + \gamma' = 3\pi - (\alpha + \beta + \gamma),$$

where α, β, γ are the internal angles of the triangle. Introducing the quantity

$$(\alpha + \beta + \gamma) - \pi,$$

called the *angular excess* of the triangle (because an ordinary triangle has angle sum π), we have

$$\int_C \kappa_g\, ds = 2\pi - \text{angular excess},$$

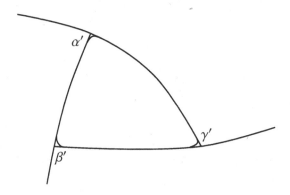

Figure 17.10: Rounding off a geodesic triangle

and the result of Gauss (1827) was that

$$\text{angular excess} = \iint_{\mathcal{R}} \kappa_1 \kappa_2 \, dA.$$

We see that the *integral* of the Gaussian curvature has a more elementary geometric meaning than the curvature $\kappa_1 \kappa_2$. It appears, in fact, that Gauss thought about angular excess first, then the curvature integral, and only last about the curvature itself. The decomposition into principal curvatures probably came later, when he reworked his geometric ideas into analytic form, reversing the order of discovery in the process. Dombrowski (1979) made a plausible reconstruction of the original approach, using clues from the unpublished work of Gauss.

The role of angular excess can be seen more plainly in the case of constant curvature $\kappa_1 \kappa_2 = c$. In this case

$$\text{angular excess} = c \times \text{area of } \mathcal{R},$$

so the angular excess gives a measure of area, a result Gauss claimed, in a letter (1846a), to have known in 1794. In fact, the special case of this result for the sphere was known to Thomas Harriot in 1603 (see Lohne (1979)). Harriot's elegant proof goes as follows (Figure 17.11).

Prolonging the sides of triangle ABC partitions the sphere into four pairs of congruent, diametrically opposite triangles (Figure 17.11a). We denote the area of ABC and its diametric opposite $A'B'C'$ by $\Delta_{\alpha\beta\gamma}$. The

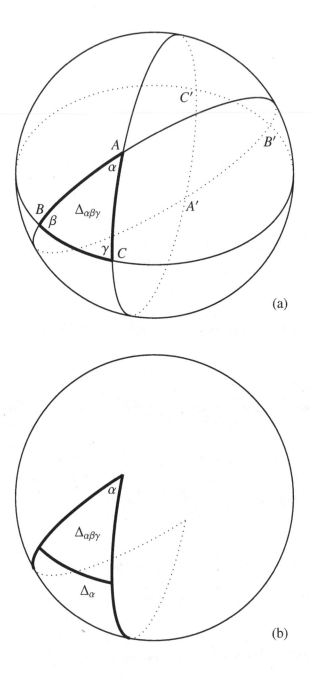

Figure 17.11: Area of a spherical triangle

other three pairs represent areas $\Delta_\alpha, \Delta_\beta, \Delta_\gamma$, which complement $\Delta_{\alpha\beta\gamma}$ in "slices" of the sphere of angles α, β, γ, respectively (Figure 17.11b).

Since the area of a slice is $2r^2$ times the angle, where r is the radius of the sphere, we have

$$\Delta_{\alpha\beta\gamma} + \Delta_\alpha = 2r^2\alpha,$$
$$\Delta_{\alpha\beta\gamma} + \Delta_\beta = 2r^2\beta,$$
$$\Delta_{\alpha\beta\gamma} + \Delta_\gamma = 2r^2\gamma,$$

whence, by addition,

$$3\Delta_{\alpha\beta\gamma} + (\Delta_\alpha + \Delta_\beta + \Delta_\gamma) = 2r^2(\alpha + \beta + \gamma). \tag{1}$$

On the other hand,

$$2(\Delta_{\alpha\beta\gamma} + \Delta_\alpha + \Delta_\beta + \Delta_\gamma) = \text{area of sphere} = 4\pi r^2,$$

and substituting this in (1) gives

$$\Delta_{\alpha\beta\gamma} = r^2(\alpha + \beta + \gamma - \pi)$$

as required, since $1/r^2$ = curvature of the sphere.

Gauss was interested in the counterpart of this result for negative curvature, in which case the angle sum of a triangle is less than π and one has angular *defect* rather than angular excess. His investigations in this case led him not only to Gaussian curvature but also to non-Euclidean geometry.

EXERCISES

It is surprising at first that area on the sphere should be measured by angles rather than lengths. However, there are general reasons (apart from the Gauss–Bonnet theorem) why area should be measured by angular excess, and this idea fails only where angular excess is zero—that is, in the Euclidean plane.

17.7.1 Consider a triangle Δ split into two triangles Δ_1 and Δ_2 by a line through a vertex. Show that

$$\text{excess}(\Delta) = \text{excess}(\Delta_1) + \text{excess}(\Delta_2).$$

17.7.2 Deduce from Exercise 17.7.1 that if any polygon Π is split into triangles Δ_i, then

$$\text{excess}(\Pi) = \text{excess}(\Delta_1) + \text{excess}(\Delta_2) + \cdots.$$

Thus the angular excess function has the same *additive* property as an area function. It can be shown that any additive function, provided it is continuous, is a constant multiple of area (see Bonola (1912), p. 46).

17.7   Biographical Notes: Harriot and Gauss

The discoveries of Thomas Harriot described in this chapter and the last seem to entitle him to a secure place in the history of mathematics, perhaps alongside others who made a few penetrating contributions, such as Desargues and Pascal. Unfortunately, Harriot's place is not yet clear. It was clouded by exaggerated claims made by 17th- and 18th-century admirers, and until recently the disorder and inaccessibility of his papers have made any claims difficult to verify. (About 140 pages, out of approximately 4000 pages that Harriot left behind, have been translated and annotated by Stedall (2003). These do not include any of his work on geometry.) In addition, Harriot was a very secretive man, and little is known about his life. He lived in the world of Sir Walter Raleigh, Christopher Marlowe, and Guy Fawkes—a lurid and fascinating world, but a very dangerous one— and probably believed that secrecy was necessary for his survival. As a result, our present understanding of Harriot, as the biography by Shirley (1983) explains, is based on a meager set of facts about him and a good deal of extrapolation from knowledge of his less discreet contemporaries.

All that we know of Harriot's early life comes from a record of his entry into Oxford University in December 1577, stating that his age was then 17 and his father "plebeian." The only other information about his family comes from his will of 1621, which mentions a sister and a cousin. It seems probable that he never had children and never married. At Oxford, Harriot gained the standard bachelor's degree in classics, but he must have picked up some Euclid and astronomy, which were offered to master's candidates. He would also have heard Richard Hakluyt, author of the famous *Voyages*, who was then just beginning to lecture on the geography of the New World opened up by 16th-century navigators.

It was probably Hakluyt who inspired Harriot to travel to London in the early 1580s and seek out Sir Walter Raleigh. Raleigh was then about 30 and the most powerful member of Queen Elizabeth's inner circle, with grand dreams of wealth through exploration. Harriot must have impressed Raleigh with his grasp of the mathematical problems of navigation, for around 1583 he joined Raleigh's household as an instructor, with considerable freedom to conduct his own research. Harriot held classes in navigation as part of Raleigh's preparations for the voyage to Virginia in 1585, led by Sir Richard Grenville, which was the first attempt at British settlement in the New World. Although the attempt was unsuccessful, it was the biggest adventure of Harriot's life. He studied Indian languages and customs and wrote a book on the settlement entitled *A Brief and True Report of the New Found Land of Virginia* (1588), the only one of Harriot's works published in his lifetime.

With Raleigh as patron, Harriot was financially secure, and he remained so for the rest of his life. However, he was also at the mercy of Raleigh's political fortunes. By 1592, the 40-year-old Raleigh was finding his role as the favorite of the nearly 60-year-old queen increasingly irksome, and he secretly married one of the queen's servants, Elizabeth Throckmorton. He may have married her as early as 1588, but at any rate the secret was out when Lady Raleigh gave birth to a son in 1592, and Raleigh was imprisoned in the Tower of London. Harriot did not suffer for his direct association with Raleigh, but through him he was linked with Christopher Marlowe, at the latter's sensational trial for atheism in 1593.

Marlowe, the dramatist, had a secret life in espionage and other unsavory activities, and any number of accusations could have been made against him, though which ones were true it is now impossible to say. Unfortunately for Harriot, the second of Marlowe's offenses against religion was said to be that "He affirmeth that Moyses was but a Jugler, and that one Heriots being Sir W. Raleighs man can do more than he." As it happened, the proceedings were terminated by the murder of Marlowe in a tavern brawl, and Harriot was not called to testify, but he was left publicly under suspicion.

Harriot did not desert Raleigh, but he was prudent enough to seek another patron, and he found one in Henry Percy, the Ninth Earl of Northumberland. Henry, known as the "Wizard Earl," was a friend of Raleigh and, like him, interested in science and philosophy. In 1593 he gave Harriot a grant, later to become an annual pension of £80. This sum was twice the salary of the best-paid teachers of the time and it enabled Harriot to maintain a house and servants on the earl's property on the Thames near London. The house, known as Sion House, remained Harriot's home and laboratory for the rest of his life.

But once again Harriot was unlucky in his choice of friends. The earl's cousin, Thomas Percy, was the man who rented the cellars under the Houses of Parliament in the famous plot to blow up King James I with gunpowder on November 5, 1605. Harriot was dragged into the investigation and imprisoned for a short time on suspicion that he had secretly cast a horoscope of the king. James I was terrified of black magic and indiscriminately viewed all mathematicians as astrologers and magicians. In the end, though, no evidence against Harriot was found, and it was the earl who suffered more, being imprisoned in the Tower from 1605 to 1621.

Meanwhile, Raleigh had fared even worse. After several spells in the Tower, he was released in 1616 to lead an expedition in search of the mythical city of gold, El Dorado. When the expedition returned in a shambles, Raleigh was rearrested and executed on an old treason charge from 1603. One of the few personal documents preserved by Harriot is his summary of the speech Raleigh made at the scene of his execution in 1618 (see Shirley (1983), p. 447).

A month after Raleigh's death, a bright comet appeared in the skies, and Harriot's observations of it were his last major scientific endeavor. He had been suffering for some years from a painful cancer of the nose and finally succumbed to it on a visit to London in 1621. He was buried at St. Christopher's Church in Threadneedle Street, later destroyed in the Great Fire of 1666. The site is now part of the Bank of England, where a replica of Harriot's original monument was installed on July 2, 1971, the three hundred and fiftieth anniversary of his death.

Figure 17.12: Thomas Harriot

Figure 17.12 is a portrait from Trinity College, Oxford. The latin inscription in the top left corner says that the portrait was made in 1602, when the sitter was aged 3 2. This does not square with the record of Harriot's entry to Oxford (which would make him 42 in 1602), but the portrait is nevertheless believed to be of him.

Carl Friedrich Gauss was born in Brunswick (Braunschweig) in 1777 and died in Göttingen in 1855. He was the only child of Gebhard Gauss and Dorothea Benze, though his father had another son from a previous marriage. Gebhard earned his living mainly from manual labor, but he also did a little accounting, and Gauss is said to have corrected an error in his father's arithmetic at the age of three. (It should be borne in mind here that stories about Gauss's youth were told by Gauss himself in old age, and in a few cases there is evidence that he was prone to exaggerate his own precocity.) Gauss was not close to his father and believed that his genius was inherited from his mother. He started school in 1784 and his teacher, Büttner, soon recognized his ability and obtained advanced books for him. Büttner's assistant, Martin Bartels (1769–1836), also gave Gauss special attention. Bartels was himself a beginning mathematician who later became professor at the University of Kazan and the teacher of Lobachevsky (see next chapter).

Gauss entered secondary school in 1788, and in 1791 he won an annual grant from the Duke of Brunswick, something like a government scholarship. He was also selected to enter the Collegium Carolinum, a new scientific academy for outstanding secondary students. In his years there, 1792–1795, Gauss studied the works of Newton, Euler, and Lagrange and began investigations of his own, mainly numerical experiments on such things as the arithmetic–geometric mean. In 1795 he left Brunswick to study at Göttingen in the adjoining state of Hannover, which was then ruled by George III of England. The duke would have preferred Gauss to remain in Brunswick, and the local university of Helmstedt, but continued his financial support nevertheless. Gauss actually chose Göttingen because of its better library and later spoke very disparagingly of its mathematics professor, Kaestner. It is true that Gauss's student achievements, which began with his construction of the regular 17-gon (Section 2.3) and culminated in his proof of the fundamental theorem of algebra (Section 14.7), dwarfed those of his teachers. Still, one wonders whether Kaestner's definition of curvature (Section 17.2) might not have been useful to Gauss when he took up differential geometry later.

Figure 17.13: Carl Friedrich Gauss

Gauss returned to Brunswick in 1798 and lived there until 1807. Figure 17.13 is a portrait of him from this period, which was the happiest and most productive of his life. Gauss published his great work on number theory, the *Disquisitiones arithmeticae*, in 1801, made a spectacular entry into astronomy in the same year by predicting the position of the asteroid Ceres, and married Johanna Osthoff in 1805. Writing to his friend Farkas Bolyai in 1804, Gauss was uncharacteristically warm and open when it came to Johanna:

> The beautiful face of a madonna, a mirror of peace of mind and health, tender, somewhat fanciful eyes, a blameless figure— this is one thing; a bright mind and an educated language— this is another; but the quiet, serene, modest and chaste soul of an angel who can do no harm to any creature—that is the best.

> Translation from Kaufmann-Bühler (1981), p. 49

If only Johanna had lived longer, Gauss might have become a very different man. But in 1809 she died, shortly after giving birth to their third child. Gauss was devastated by the blow and never quite recovered his equilibrium.

Less than a year after Johanna's death, he married Minna Waldeck, the daughter of a Göttingen professor. Unlike Johanna, who was a tanner's daughter, Minna had social status and pretensions that caused Gauss uneasiness and embarrassment. Soon after their engagement, for example, he had to tell Minna not to write to his mother, since his mother could not read. Minna also suffered from poor health, and after the couple had had three children between 1811 to 1816 she became virtually a permanent invalid. Gauss found this burden difficult to bear, and he compounded his problems by unsympathetic treatment of his children. The family drama came to a head in 1830, when their eldest son, Eugen, emigrated to America after a row with his father. The following year Minna died of tuberculosis.

During this unhappy period Gauss was less productive mathematically, but this was not due to family troubles so much as his choice of career. He had become director of the Göttingen observatory in 1807, and in 1817 substituted geodesy for some of his astronomical duties, doing arduous field work every summer from 1818 to 1825 for the geodetic survey of Hannover. Gauss appears to have seldom regretted this choice of career— he disliked teaching and thought that other mathematicians had little to teach him—but it cannot be said that his contributions to astronomy and geodesy were as great as his contributions to mathematics. Indeed, the best things to come out of his geodetic work were his theory on conformal mapping and complex functions (Section 16.2) and his intrinsic notion of curvature (Section 17.3).

In the 1830s Gauss experienced something of a rebirth with the arrival of the young physicist Wilhelm Weber in Göttingen. The two collaborated enthusiastically in the investigation of magnetism, with Gauss making contributions to both the theory and practice (the electric telegraph). However, their partnership was broken in 1837 when Weber was fired for his courageous refusal to swear an oath of allegiance to the new king of Hannover.

Among the few bright spots of Gauss's later years were his students Eisenstein and Dedekind, as well as Riemann's lecture on the foundations of geometry in 1854. After spending most of his life aloof from other mathematicians, Gauss must have been comforted to find at last that there *were* students capable of understanding his ideas and carrying them further. We can only wonder what might have been if he had made this discovery earlier.

18

Non-Euclidean Geometry

PREVIEW

Surprisingly, the geometry of curved surfaces throws light on the geometry of the plane. More than 2000 years after Euclid formulated axioms for plane geometry, differential geometry showed that *the parallel axiom does not follow from the other axioms of Euclid.*

It had long been hoped that the parallel axiom followed from the others, but no proof had ever been found. In particular, no contradiction had been derived from the contrary hypothesis, P_2, that there is *more* than one parallel to a given line through a given point. In the 1820s, Bolyai and Lobachevsky proposed that the consequences of P_2 be accepted as a new kind of geometry—*non-Euclidean* geometry.

To prove that no contradiction follows from P_2, however, one needs to find a *model* for P_2 and the other axioms of Euclid. One seeks a mathematical structure, containing objects called "points" and "lines," that satisfies Euclid's axioms with P_2 in place of the parallel axiom.

Such a structure was first found by Beltrami (1868a), in the form of a surface of constant negative curvature with geodesics as its "lines." By various mappings of this surface, Beltrami found other models, including a *projective* model in which "lines" are line segments in the unit disk, and *conformal* models in which "angles" are ordinary angles.

Finally, Poincaré (1882) showed that Beltrami's conformal models arise naturally in complex analysis. Papers had already been published with pictures of patterns of non-Euclidean "lines," most notably Schwarz (1872). Thus, non-Euclidean geometry was actually a part of existing mathematics, but a part whose geometric nature had not previously been understood.

J. Stillwell, *Mathematics and Its History*, Undergraduate Texts in Mathematics,   359
DOI 10.1007/978-1-4419-6053-5_18, © Springer Science+Business Media, LLC 2010

18.1   The Parallel Axiom

Until the 19th century, Euclid's geometry enjoyed absolute authority, both as an axiomatic system and as a description of physical space. Euclid's proofs were regarded as models of logical rigor, and his axioms were accepted as correct statements about physical space. Even today, Euclidean geometry is the simplest type of geometry, and it furnishes the simplest description of physical space for everyday purposes. Beyond the everyday world, however, lies a vast universe that can be understood only with the help of an expanded geometry. The expansion of geometric concepts initially grew from dissatisfaction with one of Euclid's axioms, the *parallel axiom*.

For our purposes, the most convenient statement of the parallel axiom is as follows:

Axiom P_1. For each straight line L and point P outside L there is exactly one line through P that does not meet L.

There are many other equivalent statements of Axiom P_1, some obviously fairly close to it, for example, Euclid's own:

> That if a straight line falling on two straight lines make the interior angles on the same side less than two right angles, the two straight lines, if produced indefinitely, meet on that side on which are the angles less than the two right angles.
>
> Heath (1925), p. 202

Other statements of Axiom P_1 are less obviously equivalent to it. For example,

 (i) The angle sum of a triangle $= \pi$ (Euclid).

 (ii) The locus of points equidistant from a straight line is a straight line. (al-Haytham, around 1000 CE).

 (iii) Similar triangles of different sizes exist (Wallis (1663); see Fauvel and Gray (1988), p. 510).

Thus a denial of the parallel axiom entails denial of (i), (ii), and (iii). A denial of (iii) means in particular that scale models would be impossible, since three points in the original object and the three corresponding points of a scale model would define similar triangles of different sizes.

Such unlikely consequences convinced many people that the parallel axiom was a logically necessary property of straight lines, already implied by the other axioms of Euclid, and so efforts were made to prove it outright.

The most tenacious attempt, entitled *Euclides ab omni naevo vindicatus* (Euclid cleared of every flaw), was made by Saccheri (1733). Saccheri's plan of attack began by subdividing the denial of the parallel axiom into two alternatives:

Axiom P_0. There is no line through P that does not meet L.

Axiom P_2. There are at least two lines through P that do not meet L.

The next step was to destroy each alternative by deducing a contradiction from it. He succeeded in deducing a contradiction from Axiom P_0, using other axioms of Euclid, such as the axiom that a straight line can be prolonged indefinitely. (Such additional assumptions are certainly necessary, since great circles on the sphere have some properties of straight lines, except that they are finite in length.)

Saccheri was less successful with Axiom P_2. The consequences he derived from it, hoping to obtain a contradiction, were as follows. Among the lines M through P that do not meet L are two extremes, M^+ or M^-, called *parallels* or *asymptotic lines* (Figure 18.1); any of these lines M strictly between M^+ and M^- has a common perpendicular with L and, moreover, the position of this perpendicular tends to infinity as M tends to M^+ or M^-. Although curious, these consequences of Axiom P_2 were not contradictory and Saccheri, sensing that the contradiction was slipping away from him, tried to overtake it by proceeding to infinity.

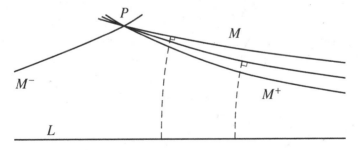

Figure 18.1: Asymptotic lines

He claimed that the asymptotic line M^+ would meet L at infinity and have a common perpendicular with it there. This was perhaps plausible, given similar arguments in projective geometry, though Euclid certainly would not have accepted it. But it *still* was not a contradiction. Saccheri merely claimed that such a conclusion was "repugnant to the nature of the straight line" (Saccheri (1733), p. 173), perhaps visualizing an intersection like Figure 18.2. But why should asymptotic lines not be tangential at infinity? History was to show that this was an appropriate resolution of Saccheri's "contradiction" (see Section 18.5). Thus Saccheri's results were not, as he thought, steps toward a proof of the parallel axiom; they were the first theorems of a *non-Euclidean* geometry in which Axiom P_2 replaces the parallel axiom.

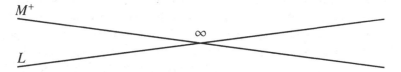

Figure 18.2: Hypothetical intersection at infinity

EXERCISES

The connection between the parallel axiom and the angle sum of a triangle is very direct and elegant.

18.1.1 Deduce, from Euclid's version of the parallel axiom, that a line falling on two *parallel* lines makes the interior angles sum to π.

18.1.2 Use Exercise 18.1.1 and the construction in Figure 18.3 (in which CD is parallel to AB) to show that $\alpha + \beta + \gamma = \pi$.

18.1.3 Deduce from Exercise 18.1.2 that the angle sum of any quadrilateral is 2π and, in particular, that squares exist.

Thus theorems mentioning squares, such as the Pythagorean theorem, can hold only when Euclid's parallel axiom is assumed.

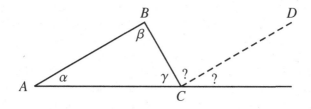

Figure 18.3: The angle sum of a triangle

18.2   Spherical Geometry

In rejecting P_0 because of its incompatibility with infinite lines, Saccheri avoided having to consider the most natural geometry in which P_0 holds, that of the sphere with great circles as "lines." Spherical geometry had been cultivated since ancient times to meet the needs of astronomers and navigators, and formulas for the side lengths and areas of spherical triangles were well known. But the sphere was considered part of Euclid's spatial geometry, so the axiomatic significance of spherical geometry was at first ignored. What did happen, though, was that the first explorations of Axiom P_2 were guided by the analogy of the sphere.

Lambert (1766) made the striking discovery that Axiom P_2 implies that the area of a triangle with angles α, β, γ is proportional to $\pi - (\alpha + \beta + \gamma)$, its angular defect. In other words,

$$\text{area} = -R^2(\alpha + \beta + \gamma - \pi)$$

for some positive constant R^2. Having rediscovered Harriot's theorem that

$$\text{area} = R^2(\alpha + \beta + \gamma - \pi)$$

for a triangle on the sphere of radius R, Lambert mused that one "could almost conclude that the new geometry would be true on a sphere of imaginary radius." What a sphere of radius iR might be was never explained, but the idea of using complex numbers to generate the formulas of a hypothetical geometry proved fruitful.

It was found that formulas derived from Axiom P_2 could also be obtained by replacing R by iR in corresponding formulas of spherical geometry. For example, Gauss (1831) deduced from Axiom P_2 that the circumference of a circle of radius r is $2\pi R \sinh r/R$. The same result follows by replacing R by iR in $2\pi R \sin r/R$, which is the circumference of a circle of radius r on the sphere (where, of course, r is measured *on* the spherical surface; see Exercise 18.2.1).

Klein (1871) called the geometry of Axiom P_2 *hyperbolic*. One reason for this is that its formulas involve hyperbolic functions, whereas those of spherical geometry involve circular functions. Lambert (1766) introduced the hyperbolic functions and noted their analogy with the circular functions, but he did not follow through with a complete translation of spherical formulas into hyperbolic formulas. This was first done by Taurinus

(1826), one of a small circle who corresponded with Gauss on geometric questions.

This gave hyperbolic geometry a second leg to stand on, but there was still nothing solid under its feet. Neither Gauss nor Taurinus seemed confident of finding a convincing interpretation of hyperbolic geometry, even though Gauss (1827) came remarkably close with the "Gauss–Bonnet" theorem. As mentioned in Section 17.6, this theorem shows that surfaces of constant negative curvature give a geometry in which angular defect is proportional to area, and Gauss knew that the pseudosphere is such a surface. Gauss's student Minding (1840) even showed that the hyperbolic formulas for triangles hold on the pseudosphere, but no one at that time commented on the likely importance of this result for hyperbolic geometry. Perhaps it was clear that the pseudosphere cannot serve as a "plane," because it is infinite in only one direction. Only in 1868, when Beltrami extended the pseudosphere to a true *hyperbolic plane*—a surface locally like the pseudosphere but infinite in all directions—was hyperbolic geometry finally placed on a firm foundation.

EXERCISES

18.2.1 Prove that the circumference of the circle C of radius r on the sphere of radius R (Figure 18.4) is $2\pi R \sin(r/R)$.

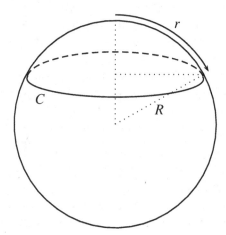

Figure 18.4: Radius and circumference on the sphere

18.2.2 Show that both $2\pi R \sin(r/R)$ and $2\pi R \sinh(r/R)$ tend to $2\pi r$ as $R \to \infty$.

These results show how even non-Euclidean geometry is "Euclidean in the small"—its formulas tend to the Euclidean formulas as size tends to zero. The same is true of the angle-sum formula for a triangle.

18.2.3 Deduce from Harriot's area formula that the angle sum of a spherical triangle tends to π as its size tends to zero.

18.3 Geometry of Bolyai and Lobachevsky

The most important contributors to hyperbolic geometry between Gauss and Beltrami were Lobachevsky and Bolyai, who published independent discoveries of the subject: Lobachevsky (1829) and János Bolyai (1832b). Because of their courage in advocating an unconventional geometry, Bolyai and Lobachevsky have been justly admired. Nevertheless, the immediate impact of their work was slight. Many of their results were already known to Gauss and his circle and could have been picked up from existing publications and personal contacts. Lambert (1766) and Taurinus (1826) were in print, and Bolyai's father, F. Bolyai, was a lifelong friend of Gauss, as was Lobachevsky's teacher Bartels. In any case their work, though more systematic and convincing than previous attempts, attracted very little attention at first. We have seen how the possibility of using differential geometry to justify hyperbolic geometry was overlooked until 1868. Up to that time, there seemed no reason to take hyperbolic geometry seriously.

In retrospect, of course, the theorems of Bolyai and Lobachevsky can be seen to unify the fragmentary results of their predecessors very nicely. They cover the basic relations between sides and angles of triangles (hyperbolic trigonometry), the measure of polygonal areas by angular defect, and formulas for circumference and area of circles. Lobachevsky (1836) broke new ground by finding volumes of polyhedra, which turn out to be far from elementary, involving the function $\int_0^{\theta} \log 2|\sin t|\, dt$.

Both Bolyai and Lobachevsky considered a three-dimensional space satisfying Axiom P_2 and made extensive use of a surface peculiar to this space, the *horosphere*. A horosphere is a "sphere with center at infinity," and it is *not* a hyperbolic plane. Wachter, a student of Gauss, observed in a letter of 1816 (published in Stäckel (1901)) that the geometry of the horosphere is in fact Euclidean. This astonishing result was rediscovered by Bolyai and Lobachevsky, and they anticipated that it would make Euclidean geometry subordinate to hyperbolic. We shall see in Section 18.5 how this view was vindicated in the work of Beltrami.

18.4   Beltrami's Projective Model

Interest in hyperbolic geometry was rekindled in the 1860s when unpublished work of Gauss, who had died in 1855, came to light. Learning that Gauss had taken hyperbolic geometry seriously, mathematicians became more receptive to non-Euclidean ideas. The works of Bolyai and Lobachevsky were rescued from obscurity and, approaching them from the viewpoint of differential geometry, Beltrami (1868a) was able to give them the concrete explanation that had eluded all his predecessors.

Beltrami was interested in the geometry of surfaces and he had found the surfaces that could be mapped onto the plane in such a way that their geodesics go to straight lines (Beltrami (1865)). They turned out to be just the surfaces of constant curvature. In the case of positive curvature, the sphere, such a mapping is central projection onto a tangent plane (Figure 18.5), though of course this maps only half the sphere onto the whole plane.

The mappings of surfaces of constant negative curvature, on the other hand, take the *whole* surface onto only *part* of the plane. Figure 18.6, from Klein (1928), shows some of these mappings (the middle one being of the pseudosphere).

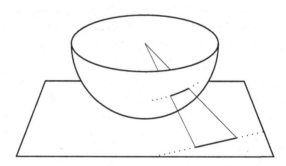

Figure 18.5: Central projection

Each negatively curved surface S is mapped onto a portion of the unit disk. Beltrami (1868a) realized that the disk can then be viewed as a natural extension of S to an "infinite plane," thus bypassing the problem of constructing "planelike" surfaces of constant negative curvature in ordinary space. Instead one takes the disk as the "plane," line segments within it as "lines," and "distance" between two points of the disk as the distance between their preimage points on the surface S. The function $d(P, Q)$, giv-

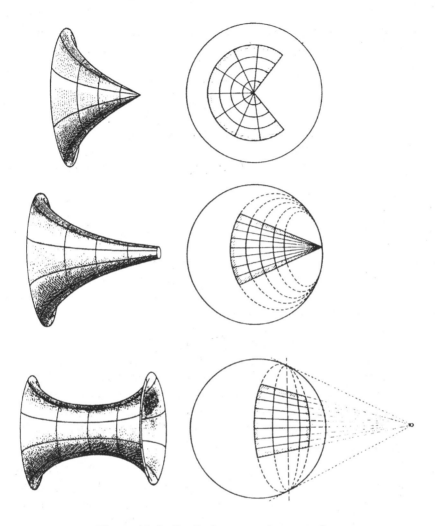

Figure 18.6: Geodesic-preserving mappings

ing "distance" between points P, Q of the disk in this way, turns out to be meaningful for all points inside the unit circle, so the notion of "distance" extends to the whole open disk. As Q approaches the unit circle, $d(P, Q)$ tends to infinity, so the "plane," and hence the "lines" in it, are indeed infinite with respect to this nonstandard "distance."

It follows that all the axioms of Euclid, except the parallel axiom, are satisfied with the new interpretation of "plane," "line," and "distance." Instead of the parallel axiom, one has of course Axiom P_2, since there is more than one "line" through a point P outside a given "line" L (Figure 18.7).

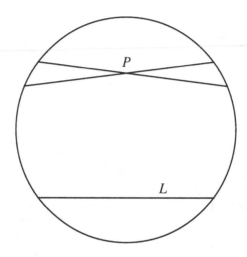

Figure 18.7: Failure of the parallel axiom

Beltrami also observed that the rigid motions of the "plane," since they preserve straight lines, are necessarily projective transformations. They are precisely those projective transformations of the plane that map the unit circle onto itself. Consequently, this model of the hyperbolic plane is often called the *projective model*. Cayley (1859) had already observed that these projective transformations could be used to define a "distance" $d(P, Q)$ in the unit disk—by saying $d(P, Q) = d(P', Q')$ if a transformation preserving the unit circle sends P to P' and Q to Q'—but he had not realized that the geometry obtained was that of Bolyai and Lobachevsky.

The pseudosphere is not entirely superseded by the projective model, since it remains the source of "real" distances and angles, whereas those in the projective model are necessarily distorted. One of the distinctive curves of the hyperbolic plane, the *horocycle*, or circle with center at infinity, is shown particularly clearly on the pseudosphere. If one imagines, following Beltrami (1868a), the pseudosphere wrapped by infinitely many turns of an infinitely thin covering, then the edge of this covering (along the rim of the pseudosphere) is a horocycle. The middle picture of Figure 18.6 shows the image of one turn of the covering, drawn solidly, and horocycles resulting from continued unwrapping are shown as dashed lines.

EXERCISES

Klein's three pictures illustrate the three types of *rigid motion* of the hyperbolic plane.

1. *Rotation*, in which one point of the plane is fixed and all other points move in *hyperbolic circles* about it. (A hyperbolic circle is the locus of a point moving at constant "distance" from a fixed point.)

2. *Limit rotation*, in which a point at infinity is fixed and all points of the plane move in horocycles centered on the fixed point at infinity.

3. *Translation*, in which a "line" moves along itself and the other points of the plane move along its *equidistant curves*. (An equidistant curve is the locus of a point moving at constant "distance" from a "line.")

18.4.1 Pick out *hyperbolic circles* and *equidistant curves* in the top and bottom pictures in Figure 18.6.

18.4.2 If the center of rotation in the top picture were not at the center of the disk, do you think the hyperbolic circles would be Euclidean circles?

18.4.3 Observe that equidistant curves at nonzero "distance" from the invariant "line" are *not* "lines." Does the translation move a point on an equidistant curve farther than a point on the invariant line?

18.4.4 Give an example of three points in the hyperbolic plane, not in a "line," that do not lie on a hyperbolic circle. (If this problem proves difficult, try it again after reading the next section.)

18.5 Beltrami's Conformal Models

The projective model of the hyperbolic plane distorts angles as well as lengths. One can see this with the asymptotic geodesics on the pseudosphere, which clearly tend to tangency at infinity yet are mapped onto lines meeting at a nonzero angle at the boundary of the unit disk (Figure 18.6). Beltrami (1868b) found that models with true angles—the so-called *conformal models*—could be obtained by sacrificing straightness of "lines." His basic conformal model is not, in fact, part of the plane but part of a hemisphere. It is erected over the projective model and its "lines" are vertical sections of the hemisphere (hence semicircles) over the "lines" of the projective model (Figure 18.8). The "distance" between points on the hemisphere is equal to the "distance" between the points beneath them in the projective model. Later we shall see that "distance" on the hemisphere also has a simple direct definition.

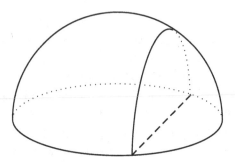

Figure 18.8: The hemisphere and the projective model

Two planar conformal models are obtained from the hemisphere model by stereographic projection, which, as we know from Section 16.2, preserves angles and sends circles to circles. The first of these is a disk (Figure 18.9) that, by change of scale, can again be taken as the unit disk. The second (Figure 18.10) is a half-plane, which we take to be the upper half-plane, $y > 0$. Since the "lines" in the hemisphere model are circular and orthogonal to the equator, "lines" in the planar conformal models are again circular, orthogonal to the boundary of the disk and half-plane, respectively, or straight lines in exceptional cases. To avoid continual mention of these exceptional cases—namely, line segments through the disk center and lines x = constant in the half-plane—we consider lines to be circles of infinite radius.

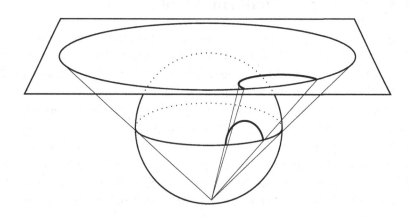

Figure 18.9: The conformal disk model

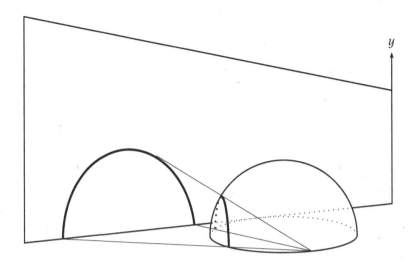

Figure 18.10: The conformal half-plane model

One of the beauties of the conformal models is that other important curves—hyperbolic "circles," horocycles, and equidistant curves—are also real circles. Each curve equidistant from a given "line" L is a circle through the endpoints of L on the boundary. Horocycles are circles tangential to the boundary and also, in the half-plane model, the lines $y = $ constant. A circle that does not meet the boundary is a hyperbolic "circle," but its "center," at equal "distance" from all its points, is not at the Euclidean center. Figure 18.11 shows some of these curves. Note also that asymptotic "lines" are tangential at "infinity" (the boundary) and that the boundary is their common perpendicular, thus resolving the situation that Saccheri (Section 18.1) thought to be contradictory.

"Distance" is particularly easy to express in the half-plane model. The "distance" ds between infinitesimally close points (x, y) and $(x + dx, y + dy)$ is

$$ds = \frac{\sqrt{dx^2 + dy^2}}{y},$$

that is, the Euclidean distance divided by y. Thus "distance" $\rightarrow \infty$ as a point approaches the boundary $y = 0$ of the half-plane, as expected. Keeping x constant, we find by integration that "distance" along a vertical line increases exponentially with respect to Euclidean distance as y decreases. For example, the "distances" between the successive points at which $x = 0$

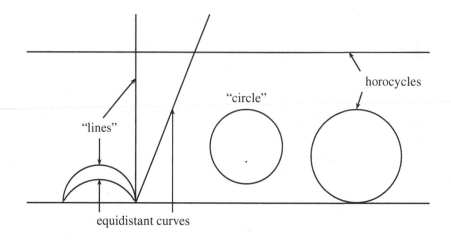

Figure 18.11: Some curves in the half-plane model

and $y = 1, \frac{1}{2}, \frac{1}{4}, \ldots$, are equal. The formula for ds was first obtained by Liouville (1850) by directly mapping the pseudosphere into the half-plane, making simplifying changes of variable. However, Liouville did not realize that the half-plane with his "distance" formula was a model of hyperbolic geometry. The "distance" formula for the conformal disk had also been obtained before Beltrami, by Riemann (1854b), but again without noticing the hyperbolic geometry.

Beltrami (1868b) not only obtained these models, in a unified way, but he also extended the idea to n dimensions. For example, he gave a model of the three-dimensional space considered by Bolyai and Lobachevsky as the upper half, $z > 0$, of ordinary (x, y, z)-space, with "distance"

$$ds = \frac{\sqrt{dx^2 + dy^2 + dz^2}}{z}.$$

"Lines" are then semicircles orthogonal to $z = 0$ and "planes" are hemispheres orthogonal to $z = 0$. Restricting the "distance" function to such a hemisphere turns out to give Beltrami's hemisphere model. Thus the hemisphere model can be viewed as a hyperbolic plane lying in hyperbolic 3-space. The horospheres of the half-space model are spheres tangential to $z = 0$, together with the planes $z = $ constant. Beltrami (1868b) pointed out that on z = constant we have

$$ds = \frac{\sqrt{dx^2 + dy^2 + dz^2}}{\text{constant}},$$

that is, "distance" is proportional to Euclidean distance. Thus he had an immediate proof of Wachter's wonderful theorem that the geometry of the horosphere is Euclidean.

EXERCISES

The mapping of the pseudosphere into the half-plane may be carried out as follows, using the parametric equations for the tractrix found in Exercise 17.5.2:

$$x = \sigma - \tanh \sigma, \quad y = \operatorname{sech} \sigma.$$

First we replace the parameter σ by the arc length τ along the tractrix.

18.5.1 Show that $\tau = \int_0^\sigma \sqrt{1 + \left(\frac{dy}{dx}\right)^2}\, dx = \log \cosh \sigma$, and hence $y = e^{-\tau}$.

Now take τ and the angle X of rotation as the coordinates on the pseudosphere obtained by rotating the tractrix about the x-axis.

18.5.2 Show that the length subtended by angle dX on a circular cross section of the pseudosphere is

$$y\, dX = e^{-\tau} dX,$$

and hence the distance between nearby points (X, τ) and $(X + dX, \tau + d\tau)$ on the pseudosphere is given by

$$ds^2 = e^{-2\tau} dX^2 + d\tau^2.$$

18.5.3 Finally, introduce the variable $Y = e^\tau$ and conclude that $ds = \frac{\sqrt{dX^2 + dY^2}}{Y}$.

Thus the pseudosphere is mapped into the (X, Y)-plane with preservation of distance, provided distance in the (X, Y) plane is defined by

$$ds = \frac{\sqrt{dX^2 + dY^2}}{Y}.$$

It follows, from what was said above, that geodesics on the pseudosphere correspond to semicircles with centers on the X-axis. This throws some light on the problem raised in Section 17.5—describing geodesics on the pseudosphere.

18.5.4 Explain why the region of the (X, Y)-plane corresponding to the pseudosphere is bounded by $X = 0$ and $X = 2\pi$ and it lies above some $Y =$ constant > 0.

18.5.5 By considering a semicircle crossing the region described in Exercise 18.5.4, show that there is no smooth closed geodesic on the pseudosphere.

18.6   The Complex Interpretations

One of the characteristics of the Euclidean plane is the existence of *regular tessellations*: tilings of the plane by regular polygons. There are three such tilings, based on the square, equilateral triangle, and regular hexagon (Figure 18.12). Associated with each tiling is a *group of rigid motions* of the plane that maps the tiling pattern onto itself. For example, the unit square pattern is mapped onto itself by unit translations parallel to the x and y axes and by the rotation of $\pi/2$ about the origin, and these three motions generate all motions of the tessellation onto itself. If we write $z = x + iy$, then these generating motions are given by the transformations

$$z \mapsto z + 1, \qquad z \mapsto z + i, \qquad z \mapsto zi.$$

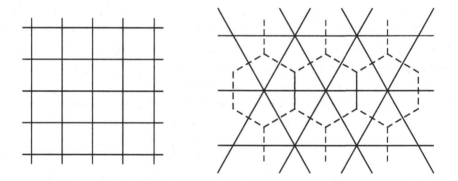

Figure 18.12: Tessellations of the Euclidean plane

The triangle and hexagon tessellations have a similar group of motions, generated by

$$z \mapsto z + 1, \quad z \mapsto z + \tau, \quad z \mapsto z\tau,$$

where $\tau = e^{i\pi/3}$ is the third vertex of the equilateral triangle whose other vertices are at 0, 1 (Figure 18.13). More generally, any motion of the Euclidean plane can be composed from translations $z \mapsto z + a$ and rotations $z \mapsto ze^{i\theta}$.

The sphere also admits a finite number of regular tessellations, obtained by central projections of the regular polyhedra (Section 2.2). Figure 18.14 shows the spherical tessellation corresponding to the icosahedron. (Each

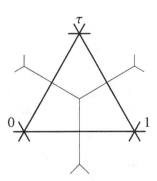

Figure 18.13: Relation between the triangle and hexagon tessellations

face has been further subdivided into six congruent triangles.) The motions that map such a tessellation onto itself can also be expressed as complex transformations by interpreting the sphere as $\mathbb{C} \cup \{\infty\}$ via stereographic projection (Section 16.2). Gauss (1819) found that any motion of the sphere can be expressed by a transformation of the form

$$z \mapsto \frac{az + b}{-\overline{b}z + \overline{a}},$$

where $a, b \in \mathbb{C}$ and an overbar denotes the complex conjugate.

The conformal models of the hyperbolic plane can be regarded as parts of $\mathbb{C}$: the unit disk $\{z : |z| < 1\}$ and the half-plane $\{z : \text{Im}(z) > 0\}$. Their rigid motions, being conformal transformations, are complex functions, and Poincaré (1882) made the beautiful discovery that they are of the form

$$z \mapsto \frac{az + b}{\overline{b}z + \overline{a}}$$

(for the disk) and

$$z \mapsto \frac{\alpha z + \beta}{\gamma z + \delta},$$

where $\alpha, \beta, \gamma, \delta \in \mathbb{R}$ (for the half-plane). Infinitely many regular tessellations are possible, since the angles of a regular n-gon can be made arbitrarily small by increasing its area. For example, there are tessellations by equilateral triangles in which n triangles meet at each vertex, for each $n \geq 7$, and similar variety is possible for other polygons (see exercises).

Figure 18.14: Icosahedral tessellation of the sphere

Some of these tessellations were known *before* Poincaré (1882) gave
the complex interpretation of hyperbolic geometry, and even before any
model of hyperbolic geometry was known at all. Figure 18.15 shows a
tessellation by equilateral triangles of angle $\pi/4$ found in unpublished, and
unfortunately undated, work of Gauss (*Werke*, vol. VIII, p. 104).

Others arise from the so-called hypergeometric differential equation
and were rediscovered in this context by Riemann (1858b) and Schwarz
(1872) (the first published example, Figure 18.16).

With his explanation of these tessellations in terms of hyperbolic geometry, Poincaré (1882) showed for the first time that hyperbolic geometry
was part of preexisting mathematics, whose geometric nature had not previously been understood.

In a subsequent paper, Poincaré (1883) explained the geometric nature
of *linear fractional transformations*,

$$z \mapsto \frac{az + b}{cz + d},$$

special cases of which, as we have seen, express the rigid motions of
the two-dimensional Euclidean, spherical, and hyperbolic geometries. He
showed that each linear fractional transformation of the plane $\mathbb{C}$ is induced
by hyperbolic motion of the three-dimensional half-space with boundary

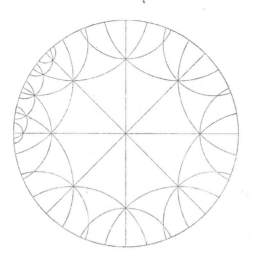

Figure 18.15: The Gauss tessellation

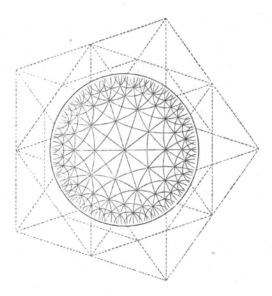

Figure 18.16: The Schwarz tessellation

plane $\mathbb{C}$; thus Poincaré's theorem embraces those of Wachter and Beltrami on the representation of two-dimensional Euclidean, spherical, and hyperbolic geometry within three-dimensional hyperbolic geometry.

18.6.1 Show that a triangle in the hyperbolic plane can have any angle sum $< \pi$.

18.6.2 Deduce that there are equilateral triangles with angle $2\pi/n$ for each $n \geq 7$.

18.6.3 Also deduce that triangles with angle zero exist, in a certain sense, and that their area is finite.

18.6.4 Find corresponding results for regular n-gons.

18.7   Biographical Notes: Bolyai and Lobachevsky

János Bolyai was born in 1802 in Kolozvár, then in the Transylvanian part of Hungary (and now Cluj, Romania) and died in Marosvásárhely in Hungary (now Tárgu-Mureş, Romania) in 1860. His father, Farkas (also known by his German name, Wolfgang), was professor of mathematics, physics, and chemistry, and his mother, Susanna von Árkos, was the daughter of a surgeon. János received his early education from his father and also studied at the Evangelic-Reformed College, where his father taught, from 1815 to 1818. Farkas and Gauss had been fellow students at Göttingen, and Farkas hoped that János would follow him there, but instead the younger Bolyai opted for a military career. He studied at the Vienna engineering academy from 1818 to 1822 and then entered the army.

In the army, János became known as an invincible duelist, but he suffered from bouts of fever and was eventually pensioned off in 1833. He returned to Marosvásárhely to live with his father, but the two did not get along, and in 1834 he moved to a small family estate. He set up house with his mistress, Rosalie von Orbán; they had three children. This could have been the start of a mathematical career, in the style of Descartes, as a leisured country gentleman. But, sad to say, Bolyai's mathematical career was already over in 1833, and it was not until after his death that the world knew he had accomplished anything.

János had inherited a passion for the foundations of geometry from his father, so much so that in 1820 Farkas tried almost desperately to steer him away from the problem of parallels: "You should not tempt the parallels in this way, I know this way until its end—I also have measured this bottomless night, I have lost in it every light, every joy of my life" (Stäckel (1913), pp. 76–77). Of course János ignored this warning, but eventually he found the way out that Farkas had missed. After unsuccessful attempts to prove the Euclidean parallel axiom, he discarded it and proceeded to derive consequences of Axiom P_2. By 1823 his results seemed so complete and elegant they somehow had to be real, and he wrote triumphantly to his father, "From nothing I have created another entirely new world."

Farkas was unwilling to accept the new geometry, but in June 1831 he agreed to send his son's results to Gauss, who did not answer for over six months (admittedly, this was the time of his wife's death). When Gauss did answer it was in the most self-serving way imaginable:

> Now something about the work of your son. You will probably be shocked for a moment when I begin by saying *that I cannot praise it*, but I cannot do anything else, since to praise it would be to praise myself. The whole content of the paper, the path that your son has taken, and the results to which he has been led, agree almost everywhere with my own meditations, which have occupied me in part already for 30–35 years.

Gauss (1832b)

Later in the letter, Gauss offered Bolyai the same backhanded thanks that he had offered to Abel (see Section 12.6) for "saving him the trouble" of writing up the results himself, and he raised the question of the volume of the tetrahedron as a problem for further research.

As we now know, Gauss *did* have many of the results of non-Euclidean geometry by this time, including the answer to the volume problem he had raised to test his young rival (see Gauss (1832a)). Nevertheless, Gauss was almost certainly wrong to imply that his understanding of non-Euclidean geometry went back 35 years. As late as 1804, when Farkas Bolyai wrote to him about the problem of parallels, Gauss could offer no help except the hope that the problem would be settled one day (see Kaufmann-Bühler (1981), p. 100).

János Bolyai was disillusioned and embittered by Gauss's reply but did not give up immediately. He published his work as an appendix to his father's book the *Tentamen* (F. Bolyai (1832a)). However, when there was no response from other mathematicians he became discouraged and never published again. He was also troubled by the possibility that there might, after all, be contradictions in his geometry. As we know, this possibility was not ruled out until 1868, and by then Gauss, Bolyai, and Lobachevsky were all dead.

Nikolai Ivanovich Lobachevsky (Figure 18.17) was born in Novgorod in 1792 and died in Kazan in 1856. He was the son of Ivan Maksimovich Lobachevsky and Praskovia Aleksandrovna. His father died when Nikolai was five years old, and his mother moved with her three sons to Kazan. By persistent efforts, she was able to secure scholarships for their education, and in 1807 Nikolai entered Kazan University, which had been founded just two years earlier. He was supervised by Martin Bartels, the former teacher of Gauss, but a link to Gauss's geometric ideas seems less likely than in the case of Bolyai, since Bartels had little contact with Gauss after his school days.

Figure 18.17: Nikolai Ivanovich Lobachevsky

Lobachevsky stayed at Kazan for the rest of his life, becoming professor in 1814 and making many contributions to the growth of the university. He married the wealthy Lady Varvara Alekseeva Moisieva in 1832 and was raised to the nobility in 1837, in recognition of his services to education. The couple had seven children.

Lobachevsky's investigation of parallels began in 1816, when he lectured on geometry, and he at first thought he could prove the Euclidean axiom. Gradually he became aware of the way in which parallels regulate other geometric properties, such as areas, and in 1832 he wrote *Geometriya*, which separated theorems not requiring an assumption about parallels from those that did. He still believed in the Euclidean axiom, however, so Bolyai was ahead of him at this stage. Lobachevsky's publications in non-Euclidean geometry began in 1829, but at first they attracted no attention, since they were in Russian and Kazan University was little known. He did gain a wider audience with an article in French in Crelle's journal in 1837, but Gauss seems to have been the only one to recognize its importance. Gauss was in fact so impressed that he collected Lobachevsky's obscure Kazan publications and taught himself Russian in order to read them. He also arranged for Lobachevsky to be elected to the Royal Academy of Sciences in Göttingen, and sent him a letter of congratulations (see Dunnington (2004), p. 187), but again he did not let his views become widely known. It is only through a letter, Gauss (1846b), published after his death that his opinion became public. As usual, Gauss's first thought was to guard his own priority, and his memory of when he discovered non-Euclidean geometry seems to have improved with age:

> Lobachevsky calls it imaginary geometry. You know that I have had the same conviction for 54 years (since 1792), with a certain later extension which I do not want to go into here. There was nothing materially new for me in Lobachevsky's paper, but he explains his theory in a way which is different from mine, and does this in a masterful way, in a truly geometric spirit.
>
> Kaufmann-Bühler (1981), p. 150

At any rate, Lobachevsky was less easily discouraged than Bolyai. Despite the silence of foreign mathematicians, opposition from mathematicians in Russia, and the handicap of blindness in his later years, he continued to refine and expand his theory. The final version of his work, *Pangéométrie*, was published in 1855–1856, the last year of his life.

19

Group Theory

PREVIEW

The next three chapters are concerned with the emergence of "modern," or *abstract*, algebra from the old algebra of equations. In the present chapter we look at group theory.

Group theory today is often described as the theory of symmetry, and indeed groups have been inherent in symmetric objects since ancient times. However, extracting algebra from a symmetric object is a highly abstract exercise, and groups first appeared in situations where some algebra was already present.

One of the first nontrivial examples was the group of integers mod p, for prime p, used by Euler (1758) to prove Fermat's little theorem. Of course, Euler had no idea that he was using a group. But he did use one of the characteristic group properties, namely, the existence of inverses.

Likewise, Lagrange (1771) was not aware of the group concept when he studied permutations of the roots of equations. But he was using the group S_n of permutations of n things, and some of its subgroups.

It was Galois (1831a) who first truly grasped the group concept, and he used it brilliantly to explain what makes an equation solvable by radicals. In particular, he was able to explain why the general quintic equation is *not* solvable by radicals. These discoveries changed the face of algebra, though few mathematicians realized it at first.

In the second half of the 19th century the group concept spread from algebra to geometry, following the observation of Klein (1872) that each geometry is characterized by a *group of transformations*. This very fruitful idea is explored further in Chapter 23.

J. Stillwell, *Mathematics and Its History*, Undergraduate Texts in Mathematics,
DOI 10.1007/978-1-4419-6053-5_19, © Springer Science+Business Media, LLC 2010

19.1   The Group Concept

The notion of group is one of the most important unifying ideas in mathematics. It draws together a wide range of mathematical structures for which a notion of combination, or "product," exists. Such products include the ordinary arithmetical product of numbers, but a more typical example is the product, or composition, of functions. If f and g are functions, then fg is the function whose value for argument x is $f(g(x))$. (Thus fg means "apply g, then f." We have to pay attention to order because in general $gf \neq fg$.)

A group G is defined formally to be a set with an operation, usually called *product* and denoted by juxtaposition, a specific element called the *identity* and written 1, and, for each $g \in G$, an element called the *inverse* of g and written g^{-1}, with the following properties:

(i) $g_1(g_2g_3) = (g_1g_2)g_3$ for all $g_1, g_2, g_3 \in G$ (associative property)

(ii) $g1 = 1g = g$ for all $g \in G$ (identity property)

(iii) $gg^{-1} = g^{-1}g = 1$ for all $g \in G$ (inverse property)

These axioms evolved over more than a century of work with particular groups, during which their essential features emerged only gradually. We look at some of the groups that played an important role in this process in the other sections of this chapter. In practice, properties (i) and (ii) are usually evident, and it is more important to ensure that the product operation is in fact *defined* for all elements of G. Many mathematical concepts have been created in response to the desire, at first unconscious, for products to exist.

For example, we saw in Section 8.2 that a perspective view of a perspective view is not, in general, a perspective view. Thus if we take the "product" fg of perspective transformations f and g to be the result of performing g then f, then fg does not always belong to the set of perspective transformations. The set of *projective* transformations is the smallest extension of the set of perspective transformations to a set on which the product is always defined, namely, the set of finite products of perspective transformations.

In other instances, concepts have arisen from the desire to have inverses. Negative numbers, for example, can be viewed as extending the set $\{0, 1, 2, 3, \ldots\}$ of natural numbers to the set $\mathbb{Z}$ of *integers*, in which each element has an inverse under the + operation. (In cases like this one, where the group operation is naturally written as +, the identity element is written 0 and the inverse of g is written $-g$.) Another example is the extension of the line $\mathbb{R}$ to the real projective line $\mathbb{RP}^1 = \mathbb{R} \cup \{\infty\}$, which ensures that each linear fractional function has an inverse. Likewise, extending the plane by points at infinity ensures that each projective transformation has an inverse, because it enables points that are projected to infinity to be projected back again.

Inverses sometimes appear unintentionally, as it were, in finite situations where repeated application of the group operation eventually produces the identity element. The simplest example is the *cyclic group* $\mathbb{Z}_n$, which consists of the numbers $0, 1, 2, \ldots, n - 1$ under "addition modulo n." Here the identity element is 0, and $n - 1$ is the inverse of 1 because their sum equals 0, modulo n. Similarly, $n - 2$ is the inverse of 2, $n - 3$ is the inverse of 3, and so on.

Perhaps the earliest nontrivial use of an inverse occurs with the operation of "multiplication modulo p," which Euler (1758) (and possibly Fermat before him) used to give an essentially group-theoretic proof of Fermat's little theorem. Recall from Section 5.1 that integers m and n are called congruent modulo p if they differ by an integer multiple of p, and from Section 5.2 that b is an *inverse of a* with respect to multiplication mod p if ab is congruent to 1 modulo p, that is, if $ab + kp = 1$ for some integer k. If p is prime and a is not a multiple of p, then such a b exists by application of the Euclidean algorithm to the relatively prime numbers a, p (Sections 3.3 and 5.2). Euler did not define a group in his proof, but it is easy for us to do so (and to rephrase his proof accordingly; see exercises). The group elements are the numbers $0, 1, 2, \ldots, p - 1$, and the product of a and b is defined to be ab mod p, where

ab mod $p =$

the number among $0, 1, 2, \ldots, p - 1$ that is congruent to ab, mod p.

Group properties (i) and (ii) follow from ordinary arithmetic; (iii), as we have seen, follows from the Euclidean algorithm.

The preceding examples illustrate the influence of geometry and number theory on the group concept. An even more decisive influence was the theory of equations, which we look at briefly in Section 19.3. But first we need to understand a little about *subgroups*—the groups within a group—and when a subgroup may be said to "divide" a group. A more detailed account of the development of the group concept may be found in Wussing (1984).

A good introduction to inverses under multiplication mod p may be had with $p = 5$. There is no need to use the Euclidean algorithm to find these inverses—just multiply by numbers < 5 until a product congruent to 1 (mod 5) is obtained.

19.1.1 Find the inverses of 2, 3, and 4 under multiplication mod 5.

Now here is the proof of Fermat's little theorem using inverses mod p. Start with the nonzero numbers, mod p,

$$1, \quad 2, \quad \ldots, \quad (p-1),$$

and multiply them all by a nonzero a (mod p).

19.1.2 Notice that if we multiply again by the *inverse* of a (mod p) we get back the numbers

$$1, \quad 2, \quad \ldots, \quad (p-1).$$

Why does this show that the numbers

$$a \cdot 1 \bmod p, \quad a \cdot 2 \bmod p, \quad \ldots, \quad a(p-1) \bmod p$$

are distinct and nonzero?

19.1.3 Deduce from Exercise 19.1.2 that if a is nonzero (mod p), then

$$\{a \cdot 1 \bmod p, \quad a \cdot 2 \bmod p, \quad \ldots, \quad a(p-1) \bmod p\}$$

is the same set as

$$\{1, \quad 2, \quad \ldots, \quad (p-1)\}.$$

19.1.4 Deduce from Exercise 19.1.3 that

$$a^{p-1} \cdot 1 \cdot 2 \cdots (p-1) \bmod p = 1 \cdot 2 \cdots (p-1) \bmod p.$$

19.1.5 Finally, deduce that

$$a^{p-1} \bmod p = 1 \bmod p,$$

that is,

$$a^{p-1} \equiv 1 \quad (\bmod\ p)$$

(Fermat's little theorem).

19.2   Subgroups and Quotients

The group concept was implicit in mathematics for a long time—arguably from the introduction of negative numbers—before it became explicit. The first substantial theorem of the subject, now known as *Lagrange's theorem*, also predates the formalization of the group concept, but to state it we will take advantage of current terminology.

A subset H of a group G is called a *subgroup* of G if H is also a group (under the same operation that makes G a group). For example, the set $\mathbb{Z}$ of integers is a subgroup of the group $\mathbb{R}$ of real numbers, under the addition operation. Lagrange's theorem concerns the number of members of a group H, which we call the *order* of H and denote by $|H|$. It states that:

If H is a subgroup of a finite group G, then $|H|$ divides $|G|$.

Lagrange (1771) proved a special case of this theorem. Jordan (1870) proved the general case and generously named it after Lagrange. The proof depends on the notion of the *cosets* of H. For each g in G we have the *left coset*

$$gH = \{gh_1, gh_2, \ldots, gh_k\}, \quad \text{where} \quad H = \{h_1, h_2, \ldots, h_k\}.$$

In words, gH is the set that results from multiplying each member of H on the left by g. (There are right cosets Hg defined similarly, but we do not need them for this proof.) The key properties of cosets are:

1. Each coset gH has $|H|$ members, because we can recover the members of H by multiplying each member of gH on the left by g^{-1}.

2. *Any two different cosets g_1H and g_2H are disjoint.* This is because, if g_1H and g_2H have a common member g, we have

$$g = g_1h_1 = g_2h_2 \quad \text{for some } h_1, h_2 \text{ in } H.$$

But then

$$g_1 = g_2h_2h_1^{-1} \quad \text{(multiplying on the right by } h_1^{-1}\text{)},$$

whence

$$g_1H = g_2(h_2h_1^{-1}H) = g_2H,$$

since $h_2h_1^{-1}$ is a member of H, and multiplying H by any one of its members gives back H.

It follows from these two properties that G can be split into disjoint sets gH, each of size $|H|$, so $|H|$ divides $|G|$.   □

Under certain conditions, it makes sense to *multiply* cosets by the rule

$$g_1 H \cdot g_2 H = g_1 g_2 H.$$

For this rule to make sense, we must get the same answer $g_1' g_2' H = g_1 g_2 H$ whenever $g_1' H = g_1 H$ and $g_2' H = g_2 H$. This happens when $gH = Hg$ for each g in G because, under this condition,

$$
\begin{aligned}
g_1' g_2' H &= g_1' H g_2' && \text{because } g_2' H = H g_2', \\
&= g_1 H g_2' && \text{because } g_1 H = g_1' H, \\
&= g_1 g_2' H && \text{because } g_2' H = H g_2', \\
&= g_1 g_2 H && \text{because } g_2' H = g_2 H. \qquad \square
\end{aligned}
$$

We call H a *normal* subgroup of G if it satisfies the condition $gH = Hg$ for each g in G, and in this case the cosets form a *group* called G/H, the *quotient* of G by H. The group properties are inherited from G, as is easy to check (see exercises).

If G has the property that $gg' = g'g$ for all g, g' in G (in which case we call G *abelian*, for reasons that will be explained in the next section), then obviously $gH = Hg$ for any subgroup H. This means that any subgroup H of an abelian group G is normal, and we can form the quotient group G/H. The concept of normal subgroup is therefore interesting only when G is *not* an abelian group. In this case, the first step towards understanding the structure of G is to look for normal subgroups.

All this was first understood and made explicit by Galois, whose work we introduce in the next section.

EXERCISES

The group properties of G/H follow from the definition of the product of cosets, $g_1 H \cdot g_2 H = g_1 g_2 H$.

19.2.1 Show that

$$g_1 H (g_2 H \cdot g_3 H) = (g_1 H \cdot g_2 H) \cdot g_3 H \quad \text{if and only if} \quad g_1 (g_2 g_3) = (g_1 g_2) g_3;$$

hence associativity in G/H follows from associativity in G.

19.2.2 Show that $H = 1H$ is the identity element of G/H.

19.2.3 What is the inverse of gH in G/H? Explain your answer.

The smallest nonabelian group is a group of six elements that may be viewed as the "symmetries" of the equilateral triangle. If we fix a position of the triangle, then there are six motions of it (including the "motion" that does not change it at all) leading to a position where it looks the same as it did before. These motions can be identified by where they send the vertices A, B, and C (Figure 19.1).

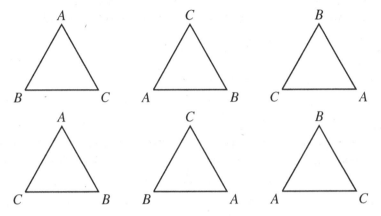

Figure 19.1: The symmetries of the equilateral triangle

The six motions form a group (called S_3 for reasons that will be given in the next section) under the operation of combining motions. We combine motions by viewing each motion as a function $f(P)$ of points P in the triangle, so "do f, then g" means to form the function $gf(P)$, as mentioned in Section 19.1.

19.2.4 Why are there only six motions leading to positions that look the same? Why is this group not abelian?

19.2.5 A subgroup H of S_3 consists of three rotations, through $0°$, $120°$, and $240°$, represented by the pictures in the top row.

19.2.6 The bottom row of the picture represents a coset gH for some g in S_3. Describe the motion g, and verify that Hg is the same set as gH.

19.2.7 Show that any subgroup H with only two cosets in a group G is a normal subgroup.

19.3   Permutations and Theory of Equations

We saw in Section 11.1 that, as early as 1321, Levi ben Gershon found that there are $n!$ permutations of n things. These permutations are invertible functions that form a group S_n (called the *symmetric* group) under composition, though their behavior under composition was not considered until the 18th century. It was when the idea of permutation was applied to the roots of polynomial equations, by Vandermonde (1771) and Lagrange (1771), that the first truly group-theoretic properties of permutations were discovered. At the same time, Vandermonde and Lagrange discovered the key to understanding the solution of equations by radicals.

They began with the observation that if an equation

$$x^n + a_1 x^{n-1} + \cdots + a_{n-1} x + a_n = 0 \tag{1}$$

has roots $x_1, x_2, \ldots, x_n$, then

$$x^n + a_1 x^{n-1} + \cdots + a_{n-1} x + a_n = (x - x_1)(x - x_2) \cdots (x - x_n), \tag{2}$$

and by multiplying out the right-hand side and comparing coefficients one finds that the a_i are certain functions of $x_1, x_2, \ldots, x_n$. For example,

$$a_n = (-1)^n x_1 x_2 \cdots x_n,$$
$$a_1 = -(x_1 + x_2 + \cdots + x_n).$$

These functions are *symmetric*, that is, unaltered by any permutation of $x_1, x_2, \ldots, x_n$, since the right-hand side of (2) is unaltered by such permutations. Consequently, any rational function of $a_1, a_2, \ldots, a_n$ is symmetric as a function of $x_1, x_2, \ldots, x_n$. Now the object of solution by radicals is to apply rational operations *and radicals* to $a_1, a_2, \ldots, a_n$ so as to obtain the roots, that is, the completely *asymmetric* functions x_i.

Radicals must therefore reduce symmetry in some way, and one can see that they do in the quadratic case. The roots of

$$x^2 + a_1 x + a_2 = (x - x_1)(x - x_2) = 0$$

are

$$x_1, x_2 = \frac{-a_1 \pm \sqrt{a_1^2 - 4a_2}}{2} = \frac{(x_1 + x_2) \pm \sqrt{x_1^2 - 2x_1 x_2 + x_2^2}}{2},$$

and we notice that the symmetric functions $x_1 + x_2$ and $x_1^2 - 2x_1 x_2 + x_2^2$ yield the two asymmetric functions x_1, x_2 when the two-valued radical $\sqrt{}$ is introduced. In general, introduction of radicals $\sqrt[p]{}$ multiplies the number of values of the function by p and divides symmetry by p, in the sense that the group of permutations leaving the function unaltered is reduced to $1/p$ of its previous size.

Vandermonde and Lagrange found that they could explain the previous solutions of cubic and quartic equations in terms of such symmetry reduction in the corresponding permutation groups, S_3 and S_4. They also found some properties of subgroups. For example, Lagrange found a special case of the result now known as "Lagrange's theorem": the order of a subgroup divides the order of the group. However, they were unable to obtain sufficient understanding of the relation between radicals and subgroups of S_n to settle the equations of degree ≥ 5. Ruffini (1799) and Abel (1826) made enough progress with S_5 to be able to prove the unsolvability of the quintic, but none of these authors had a firm enough grip on the relation between radicals and permutations to handle arbitrary equations. They were not, in fact, conscious of the group concept, and it is only with hindsight that we can interpret their results in group-theoretic terms.

The concept, and indeed the word "group," is due to Galois (1831b). Along with it, Galois introduced the concept of normal subgroup, which finally unlocks the secret of solvability by radicals. Galois showed that each equation E has a group G_E consisting of the permutations of the roots that leave rational functions of the coefficients unaltered, and that the reduction of symmetry caused by introduction of a radical corresponds to formation of a normal subgroup. More precisely, *if E is an equation solvable by radicals then there is a chain of subgroups*

$$G_E = H_1 \supseteq H_2 \supseteq \cdots \supseteq H_k = \{1\}$$

such that each H_{i+1} is a normal subgroup of H_i and H_i/H_{i+1} is cyclic. (Moreover, if H_i/H_{i+1} is cyclic of order n then the step from H_i to H_{i+1} corresponds to introduction of an nth root.) Such a group G_E is now called *solvable*.

Examples of solvable groups are S_3 and S_4, as one would expect from the known solvability of the corresponding equations. Also, it is easy to see that all finite abelian groups are solvable, so each equation with an abelian group is solvable by radicals—a result due to Abel (1829). This is why we call such groups "abelian." If E is the general equation of degree n, then $G_E = S_n$ and the theorem of Ruffini and Abel is recovered by showing that S_n is *not* solvable for $n \geq 5$ (see, for example, Dickson (1903)).

This brief sketch of Galois's ideas covers only a part of his theory. Another part is his theory of *fields*, which is needed to clarify the notion of rational function. The group theory and the field theory make up what is currently known as "Galois theory" (see, for example, Edwards (1984)).

What one might consider to be the summit of Galois's theory, rising above the confines of algebra, is currently neglected. This is the solution of equations by elliptic and related functions, for which one must consult earlier books such as Jordan (1870) and Klein (1884). The greatest triumph of this theory was the solution of the general quintic equation by elliptic modular functions in Hermite (1858), following a hint in Galois (1831a) (see also Sections 6.6 and 19.8).

EXERCISES

The simplest type of permutation is a *transposition*, which swaps two things and leaves the others fixed.

19.3.1 Show that any permutation is a product of transpositions, that is, any arrangement of n things may be achieved by repeated swaps.

The group S_n of all permutations of n things has an important subgroup A_n, consisting of the permutations that are *even* in the following sense.

An *even permutation* f of $\{1, 2, \ldots, n\}$ is one with an even number of *inversions*, that is, pairs (i, j) for which $i < j$ and $f(i) > f(j)$ (Cramer (1750), p. 658). This can be visualized by placing the numbers $1, 2, \ldots, n$ in two rows, one above the other, and drawing a line from k in the top row to $f(k)$ in the bottom row. Figure 19.2 illustrates the permutation $f(1) = 2$, $f(2) = 3$, $f(3) = 1$ in this way.

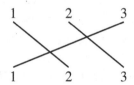

Figure 19.2: A permutation diagram

19.3.2 Explain why a permutation is even if and only if its diagram has an even number of crossings.

19.3.3 Show that the product of even permutations is even, and hence that the even permutations of $\{1, 2, \ldots, n\}$ form a group A_n. (It is called the *alternating* group.)

19.3.4 Show that evenness does not depend on how the numbers $1, 2, \ldots, n$ are assigned to the n things. (Hint: if the numbers are permuted by g, show that the permutation f is replaced by the permutation $g^{-1}fg$.)

19.3.5 If g is an odd permutation, that is, $g \in S_n - A_n$, show that the set $gA_n = \{gf : f \in A_n\}$ is all the odd permutations in S_n; hence A_n contains exactly half the members of S_n.

It follows from Exercise 19.3.5, and Exercise 19.2.7, that A_n is a normal subgroup of S_n; hence we can always form the cyclic quotient $S_n/A_n = \mathbb{Z}_2$. Thus the real problem in solving the general equation of degree n is to "solve" A_n by finding normal subgroups inside it.

The group S_3 turns out to be solvable because its normal subgroup A_3 is already cyclic. This can be seen by studying the permutations in A_3, but more easily by interpreting S_3 geometrically.

19.3.6 Interpret the symmetry group of the equilateral triangle, discussed in the previous exercise set, as the group S_3 of all permutations of three things.

19.3.7 Show that, under this interpretation, the cyclic subgroup of rotations is A_3.

The "interpretation" we speak of here is an example of what is technically called an *isomorphism* between the triangle symmetry group and S_3. An isomorphism is a one-to-one correspondence between the two groups that preserves the group operation, thus establishing that the groups have the "same form." (We used this expression in the same sense in Section 16.5.) In calling the rotation subgroup "cyclic" we also imply an isomorphism, namely, the one that pairs the rotations through $0°$, $120°$, $240°$ with the members 0, 1, 2 of $\mathbb{Z}_3$ respectively.

19.3.8 Show that there is an isomorphism between the symmetry group of the regular tetrahedron and S_4. To which symmetries do the members of A_4 correspond?

19.4   Permutation Groups

Galois understood "group" to mean a group of permutations of a finite set, so his definition stated only that the product of two permutations in the group must again be a member of the group. Associativity, identity, and inverses were consequences of his assumptions, and indeed too obvious to be considered important from his point of view. Galois's work was published only in 1846, and by that time the theory of finite permutation groups had been taken up and systematized by Cauchy (1844). Cauchy likewise required only closure under product in his definition of group, but he recognized the importance of identity and inverses by introducing the notation of 1 for the identity and f^{-1} for the inverse of f.

Cayley (1854) was the first to consider the possibility of more abstract group elements, and with it the need to postulate associativity. (Incidentally, a group operation for which associativity is not obvious is that defined by the chord construction on a cubic curve: see Sections 11.6 and 16.5.) He took group elements to be simply "symbols," with a "product" of A and B written $A \cdot B$ and subject to the law $A \cdot (B \cdot C) = (A \cdot B) \cdot C$, and a unique element 1 subject to the laws $A \cdot 1 = 1 \cdot A = A$. He still assumed that each group was finite, however. This meant that the existence of inverses did not have to be postulated, only the validity of cancellation.

The existence of inverses in a finite group G, as defined by Cayley, follows from an argument used by Cauchy (1815) and developed more fully in Cauchy (1844). If $A \in G$, then the powers $A^2, A^3, \ldots$ all belong to G and hence they eventually include a recurrence of the same element:

$$A^m = A^n \quad \text{where } m < n.$$

Then, assuming that it is valid to cancel A^m from both sides, A^{n-m} is the identity element 1 and A^{n-m-1} is the inverse of A.

The need to postulate inverses first arises with infinite groups, where this argument no longer holds. Geometry was historically the most important source of infinite groups, as we shall see in Section 19.6. It was in extending Cayley's abstract group theory to cover the symmetry groups of infinite tessellations that Dyck (1883) made first mention of inverses in the definition of group. We shall return to Dyck's concept of group in Section 19.7.

A theorem of Cayley (1878) shows that abstraction of the group concept is, in a sense, empty, because every group is essentially the same as a group of permutations. Cayley proved the theorem for finite groups only, where it is more valuable, but the proof easily extends to arbitrary groups (see exercises).

EXERCISES

The proof of Cayley's theorem goes as follows. Given any group G, associate any g in G with the function $\times g$ that sends each $h \in G$ to hg.

19.4.1 Show that function $\times g$ is a permutation of G, by showing that its effect can be undone by the function $\times g^{-1}$.

19.4.2 Show that different group elements g_1, g_2 give different functions $\times g_1$, $\times g_2$, and hence that there is a one-to-one correspondence between the elements g *in* G and the permutations $\times g$ *of* G.

19.4.3 Show that the permutation of G obtained by applying $\times g_1$, then $\times g_2$, is the permutation obtained by applying $\times g_1 g_2$.

Thus the group of permutations $\times g$ is *isomorphic* to the group G, in the sense described in the previous exercise set. This is the precise way of saying that G is "essentially the same" as a group of permutations.

19.5 Polyhedral Groups

A beautiful illustration of Cayley's theorem that every group is a permutation group is provided by the regular polyhedra, whose rotation groups turn out to be important subgroups of S_4 and S_5. If we imagine a polyhedron P occupying a region R in space, the rotations of P can be viewed as the different ways of fitting P into R.

We begin with the rotations of the tetrahedron T: T has four vertices, V_1, V_2, V_3, V_4, so each rotation of T is determined by a permutation of the four things V_1, V_2, V_3, V_4. There are $4 \times 3 = 12$ rotations, because V_1 can be put at any of the four vertices of R, after which three choices remain for the remaining triangle of vertices V_2, V_3, V_4. One can check, using the fact that a permutation that leaves one element fixed and rotates the other three is even, that all the symmetries of T are even permutations of V_1, V_2, V_3, V_4. But the subgroup A_4 of *all* even permutations in S_4 has $\frac{1}{2} \times 4! = 12$ elements by the exercises in Section 19.3, so the rotation group of T is precisely A_4.

The full permutation group S_4 can be realized by the rotations of the cube. The four elements of the cube that are permuted are the long diagonals AA', BB', CC', DD' (Figure 19.3).

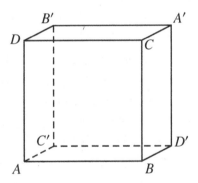

Figure 19.3: The cube and its diagonals

One has to check, first, that each permutation of the diagonals is actually realizable. While doing this, it will become apparent that the position of the diagonals (bearing in mind that endpoints could be swapped) really determines the position of the cube (Exercise 19.5.1). S_4 is also the rotation group of the octahedron, because of the dual relationship between cube and octahedron seen in Figure 19.4. Each rotation of the cube is clearly a rotation of its dual octahedron, and conversely.

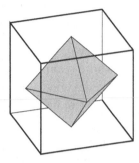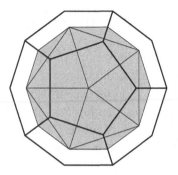

Figure 19.4: Dual polyhedra

Likewise, the dual relationship between dodecahedron and icosahedron (Figure 19.4) shows that they have the same rotation group. This group turns out to be A_5, the subgroup of even permutations in S_5. The five elements of the dodecahedron whose even permutations determine these rotations are tetrahedra formed from sets of four vertices (see Figure 19.5).

Figure 19.5: The tetrahedra in a dodecahedron

For more information on the polyhedral groups, see Klein (1884). This book relates the theory of equations to the rotations of the regular polyhedra and functions of a complex variable. The complex variable makes its appearance when the regular polyhedra are replaced by regular tessellations of the sphere $\mathbb{C} \cup \{\infty\}$, and their rotations by linear fractional transformations, as in Section 18.6. Klein (1876) showed that, with trivial exceptions, *all* finite groups of linear fractional transformations come from the rotations of the regular polyhedra in this way.

The regular polyhedra were also the source of another approach to groups: *presentation by generators and relations*. Hamilton (1856) showed that the icosahedral group can be generated by three elements ι, χ, λ subject to the relations

$$\iota^2 = \chi^3 = \lambda^5 = 1, \quad \lambda = \iota\chi. \tag{1}$$

This means that any element of the icosahedral group is a product (possibly with repetitions) of ι, χ, λ and that any relation between ι, χ, λ follows from the relations (1). Dyck (1882) gave similar presentations of the cube and tetrahedron groups, and for the groups of certain finite tessellations, as part of the first general discussion of generators and relations. We return to this in Section 19.7.

EXERCISES

19.5.1 Show that each permutation of the diagonals of a cube is realizable, for example by showing that each transposition is realizable.

19.5.2 Show that a permutation of the diagonals uniquely determines the position of the cube.

Now consider the following rotations of the cube:

$\iota = 180°$ rotation about a line through the midpoints of opposite edges,

$\chi = 120°$ rotation about a diagonal.

These obviously satisfy $\iota^2 = \chi^3 = 1$.

19.5.3 Show that $\iota\chi = \lambda$, where

$\lambda = 90°$ rotation about a line through the centers of opposite faces,

where the lines are, for example, those shown in Figure 19.6 (these lines are fixed in space, not in the cube).

19.5.4 Deduce from Exercise 19.5.3 that $\iota^2 = \chi^3 = (\iota\chi)^4 = 1$ for the cube.

19.5.5 Show that the analogous ι, χ for the tetrahedron satisfy

$$\iota^2 = \chi^3 = (\iota\chi)^3 = 1,$$

and the analogous ι, χ for the dodecahedron satisfy

$$\iota^2 = \chi^3 = (\iota\chi)^5 = 1.$$

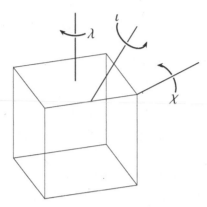

Figure 19.6: Rotations of the cube

19.6   Groups and Geometries

As the regular polyhedra show, geometric symmetry is fundamentally a group-theoretic notion. More generally, many notions of "equivalence" in geometry can be explained as properties that are preserved by certain groups of transformations. However, some revision of classical notions was necessary before geometry could benefit from group-theoretic ideas.

The oldest notion of geometric equivalence is that of *congruence*. The Greeks understood figures F_1 and F_2 to be congruent if there is a rigid motion of F_1 that carries it into F_2. But this concept of motion had meaning only for the individual figure. The "product" of motions of different figures was meaningless, so one did not have a group of motions.

The step that paved the way for group theory in geometry was the extension of the idea of motion to the whole plane by Möbius (1827). This gave meaning to the product of motions. In fact, Möbius considered all continuous transformations of the plane that preserve straightness of lines, and he picked out several subclasses of these transformations: those that preserve length (congruences), shape (similarities), and parallelism (affinities). He showed that the most general continuous transformations preserving straightness are just the projective transformations, so in one stroke Möbius defined the notions of congruence, similarity, affinity, and projective equivalence as properties invariant under certain classes of transformations. That the classes in question are groups was obvious as soon as one recognized the concept of group. It is an indication of the slowness with which the group concept was recognized that the restatement of Möbius's ideas in terms of groups occurred only with Klein (1872).

Klein's formulation became known as the *Erlanger Programm* because he announced it at the University of Erlangen. His idea is to associate each geometry with a group of transformations that preserve its characteristic properties. Thus, characteristic properties show up as *invariants of the group*. For example, the group of plane Euclidean geometry is the group of *Euclidean rigid motions*—transformations of $\mathbb{R}^2$ that preserve the Euclidean distance $\sqrt{(x_2 - x_1)^2 + (y_2 - y_1)^2}$ between points (x_1, y_1) and (x_2, y_2). Euclidean distance is therefore an invariant, by the very definition of the group.

A more interesting example is the group of the real projective line $\mathbb{RP}^1$, studied in Section 8.6. In this case we start with the group, the group of linear fractional transformations, and *discover* its invariant, the cross-ratio, which is not at all obvious visually. Plane projective geometry is similarly associated with the group of projective transformations of $\mathbb{RP}^2$, and its fundamental invariant is likewise the cross-ratio.

Plane hyperbolic geometry, in view of the projective model, can be defined by the group of projective transformations that map the unit circle onto itself. An important influence on the Erlanger Programm was indeed Cayley (1859), where this group was first shown to determine a geometry, and the subsequent realization of Klein (1871) that the elements of this group are the rigid motions of hyperbolic geometry. Not surprisingly, its fundamental invariant is the hyperbolic distance, and this turns out to be a function of the cross-ratio.

When geometry is reformulated in terms of groups, certain geometric questions become natural questions about groups. A regular tessellation, for example, corresponds to a subgroup of the full group of motions, consisting of those motions that map the tessellation onto itself. In the case of hyperbolic geometry, where the problem of classifying tessellations is formidable, the interplay between geometric and group-theoretic ideas proved to be very fruitful. In the work of Poincaré (1882, 1883) and Klein (1882b), group theory is the catalyst for a new synthesis of geometric, topological, and combinatorial ideas, which are described in Sections 19.7 and 22.7.

EXERCISES

If we view geometric objects (points, lines, curves, and so on) as subsets X of a space S, then relations such as congruence arise from groups of transformations of S in the following way. There is a group G of maps $g : S \to S$, and each geometric object X has a *G-orbit* $\{g(X) : g \in G\}$, consisting of the objects onto which X is mapped by elements of G.

For example, if Δ is a triangle in the plane $\mathbb{R}^2$, and G consists of all transformations of $\mathbb{R}^2$ that preserve length, then $\{g(\Delta) : g \in G\}$ consists of all triangles congruent to Δ. This example shows that members of the same G-orbit are "equivalent" in a sense that depends on G. In fact, we always get an equivalence relation from a group in this way. Here is another example.

19.6.1 If $G = \{$similarities of $\mathbb{R}^2\}$, what is $\{g(\Delta) : g \in G\}$ for a triangle Δ?

For any group G of transformations, we define a relation $X \cong_G Y$ ("X is G-equivalent to Y") between subsets X, Y of S by

$$X \cong_G Y \quad \Longleftrightarrow \quad X \text{ is in the } G\text{-orbit of } Y.$$

Then the group properties of G imply the following properties of the relation $\cong_G$.

19.6.2 Show that the relation $\cong_G$ has the properties

$$X \cong_G X \qquad \qquad \text{(reflexive)}$$
$$X \cong_G Y \quad \Longrightarrow \quad Y \cong_G X \qquad \qquad \text{(symmetric)}$$
$$X \cong_G Y \text{ and } Y \cong_G Z \quad \Longrightarrow \quad X \cong_G Z \qquad \qquad \text{(transitive)}$$

19.6.3 At which points does your solution of Exercise 19.5.2 involve the existence of an identity, existence of inverses, and existence of products in G?

The properties in Exercise 19.6.2 show that $\cong_G$ is an *equivalence relation*, according to the definition in the exercises for Section 2.1. It was also noted there that the reflexive and transitive properties actually imply symmetry, provided that transitivity is stated in the manner of Euclid's Common Notion 1: "Things equivalent to the same thing are equivalent to each other."

19.6.4 Prove Common Notion 1 for $\cong_G$:

$$X \cong_G Y \text{ and } Z \cong_G Y \quad \Longrightarrow \quad X \cong_G Z.$$

You will see that this proof involves inverses, which previously were needed only to prove symmetry. This confirms that Euclid's Common Notion 1 is in some sense a combination of both transitivity and symmetry.

Returning to a particular group and its invariants, here is an example of the way in which an invariant can throw light on its group.

19.6.5 Given three points A, B, C on $\mathbb{RP}^1$, show that there is a unique fourth point x such that the cross-ratio

$$\frac{(C - A)(x - B)}{(C - B)(x - A)}$$

has a given value y.

19.6.6 Deduce from Exercise 19.6.5 that each linear fractional transformation of $\mathbb{RP}^1$ is determined by its values on any three points A, B, C.

19.7   Combinatorial Group Theory

As mentioned in Section 19.5, the groups of the regular polyhedra were
the first to be defined in terms of generators and relations. With finite
groups such as these, however, one is concerned mainly with the simplicity
and elegance of a presentation; the question of *existence* does not arise.
For any finite group G one can trivially obtain a finite set of generators
(namely, *all* the elements $g_1, \ldots, g_n$ of G) and defining relations (namely,
all equations $g_i g_j = g_k$ holding among the generators). Of course the same
argument gives an infinite set of generators and defining relations for an
infinite group, but this is also not interesting. The real problem is to find
finite sets of generators and defining relations for infinite groups where
possible.

This problem was first solved for the symmetry groups of certain regular tessellations, and such examples were the basis of the first systematic
study of generators and relations, by Klein's student Dyck. Dyck's papers
(1882, 1883) laid the foundations of this approach to group theory, now
called *combinatorial*. For more technical information, as well as detailed
history of the development of combinatorial group theory, see Chandler
and Magnus (1982).

Figure 19.7 illustrates how generators and relations arise naturally from
tessellations. This tessellation is based on the regular tessellation of the
Euclidean plane by unit squares, but each square has been subdivided into
black and white triangles to eliminate symmetries by rotation and reflection. The symmetries that remain are generated by

1. horizontal translation of length 1

2. vertical translation of length 1

These generators are subject to the obvious relation

$$ab = ba,$$

which implies that any element of the group can be written in the form
$a^m b^n$. If $g = a^{m_1} b^{n_1}$ and $h = a^{m_2} b^{n_2}$, then $g = h$ only if $m_1 = m_2$ and
$n_1 = n_2$, that is, only if $g = h$ is a *consequence* of the relation $ab = ba$.
Thus all relations $g = h$ in the group follow from $ab = ba$, which means
that the latter relation is a defining relation of the group.

The obviousness of the defining relation in this case blinds us to a fact
that becomes more evident with tessellations of the hyperbolic plane: *the*

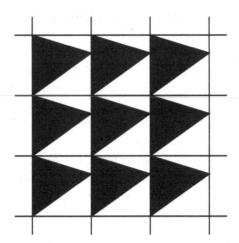

Figure 19.7: A tessellation of the plane

generators and relations can be read off the tessellation. Group elements correspond to cells in the tessellation, squares in the present example. If we fix the square corresponding to the identity element 1, then the square to which square 1 is sent by the group element g may be called square g. The generators $a^{\pm 1}$, $b^{\pm 1}$ are the elements that send square 1 to adjacent squares. They generate the group because square 1 can be sent to any other square by a series of moves from square to adjacent square. Relations correspond to equal sequences of moves or, what amounts to the same thing, to sequences of moves that return square 1 to its starting position. These sequences can all be derived from a circuit around a vertex (Figure 19.8), that is, the sequence $aba^{-1}b^{-1}$. Thus all relations are derived from $aba^{-1}b^{-1} = 1$, or, equivalently, $ab = ba$.

Generalizing these ideas, Poincaré (1882) showed that the symmetry groups of all regular tessellations, whether of the sphere, Euclidean plane, or hyperbolic plane, can be represented by finitely many generators and relations. Generators correspond to moves of the basic cell to adjacent cells, and hence to the sides of the basic cell; defining relations correspond to its vertices. These results are also important for topology, as we shall see in Chapter 22.

The notion of group abstracted from such examples was expressed in a somewhat technical way, involving normal subgroups, by Dyck (1882). The following, simpler, approach was worked out by Dehn and used by

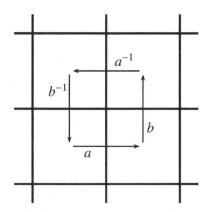

Figure 19.8: Circuit around a vertex

Dehn's student Magnus (1930). A group G is defined by a set $\{a_1, a_2, \ldots\}$ of *generators* and a set $\{W_1 = W_1', W_2 = W_2', \ldots\}$ of *defining relations*. Each generator a_i is called a *letter*; a_i has an *inverse* a_i^{-1}, and arbitrary finite sequences ("products") of letters and inverse letters are called *words*.

Words W, W' are called *equivalent* if $W = W'$ is a consequence of the defining relations, that is, if W may be converted to W' by a sequence of replacements of subwords W_i by W_i' (or vice versa) and cancellation (or insertion) of subwords $a_i a_i^{-1}$, $a_i^{-1} a_i$. The elements of G are the equivalence classes

$$[W] = \{W' : W' \text{ is equivalent to } W\},$$

and the product of elements $[U], [V]$ is defined by

$$[U][V] = [UV],$$

where UV denotes the result of concatenating the words U, V. It has to be checked that this product is well defined, but once this is done, the group properties (i), (ii), and (iii) of Section 19.1 follow easily.

EXERCISES

Here is how one verifies that the classes $[W]$ have the group properties.

19.7.1 If U is equivalent to U', show that UV is equivalent to $U'V$. Conclude, using this and a similar result for V', that the product $[U][V]$ is independent of the choice of representatives for $[U], [V]$.

19.7.2 $[U]([V][W]) = ([U][V])[W]$ is trivial. Why?

19.7.3 Show that 1 = equivalence class of the empty word.

19.7.4 Show that $[W]^{-1} = [W^{-1}]$, where W^{-1} is the result of writing W backward and changing the sign of each exponent.

The smallest nonabelian group S_3 is also the smallest group with interesting defining relations. We take S_3 to be the group of symmetries of the equilateral triangle, as in the exercises to Section 19.2.

19.7.5 Show that S_3 is generated by a 120° rotation r about its center, and a 180° rotation s about the vertical axis of symmetry. Also show that r and s satisfy the relations
$$r^3 = s^2 = 1, \quad r^2 s = sr.$$

19.7.6 Deduce from Exercise 19.7.5 that each element of S_3 can be written in the form
$$r^m s^n, \quad \text{where } m = 0, 1, 2 \text{ and } n = 0, 1.$$

19.7.7 Conclude from Exercise 19.7.6 that $r^3 = s^2 = 1$ and $r^2 s = sr$ are defining relations for S_3.

19.7.8 By a similar argument, show the group of symmetries of a regular n-gon has defining relations $r^n = s^2 = 1$, $r^{n-1} s = sr$.

19.8 Finite Simple Groups

A group is called *simple* if it has no normal subgroups other than itself and the group $\{1\}$ whose only member is the identity element. The reason for the name is that a simple group cannot be "simplified" by forming the quotient G/H by a normal subgroup H. In this sense of simplicity, simple groups are like prime numbers, which cannot be "simplified" by dividing them by smaller integers. We do not claim that simple groups or prime numbers are not complicated!

The most obvious examples of finite simple groups *are* in fact the prime numbers, or more precisely the cyclic groups $\mathbb{Z}_p$ for prime numbers p. $\mathbb{Z}_p$ is simple because it has no subgroups whatever except itself and $\{1\}$ (thanks to Lagrange's theorem that the size of a subgroup divides the size of the group). In fact, these are the only *abelian* simple groups, and we will ignore them from now on. The interesting simple groups are those that are not abelian, and the first examples were discovered by Galois in his study of polynomial equations.

The smallest nonabelian simple group is A_5, the group of the 60 even permutations of five things. The simplicity of A_5 is the obstruction to the solution of the quintic equation by radicals. As we saw in Section 19.3, the group of the quintic equation is S_5, the group of all 120 permutations of five things. Solving the quintic equation by radicals is equivalent to finding a chain of subgroups

$$S_5 \supseteq H_1 \supseteq H_2 \supseteq \cdots \supseteq \{1\}$$

such that the quotient of each group by the next is cyclic. We can make a first step,

$$S_5 \supseteq A_5,$$

but we can go no further because S_5 has no other nontrivial normal subgroup and A_5 is simple.

The proof that A_5 is simple (see exercises below) can be extended to show that A_n is simple for all $n \geq 5$, so Galois actually discovered a whole infinite family of simple groups. He also found three remarkable simple groups in the study of *modular equations*, which arise in the theory of elliptic functions. The starting point of these investigations was the Fagnano (1718) formula for doubling the arc length of the lemniscate (Section 12.4):

$$2 \int_0^x \frac{dt}{\sqrt{1 - t^4}} = \int_0^y \frac{dt}{\sqrt{1 - t^4}}, \quad \text{where} \quad y = \frac{2x\sqrt{1 - x^4}}{1 + x^4}.$$

This gives the polynomial equation between x and y, of degree 2 in y:

$$y^2(1 + x^4)^2 = 4x(1 - x^4).$$

In the early 19th century, Fagnano's discovery was generalized to other elliptic integrals and to n-tupling instead of doubling, by Legendre, Gauss, Abel, and Jacobi. Galois left only some cryptic remarks about multiplication by 5, 7, and 11 (implying that they yield equations of degree 5, 7, and 11) in a letter that he wrote just before his death.

It turns out that the modular equation of degree 5 is equivalent to the general quintic equation, which is why Hermite (1858) was able to solve the general quintic equation by elliptic modular functions. However, the modular equations of degree 7 and 11 have groups of size 336 and 1320 respectively, so they are *not* symmetric groups S_n. The nature of these new groups was revealed by Jordan (1870). They can be viewed as (what we would now call) *transformation groups of finite projective lines*.

What is a finite projective line? It is like the real projective line $\mathbb{RP}^1 = \mathbb{R} \cup \{\infty\}$ discussed in Section 8.6, except that $\mathbb{R}$ is replaced by a *finite field.* Finite fields were discovered by Galois, and we met some of them in Section 19.1 when we discussed addition and multiplication mod p. Since the latter operations have the same behavior as ordinary addition and multiplication—in particular, each nonzero number has an inverse—we can operate on the set $\mathbb{F}_p = \{0, 1, 2, \ldots, p - 1\}$ as we normally do, to solve equations and so on. Moroever, *linear fractional functions* make sense on $\mathbb{F}_p \cup \{\infty\}$, if we agree as usual that

$$1/0 = \infty \quad \text{and} \quad 1/\infty = 0.$$

So we can view $\mathbb{F}_p \cup \{\infty\}$ as a finite projective line, and its linear fractional functions as "projections." Moreover, the cross-ratio makes sense on $\mathbb{F}_p \cup \{\infty\}$, and it is invariant under linear fractional functions by the same argument as in Section 8.6.

For this reason, the group of functions

$$x \mapsto \frac{ax + b}{cx + d}, \quad \text{where} \quad a, b, c, d \in \mathbb{F}_p \quad \text{and} \quad ad - bc \neq 0,$$

is called the *projective general linear group*, PGL$(2, p)$. (The reason for the 2 is that the coefficients a, b, c, d behave like the 2×2 matrix $\left(\begin{smallmatrix} a & b \\ c & d \end{smallmatrix}\right)$— see Section 23.1.) It turns out that PGL$(2, 5)$, PGL$(2, 7)$, and PGL$(2, 11)$ are the groups of the modular equations of degree 5, 7, and 11 respectively. Moreover, each of these groups PGL$(2, p)$ contains a simple subgroup, called PSL$(2, p)$, which is half of its size. This was shown by Jordan (1870).

We will not prove that PSL$(2, p)$ is simple here, nor that PSL$(2, 5)$ is the same as A_5, but we can confirm that PSL$(2, 5)$ has 60 elements by interpreting its elements as transformations of the projective line

$$\mathbb{F}_5 \cup \{\infty\} = \{0, 1, 2, 3, 4, \infty\}.$$

The key observation is that any linear fractional function $f(x) = \frac{ax+b}{cx+d}$ on a projective line is determined by its values on three points, say $f(0)$, $f(1)$, $f(\infty)$. This is because any fourth point x has a certain cross-ratio with 0, 1, ∞, and $f(x)$ has the same cross-ratio with $f(0), f(1), f(\infty)$, which determines $f(x)$ uniquely. Also, $f(0), f(1), f(\infty)$ can be *any* triple of distinct values among $0, 1, 2, 3, 4, \infty$ because we can solve the equations

$$\frac{a0 + b}{c0 + d} = f(0), \quad \frac{a1 + b}{c1 + d} = f(1), \quad \frac{a\infty + b}{c\infty + d} = f(\infty)$$

for a, b, c, d. Thus, the number of elements in PGL(2, 5) is the number of 3-element sequences of distinct elements among $0,1,2,3,4,\infty$, namely

$$6 \cdot 5 \cdot 4 = 120.$$

Now, the reason that PGL(2, 5) is not a simple group is that it has an "obvious" normal subgroup when we view linear fractional functions as permutations of the set $\{0, 1, 2, 3, 4, \infty\}$. For example, the function $f(x) = x + 1$ is the function

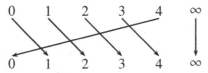

—which is an even permutation. On the other hand, the function $g(x) = 2x$ is the function

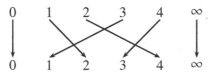

—which is an odd permutation. Therefore, the subgroup of even permutations, which is PSL(2, 5), is not the whole of PGL(2, 5). In fact it is half the size of PGL(2, 5), so it has 60 elements. It also turns out to be a normal subgroup, simple, and isomorphic to A_5.

We can similarly see that PGL(2, 7) has $8 \cdot 7 \cdot 6 = 336$ elements, half of which are even permutations. The subgroup PSL(2, 7) of even permutations therefore has 168 elements, and it turns out to be a simple normal subgroup. The third group considered by Galois, PSL(2, 11), likewise turns out to be simple, with 660 elements. It so happens that PSL(2, 7) is the smallest nonabelian simple group, other than PSL(2, 5) $= A_5$. PSL(2, 7) makes several other spectacular appearances in geometry, which may seen in the article Gray (1982).

These examples give only the tiniest glimpse of the world of simple groups. Nevertheless, they hint at one of its most fascinating features— there are meaningful finite analogues of infinite structures such as the real projective line. We reveal more about this world in Chapter 23.

EXERCISES

A_5 is simple for quite elementary reasons, which can be understood with only slight knowledge of permutations. This includes the nature of even permutations, explored in the exercises to Section 19.3, and the decomposition of permutations into *cycles*, which we explore here.

We say that $(a_1, a_2, \ldots, a_k)$ is a *k-cycle* of a permutation f of $\{1, 2, \ldots, n\}$ if

$$f(a_1) = a_2, \quad f(a_2) = a_3, \quad \ldots, \quad f(a_k) = a_1$$

for distinct numbers $a_1, a_2, \ldots, a_k$. Each number in $\{1, 2, \ldots, n\}$ belongs to some *k*-cycle of f, so f is a product of disjoint cycles. For example, if f is

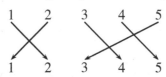

then $f = (1, 2)(3, 4, 5)$. It follows from the Exercises in Section 19.3 that the only even *k*-cycles among the even permutations of $\{1, 2, 3, 4, 5\}$ are the 3-cycles and the 5-cycles.

19.8.1 Omitting 1-cycles from the cycle decomposition, show that the only possible types of cycle decomposition for (nonidentity) members of A_5 are (a, b, c), $(a, b)(c, d)$, and (a, b, c, d, e).

19.8.2 Recalling that $g \cdot f$ means "f, then g," check that

(i) $(1, 2, 3, 4, 5) \cdot (2, 1, 4, 3, 5) = (1, 5, 3)$.

(ii) $(1, 2)(3, 4) \cdot (1, 2)(4, 5) = (3, 4, 5)$.

The preceding exercises show that a subgroup H of A_5 with "enough" elements of type $(a, b)(c, d)$ of (a, b, c, d, e) also contains a 3-cycle. We now study what happens when H is normal and not equal to $\{1\}$, and show that such an H contains "enough" elements to ensure that 3-cycles are present.

Recall from Section 19.2 that a normal subgroup H of A_5 satisfies $gH = Hg$ for each g in A_5. It follows that $gHg^{-1} = H$, that is, if h is in H, so is ghg^{-1} for any g in A_5.

19.8.3 Show that if H contains a 5-cycle (a, b, c, d, e) then it also contains the 5-cycle $(g(a), g(b), g(c), g(d), g(e))$ for each g in A_5.

19.8.4 Show that if H contains a product of 2-cycles $(a, b)(c, d)$ then it also contains the product of 2-cycles $(g(a), g(b))(g(c), g(d))$ for each g in A_5.

19.8.5 Deduce from Exercises 19.8.3 and 19.8.4, and calculations like those in Exercise 19.8.2, that H contains a 3-cycle.

19.8.6 Deduce from the preceding exercises that H contains all 3-cycles.

To prove that A_5 is simple, it now remains to prove that the normal subgroup $H \neq \{1\}$ in fact contains all members of A_5.

19.8.7 By using 3-cycles to produce other elements of A_5, show that $H = A_5$.

19.9   Biographical Notes: Galois

Evariste Galois (Figure 19.9) was born in the town of Bourg-la-Reine, near
Paris, in 1811, and died, from wounds received in a duel, in Paris in 1832.
The tragedy and mystery of his short life make him the most romantic
figure in mathematics, and several biographers have been tempted to cast
Galois in the role of misunderstood genius and victim of the Establishment.
However, it has been amply documented by Rothman (1982) that Galois
does not easily fit this role. Though the known facts of his life should
satisfy anyone's appetite for drama, his tragedy is of the more classical
kind, whose seeds lie in the character of the victim himself.

Figure 19.9: Evariste Galois at the age of 15

Galois was the second of three children of Nicholas-Gabriel Galois,
the director of a boarding school and later mayor of Bourg-la-Reine, and
Adelaïde-Marie Demante, who came from a family of jurists. Both parents were well educated, and Galois seems to have had a happy, if unconventional, childhood. Up to the age of 12, he was educated entirely by
his mother, a severe classicist who instilled in him a knowledge of Latin
and Greek and a respect for Stoic morality. His father was far less of a
stoic, but unconventional in a different way, being a republican at a time

when France was returning to the monarchy. In October 1823 Evariste entered the Lycée of Louis-le-Grand, a well-known school whose pupils had included Robespierre and Victor Hugo and would later include the mathematician Charles Hermite, who found the transcendental solution to the quintic equation. There does not seem to have been any mathematics in Galois's family background, and he did not begin studying it at school until February 1827. To make the progress he did, he must have devoured mathematics at a greater rate than anyone in history, except perhaps Newton in 1665–1666, so it is no wonder that for the first time his school reports noted unsatisfactory progress in other subjects. Comments about his character being "singular" and "closed" but "original" also began to be made. At this stage Galois was studying Legendre's *Geometry* and Lagrange's works on the theory of equations and analytic functions. He believed he was ready to enter the École Polytechnique but, due to his lack of preparation in the standard syllabus, failed the entrance examination.

In 1828 he had the good fortune to study mathematics under a teacher who recognized his genius, Louis-Paul-Emile Richard. This led to Galois's first publication, a paper on continued fractions that appeared in the *Annales* of Gergonne in March 1829. Thanks to Richard, many pieces of Galois's early work still exist and have been published in Bourgne and Azra (1962). They include class papers saved by Richard and later preserved for posterity by Hermite. One fragment from 1828 shows that Galois, like Abel, initially believed he could solve the quintic equation.

One might think that publication in a respected journal was reasonable encouragement for a 17-year-old mathematician, but it was not enough for Galois. He nursed a grudge against the examiners of the École Polytechnique for failing him, and he was supported by Richard, who declared that he should be admitted without examination. Needless to say, this did not happen, but worse disappointments were to follow.

Galois had already begun working on his theory of equations and submitted his first paper on the subject to the Paris Academy in May 1829. Cauchy was referee and even seemed to be favorably impressed (see Rothman (1982), p. 89), but months went by and the paper failed to appear. Then, in July 1829, Galois's father committed suicide. The cause was trivial, even childish—a spiteful attack on him by the priest of Bourg-la-Reine—but it unleashed political passions with which Galois senior could not cope. Nor could Evariste cope with the loss of his father. His distrust of the political and educational establishment deepened into paranoia, and the sacrifice of his own life must suddenly have seemed a real possibility. It was almost the last straw when, a few days after his father's death, he failed the entrance examination for the École Polytechnique a second time.

Despite these crushing blows, Galois persevered with examinations and succeeded in entering the less prestigious École Normale in November 1829. In early 1830 he got his theory of equations into print (though not through the Academy) with the publication of three papers. The more decisive event of 1830, however, was the July revolution against the Bourbon monarchy. It gave Galois the ideal focus for his rage over the death of his father and his own humiliations, and he emerged as a republican firebrand. He made friends with the republican leaders Blanqui and Raspail and began political agitation at the École Normale—until he was expelled in December 1830 for an article he wrote against its director. In the same month, the Bourbons fled France and, as mentioned in Section 16.7, Cauchy fled with them.

Immediately after leaving the École Normale, Galois joined the Artillery of the National Guard, a republican stronghold, to concentrate on revolutionary activity. At a republican banquet on May 9, 1831, he proposed a toast with a dagger in his hand, implying a threat against the life of the new king, Louis-Philippe. Galois was arrested the following day and held until June 15 in Sainte-Pélagie prison. He was then tried for threatening the life of the king, but he was acquitted almost immediately, evidently on the grounds that he was young and foolish. The acquittal was an act of considerable leniency, for Galois gave full vent to his opinions during the trial. He admitted that he still intended to kill the king "if he betrays" and added his view that the king "will soon turn traitor if he has not done so already."

Galois was arrested a second time on Bastille Day 1831, for illegal possession of weapons and for wearing the uniform of the Artillery Guard (which had been disbanded at the end of 1830). He was held in Sainte-Pélagie prison until October and then sentenced to a further six months. Galois became very despondent and once, thinking of his father, attempted suicide. Thus he was not in a receptive mood when he finally heard from the Academy—that they were returning his manuscript—even though he was invited to submit a more complete account of his theory. Galois did in fact begin to revise his work, but he poured most of his energy into the preface, a scorching condemnation of the scientific establishment and Academicians in particular "who already have the death of Abel on their consciences." The last six weeks of his imprisonment were spent in a nursing home. Some prisoners were transferred there as a measure against cholera, which was then epidemic in Paris. In these relatively pleasant surroundings, Galois resumed his research and managed to write a few philosophical essays.

He was released on April 29, 1832. Frustratingly little is known about the next, final, month of his life. He wrote to his friend Chevalier on May 25, expressing his complete disenchantment with life and hinting that a broken love affair was the reason. It appears that the woman was Stéphanie Dumotel, daughter of the resident physician at the nursing home. Two letters from her to Galois exist, though they are defaced (presumably by Galois himself) so as to be only partly readable. One, dated May 14, says "Please let us break up this affair." The other mentions sorrows someone else had caused her, in such a way that Galois might have felt obliged to come to her defense. Whether this was the cause of the fatal duel we do not know. It is also possible that Galois felt that the duel had been hanging over his head for a long time. When he first went to prison in 1831, one of his comrades was Raspail, who, in a letter from prison on July 25 that year, quoted Galois as follows: "And I tell you, I will die in a duel over some low class coquette. Why? Because she will invite me to avenge her honour which another has compromised" (Raspail (1839), p. 89). In letters he wrote to friends on the night before the duel, Galois again spoke of an "infamous coquette."

He also wrote, "Forgive those who kill me for they are of good faith." His opponent was in fact a fellow republican, Pescheux d'Herbinville. Authors who like conspiracy theories have since conjectured that d'Herbinville was really a police agent, but no evidence exists for this. His revolutionary credentials were as good as those of Galois. The police agent theory seems rather to reflect 20th-century bafflement over dueling, something we no longer understand or sympathize with (though we still applaud successful duelists, such as Bolyai and Weierstrass). There may be *no* rational explanation for the duel, but no doubt the suicide of his father and Galois's own self-destructive tendencies were among the conditions that made it possible. Galois was convinced he was going to die over something small and contemptible, and the tragedy is that he let it happen.

The tragedy for mathematics was the incompleteness of Galois's work at the time of his death. The night before the duel, he wrote a long letter to Chevalier outlining his discoveries and hoping "some men will find it profitable to sort out this mess." Chevalier and Alfred Galois (Evariste's younger brother) later copied the mathematical papers and sent them to Gauss and Jacobi, but there was no response. The first to study them conscientiously was Liouville, who became convinced of their importance in 1843 and arranged to have them published. They finally appeared in 1846, and by the 1850s the algebraic part of the theory began to creep into textbooks. But, as mentioned in Section 19.3, there was more. Galois also talked of connections between algebraic equations and transcendental functions and made a cryptic reference to a "theory of ambiguity." The latter probably concerned the many-valuedness of algebraic functions, and we may be fairly sure that whatever Galois did was later superseded by Riemann. As for the transcendental functions, we also know that Hermite (1858) successfully completed one of Galois's investigations into solving the quintic equation by means of elliptic modular functions, and that Jordan (1870) exposed the group theory governing the behavior of such functions. However, these results only scratch the surface, and it is still possible that a bigger "Galois theory" remains to be discovered.

20

Hypercomplex Numbers

PREVIEW

This chapter is the story of a generalization with an unexpected outcome. In trying to generalize the concept of real number to n dimensions, we find only four dimensions where the idea works: $n = 1, 2, 4, 8$. "Numberlike" behavior in $\mathbb{R}^n$, far from being common, is a rare and interesting exception.

Our idea of "numberlike" behavior is motivated by the cases $n = 1, 2$ that we already know: the real numbers $\mathbb{R}$ and the complex numbers $\mathbb{C}$. The number systems $\mathbb{R}$ and $\mathbb{C}$ have both algebraic and geometric properties in common.

The common algebraic property is that of being a *field*, and it is captured by nine laws governing addition and multiplication, such as $ab = ba$ and $a(bc) = (ab)c$ (commutative and associative laws for multiplication). The common geometric property is the existence of an *absolute value*, $|u|$, which measures the distance of u from O and is *multiplicative*: $|uv| = |u||v|$.

In the 1830s and 1840s, Hamilton and Graves searched long and hard for "numberlike" behavior in $\mathbb{R}^n$, but they came up short. Beyond $\mathbb{R}$ and $\mathbb{C}$, only two *hypercomplex number* systems even come close: for $n = 4$ the *quaternion* algebra $\mathbb{H}$, which has all the required properties except commutative multiplication, and for $n = 8$ the *octonion* algebra $\mathbb{O}$, which has all the required properties except commutative and associative multiplication.

Despite lacking some of the field properties, $\mathbb{H}$ and $\mathbb{O}$ can serve as coordinates for projective planes. In this setting, the missing field properties have a remarkable geometric meaning. Failure of the commutative law corresponds to failure of the Pappus theorem, and failure of the associative law corresponds to failure of the Desargues theorem.

J. Stillwell, *Mathematics and Its History*, Undergraduate Texts in Mathematics, DOI 10.1007/978-1-4419-6053-5_20, © Springer Science+Business Media, LLC 2010

20.1   Complex Numbers in Hindsight

We saw in Chapter 14 how a need for complex numbers was first recognized in the 16th century, with the solution of cubic equations. Mathematicians were forced to include $\sqrt{-1}$ among the numbers, in order to
reconcile obvious real solutions of cubic equations with solutions given by
the Cardano formula. As time went by, complex numbers were also found
to be indispensable in geometry and analysis, as we saw in Chapters 15
and 16. With hindsight, one realizes that there is nothing "impossible" or
"imaginary" about complex numbers. They are just as real as the so-called
"real" numbers, because two dimensions are just as real as one. And they
have just as much right to be called "numbers," because they have the same
arithmetic behavior as the real numbers.

But if complex numbers are so real—and not merely a lucky side effect
of the Cardano formula—they should have been observed independently,
and earlier, in the history of mathematics. There is a comparable situation
in the history of astronomy that may help to make this point. The planet
Neptune was discovered through the calculations of Adams and Leverrier
in 1846, as we know from Section 13.4. But of course, Neptune had always been there, so it could have been observed earlier, before its special
importance was understood. This actually happened! A check of Galileo's
records by Kowal and Drake (1980) showed that he had observed Neptune
in 1612, without realizing that it was a planet. (He even observed its apparent movement with respect to the fixed stars, but presumably put this down
to experimental error.)

There was an analogous "observation" of complex numbers, without
recognizing all their properties, by Diophantus. He gave no thought to
$i = \sqrt{-1}$, which we tend to regard as the starting point of complex numbers today, but he did something else that is equally crucial—he operated
on *pairs* of ordinary numbers. This happens in his work on sums of two
squares, and it is significant because similar observations on sums of four
and eight squares foreshadowed the discovery of the four-dimensional and
eight-dimensional "numbers" that are the main subject of this chapter. Because these "numbers" have higher dimension than the complex numbers,
they are called *hypercomplex*. We shall see how much they deserve to be
called "numbers," but first it will be helpful to set the scene by recounting
the discovery made by Diophantus.

20.2   The Arithmetic of Pairs

In Book III, Problem 19 of his *Arithmetica*, Diophantus remarks that

> 65 is naturally divided into two squares in two ways, namely into $7^2 + 4^2$ and $8^2 + 1^2$, which is due to the fact that 65 is the product of 13 and 5, each of which is the sum of two squares.

Apparently, he knows that the product of sums of two squares is itself the sum of two squares, because of the identity

$$(a_1^2 + b_1^2)(a_2^2 + b_2^2) = (a_1 a_2 \mp b_1 b_2)^2 + (b_1 a_2 \pm a_1 b_2)^2.$$

As usual, Diophantus merely illustrates the general result, in this case taking $a_1 = 3$, $b_1 = 2$, $a_2 = 2$, and $b_2 = 1$. But later mathematicians realized what he was driving at: the general identity was observed by al-Khazin around 950 CE, commenting on this very problem in Diophantus, and it was proved in Fibonacci's *Book of Squares* in 1225.

Although Diophantus talks in terms of products of sums of squares $a^2 + b^2$, he is really operating on *pairs* (a, b), because he regards $a^2 + b^2$ as the square on the hypotenuse of the right-angled triangle with the pair of sides a, b. Taking the upper signs in his identity, he is describing a rule for taking two triangles, (a_1, b_1) and (a_2, b_2), and producing a third triangle, $(a_1 a_2 - b_1 b_2, b_1 a_2 + a_1 b_2)$, whose hypotenuse is the product of the hypotenuses of the two triangles given initially.

Now if we interpret (a, b) as $a + ib$ instead of a triangle, Diophantus's rule is nothing but the rule for *multiplying complex numbers*, because

$$(a_1 + ib_1)(a_2 + ib_2) = (a_1 a_2 - b_1 b_2) + i(b_1 a_2 + a_1 b_2).$$

His hypotenuse $\sqrt{a^2 + b^2}$ is what we call the *absolute value* $|a+ib|$ of $a+ib$, and his identity (with the upper signs) is the *multiplicative property of the absolute value*:

$$|a_1 + ib_1||a_2 + ib_2| = |(a_1 + ib_1)(a_2 + ib_2)|.$$

Thus in some sense Diophantus "observed" the rule for multiplying complex numbers, and also the multiplicative property it implies for the absolute value. Admittedly, there is no *addition rule*, which takes the pairs (a_1, b_1), (a_2, b_2) and produces the pair $(a_1 + a_2, b_1 + b_2)$, so Diophantus had no real arithmetic of pairs—but this could wait.

The concept of complex number had to emerge in algebra, and take charge of geometry and analysis, before mathematicians felt compelled to ask, what *is* a complex number? The definitive answer was given by Hamilton (1835): *a complex number is an ordered pair (a, b) of real numbers, and these pairs are added and multiplied according to the rules*

$$(a_1, b_1) + (a_2, b_2) = (a_1 + a_2, b_1 + b_2),$$
$$(a_1, b_1) \times (a_2, b_2) = (a_1 a_2 - b_1 b_2, b_1 a_2 + a_1 b_2).$$

The reason for replacing $a + ib$ by the pair of real numbers (a, b), of course, is to remove the controversial object $i = \sqrt{-1}$. Once this is done, it is easy to find the rules for adding and multiplying (a_1, b_1) and (a_2, b_2). Just rewrite the rules for adding and multiplying $a_1 + ib_1$ and $a_2 + ib_2$ in terms of pairs. This seems like a sly trick—using $i^2 = -1$ to find the multiplication rule, then removing the i—until we remember that Diophantus found the multiplication rule without any help from $\sqrt{-1}$.

Hamilton realized that multiplying pairs of real numbers was an important question in its own right. In fact, he was interested in the bigger question of multiplying triples, quadruples, and so on. There is an obvious way to add triples for example, the *vector addition*

$$(a_1, b_1, c_1) + (a_2, b_2, c_2) = (a_1 + a_2, b_1 + b_2, c_1 + c_2),$$

which generalizes to n-tuples for any n. But what would it mean to multiply triples? The multiplication rule for pairs does not generalize in any obvious way. Hamilton was tormented by this problem for years, and for a long time his arithmetic of pairs was the only progress he had to report. As we shall see, it played an important role in clarifying what arithmetic is in one and two dimensions, and what it should be in higher dimensions.

EXERCISES

In case there is any doubt that multiplication of complex numbers could be observed before the numbers themselves were recognized, here is another sighting, by Viète in his *Genesis triangulorum* from around 1590.

Viète independently discovered the rule of Diophantus that takes two triangles and produces a third, but Viète used it for an entirely different purpose. Instead of multiplying hypotenuses, he wanted to *add angles*.

20.2.1 Suppose the right-angled triangle with sides a_1, b_1 has angle θ_1 opposite the side b_1, and the right-angled triangle with sides a_2, b_2 has angle θ_2 opposite the side b_2. Write down $\tan \theta_1$, $\tan \theta_2$, and $\tan(\theta_1 + \theta_2)$.

20.2.2 Deduce from Question 20.2.1 that the right-angled triangle with sides $a_1a_2 - b_1b_2$, $b_1a_2 + a_1b_2$ has angle $\theta_1 + \theta_2$. (Opposite which side?)

20.2.3 Interpret the results of Diophantus and Viète in terms of the *polar form* $r(\cos\theta + i\sin\theta)$ of the complex number $a + ib$.

It has even been speculated that multiplication of complex numbers, at least "multiplication of pairs," lies behind the mysterious collection of Pythagorean triples in Plimpton 322 (Section 1.2).

To explore this speculation more fully one needs to have the complete triples (a, b, c) from Exercise 1.2.1. It turns out that every side pair (a, b) is of the form $(a_1a_2 - b_1b_2, b_1a_2 + a_1b_2)$ for some smaller integer pairs (a_1, b_1) and (a_2, b_2). That is, $a + ib = (a_1 + ib_1)(a_2 + ib_2)$. Even more amazing, with the exception of the multiple $(45, 60, 75)$ of $(3, 4, 5)$, *every $a + ib$ is a perfect square*, up to a factor of $\pm i$. Here are some for which this is not hard to verify.

20.2.4 For $(a, c) = (119, 169)$ show that $b = 120$ and that $119 + 120i$ is a perfect square. *Hint*: Observe that $169 = 13^2 = \text{hypotenuse}^2$.

20.2.5 Show that a similar result holds for $(a, c) = (161, 289)$.

20.3 Properties of + and ×

During the 1830s, Hamilton and his colleagues Peacock, De Morgan, and John Graves pursued the idea of extending the concept of number. The existing concept of number was already the result of a series of extensions— from natural and rational numbers to real and complex numbers—and Peacock observed that some *principle of permanence* was involved. It was tacitly agreed that certain properties of addition and multiplication should continue to hold with each extension of the number concept.

The "permanent" properties were not completely clear at the time, but most of them crystallized in the definition of a *field* given by Dedekind (1871). This concept had an independent origin, also around 1830, in the work of Galois on the theory of equations. So for convenience we start with the definition of a field and then explain its role in Hamilton's search for an arithmetic of n-tuples.

A field is a set of objects on which operations + and × are defined, with certain properties or "laws." To state these properties concisely, we also use the − operation. Notice that − is interpreted as the operator that turns a natural number a into its *negative*, or *additive inverse*, $-a$. The negative of a negative is defined so that $- - a = a$ always, and the *difference $a - b$*

is defined to be $a + (-b)$. Then the properties of $+$ and $-$ are as follows:

$$a + (b + c) = (a + b) + c \qquad \text{(associative law)}$$
$$a + b = b + a \qquad \text{(commutative law)}$$
$$a + (-a) = 0 \qquad \text{(additive inverse property)}$$
$$a + 0 = a \qquad \text{(property of 0)}$$

There is a similar set of properties describing the behavior of $\times$:

$$a \times (b \times c) = (a \times b) \times c \qquad \text{(associative law)}$$
$$a \times b = b \times a \qquad \text{(commutative law)}$$
$$a \times 1 = a \qquad \text{(property of 1)}$$
$$a \times 0 = 0 \qquad \text{(property of 0)}$$

and a rule for the interaction of $+$ and $\times$:

$$a \times (b + c) = a \times b + a \times c \qquad \text{(distributive law)}$$

The properties so far define what is called a *commutative ring with unit*, a typical example of which is the set $\mathbb{Z}$ of integers.

The defining properties of a *field* are those above, together with existence of the *multiplicative inverse* a^{-1}, which is defined for each $a \neq 0$ and satisfies

$$a \times a^{-1} = 1 \quad \text{(multiplicative inverse property)}$$

Typical examples of fields are the number systems $\mathbb{Q}$ of rationals, $\mathbb{R}$ of reals, and $\mathbb{C}$ of complex numbers.

In trying to see beyond these systems, Hamilton was guided by one more property they all have in common: the existence of a *multiplicative absolute value*, a real-valued function $|\ |$ with the properties

$$a \neq 0 \Rightarrow |a| > 0, \quad |ab| = |a||b|.$$

As we have seen in Section 20.2, the multiplicative absolute value for complex numbers was essentially discovered by Diophantus, long before the discovery of complex numbers themselves. Hamilton was unaware of this, because he had not studied number theory, and he was *blissfully* unaware of what number theory had to say about a multiplicative absolute value for triples. The subsequent history of hypercomplex numbers might have been very different had he known what he was up against.

20.4   Arithmetic of Triples and Quadruples

Diophantus's *Arithmetica* contains many results about sums of two squares. This is natural, because of the long history of Pythagorean triples, and because of Diophantus's own contribution to the subject in showing that sums of two squares could be "multiplied." There are also some results on sums of four squares, which led Bachet de Méziriac (1621) to the conjecture that every positive integer is the sum of four squares, and the eventual proof of this conjecture by Lagrange (1770). However, Diophantus has nothing much to say about sums of *three* squares, and it was probably obvious to him that sums of three squares could *not* be multiplied.

For example, $3 = 1^2 + 1^2 + 1^2$ and $5 = 0^2 + 1^2 + 2^2$ are both sums of three squares, but their product, 15, is not. It follows that there can be no identity of the form

$$(a_1^2 + b_1^2 + c_1^2)(a_2^2 + b_2^2 + c_2^2) = A^2 + B^2 + C^2,$$

where A, B, and C are combinations of the a_m, b_m, and c_m with integer coefficients. This means in turn that there cannot be a product of triples

$$(a_1, b_1, c_1)(a_2, b_2, c_2) = (A, B, C)$$

with multiplicative absolute value, at least if A, B, and C are such combinations of the a_m, b_m, and c_m.

In one of the most extraordinary oversights in the history of mathematics, Hamilton failed to notice this or any other evidence, and persisted to search for a product of triples for at least 13 years (from 1830 to 1843). For most of this time, he was hoping to achieve all the field properties listed above, together with a multiplicative absolute value.

Following the example of the complex numbers, he wrote the triple (a, b, c) as $a + ib + jc$, thus reducing the problem of multiplication to determining the products i^2, j^2, and ij. He wanted $i^2 = j^2 = -1$, so it remained only to find real coefficients α, β, γ such that $ij = \alpha + i\beta + j\gamma$. But nothing worked. In particular, it seemed impossible to reconcile the distributive law with the commutative law for multiplication. In 1843, he briefly considered making $ij = 0$ (which would violate the multiplicative absolute value), but then

> made what appeared to me a *less harsh* supposition, namely the supposition ... that
>
> $$ij = -ji : \quad \text{or that} \quad ij = +k, \, ji = -k,$$

the value of the product k being still left undetermined ....
This led me to conceive that perhaps instead of seeking to
confine ourselves to *triplets*, such as $a + ib + jc$ or (a, b, c),
we ought to regard these as only *imperfect forms of* QUATER-
NIONS, such as $a + ib + jc + kd$ or (a, b, c, d), the symbol k being
some new sort of unit operator.

Hamilton (1853), pp. 143–144

Thus Hamilton abandoned commutative multiplication, but everything
else fell into place. This is how he described it later, in a letter to his son:

> But on the 16th day of the month [namely, October 1843]
> which happened to be a Monday and a council day of the
> Royal Irish Academy—I was walking along to preside, and
> your mother was walking with me, along the Royal Canal
> ...and although she talked with me now and then, yet an un-
> dercurrent of thought was going on in my mind, which gave at
> last a result ... An electric current seemed to close, and a spark
> flashed forth, the herald (as I foresaw immediately) of many
> long years to come of definitely directed thought and work ...
>
> I pulled out on the spot a pocket-book, which still exists, and
> made an entry there and then. Nor could I resist the impulse—
> unphilosophical as it may have been—to cut with a knife on a
> stone of Brougham Bridge the fundamental formula with the
> symbols i, j, k:
>
> $$i^2 = j^2 = k^2 = ijk = -1,$$
>
> which contains the solution of the Problem, but of course, as
> an inscription it has long since mouldered away.

Hamilton (1865)

The pocket-book contains not only the values of ij, ji, jk, kj, ki, ik,
which follow from the fundamental formula, but also the four components
of the general product of quaternions:

$$
\begin{aligned}
(a + ib + jc + kd)(\alpha + i\beta + j\gamma + k\delta) = \quad & (a\alpha - b\beta - c\gamma - d\delta) \\
& + i\,(a\beta + b\alpha + c\delta - d\gamma) \\
& + j\,(a\gamma - b\delta + c\alpha + d\beta) \\
& + k\,(a\delta + b\gamma - c\beta + d\alpha).
\end{aligned}
$$

As in all his previous attempts, Hamilton's starting point was the multiplicative property of the absolute value, or as he put it: "the modulus of a product is equal to the product of the moduli of the factors." This generalizes the multiplicative property of the absolute value for complex numbers, and shows that the product of two nonzero quaternions is nonzero.

The square of the absolute value of the quaternion $\alpha + \beta i + \gamma j + \delta k$ is $\alpha^2 + \beta^2 + \gamma^2 + \delta^2$, so the quaternion product gives the following identity, showing that the product of sums of four squares is a sum of four squares:

$$
\begin{aligned}
(a^2 + b^2 + c^2 + d^2)(\alpha^2 + \beta^2 + \gamma^2 + \delta^2) = \ & (a\alpha - b\beta - c\gamma - d\delta)^2 \\
& + (a\beta + b\alpha + c\delta - d\gamma)^2 \\
& + (a\gamma - b\delta + c\alpha + d\beta)^2 \\
& + (a\delta + b\gamma - c\beta + d\alpha)^2.
\end{aligned}
$$

If Hamilton had studied number theory he would have known this, because the identity was discovered by Euler (1748c), and used by Euler and Lagrange to prove that every natural number is the sum of four squares.

Hamilton thought at first that his four-square identity was original, but in the months following the discovery of quaternions he and his friend John Graves caught up with the news on three and four squares. It dawned on Graves that they should never have expected a three-square identity, because $3 = 1^2 + 1^2 + 1^2$ and $21 = 1^2 + 2^2 + 4^2$ are sums of three squares, but their product 63 is not. He then consulted the literature and

> On Friday last I looked into Lagrange's [he meant Legendre] Théorie des Nombres and found for the first time that I had lately been on the track of former mathematicians. For example, the mode by which I satisfied myself that a general theorem
>
> $$(x_1^2 + x_2^2 + x_3^2)(y_1^2 + y_2^2 + y_3^2) = z_1^2 + z_2^2 + z_3^2$$
>
> was impossible was the very mode mentioned by Legendre, who gives the very example that occurred to me, viz., $3 \times 21 = 63$, it being impossible to compound 63 of three squares.
>
> I then learned that the theorem
>
> $$(x_1^2 + x_2^2 + x_3^2 + x_4^2)(y_1^2 + y_2^2 + y_3^2 + y_4^2) = z_1^2 + z_2^2 + z_3^2 + z_4^2$$
>
> was Euler's.
>
> Graves (1844) letter to Hamilton

It is tempting to think that Hamilton could have discovered quaternions much more easily had he known there was an identity for sums of four squares, and none for sums of three squares. But the course of mathematical discovery is seldom so smooth. Perhaps the hopeless struggle with triples was good for him, because he did not want it to be in vain—he may not have been willing to abandon commutative multiplication otherwise.

EXERCISES

It can be checked that 15 is not a sum of three (integer) squares by trying all possible sums of the squares 0, 1, 4, 9 that are less than 15. However, a much more general result is possible. As Exercises 3.2.1 and 3.2.2 show, no natural number of the form $8n + 7$ is a sum of three squares.

With an infinite supply of such numbers on tap, it becomes easy to understand how both Legendre and Graves stumbled on the example $3 \times 21 = 63$.

20.4.1 Find the smallest number of the form $8n + 7$ (hence not a sum of three squares) that is the product of sums of three *nonzero* squares.

We can also improve the result of Exercise 3.2.2 to one about sums of *rational* squares (which would have been more interesting to Diophantus).

20.4.2 Show that if there are rationals x, y, and z such that $x^2 + y^2 + z^2 = 7$, then $7s^2$ is a sum of three integer squares, for some integer s. Show that the latter is impossible.

20.4.3 Generalize the argument of Question 20.4.2 to show that $8n + 7$ is not the sum of three rational squares, for any integer n.

It is interesting that Diophantus actually remarked (in his Book VI, Problem 14) that 15 is not a sum of two rational squares. Question 20.4.3 shows that 15 is not even the sum of three rational squares—a result Diophantus may also have known, since the most obvious proofs of the two results are similar. (To prove that 15 is not a sum of two rational squares it suffices to use remainders on division by 4. Try it!)

20.5   Quaternions, Geometry, and Physics

Hamilton may have seen, at the instant of discovery, that quaternions would be worth his attention for the rest of his life, but even his best friends were skeptical at first. On 26 October 1843, John Graves wrote to him:

> You must have been in a very bold mood to start the happy idea that ij might be different from ji . . . Have you any

inkling of the existence in nature of processes, or operations, or phenomena, or conceptions analogous to the circuit

$$ij = -ji = k$$
$$jk = -kj = i$$
$$ki = -ik = j?$$

And after receiving a letter from Hamilton hinting at applications to physics, and announcing that quaternions could certainly be used to derive theorems of spherical trigonometry, Graves replied:

> There is still something in the system that gravels me. I have not yet any clear views as to the extent to which we are at liberty arbitrarily to create imaginaries, and to endow them with supernatural properties . . . . But supposing that your symbols have their physical antitypes, which might have led to your quaternions, what right have you to such luck, getting at your system by such an *inventive* mode as yours?

(For more of these letters, see the biography of Hamilton by Graves's brother Robert: Graves (1975), vol. 3, p. 443.)

Of course, Graves's question about luck was tongue-in-cheek, but it is still a good question. Many mathematicians and physicists have marveled at the capacity of pure mathematics to become applied, for number theory and algebra to become geometry and physics. In the case of quaternions, more surprises were in store.

Not only was it true that quaternions had implications for spherical trigonometry, their geometric aspect had already been discovered twice before! The first discovery was the unpublished work of Gauss (1819) on rotations of the sphere, which Hamilton could not have known about; the second was a publication by Rodrigues (1840) that (typically) escaped his attention.

The result of Gauss is easiest to explain, because we have already mentioned it in Section 18.6: every rotation of the sphere can be expressed by a complex function of the form

$$f(z) = \frac{az + b}{-\overline{b}z + \overline{a}}.$$

Any such function can be represented by the matrix of its coefficients,

$$\begin{pmatrix} a & b \\ -\overline{b} & \overline{a} \end{pmatrix},$$

and it is easily checked that the matrix of $f_1 f_2$ is the product of the matrices for f_1 and f_2. Thus rotations of the sphere can be studied via products of matrices of the above type, involving pairs of complex numbers a, b. Such a matrix can also be written in terms of four *real* parameters $\alpha, \beta, \gamma, \delta$ if we set

$$a = \alpha + i\beta, \quad b = \gamma + i\delta.$$

And we can then write the resulting matrix as a linear combination of four special matrices with coefficients $\alpha, \beta, \gamma, \delta$:

$$\begin{pmatrix} a & b \\ -\overline{b} & \overline{a} \end{pmatrix} = \begin{pmatrix} \alpha + i\beta & \gamma + i\delta \\ -\gamma + i\delta & \alpha - i\beta \end{pmatrix}$$

$$= \alpha \begin{pmatrix} 1 & 0 \\ 0 & 1 \end{pmatrix} + \beta \begin{pmatrix} i & 0 \\ 0 & -i \end{pmatrix} + \gamma \begin{pmatrix} 0 & 1 \\ -1 & 0 \end{pmatrix} + \delta \begin{pmatrix} 0 & i \\ i & 0 \end{pmatrix}$$

$$= \alpha\mathbf{1} + \beta\mathbf{i} + \gamma\mathbf{j} + \delta\mathbf{k}.$$

The four special matrices $\mathbf{1}, \mathbf{i}, \mathbf{j}, \mathbf{k}$ play the role of $1, i, j, k$ in the quaternions, because

$$\mathbf{i}^2 = \mathbf{j}^2 = \mathbf{k}^2 = \mathbf{ijk} = -1.$$

In fact, the same matrices were discovered by Cayley (1858), who proposed them as a new realization of the quaternions. Today, they are often known as *Pauli* matrices, particularly in physics. They were rediscovered in quantum theory, where the rotations of the sphere are also important.

EXERCISES

The Cayley matrices

$$\mathbf{1} = \begin{pmatrix} 1 & 0 \\ 0 & 1 \end{pmatrix}, \quad \mathbf{i} = \begin{pmatrix} i & 0 \\ 0 & -i \end{pmatrix}, \quad \mathbf{j} = \begin{pmatrix} 0 & 1 \\ -1 & 0 \end{pmatrix}, \quad \mathbf{k} = \begin{pmatrix} 0 & i \\ i & 0 \end{pmatrix}$$

make it easy to prove the basic properties of quaternions.

20.5.1 Show that $\mathbf{i}^2 = \mathbf{j}^2 = \mathbf{k}^2 = \mathbf{ijk} = -1$, and also that $\mathbf{ij} = \mathbf{k}$, etc.

It follows that the arbitrary quaternion $\alpha + \beta i + \gamma j + \delta k$ is represented by the complex matrix

$$\alpha\mathbf{1} + \beta\mathbf{i} + \gamma\mathbf{j} + \delta\mathbf{k} = \begin{pmatrix} \alpha + i\beta & \gamma + i\delta \\ -\gamma + i\delta & \alpha - i\beta \end{pmatrix}.$$

A nice feature of this representation is that the square of the absolute value of a quaternion is simply the determinant of the corresponding matrix. Since the square of the absolute value turns up so often, it is also given a name: the *norm*.

20.5.2 Show that

$$\det \begin{pmatrix} \alpha + i\beta & \gamma + i\delta \\ -\gamma + i\delta & \alpha - i\beta \end{pmatrix} = \alpha^2 + \beta^2 + \gamma^2 + \delta^2.$$

The multiplicative property of the norm then follows from the multiplicative property of determinants: $\det AB = \det A \det B$ for any 2×2 matrices A and B. The other algebraic properties of quaternions also follow from properties of matrices that are familiar today: addition is associative and commutative; multiplication is associative but not commutative; the distributive law holds, and every matrix with nonzero determinant has a multiplicative inverse.

The quaternions have a *conjugation* operation analogous to conjugation in $\mathbb{C}$. The conjugate of $q = \alpha + \beta i + \gamma j + \delta k$ is defined to be $\bar{q} = \alpha - \beta i - \gamma j - \delta k$.

20.5.3 Show that $q\bar{q} = \alpha^2 + \beta^2 + \gamma^2 + \delta^2$ (the norm $|q|^2$ of q), and hence express the multiplicative inverse of q in terms of $\bar{q}$ and $|q|$.

The quaternion product is related to two well-known product operations on the space of 3-dimensional vectors: the *scalar product* $u \cdot v$ and the *vector product* $u \times v$. If we write 3-dimensional vectors as

$$u = u_1 i + u_2 j + u_3 k, \quad v = v_1 i + v_2 j + v_3 k,$$

then the scalar and vector products are defined by

$$u \cdot v = u_1 v_1 + u_2 v_2 + u_3 v_3$$

and

$$u \times v = \begin{vmatrix} i & j & k \\ u_1 & u_2 & u_3 \\ v_1 & v_2 & v_3 \end{vmatrix} = (u_2 v_3 - u_3 v_2)i - (u_1 v_3 - u_3 v_1)j + (u_1 v_2 - u_2 v_2)k.$$

20.5.4 Show that if u and v are quaternions with zero real part ("pure imaginary quaternions"), then

$$uv = -u \cdot v + u \times v.$$

20.5.5 Deduce from Exercise 20.5.4 that $u^2 = -1$ for any unit pure imaginary quaternion, and that uv is pure imaginary if and only if u is perpendicular to v.

20.6   Octonions

Hamilton and his friend John Graves had long discussed the problem of defining multiplication for triples and other n-tuples of real numbers. The discovery of quaternions evidently catalyzed Graves's own thinking about n-tuples, because by December 1843 he was able to tell Hamilton of an interesting discovery of his own: a system of octuples with a multiplicative absolute value, which he called the *octaves*. Hamilton congratulated Graves on his discovery, but pointed out that octaves were not quite as nice as quaternions, because their multiplication was not only noncommutative, but also nonassociative. He agreed to arrange for publication of Graves's discovery, but failed to follow up, with the result that the octaves were rediscovered by Cayley (1845b) before Graves's priority was generally known. As a consequence, they have often been called *Cayley numbers* or *Cayley–Graves* numbers. Today they are generally called the *octonions*, and the set of them is called $\mathbb{O}$.

Octonions are octuples of real numbers with the usual vector addition and scalar multiplication. The standard basis vectors $(1, 0, 0, 0, 0, 0, 0, 0)$, $(0, 1, 0, 0, 0, 0, 0, 0)$, ..., $(0, 0, 0, 0, 0, 0, 0, 1)$ are called *1, i, j, k, l, m, n, o*, respectively, so any octonion can be written in the form

$$\alpha + \beta i + \gamma j + \delta k + \varepsilon l + \zeta m + \eta n + \theta o.$$

They satisfy the distributive axiom, so the value of any octonion product is determined by the products of the "imaginary units" i, j, k, l, m, n, o. The square of each imaginary unit is -1, and Figure 20.1 gives a description of all products of distinct basis vectors. The product of any two basis vectors

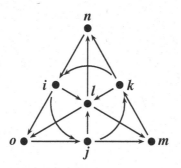

Figure 20.1: Products of octonion basis vectors

is the third vector in the "line" containing them, with a plus or minus sign determined by the arrow and the position of the two vectors in the product. The "lines" include the circle through i, j, and k, and in fact all the "lines" are supposed to be like this—you should imagine adding a third segment to each of them, joining the endpoints.

A much simpler description of octonion multiplication was given by Dickson (1914), p. 15. Dickson's description is a generalization of Hamilton's definition of multiplication of pairs, and in fact it shows that the *same construction* produces $\mathbb{C}$ from $\mathbb{R}$, $\mathbb{H}$ from $\mathbb{C}$, and $\mathbb{O}$ from $\mathbb{H}$. Each system consists of ordered pairs (a, b) from the previous system, and pairs are multiplied by the rule

$$(a_1, b_1) \times (a_2, b_2) = (a_1 a_2 - \overline{b_2} b_1, b_2 a_1 + b_1 \overline{a_2}),$$

where ⁻ denotes the conjugation operation, which changes the sign of all the imaginary units. (Thus conjugation has no effect on a real number.) In particular, octonions can be viewed as pairs (a, b) of quaternions a and b. In this case it is important to observe the precise order of products in the definition, because the quaternion product is generally not commutative.

The octonion $p = \alpha + \beta i + \gamma j + \delta k + \varepsilon l + \zeta m + \eta n + \theta o$ has the square of its absolute value equal to $p\overline{p} = \alpha^2 + \beta^2 + \gamma^2 + \delta^2 + \varepsilon^2 + \zeta^2 + \eta^2 + \theta^2$, so the multiplicative property of absolute value gives an identity expressing the product of two sums of eight squares as a sum of eight squares. After discovering this, Graves searched the literature for such identities, and uncovered Euler's four-square identity from 1748 (though actually a later appearance of it), and also his own identity in a paper of Degen (1822). Thus the octonions, like the complex numbers and quaternions, gave the first intimation of their existence in the theory of sums of squares.

EXERCISES

The Dickson formula

$$(a_1, b_1) \times (a_2, b_2) = (a_1 a_2 - \overline{b_2} b_1, b_2 a_1 + b_1 \overline{a_2})$$

can be taken as the definition of multiplication for the octonions, but first we should check that the formula gives the correct definition of multiplication for quaternions. This can be done using Cayley's representation of the quaternions by 2×2 matrices of complex numbers (Exercises 20.5.1 and 20.5.2). Each quaternion $\alpha + \beta i + \gamma j + \delta k$ is represented by the complex matrix

$$\begin{pmatrix} \alpha + i\beta & \gamma + i\delta \\ -\gamma + i\delta & \alpha - i\beta \end{pmatrix},$$

which we may call $M(\alpha+i\beta, \gamma+i\delta)$, because it is determined by the pair of complex numbers $\alpha + i\beta$, $\gamma + i\delta$. Then if we write this pair more simply as

$$a = \alpha + i\beta,$$
$$b = \gamma + i\delta,$$

it suffices to prove that the product according to Dickson,

$$(a_1, b_1) \times (a_2, b_2) = (a_1 a_2 - \overline{b_2} b_1, b_2 a_1 + b_1 \overline{a_2}),$$

corresponds to the product according to Cayley's matrices. That is, we have to show that

$$M(a_1, b_1)M(a_2, b_2) = M(a_1 a_2 - \overline{b_2} b_1, b_2 a_1 + b_1 \overline{a_2}).$$

20.6.1 Show that

$$M(a, b) = \begin{pmatrix} a & b \\ -\overline{b} & \overline{a} \end{pmatrix}.$$

Hence compute $M(a_1, b_1)M(a_2, b_2)$ for any complex numbers a_1, b_1, a_2, b_2, and show that it equals $M(a_1 a_2 - \overline{b_2} b_1, b_2 a_1 + b_1 \overline{a_2})$.

Figure 20.1 for the products of the octonion units is due to Freudenthal (1951), and it shows that $ij = k$ as we would expect, because these i, j, k behave the same as the quaternion units. Since the "line" $i \to j \to k$ is closed by an arrow from k to i, it also shows that $kj = i$, and likewise (using the invisible arrow from m to o) that $jm = o$ and $mo = j$.

20.6.2 Check that the same products result from Dickson's multiplication formula when i, j, k, l, m, n, o are defined in terms of the quaternion units $\mathbf{i}, \mathbf{j}, \mathbf{k}$ by

$$l = (0, 1),$$
$$i = (\mathbf{i}, 0), \quad m = (0, \mathbf{i}),$$
$$j = (\mathbf{j}, 0), \quad n = (0, \mathbf{j}),$$
$$k = (\mathbf{k}, 0), \quad o = (0, \mathbf{k}).$$

20.7   Why $\mathbb{C}$, $\mathbb{H}$, and $\mathbb{O}$ Are Special

The pre-established harmony between the two-square, four-square, and eight-square identities and the norms on $\mathbb{C}$, $\mathbb{H}$, and $\mathbb{O}$ suggest that $\mathbb{C}$, $\mathbb{H}$, and $\mathbb{O}$ are not just random curiosities, but actually very special structures. In fact, they are unique. If we define a *hypercomplex number system* to consist of n-tuples of real numbers ($n \geq 2$) with vector addition, a distributive multiplication and a multiplicative absolute value, then

- $\mathbb{C}$ is the only hypercomplex number system for which the multiplication is commutative and associative. This was proved by Weierstrass (1884).

- $\mathbb{H}$ is the only other hypercomplex number system for which the multiplication is associative. This was proved by Frobenius (1878).

- $\mathbb{O}$ is the only other hypercomplex number system. This was proved by Hurwitz (1898). (In the process, Hurwitz proved that there are no n-square identities except for $n = 1, 2, 4, 8$.)

Since that time, it has been found that $\mathbb{C}$, $\mathbb{H}$, and $\mathbb{O}$ have relationships with many other "exceptional" structures in mathematics. One of the most remarkable is their relationship with projective geometry, via the theorems of Pappus and Desargues.

The theorem of Pappus is a theorem of classical geometry that belongs to projective geometry, seemingly by accident. As mentioned in the exercises to Section 8.8, it states that *if the vertices of a hexagon ABCDEF lie alternately on two straight lines, then the intersections of opposite sides of the hexagon lie on a line* (Figure 20.2).

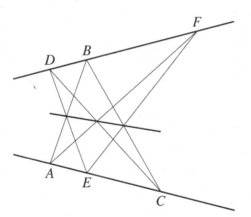

Figure 20.2: Pappus's theorem

This theorem is meaningful in projective geometry, because it involves only points and lines and whether they meet or not, yet its *proof* involves the concept of distance. Desargues's theorem in the plane is like this too, as mentioned in Section 8.3; it is a projective theorem without a projective proof, and this is even more puzzling because Desargues's theorem in space *does* have a projective proof.

An amazing explanation of these phenomena was uncovered by the work of von Staudt (1847) and Hilbert (1899). In 1847 von Staudt gave geometric constructions of + and ×, allowing each projective plane to be "coordinatized" by hypercomplex numbers. Then in 1899 Hilbert made the wonderful discovery that the geometry of a projective plane is tied to the algebra of the corresponding hypercomplex number system:

- *Pappus's theorem holds ⇔ the system is commutative.*

- *Desargues's theorem holds ⇔ the system is associative.*

Conversely, any hypercomplex number system $\mathbb{K}$ yields a projective plane $\mathbb{K}P^2$, by construction of homogeneous coordinates essentially as in Section 8.5. Then by Hilbert's theorem,

- $\mathbb{R}P^2$ and $\mathbb{C}P^2$ satisfy Pappus,

- $\mathbb{H}P^2$ satisfies Desargues but not Pappus, and

- $\mathbb{O}P^2$ satisfies neither.

The results of Hilbert explain why the theorems of Pappus and Desargues do not have projective proofs. It is because these theorems *do not hold for all projective planes*, only for those with enough algebraic structure. This is a remarkable contribution of algebra to geometry, but it also gives insight in the opposite direction. It can seriously be said that Pappus's theorem "explains" why $\mathbb{R}$ and $\mathbb{C}$ have commutative multiplication, because it is simpler (takes fewer axioms) to describe a projective plane satisfying Pappus's theorem than to describe a field. This is possibly the most remarkable aspect of Hilbert's work on the foundations of geometry. It shows that the long historical trend of turning geometry into algebra—which began with Fermat and Descartes—may conceivably be coming to an end.

EXERCISES

Freudenthal's diagram of octonion units (Figure 20.1) itself has the structure of a projective plane, which is why we used the name "lines" for the collinear (or cocircular) triples of points in it.

20.7.1 Check that the seven "points" (octonion units) and seven "lines" of Freudenthal's diagram have the following properties.

- Through any two "points" there is exactly one "line."
- Any two "lines" have exactly one "point" in common.

Such a structure is called a *finite projective plane*, and this one is often called the *Fano* plane after its discoverer. The diagram makes it easy to show that $\mathbb{O}$ is not associative.

20.7.2 Find a triple of octonion units a, b, c such that $a(bc) \neq (ab)c$.

The weakening of multiplicative structure as we construct hypercomplex number systems of higher dimension (losing commutativity with $\mathbb{H}$ and associativity with $\mathbb{O}$) is a hint that we can't go on constructing hypercomplex systems indefinitely. In fact, the 16-dimensional system of pairs of octonions, with the Dickson multiplication rule, does not have a multiplicative absolute value. This is because it includes "zero divisors"—nonzero elements whose product is the zero element $(0, 0)$.

20.7.3 Show that the nonzero pairs (i, n), (k, l) of octonion units have Dickson product $(0, 0)$. Also, find another pair (a, b) of octonion units such that $(i, n)(a, b) = (0, 0)$.

20.7.4 Show that in any system with a multiplicative absolute value $|\ |$, $x \neq 0$ and $y \neq 0$ imply $xy \neq 0$. (Hence the system of pairs of octonions does not have a multiplicative absolute value.)

20.8 Biographical Notes: Hamilton

The world of mathematics is one of logic and order, so mathematicians tend to look for order in their personal lives. Usually they find it (it is hard to do mathematics otherwise!), even though the human world is not very orderly. But sometimes they don't, and the result can be both a mathematical and human tragedy. One such case was Galois; another was Hamilton.

William Rowan Hamilton (Figure 20.3) was born in Dublin at midnight, between August 3 and 4, in 1805. His father, Archibald, a lawyer, and his mother, Sarah, cared for him until he was three years old, but then they got into financial difficulties, and young William was sent to live with Archibald's brother James and his wife, Sydney. Uncle James Hamilton was an Anglican curate and schoolmaster in Trim, about 40 miles from Dublin, a devoted father figure and educator, but with highly eccentric methods of instruction. Here is how he taught William to spell at age three:

James printed on cards every word he has yet spelled; he began with every monosyllable in which *A* is the principal letter, and so on alphabetically, never beginning a new set till he could spell them off book and on book; every spelling book and dictionary was searched . . . so that he is now completely grounded in words that most children are very deficient in, and indeed many grown people . . . he is going through them now for the last time, and James is now preparing words of two syllables.

Letter from Sydney to Sarah Hamilton, 17 October 1808, in Graves (1975), vol. 1, p. 31

At this time, William was also taught addition, subtraction, and multiplication of numbers up to 10, but mathematics did not play a big part in his childhood. Uncle James was primarily a classicist with an interest in Asian languages, and William was the ideal pupil. He began learning Hebrew at three, followed by Latin and Greek by age five, Italian and French at eight, and Arabic, Sanskrit, and Persian by age ten. Only then do we hear of mathematics again, when William reports in a letter to his sister Grace that "I have done near half the first book of Euclid with uncle"—a pretty ordinary accomplishment by the standards of the time.

Hamilton reached a turning point in his intellectual life at age 13. He seems to have decided that he knew enough languages, because he stopped picking up new ones and wrote a small book on Syriac grammar for the benefit of other learners. At the same time, he met another boy who could beat him in an intellectual contest, the American calculating prodigy Zerah Colburn. Hamilton was consistently outclassed by Colburn's feats, such as calculating the number of minutes in 1811 years, and factorizing numbers in the billions. But far from being discouraged by the experience, he wanted to know more. When Colburn retired from the mental arithmetic game and returned two years later as an actor, Hamilton asked him about his computational methods, and found he was able to simplify them. This was probably his first mathematical research.

In 1823 Hamilton entered Trinity College, Dublin, beginning an academic career of extraordinary distinction in both science and classics. Over the next three years he laid the foundations of his brilliant mathematical life but also, alas, for his miserable personal life. Hamilton was a romantic—he loved *Romeo and Juliet* and the poetry of Wordsworth—and on 17 August 1824 he met the lady of his dreams, Catherine Disney.

Figure 20.3: Sir William Rowan Hamilton

Her family were friends of his uncle James, and some of her brothers in fact became friends of Hamilton at Trinity College. Hamilton fell in love with Catherine at first sight, and she apparently reciprocated his feeling; but the boy who knew all the words, in all the languages, did not manage to convey his love to her. Perhaps he thought it improper to express such feelings before he had any prospect of marriage, or before he was sure how she felt; but at any rate his hesitation was fatal. In February 1825 Catherine became engaged to an older and wealthier suitor, encouraged by her family, and on 25 May they were married. Hamilton despaired almost to the point of suicide, and never really recovered. Only his mathematical spirit was not crushed.

On this occasion he rebounded with his first important mathematical paper, *Theory of Systems of Rays*, presented to the Royal Irish Academy in 1827. This paper led to his appointment as Professor of Astronomy and Director of Dunsink Observatory, an amazing achievement for a 22-year-old. His fame grew, and over the next few years he became friends with several men who were to influence his intellectual life: the poets Wordsworth and Coleridge, the mathematicians John and Charles Graves, and their brother

Robert, who eventually wrote Hamilton's biography.

The scene was also set for his next disaster of the heart. Among Hamilton's students at the observatory in 1830 was a young aristocrat and astronomy enthusiast named Lord Adare. From time to time he invited Hamilton to his family home, Adare Manor in County Limerick. There in 1831 Hamilton met the second love of his life, Ellen de Vere, a beautiful and intelligent 18-year-old whose appreciation of romantic poetry surpassed even his own.

They seemed perfect for each other, and this time he had money, position, and the support of her family. How could he fail? Only by giving up at the first sign of difficulty! Ellen dropped a casual remark that "she could not live happily anywhere but at Curragh" (her home). Hamilton took this as a polite but firm rebuff—and that was the end of the courtship. He retired to nurse his broken heart again, writing an excruciating sonnet entitled *To E. de V. On her saying that she could not live happily anywhere but at Curragh.* In due course Ellen married another, and of course left Curragh.

Hamilton returned to mathematics to ease the pain, and in 1832 lifted his theory of optics to a new level. A supplement to his *Theory of Systems of Rays* in 1832 presented a sensational and unprecedented discovery: a new physical phenomenon predicted by pure mathematics. This was the previously unobserved *conical refraction,* in which a single ray of light entering a slab of suitable crystalline material diverges as a hollow cone. Hamilton's prediction was verified experimentally by Humphrey Lloyd at Trinity College, and was the first of many such predictions. Two of the best-known ones are the prediction of electromagnetic waves from Maxwell's equations of 1864, and the bending of light predicted by Einstein's general theory of relativity in 1915. As in the latter cases, Hamilton's success was no fluke. It was based on a deep and powerful mathematical theory that generalizes to other situations, and is now known as *Hamiltonian dynamics.*

Having regained some self-confidence, Hamilton in 1832 found what he called a "dim perspective of possible marriage" in Helen Bayly, who lived near him and was two years his senior. Dim it was, but this time he steeled himself to resist all opposition. Despite Helen's fragile health (which she warned him about herself) and the total opposition of his family, they were married on 9 April 1833. They spent their honeymoon at the cottage of Helen's widowed mother, where Hamilton continued working on his mathematical papers.

When they returned to his home at Dunsink observatory, Hamilton's sisters, who had previously kept house for him, had moved out. His domestic life descended into chaos, as Helen was frequently ill or absent entirely, and Hamilton came to depend on alcohol for consolation. Despite this, his mathematical work continued unabated. He was knighted in 1835, elected president of the Royal Irish Academy in 1837, and (as we know) discovered quaternions in 1843.

It is probably true that Hamilton spent too much time on quaternions. He did little else until his death in 1865, and few mathematicians shared his enthusiasm. Nevertheless, quaternions changed the course of mathematics, though not in the way Hamilton intended. In the 1880s, Josiah Willard Gibbs and Oliver Heaviside created what we now know as vector analysis, essentially by separating the real ("scalar") part of a quaternion from its imaginary ("vector") part. Hamilton's followers were outraged to see the simple and elegant quaternions torn limb from limb, but the idea caught on in physics and engineering, where it still holds sway today.

There are at least three biographies of Hamilton, all worth reading. Graves's three volumes (Graves (1975)) are still valuable, if only for the large amount of correspondence they contain. Hankins (1980) is entertaining and authoritative, with good coverage of the mathematics. O'Donnell (1983) throws more light on Hamilton's psychology and is refreshingly skeptical about his childhood precocity with languages. For more on the remarkable metamorphosis of quaternions into vector analysis, see Crowe (1967).

21

Algebraic Number Theory

PREVIEW

Another concept of abstract algebra that emerged from the old algebra of equations was that of *ring*, which arose from attempts to find *integer* solutions of equations. The first steps towards the ring concept were taken by Euler (1770b), who discovered equations whose integer solutions are most easily found with the help of irrational or imaginary numbers.

Gauss realized that these auxiliary numbers work because they *behave like* integers. In particular, they admit a concept of "prime" for which unique prime factorization holds.

In the 1840s and 1850s the idea of "algebraic integers" was pushed further by various mathematicians, and it reached maturity when Dedekind (1871) defined the concept of *algebraic integer* in a *number field of finite degree*. By this time, considerable experience with number fields had been acquired, and Kummer had noticed that such fields do *not* always admit unique prime factorization.

Kummer found a way around this difficulty by introducing new objects that he called *ideal numbers* (in analogy with "ideal" objects in geometry, such as points at infinity). Dedekind replaced Kummer's undefined "ideal numbers" by concrete sets of numbers that he called *ideals*. He was then able to restore unique prime factorization by proving that it holds for ideals.

Ring theory as we know it today is largely the result of building a general setting for Dedekind's theory of ideals. It owes its existence to Emmy Noether, who used to say that "it's already in Dedekind."

J. Stillwell, *Mathematics and Its History*, Undergraduate Texts in Mathematics,   439
DOI 10.1007/978-1-4419-6053-5_21, © Springer Science+Business Media, LLC 2010

21.1 Algebraic Numbers

The integers are the simplest objects in mathematics but, as history shows, their secrets are deeply hidden. A vast range of mathematical disciplines—such as geometry, algebra, and analysis—has been called upon to clarify the apparently simple concept of integer. In particular, a broader *concept of integer* itself seems to be useful. We have seen in Section 5.4, for example, how integer solutions of the Pell equation $x^2 - Ny^2 = 1$ can be produced with the help of irrational numbers of the form $a + b\sqrt{N}$, and in Section 10.6 how the number $(1 + \sqrt{5})/2$ helps explain the mysterious sequence of Fibonacci numbers. These are examples of the way *algebraic* numbers help elucidate the behavior of integers.

In the 19th century, a powerful theory of algebraic numbers was developed, with the aim of throwing more light on ordinary number theory. It was very successful in this respect, but it also developed a life of its own, and in the 20th century its concepts were appropriated by the abstract theories of rings, fields, and vector spaces. Later in the chapter we sketch how this happened, but our main goal is to explain algebraic number theory itself, the inspiration for this whole development.

First we should state the definition: an *algebraic number* is one that satisfies an equation of the form

$$a_n x^n + a_{n-1} x^{n-1} + \cdots + a_1 x + a_0 = 0, \quad \text{where } a_0, a_1, \ldots, a_n \in \mathbb{Z}.$$

The symbol $\mathbb{Z}$ for integers comes from the German word "Zahlen," meaning "numbers." We sometimes call these integers the "ordinary," or *rational*, integers, to avoid confusion with the algebraic integers defined in Section 21.3.

The algebraic numbers obviously include $\sqrt{2}$ (a solution of $x^2 - 2 = 0$), $\sqrt[3]{2}$ (a solution of $x^3 - 2 = 0$), and less obviously $\sqrt{2} + \sqrt{3}$ (see Exercise 21.1.1). The first mathematicians to use algebraic numbers systematically in number theory were Lagrange and Euler around 1770. A spectacular example was given by Euler (1770b), when he used the algebraic number $\sqrt{-2}$ to prove the following claim of Fermat: $x = 5$ *and* $y = 3$ *is the only positive integer solution of* $y^3 = x^2 + 2$. (The equation in fact goes back to Diophantus, who mentioned its integer solution in his Book VI, Problem 17.)

Euler's argument is incomplete but essentially correct, and we complete it later by closer study of the set $\mathbb{Z}[\sqrt{-2}]$ of numbers $a + b\sqrt{-2}$, where $a, b \in \mathbb{Z}$. It goes as follows.

Suppose x and y are integers such that $y^3 = x^2 + 2$. Then

$$y^3 = (x + \sqrt{-2})(x - \sqrt{-2}).$$

Assuming that numbers of the form $a + b\sqrt{-2}$ "behave like" ordinary integers, we can conclude that $x + \sqrt{-2}$ and $x - \sqrt{-2}$ are *cubes* (since their product is the cube y^3). That is, there are $a, b \in \mathbb{Z}$ such that

$$\begin{aligned} x + \sqrt{-2} &= (a + b\sqrt{-2})^3 \\ &= a^3 + 3a^2 b \sqrt{-2} + 3ab^2(-2) + b^3(-2\sqrt{-2}) \\ &= a^3 - 6ab^2 + (3a^2 b - 2b^3)\sqrt{-2}. \end{aligned}$$

Equating real and imaginary parts, we get

$$\begin{aligned} x &= a^3 - 6ab^2, \\ 1 &= 3a^2 b - 2b^3 = b(3a^2 - 2b^2) \quad \text{for some } a, b \in \mathbb{Z}. \end{aligned}$$

Now the only integer products equal to 1 are 1×1 and $(-1) \times (-1)$; hence $b = \pm 1$, and therefore $a = \pm 1$, from the second equation. Then the only positive solution for x occurs with $a = -1$, $b = \pm 1$, in which case $x = 5$ and hence $y = 3$.   □

This wonderful flight of fancy, that the numbers $a + b\sqrt{-2}$ "behave like" ordinary integers, can actually be justified. It depends on the theory of divisibility in $\mathbb{Z}[\sqrt{-2}]$, which turns out to be similar to divisibility in $\mathbb{Z}$, already studied in Section 3.3.

EXERCISES

21.1.1 Show that the number $\sqrt{2} + \sqrt{3}$ satisfies the equation $x^4 - 10x^2 + 1 = 0$.

Before starting to investigate divisibility in $\mathbb{Z}[\sqrt{-2}]$, it will be useful to renew our acquaintance with $\mathbb{Z}$, particularly with regard to the behavior of squares, cubes, and their divisors.   •

21.1.2 Use unique prime factorization to show that a positive integer n is a square if and only if each prime in the prime factorization of n occurs to an even power.

21.1.3 If l and m are positive integers with no common prime divisor, and lm is a square, use Exercise 21.1.2 to show that l and m are both squares.

21.1.4 Show similarly that if l and m are integers with no common prime divisor, and if lm is a cube, then l and m are both cubes.

Thus to prove such results about the numbers $x + \sqrt{-2}$ and $x - \sqrt{-2}$ we need to know, first, that they have no common prime divisor. In the next section we introduce the concept of *norm*, which reduces such divisibility questions to questions about divisibility in the ordinary integers.

21.2 Gaussian Integers

Beyond $\mathbb{Z}$ itself, the simplest set to "behave like" integers is $\mathbb{Z}[i]$, the set of numbers of the form $a + bi$, where $a, b \in \mathbb{Z}$. These are called the *Gaussian integers*, because Gauss (1832c) was the first to study them and prove their basic properties. $\mathbb{Z}[i]$ is like $\mathbb{Z}$ in being closed under the operations $+$, $-$, and $\times$, but also in having primes and unique prime factorization.

An ordinary prime may be defined as an integer of size >1 that is not the product of integers of smaller size. A *Gaussian prime* may be defined in the same way, provided we make a sensible definition of "size." The ordinary absolute value $|a + bi| = \sqrt{a^2 + b^2}$ is a suitable measure, so we say that a Gaussian integer α is a Gaussian prime if $|\alpha| > 1$ but α is not the product of Gaussian integers of smaller absolute value.

An equivalent definition of Gaussian primes is in terms of the *square* of the absolute value, the *norm* of α, $N(\alpha)$. Namely, α is a Gaussian prime if $N(\alpha) > 1$ and α is not the product of Gaussian integers of smaller norm.

The norm has the advantage that $N(a + ib) = a^2 + b^2$ is an ordinary positive integer, so we can exploit the known properties of integers. For example, we can see immediately why *every Gaussian integer has a Gaussian prime factorization*. Namely, if α is not itself a Gaussian prime, then $\alpha = \beta\gamma$, where $N(\beta), N(\gamma) < N(\alpha)$. If β, γ are Gaussian primes, then we have a Gaussian prime factorization of α; if not, at least one of them factorizes into Gaussian integers of smaller norm, and so on. *This process must terminate*, because norms are ordinary nonnegative integers and hence they cannot decrease in size indefinitely. At termination, we have a Gaussian prime factorization of α.

The *uniqueness* of this prime factorization is a deeper result, for which it is convenient to revert to the absolute value measure of size and interpret $|a + ib|$ as the distance of $a + ib$ from O. This gives a surprisingly geometric proof that Gaussian integers have "division with remainder."

Division property of $\mathbb{Z}[i]$. *For any α and $\beta \neq 0$ in $\mathbb{Z}[i]$, there are μ and ρ in $\mathbb{Z}[i]$ such that*

$$\alpha = \mu\beta + \rho \quad \text{with} \quad |\rho| < |\beta|.$$

Proof. The multiples $\mu\beta$ for $\mu \in \mathbb{Z}[i]$ are sums of terms $\pm\beta$ and $\pm i\beta$. It follows, since the lines from O to β and $i\beta$ are perpendicular, that the numbers $\mu\beta$ lie at the corners of a lattice of squares of side $|\beta|$, as in Figure 21.1.

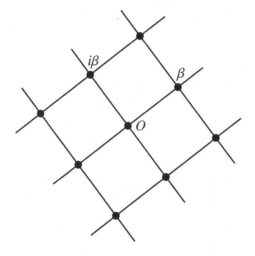

Figure 21.1: Multiples of β in $\mathbb{Z}[i]$

Now α lies in one of these squares, and if we let

$$\rho = \alpha - \text{nearest corner } \mu\beta,$$

it follows that the perpendiculars from α to the nearest sides are of length $\leq |\beta|/2$ (draw a picture). Therefore, since two sides of a triangle have total length greater than the third, we have

$$|\rho| < \frac{|\beta|}{2} + \frac{|\beta|}{2} = |\beta|,$$

as required.   □

The division property of $\mathbb{Z}[i]$ has the following consequences, parallel to those for natural numbers described in Section 3.3.

1. There is a *Euclidean algorithm* for $\mathbb{Z}[i]$, which takes any $\alpha, \beta \in \mathbb{Z}[i]$ and repeatedly divides the larger of the pair by the smaller, keeping the smaller number and the remainder. It ends by finding $\gcd(\alpha, \beta)$, a common divisor of α, β that is greatest in norm.

2. There are $\mu, \nu \in \mathbb{Z}[i]$ such that $\gcd(\alpha, \beta) = \mu\alpha + \nu\beta$.

3. If ϖ is a Gaussian prime that divides $\alpha\beta$, then ϖ divides α or β.

4. The *Gaussian prime factorization of a Gaussian integer is unique*, up to the order of factors and factors of norm 1 (that is, factors ± 1, $\pm i$).

EXERCISES

We know from Section 20.2 that the absolute value is multiplicative, and hence so is the norm: $N(\alpha\beta) = N(\alpha)N(\beta)$. Indeed, this is just a restatement of Diophantus's identity. It follows that *if α divides γ* (that is, if $\gamma = \alpha\beta$ for some β), *then $N(\alpha)$ divides $N(\gamma)$* [because $N(\gamma) = N(\alpha)N(\beta)$].

Thus we have a criterion for divisibility in the Gaussian integers based on divisibility in the ordinary integers. Among other things, this enables us to show that certain Gaussian integers are Gaussian primes.

21.2.1 By considering $N(4 + i)$, show that $4 + i$ is a Gaussian prime.

21.2.2 Show that an ordinary prime of the form $a^2 + b^2$ is *not* a Gaussian prime, and find its Gaussian prime factorization.

Now we modify the above argument for the division property of $\mathbb{Z}[i]$ to show that $\mathbb{Z}[\sqrt{-2}]$ also has it. That is, *if α and $\beta \neq 0$ are in $\mathbb{Z}[\sqrt{-2}]$, then there are μ and ρ in $\mathbb{Z}[\sqrt{-2}]$ such that*

$$\alpha = \mu\beta + \rho \quad \text{with} \quad |\rho| < |\beta|.$$

21.2.3 Show that the multiples $\mu\beta$ of any $\beta \in \mathbb{Z}[\sqrt{-2}]$ lie at the corners of a grid of rectangles, each of which has sides of length $|\beta|$ and $\sqrt{2}|\beta|$.

21.2.4 Deduce from Exercise 21.2.3 and the Pythagorean theorem that any α lies at distance $< |\beta|$ from the nearest multiple $\mu\beta$ of $\beta \neq 0$, and hence that $\mathbb{Z}[\sqrt{-2}]$ has the division property.

As in $\mathbb{Z}[i]$, the division property leads to a Euclidean algorithm for gcd, and eventually to unique prime factorization in $\mathbb{Z}[\sqrt{-2}]$. This enables us to fill in the gaps of Euler's argument in the previous section, as soon as we have checked that $\gcd(x + \sqrt{-2}, x - \sqrt{-2}) = 1$ when $y^3 = x^2 + 2$.

21.2.5 Show that if x and y are ordinary integers with $y^3 = x^2 + 2$, then x is odd.

Finally, we invoke the norm in $\mathbb{Z}[\sqrt{-2}]$,

$$N(a + b\sqrt{-2}) = |a + b\sqrt{-2}|^2 = a^2 + 2b^2.$$

21.2.6 Show that $N(x + \sqrt{-2})$ is odd, whereas $N(2\sqrt{-2}) = 2^3$, and hence that

$$1 = \gcd(x + \sqrt{-2}, 2\sqrt{-2}) = \gcd(x + \sqrt{-2}, x - \sqrt{-2}).$$

Unique prime factorization in $\mathbb{Z}[\sqrt{-2}]$ gives an easy proof of one of the results of Fermat proved in Section 11.4. This was pointed out to me by Lin Tan.

21.2.7 Suppose that $t^2 = u^2 + 2s^2$, for ordinary integers s, t, u. By considering the prime factorization of both sides in $\mathbb{Z}[\sqrt{-2}]$, show that t is also of the form $p^2 + 2q^2$, for ordinary integers p, q.

21.3   Algebraic Integers

The Gaussian integers are an excellent example of algebraic numbers that "behave like" integers, but it is not yet clear what the general concept of "integer" should be. After a period of exploration by Dirichlet, Kummer, Eisenstein, Hermite, and Kronecker in the 1840s and 1850s, the following definition was proposed by Dedekind (1871): an *algebraic integer* is a root of an equation of the form

$$x^n + a_{n-1}x^{n-1} + \cdots + a_1 x + a_0 = 0, \quad \text{where } a_0, a_1, \ldots, a_{n-1} \in \mathbb{Z}. \quad (*)$$

Thus the definition of algebraic integer results from the definition of algebraic number (Section 21.1) by restricting the polynomials to those with leading coefficient 1, or *monic* polynomials as they are often called.

One reason that this definition suggested itself was a result proved by Eisenstein (1850) that the numbers satisfying such equations are closed under $+$, $-$, and $\times$. It follows, since algebraic numbers inherit the properties of $+$, $-$, and $\times$ from $\mathbb{C}$, that algebraic integers form a *commutative ring with unit*, as defined in Section 20.3.

Another reason for the restriction to monic polynomials is that the *rational* algebraic integers are precisely the ordinary integers. This property of monic polynomials was pointed out by Gauss (1801), Article 11, and it is quite easy to prove. We suppose that the equation $(*)$ has a rational solution that is not an ordinary integer. Then we may assume that the solution is of the form $x = r/pq$, where p, q, r are ordinary integers and p is a prime not dividing r. Substituting this value for x in $(*)$, and multiplying through by $(pq)^n$, we get

$$r^n = -a_{n-1}r^{n-1}(pq) - \cdots - a_1 r(pq)^{n-1} - a_0(pq)^n.$$

However, this is impossible, because p divides the right-hand side but not the left.

In practice, it is difficult to work in the ring of all algebraic integers, and we prefer to work in smaller rings such as $\mathbb{Z}[i]$ or $\mathbb{Z}[\sqrt{-2}]$. The exercises in the previous section show that $\mathbb{Z}[\sqrt{-2}]$ is the perfect setting for Euler's proof that $y^3 = x^2 + 2$ has only one positive solution in $\mathbb{Z}$.

The advantage of rings such as $\mathbb{Z}[i]$ or $\mathbb{Z}[\sqrt{-2}]$ is that they have the concept of norm, which allows us to define the concept of prime and to show that each element of the ring has a prime factorization. However, the *uniqueness* of prime factorization is not guaranteed, and in a sense we were lucky to find it in $\mathbb{Z}[i]$ and $\mathbb{Z}[\sqrt{-2}]$.

A more typical ring of algebraic integers is

$$\mathbb{Z}[\sqrt{-5}] = \{a + b\sqrt{-5} : a, b \in \mathbb{Z}\}.$$

In this ring $|a + b\sqrt{-5}| = \sqrt{a^2 + 5b^2}$, and hence the norm is

$$N(a + b\sqrt{-5}) = a^2 + 5b^2.$$

As before, we define a *prime* to be a number of norm >1 that is not the product of numbers of smaller norm, and it follows as in $\mathbb{Z}[i]$ that every member of $\mathbb{Z}[\sqrt{-5}]$ factorizes into primes of $\mathbb{Z}[\sqrt{-5}]$.

It is likewise true that if β divides α in $\mathbb{Z}[\sqrt{-5}]$, then $N(\beta)$ divides $N(\alpha)$ in $\mathbb{Z}$. Hence α is a prime of $\mathbb{Z}[\sqrt{-5}]$ if $N(\alpha)$ is not divisible by any smaller norm $\neq 1$, that is, by any smaller integer of the form $a^2 + 5b^2 \neq 1$. Examples of primes in $\mathbb{Z}[\sqrt{-5}]$ are

$$2, \quad \text{because } N(2) = 4,$$
$$3, \quad \text{because } N(3) = 9,$$
$$1 + \sqrt{-5}, \quad \text{because } N(1 + \sqrt{-5}) = 6,$$
$$1 - \sqrt{-5}, \quad \text{because } N(1 - \sqrt{-5}) = 6.$$

Hence it follows that 6 has *two different prime factorizations* in $\mathbb{Z}[\sqrt{-5}]$:

$$6 = 2 \cdot 3 = (1 + \sqrt{-5})(1 - \sqrt{-5}).$$

In the 1840s Kummer noticed examples of the failure of unique prime factorization, and he realized that it is a serious problem. He wrote:

> It is greatly to be lamented that this virtue of the real num-
> bers [that is, of the ordinary integers] to be decomposable
> into prime factors, always the same ones for a given num-
> ber, does not also belong to the complex numbers [that is,
> the algebraic integers]; were this the case, the whole theory,

which is still laboring under such difficulties, could easily be brought to a conclusion. For this reason, the complex numbers we have been considering seem imperfect, and one may well ask whether one ought not to look for another kind which would preserve the analogy with the real numbers with respect to such a fundamental property.

Translation by Weil (1975) from Kummer (1844)

Kummer found "another kind of number" that preserved the property of unique prime factorization, and he called them *ideal numbers*. Today we know them under the name of *ideals*.

EXERCISES

Although ordinary fractions, such as 1/2, are not algebraic integers, some "algebraic fractions" are.

21.3.1 Show that the golden ratio $(1 + \sqrt{5})/2$ is an algebraic integer.

21.3.2 Find the three algebraic integers that satisfy the equation $x^3 - 1 = 0$.

Eisenstein's theorem that the algebraic integers are closed under +, −, and × was given a new proof by Dedekind (1871) using linear algebra.

21.3.3 Suppose that α and β are algebraic integers satisfying the equations

$$\alpha^a + p_{a-1}\alpha^{a-1} + \cdots + p_1\alpha + p_0 = 0,$$
$$\beta^b + q_{b-1}\beta^{b-1} + \cdots + q_1\beta + q_0 = 0.$$

Deduce from these that any power $\alpha^{a'}$ may be written as a linear combination of $1, \alpha, \alpha^2, \ldots, \alpha^{a-1}$ with ordinary integer coefficients, and any power $\beta^{b'}$ as a linear combination of $1, \beta, \beta^2, \ldots, \beta^{b-1}$ with ordinary integer coefficients.

21.3.4 Now let $\omega_1, \omega_2, \ldots, \omega_n$ denote the $n = ab$ products of the form $\alpha^{a'}\beta^{b'}$, where $a' < a$ and $b' < b$. Show that, if ω denotes any one of $\alpha + \beta$, $\alpha - \beta$, or $\alpha\beta$, then we have n equations with ordinary integer coefficients $k_j^{(i)}$:

$$\omega\omega_1 = k_1'\omega_1 + k_2'\omega_2 + \cdots + k_n'\omega_n,$$
$$\omega\omega_2 = k_1''\omega_1 + k_2''\omega_2 + \cdots + k_n''\omega_n,$$
$$\vdots$$
$$\omega\omega_n = k_1^{(n)}\omega_1 + k_2^{(n)}\omega_2 + \cdots + k_n^{(n)}\omega_n.$$

21.3.5 Explain why the equations in Exercise 21.3.4 have a nonzero solution for $\omega_1, \omega_2, \ldots, \omega_n$, and hence that

$$\begin{vmatrix} k_1' - \omega & k_2' & \cdots & k_n' \\ k_1'' & k_2'' - \omega & \cdots & k_n'' \\ \cdots & \cdots & \cdots & \cdots \\ k_1^{(n)} & k_2^{(n)} & \cdots & k_n^{(n)} - \omega \end{vmatrix} = 0.$$

Also explain why this is a monic equation, with ordinary integer coefficients, for $\omega = \alpha + \beta, \alpha - \beta$, or $\alpha\beta$.

21.4   Ideals

Kummer did not explicitly define his "ideal numbers." Rather, he observed that prime algebraic integers sometimes behave *as if* they were nontrivial products, and from their behavior he inferred the behavior of their "ideal factors." Dedekind (1871) showed that "ideal factors" could be realized by sets of actual numbers, and he called these sets *ideals*. In his (1877) work he used the numbers in $\mathbb{Z}[\sqrt{-5}]$ to illustrate his method, showing that 2 and 3 behave as if they were products of primes—$2 = \alpha^2$ and $3 = \beta_1\beta_2$—and then showing how α, β_1, and β_2 may be realized as ideals.

Here we shall take a slightly different route to the same goal: using ideals first to rewrite the theory of divisibility and gcd in $\mathbb{Z}$ and $\mathbb{Z}[i]$, then using them to *introduce* the gcd in $\mathbb{Z}[\sqrt{-5}]$. The ideals realizing α, β_1, and β_2 turn out to be gcds of algebraic integers.

Ideals in $\mathbb{Z}$

In $\mathbb{Z}$ we have the commonplace facts that

$$2 \text{ divides } 6, \quad 3 \text{ divides } 6, \quad \gcd(2, 3) = 1.$$

These facts can be rewritten in terms of the sets

$$(2) = \{\text{multiples of 2}\}, \quad (3) = \{\text{multiples of 3}\}, \quad (6) = \{\text{multiples of 6}\},$$

which are examples of ideals. The equivalents of the first two facts are

$$(2) \text{ contains } (6), \quad (3) \text{ contains } (6),$$

which may be summed up by the slogan *to divide is to contain*. To express the third fact we consider another ideal, the *sum* of (2) and (3):

$$(2) + (3) = \{a + b : a \in (2), b \in (3)\}.$$

It is clear that gcd(2, 3) divides any member of the set (2) + (3), and in fact it is not hard to show that

$$(2) + (3) = \{\text{multiples of } 1\} = (1) = (\gcd(2, 3)).$$

In general, we call a subset I of a ring R an *ideal* if

- $a \in I$ and $b \in I \Longrightarrow a + b \in I$,

- $a \in I$ and $m \in R \Longrightarrow am \in I$.

Then, for any $a \in \mathbb{Z}$, the set $(a) = \{\text{multiples of } a\}$ is obviously an ideal, called the *principal ideal* generated by a. It is not hard to prove (see the subsection below and the exercises) that

- every ideal in $\mathbb{Z}$ is (a) for some a,

- a divides $b \Longleftrightarrow (a)$ contains (b),

- $(a) + (b) = (\gcd(a, b))$.

Since ideals in $\mathbb{Z}$ correspond to numbers in $\mathbb{Z}$, the language of ideals tells us nothing we do not already know. However, the *concept* of ideal generalizes to other rings where it might conceivably give us new insight.

Ideals in $\mathbb{Z}[i]$

We know from Section 21.2 that $\mathbb{Z}[i]$ has many similarities to $\mathbb{Z}$, because they both have the division property. These similarities extend to properties of ideals in $\mathbb{Z}[i]$, and the division property explains why. In particular, it explains why every ideal in $\mathbb{Z}[i]$ is of the form $(\beta) = \{\text{multiples of } \beta\}$.

Suppose that I is an ideal of $\mathbb{Z}[i]$, and consider a nonzero element $\beta \in I$ of minimal norm. Then I contains the set (β) of multiples of β, since an ideal contains all multiples of any element. Also, I cannot contain any $\alpha \notin (\beta)$ by the division property: if such an α exists, there is a multiple $\mu\beta$ with $0 < |\alpha - \mu\beta| < |\beta|$. But $-\mu\beta \in I$ and hence $\alpha - \mu\beta \in I$ also, which contradicts the choice of β as a nonzero element of I of minimal norm.

Thus any ideal of $\mathbb{Z}[i]$ consists of all the multiples of some $\beta \in \mathbb{Z}[i]$, which, as we saw in Figure 21.1, is a set with the same shape as $\mathbb{Z}[i]$. The same is true for principal ideals in any $\mathbb{Z}[\sqrt{-n}]$: *they all have the same (rectangular) shape*. In fact, the set (β) of multiples of β consists of sums of the elements β and $\beta\sqrt{-n}$, which define a rectangle of the same shape as the rectangle defined by the generating elements 1 and $\sqrt{-n}$ of $\mathbb{Z}[\sqrt{-n}]$.

Ideals in $\mathbb{Z}[\sqrt{-5}]$

The ring $\mathbb{Z}[\sqrt{-5}]$ contains an ideal that is *not* the same shape as $\mathbb{Z}[\sqrt{-5}]$ itself. We expect this, since unique prime factorization fails in $\mathbb{Z}[\sqrt{-5}]$, and so the division property fails too; however, it is satisfying to make this failure visible.

One such ideal is the sum I of the principal ideals (2) and $(1 + \sqrt{-5})$,

$$(2) + (1 + \sqrt{-5}) = \{2m + (1 + \sqrt{-5})n : m, n \in \mathbb{Z}\},$$

part of which is shown in Figure 21.2.

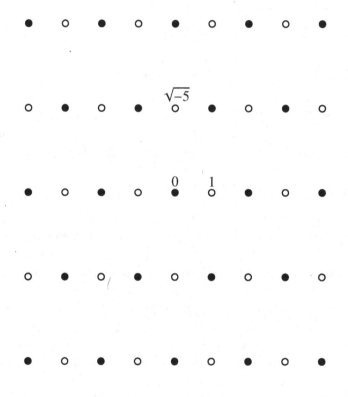

Figure 21.2: The nonprincipal ideal $(2) + (1 + \sqrt{-5})$ in $\mathbb{Z}[\sqrt{-5}]$

It is clear from the figure that I (consisting of the black dots) is *not* rectangular in shape like $\mathbb{Z}[\sqrt{-5}]$ (consisting of the black and white dots)—the black neighbors of any black dot do not include any two in perpendicular directions.

Thus the members of I are not the multiples of any one $\beta \in \mathbb{Z}[\sqrt{-5}]$. They are, if you like, the multiples of an "ideal number"—a number that is outside $\mathbb{Z}[\sqrt{-5}]$.

EXERCISES

Implicit in the discussion above is the following *definition* of the sum of ideals: if A and B are ideals, then

$$A + B = \{a + b : a \in A, b \in B\}.$$

It should also be checked that $A + B$ thus defined is itself an ideal.

21.4.1 Check that $A + B$ has the two defining properties of an ideal.

In $\mathbb{Z}$, we know that $\gcd(a, b) = ma + nb$ for some m and n. This makes it easy to describe the sum of principal ideals $(a) + (b)$ in terms of the gcd.

21.4.2 Show that $(a) + (b) = (\gcd(a, b))$ in $\mathbb{Z}$.

We take up this idea in the next section to find the gcd of any ideals. For the moment, we continue to explore nonprincipal ideals in $\mathbb{Z}[\sqrt{-5}]$, arising as sums of principal ideals.

21.4.3 Show that the vectors from O to 2 and $1 + \sqrt{-5}$ define a parallelogram of the same shape as the vectors from O to 3 and $1 - \sqrt{-5}$. *Hint*: Consider quotients of complex numbers and what they say about the ratio of side lengths, and the angle between the sides. (The same idea occurs in the exercises for Section 16.5.)

21.4.4 Deduce from Exercise 21.4.3 that the ideal $(3) + (1 - \sqrt{-5})$ has the same shape as the ideal $(2) + (1 + \sqrt{-5})$.

21.4.5 Show also that the ideal $(3) + (1 - \sqrt{-5})$ has the same shape as the ideal $(3) + (1 + \sqrt{-5})$.

Thus we have found so far only two different shapes of ideals in $\mathbb{Z}[\sqrt{-5}]$: the shape of $\mathbb{Z}[\sqrt{-5}]$ itself, which is the shape of all principal ideals, and the shape of the nonprincipal ideal $(2) + (1 + \sqrt{-5})$.

It can be shown that any ideal in $\mathbb{Z}[\sqrt{-5}]$ has one of these two shapes, which represent what Dedekind called the ideal *classes* of $\mathbb{Z}[\sqrt{-5}]$. This term goes back to the older theory of quadratic forms, where forms $ax^2 + bxy + cy^2$ with the same discriminant $b^2 - 4ac$ were divided into a number of equivalence classes, the number of which was called the *class number*. Lagrange (1773a) showed that any form with discriminant -20 is equivalent to either $x^2 + 5y^2$ (the norm of $x + y\sqrt{-5}$) or $2x^2 + 2xy + 3y^2$. These two forms correspond to the two ideal classes of $\mathbb{Z}[\sqrt{-5}]$. For more on classes of quadratic forms, see Section 21.6.

21.5   Ideal Factorization

In $\mathbb{Z}$ we saw that "to divide is to contain," because

$$a \text{ divides } b \quad \Longleftrightarrow \quad (a) \text{ contains } (b).$$

In $\mathbb{Z}[\sqrt{-5}]$, we can then say that the nonprincipal ideal $(2) + (1 + \sqrt{-5})$ behaves like a common divisor of 2 and $1 + \sqrt{-5}$, because

$$(2) + (1 + \sqrt{-5}) \text{ contains } (2), \quad (2) + (1 + \sqrt{-5}) \text{ contains } (1 + \sqrt{-5}).$$

Indeed, we can expect that $(2) + (1 + \sqrt{-5})$ is the *greatest common divisor* of 2 and $1 + \sqrt{-5}$ in $\mathbb{Z}[\sqrt{-5}]$, since in $\mathbb{Z}$ it is always true that $(a) + (b) = (\gcd(a, b))$.

Not only that, we can expect that $(2) + (1 + \sqrt{-5})$ is *prime*. In $\mathbb{Z}$ we notice that p is prime if and only the ideal (p) is *maximal*; that is, the only ideal properly containing (p) is $\mathbb{Z}$ itself. This is because any $a \notin (p)$ is relatively prime to p; hence $ma + np = 1$ for some m and n, so 1 is in any ideal containing both a and p.

To prove that $(2) + (1 + \sqrt{-5})$ is maximal is even easier. We suppose that $a = m + n\sqrt{-5} \notin (2) + (1 + \sqrt{-5})$, which means that m is even. But then $a - 1 \in (2) + (1 + \sqrt{-5})$; hence 1 is in any ideal containing both a and $(2) + (1 + \sqrt{-5})$. Such an ideal is therefore $\mathbb{Z}[\sqrt{-5}]$ itself.

To sum up: if ideals in $\mathbb{Z}[\sqrt{-5}]$ have divisibility properties like those in $\mathbb{Z}$, then $(2) + (1 + \sqrt{-5})$ is the gcd of 2 and $1 + \sqrt{-5}$, and it is prime. Dedekind (1871) defined the product of ideals so that divisibility behaves as expected.

Definition. If A and B are ideals, then

$$AB = \{a_1 b_1 + a_2 b_2 + \cdots + a_k b_k : a_1, a_2, \ldots, a_k \in A, \ b_1, b_2, \ldots, b_k \in B\}.$$

It is easily checked that AB is an ideal and (with greater difficulty) that the containment concept of divisibility agrees with the usual concept: B divides A if there is an ideal C such that $A = BC$. However, what is really delightful is that *the product of ideals explains the nonunique prime factorization of 6 in $\mathbb{Z}[\sqrt{-5}]$,*

$$6 = 2 \cdot 3 = (1 + \sqrt{-5})(1 - \sqrt{-5}),$$

by resolving both sides into the same product of prime ideals. In fact, we have

- (2) is the square of the prime ideal $(2) + (1 + \sqrt{-5})$,
- (3) is the product of ideals $(3) + (1 + \sqrt{-5})$ and $(3) + (1 - \sqrt{-5})$, which are prime,
- $(1 + \sqrt{-5})$ is the product of $(2) + (1 + \sqrt{-5})$ and $(3) + (1 + \sqrt{-5})$,
- $(1 - \sqrt{-5})$ is the product of $(2) + (1 + \sqrt{-5})$ and $(3) + (1 - \sqrt{-5})$.

As an example, we prove the first of these claims.

The ideal factorization of 2: $(2) = [(2) + (1 + \sqrt{-5})]^2$.

It follows from the definition of product of ideals that

$$4 = 2 \times 2 \in [(2) + (1 + \sqrt{-5})]^2,$$
$$2 + 2\sqrt{-5} = 2 \times (1 + \sqrt{-5}) \in [(2) + (1 + \sqrt{-5})]^2,$$
$$-4 + 2\sqrt{-5} = (1 + \sqrt{-5})^2 \in [(2) + (1 + \sqrt{-5})]^2.$$

Adding the elements 4, $2 + 2\sqrt{-5}$, and $-4 + 2\sqrt{-5}$ of $[(2) + (1 + \sqrt{-5})]^2$, we find that $2 \in [(2) + (1 + \sqrt{-5})]^2$. It follows that all multiples of 2 are in $[(2) + (1 + \sqrt{-5})]^2$, that is, $[(2) + (1 + \sqrt{-5})]^2$ contains (2).

Conversely, any element of $[(2) + (1 + \sqrt{-5})]^2$ is a sum of products of terms $2m$ and $(1 + \sqrt{-5})n$. Any product involving $2m$ is a multiple of 2, and so is any product involving $(1 + \sqrt{-5})^2 = -4 + 2\sqrt{-5}$. Thus any element of $[(2) + (1 + \sqrt{-5})]^2$ is a multiple of 2; hence $[(2) + (1 + \sqrt{-5})]^2$ is contained in (2), as required.   □

EXERCISES

The other ideal factorizations claimed above, and proofs that the factors are maximal ideals, go along the same lines as the examples just worked out.

21.5.1 Show in turn that 9, 6, and hence 3 belong to the product of ideals

$$[(3) + (1 + \sqrt{-5})][(3) + (1 - \sqrt{-5})],$$

so $[(3) + (1 + \sqrt{-5})][(3) + (1 - \sqrt{-5})]$ contains the ideal (3).

21.5.2 Show that an element of $(3) + (1 + \sqrt{-5})$ times one of $(3) + (1 - \sqrt{-5})$ is a multiple of 3, so that (3) contains $[(3) + (1 + \sqrt{-5})][(3) + (1 - \sqrt{-5})]$.

21.5.3 Consider an ideal A containing $(3) + (1 + \sqrt{-5})$ and an element a outside $(3) + (1 + \sqrt{-5})$. Show that A contains either 1 or 2, and in the latter case A also contains 1.

21.5.4 Deduce from Exercise 21.5.3 that $(3) + (1 + \sqrt{-5})$ is a maximal ideal in $\mathbb{Z}[\sqrt{-5}]$, and show that $(3) + (1 - \sqrt{-5})$ is maximal similarly.

21.6   Sums of Squares Revisited

Algebraic number theory has a very long pedigree, which can plausibly be traced back to the Babylonian discovery of Pythagorean triples around 1800 BCE. It is still mysterious how the Babylonians were able to generate triples, seemingly at will, but a method of generation can be clearly recognized in the work of Diophantus. It lies in the Diophantus two-square identity from Section 20.2:

$$(a_1^2 + b_1^2)(a_2^2 + b_2^2) = (a_1a_2 - b_1b_2)^2 + (a_1b_2 + b_1a_2)^2.$$

This identity allows us to "compose" two Pythagorean triples, (a_1, b_1, c_1) and (a_2, b_2, c_2), to obtain a third triple, $(a_1a_2 - b_1b_2, a_1b_2 + b_1a_2, c_1c_2)$.

But with Diophantus the focus shifts from the triples (a, b, c) to the pairs (a, b), and particularly to the sums $a^2 + b^2$. As Diophantus said (Section 20.2), 65 is the sum of two squares *because* $65 = 5 \times 13$, and because 5 and 13 are also sums of two squares. To understand which numbers are sums of two squares, we evidently need to look at their factors, and hence the problem boils down to knowing which *primes* are sums of two squares. Apparently Fermat was the first to see that this was the ultimate question about sums of two squares. At any rate, Fermat (1640b) was the first to answer it: *an odd prime p is the sum of two squares if and only if p is of the form $4n + 1$.*

Fermat, in his usual manner, stated this theorem without proof. The first published proof was given by Euler (1749), and a series of increasingly elegant proofs was given by illustrious mathematicians, usually when they had new methods to show off: for example, Lagrange (1773b) (theory of quadratic forms), Gauss (1832c) (Gaussian integers), and Dedekind (1877) (ideal theory).

Lagrange's theory of quadratic forms was in fact a precursor of algebraic number theory, stimulated by a trio of theorems stated by Fermat, and by a problem that Fermat was unable to solve. The three theorems are about odd primes p of the forms $x^2 + y^2$ (the one inspired by Diophantus), $x^2 + 2y^2$, and $x^2 + 3y^2$, and they may be stated as follows.

$$p = x^2 + y^2 \iff p \equiv 1 \pmod 4 \qquad \text{(Fermat (1640b))}$$
$$p = x^2 + 2y^2 \iff p \equiv 1 \text{ or } 3 \pmod 8 \qquad \text{(Fermat (1654))}$$
$$p = x^2 + 3y^2 \iff p \equiv 1 \pmod 3 \qquad \text{(Fermat (1654))}$$

The problem Fermat was unable to solve was to characterize odd primes of the form $x^2 + 5y^2$. Here there was a puzzling new phenomenon: primes *not* of the form $x^2 + 5y^2$, such as 3 and 7, whose product *is* of the form $x^2 + 5y^2$.

Lagrange (1773b) was able to prove Fermat's three theorems, and to explain the anomalous behavior of $x^2 + 5y^2$, by his theory of *equivalence of quadratic forms*. If we are interested in the numbers represented by a form $ax^2 + bxy + cy^2$, then we also need to survey the forms $a'x'^2 + b'x'y' + c'y'^2$ obtainable from $ax^2 + bxy + cy^2$ by a change of variables

$$x' = px + qy, \quad y' = rx + sy, \quad \text{where } p, q, r, s \in \mathbb{Z} \text{ and } ps - qr = \pm 1,$$

because such a change of variables $(x, y) \mapsto (x'y')$ is a one-to-one map of $\mathbb{Z} \times \mathbb{Z}$, and so the new form represents exactly the same numbers as the old.

Lagrange called such forms *equivalent* and observed that they have the same *discriminant*: $b^2 - 4ac = b'^2 - 4a'c'$. Moreover, he found that

all forms with discriminant -4 are equivalent to $x^2 + y^2$,

all forms with discriminant -8 are equivalent to $x^2 + 2y^2$,

all forms with discriminant -12 are equivalent to $x^2 + 3y^2$,

but *there are two inequivalent forms with discriminant* -20: namely, the forms $x^2 + 5y^2$ and $2x^2 + 2xy + 3y^2$. By exposing the "invisible companion" $2x^2 + 2xy + 3y^2$ of $x^2 + 5y^2$, Lagrange explained the behavior of numbers of the form $x^2 + 5y^2$. They cannot be understood in isolation, but only as a class that interacts with numbers of the form $2x^2 + 2xy + 3y^2$. In fact, the primes of the form $x^2 + 5y^2$ are those $\equiv 1$ or $9 \pmod{20}$, while the primes of the form $2x^2 + 2xy + 3y^2$ are those $\equiv 3$ or $7 \pmod{20}$. And products of the latter primes are $\equiv 1$ or $9 \pmod{20}$ and of the form $x^2 + 5y^2$.

It appears that Gauss was aware that the theory of quadratic forms could be replaced, at least up to a point, by a theory of "quadratic integers." His theory of $\mathbb{Z}[i]$ is indeed a replacement for Lagrange's theory of the quadratic form $x^2 + y^2$. But Gauss was also aware that in some cases the corresponding quadratic integers failed to have unique prime factorization (which is perhaps why he was the first to recognize the importance of unique prime factorization elsewhere). He was unable to see a way around this obstacle, so Kummer's creation of ideal numbers can be regarded as the solution to a problem that had baffled even the great Gauss.

We do not know how far Kummer developed the theory of ideal numbers in rings of quadratic integers such as $\mathbb{Z}[\sqrt{-5}]$, because he was actually interested in algebraic integers of higher degree, the so-called *cyclotomic integers*. As their name suggests, these arise from circle division (Sections 2.3 and 14.5), where the solutions $1, \zeta_n, \zeta_n^2, \ldots, \zeta_n^{n-1}$ of the equation

$$x^n - 1 = 0$$

represent n equally spaced points on the unit circle. The numbers

$$a_0 + a_1\zeta_1 + a_2\zeta_n^2 + \cdots + a_{n-1}\zeta_n^{n-1}, \quad \text{where } a_0, a_1, \ldots, a_{n-1} \in \mathbb{Z},$$

form a ring $\mathbb{Z}[\zeta_n]$ of *cyclotomic integers*.

In Kummer's time it was thought that $\mathbb{Z}[\zeta_n]$ was the key to Fermat's last theorem, because if $a, b, c \in \mathbb{Z}$ are such that $a^n + b^n = c^n$, then the nth power $a^n + b^n$ factorizes into n linear factors in $\mathbb{Z}[\zeta_n]$. In fact, this was the basis of a mistaken "proof" by Lamé (1847). However, Kummer noticed that such arguments break down, precisely because *unique prime factorization fails in* $\mathbb{Z}[\zeta_n]$. Kummer showed that this happens for $n \geq 23$, and he created the theory of ideal numbers in an attempt to repair the damage. In this respect, ideal numbers were only partially successful (not that it matters, now that we have Wiles's proof of Fermat's last theorem), but they proved their worth elsewhere. Dedekind's revision of Kummer's idea gave us the concept of ideal, which is indispensable in algebra today.

For a treatment of primes of the form $x^2 + 5y^2$ using ideals, see Artin (1991) or Stillwell (2003), and for more on the history of $x^2 + ny^2$, see the introduction to Dedekind (1877), and Cox (1989). The latter pursues another remarkable thread in number theory—the modular function. As mentioned in the exercises to Section 16.5, the modular function is a function of lattice shapes, which is why it can reflect ideals of imaginary quadratic integers. For more, see Cox's book, or McKean and Moll (1997).

EXERCISES

There is an "easy direction" of Fermat's theorems about $x^2 + y^2$, $x^2 + 2y^2$, and $x^2 + 3y^2$ that can be proved with the help of congruences. This direction shows that primes are *not* representable in the given forms if they have the wrong remainders on division by 4, 8, and 3, respectively. (Compare with Exercises 1.5.2 and 3.2.1.)

21.6.1 Show that

1. An odd prime $x^2 + y^2 \not\equiv 3 \pmod{4}$.

2. An odd prime $x^2 + 2y^2 \not\equiv 5$ or $7 \pmod{8}$.

3. An odd prime $x^2 + 3y^2 \not\equiv 2 \pmod{3}$.

The "hard direction" of Fermat's theorems, finding the x^2 and y^2 to represent primes with the right remainders, involves more than we can cover completely here. However, for $x^2 + y^2$ and $x^2 + 2y^2$ it involves unique prime factorization in $\mathbb{Z}[i]$ and $\mathbb{Z}[\sqrt{-2}]$, both of which were discussed earlier in this chapter.

For $x^2 + 3y^2$, the proof involves not so much $\mathbb{Z}[\sqrt{-3}]$ as the larger ring

$$\mathbb{Z}\left[\frac{1 + \sqrt{-3}}{2}\right] = \left\{ m + \frac{1 + \sqrt{-3}}{2}n : m, n \in \mathbb{Z} \right\}.$$

21.6.2 Show that $(1 + \sqrt{-3})/2$ is an algebraic integer and that $\mathbb{Z}[(1 + \sqrt{-3})/2]$ contains $\mathbb{Z}[\sqrt{-3}]$.

21.6.3 Show that 2, $1 + \sqrt{-3}$, and $1 - \sqrt{-3}$ are primes of $\mathbb{Z}[\sqrt{-3}]$, and deduce that 4 has two distinct prime factorizations in $\mathbb{Z}[\sqrt{-3}]$.

21.6.4 By a geometric argument like those used for $\mathbb{Z}[i]$ and $\mathbb{Z}[\sqrt{-2}]$, show that $\mathbb{Z}[(1 + \sqrt{-3})/2]$ has unique prime factorization.

21.7   Rings and Fields

Kronecker is famous for saying "God made the natural numbers, the rest is the work of man." (This is reported, for example, in his obituary by Weber (1892).) Algebraic number theory was very much what he had in mind, because Kronecker, like Dedekind, saw number theory as the source of the most interesting problems, and the inspiration for all mathematical concepts. We can at least agree that number theory was the inspiration for two of the most important *algebraic* concepts: rings and fields.

Perhaps the first step toward abstract algebra was the introduction of negative numbers, creating the ring $\mathbb{Z}$ of integers from the natural numbers. This seems to have been a very difficult step, because mathematicians for many centuries (say, from the time of Diophantus to Descartes) lived in a halfway house where negative numbers were only partially accepted—sometimes being admitted in intermediate calculations, but not allowed as answers. Likewise, it was a long time before the "ratios" of the Greeks became the *field* $\mathbb{Q}$ of rational numbers.

Thus the first level of abstraction, the creation of inverses for addition and multiplication, took place unconsciously over thousands of years. The next level, identifying *axioms* for rings and fields, took place in the 19th century, mainly under the influence of algebraic number theory. The ring axioms are essentially the result of writing down the properties of + and ×

that algebraic integers share with the ordinary integers, and the field axioms are the properties that algebraic numbers share with rational numbers.

The concept of field was implicit in the work of Abel and Galois in the theory of equations, but it became explicit when Dedekind introduced *number fields of finite degree* as the setting for algebraic number theory. He saw that the ring of all algebraic integers is not a convenient ring, because it has no "primes." This is because $\sqrt{\alpha}$ is an algebraic integer if α is, so there is always a nontrivial factorization $\alpha = \sqrt{\alpha}\sqrt{\alpha}$ in the ring of all algebraic integers. On the other hand, the algebraic integers in a field generated from a single algebraic number α of degree n,

$$\mathbb{Q}(\alpha) = \{a_0 + a_1\alpha + \cdots + a_{n-1}\alpha^{n-1} : a_0, a_1, \ldots, a_{n-1} \in \mathbb{Q}\},$$

have better behavior. The algebraic integers β in $\mathbb{Q}(\alpha)$ have a norm $N(\beta)$ that is an ordinary integer, and this guarantees the existence of primes, as we have seen in special cases like $\mathbb{Z}[i]$ and $\mathbb{Z}[\sqrt{-2}]$, which are the algebraic integers in the fields $\mathbb{Q}(i)$ and $\mathbb{Q}(\sqrt{-2})$ of degree 2.

By drawing attention to the field $\mathbb{Q}(\alpha)$ of degree n, Dedekind also brought to light some *vector space* structure: the *basis* $1, \alpha, \alpha^2, \ldots, \alpha^{n-1}$ of $\mathbb{Q}[\alpha]$, the *linear independence* of these basis elements over $\mathbb{Q}$, and the *dimension* (equal to the degree) of $\mathbb{Q}[\alpha]$ over $\mathbb{Q}$. Despite the long history of linear algebra, dating back 2000 years in China at least, again it was the greater generality afforded by algebraic number theory that finally brought its fundamental concepts to light.

The next level of abstraction was reached in the 20th century and it was (in a new twist to Kronecker's words) the work of a woman, Emmy Noether. In the 1920s she developed concepts for discussing common properties of different algebraic structures, such as groups and rings. One of the things groups and rings have in common is *homomorphisms*, or structure-preserving maps. A map $\varphi : G \to G'$ is a *homomorphism of groups* if $\varphi(gh) = \varphi(g)\varphi(h)$ for any $g, h \in G$. Similarly, a map $\varphi : R \to R'$ is a homomorphism of rings if $\varphi(r + s) = \varphi(r) + \varphi(s)$ and $\varphi(rs) = \varphi(r)\varphi(s)$ for any $r, s \in R$. From this higher vantage point, normal subgroups (Section 19.2) and ideals can be seen as instances of the same concept. Each is the *kernel* of a homomorphism φ: the set of elements mapped by φ to the identity element (1 for a group, 0 for a ring).

EXERCISES

It is not clear that $\mathbb{Q}(\alpha)$ (as defined above) is a field for any algebraic number α. The hardest part is to prove that the quotient of any two of its elements is also an element. Some inkling of the difficulty may be grasped by working out the special case of $\mathbb{Q}(i)$.

21.7.1 Show that, if $a_1, b_1, a_2, b_2 \in \mathbb{Q}$, then $\frac{a_1 + ib_1}{a_2 + ib_2}$ is of the form $a + ib$, where $a, b \in \mathbb{Q}$.

It is also not obvious that the kernel of a group homomorphism is a normal subgroup, partly because the definition of normal subgroup in Section 19.2 is not the most convenient for this purpose. It is easier to prove that the kernel of a ring homomorphism is an ideal, using the definition of an ideal given in Section 21.4.

21.7.2 Suppose that R is a ring and φ maps R into another ring in such a way that $\varphi(r + s) = \varphi(r) + \varphi(s)$ and $\varphi(rs) = \varphi(s)\varphi(s)$ for any $r, s \in R$. Show that the set

$$\{r : \varphi(r) = 0\}$$

has the two defining properties of an ideal.

The equivalence of kernels and ideals may be illustrated in $\mathbb{Z}$ by the ideal (3) of multiples of 3.

21.7.3 Find a homomorphism of $\mathbb{Z}$ whose kernel is (3).

21.8 Biographical Notes: Dedekind, Hilbert, and Noether

Richard Dedekind (Figure 21.3) was born in 1831 in Brunswick, the home town of Gauss, into an academic family. His father, Julius, was professor of law at the Collegium Carolinum, and his mother, Caroline Emperius, was the daughter of another professor there. Richard was the youngest of four children in a close-knit family. They remained in Brunswick for most of their lives, and Richard lived with his sister Julie (both of them being unmarried) until 1914. Sounds dull, but this seemingly eventless life was the background to revolutionary activity in mathematics, in its way as provocative as the work of Galois.

Dedekind became interested in mathematics in high school, after coming to the conclusion that chemistry and physics were not sufficiently logical. He attended the Collegium Carolinum, the scientific academy that Gauss also attended, before entering Göttingen University in 1850. There he became friends with Riemann and made rapid academic progress, completing a thesis under Gauss's supervision in 1852. After the death of Gauss in 1855, Dirichlet was appointed to Gauss's chair, and he became the third major influence on Dedekind's career. After a brief period at the

Figure 21.3: Richard Dedekind

Polytechnikum in Zurich (now known as the ETH), a position that he won in competition with Riemann, Dedekind returned to the Polytechnikum in Brunswick, where he remained for the rest of his life. It was not a prestigious position, but the home comforts enabled him to concentrate on mathematics.

Dedekind was the last student of Gauss, and Gauss's number theory was the inspiration for much of Dedekind's work, as it was for many of the great German mathematicians of the 19th century. When Dedekind started, the new generation of Eisenstein, Dirichlet, and Kronecker was beginning to understand Gauss's ideas, and making further progress. Dirichlet in particular made Gauss more approachable with his elegant and readable *Vorlesungen über Zahlentheorie* (Lectures on Number Theory, Dirichlet (1863)), which simplified much of Gauss's difficult theory of quadratic forms and added stunning new results and proofs of his own. The climax of Dirichlet's lectures is a *class number formula*, giving a uniform description of the number of inequivalent quadratic forms with given discriminant. The lectures were edited by Dedekind and first published in 1863, four years after Dirichlet's death. Dedekind took this project very seriously and made it virtually his life's work, bringing out further editions in 1871, 1879, and 1894, each time adding supplementary material, until the supplements amounted to more than Dirichlet's book itself. The theory of ideals made its first appearance in the 1871 edition, and was expanded and deepened in 1879 and 1894, eventually including a lot of Galois theory as well.

However, Dedekind was disappointed in the low enthusiasm for ideals shown by other mathematicians, and in 1877 he attempted a more popular approach. Dedekind (1877) is nearly perfect for the modern reader—clear, concise, and well motivated—but apparently it was still too abstract for his contemporaries. The theory of ideals did not really catch on until it was given a new exposition by Hilbert (1897), as we shall see below.

In the meantime, Dedekind had made several other great contributions to mathematics that were slowly taking root:

- the theory of real numbers as "Dedekind cuts,"

- the theory of Riemann surfaces as algebraic function fields,

- the characterization of natural numbers as an "inductive set."

What these contributions had in common, and what made them hard for Dedekind's contemporaries to grasp, was the idea of treating *infinite sets* as mathematical objects. Dedekind actually started doing this in 1857, when he treated congruence modulo n as the arithmetic of congruence classes

$$0 \bmod n = \{0, \pm n, \pm 2n, \ldots\},$$
$$1 \bmod n = \{1, 1 \pm n, 1 \pm 2n, \ldots\},$$
$$\vdots$$
$$n - 1 \bmod n = \{n - 1, n - 1 \pm n, n - 1 \pm 2n, \ldots\},$$

which are added and multiplied according to the rules

$$(i \bmod n) + (j \bmod n) = (i + j) \bmod n,$$
$$(i \bmod n)(j \bmod n) = (i \cdot j) \bmod n.$$

(We mentioned multiplication mod n in Section 19.1, but without mention of congruence classes.)

The idea of adding or multiplying sets by adding or multiplying *representatives* transfers directly to Dedekind cuts and, with some modification, to ideals and Riemann surfaces. Dedekind hoped that this cornucopia of applications would convince his colleagues of the value of the idea that "mathematical objects are sets," but it was a hard idea to sell. At first he was joined only by Cantor, who took up the theory of infinite sets as enthusiastically as Dedekind took up the applications (see Chapter 24).

Dedekind had to wait decades before his ideas entered the mainstream (and in some cases after they had been rediscovered by others—for example, his theory of natural numbers became the "Peano axioms"), but fortunately he lived long enough. He died in 1916 at the age of 84.

David Hilbert (Figure 21.4) was born in 1862 in Königsberg and died in Göttingen in 1943. His father, Otto, was a judge, and David may have inherited his mathematical ability from his mother, about whom we know little except that her maiden name was Erdtmann. Königsberg was in the remote eastern part of Prussia (it is now Kaliningrad, a small, disconnected piece of Russia), but with a strong mathematical tradition dating back to Jacobi. When Hilbert attended university there in the 1880s he became friends with Hermann Minkowski, a former child mathematical prodigy two years his junior, and Adolf Hurwitz, who was three years older and a professor in Königsberg from 1884. The three used to discuss mathematics on long walks, and Hilbert seems to have picked up his basic mathematical education in this way. In later life he made "mathematical walks" an important part of the education of his own students.

Figure 21.4: David Hilbert

Hilbert's first research interest was in the theory of invariants, an algebraic topic then held in high esteem. An elementary example of an invariant is the discriminant $b^2 - 4ac$ of a quadratic form, which Lagrange (1773b) noticed is invariant when the form is transformed into an equivalent form (Section 21.6). By Hilbert's time, invariant theory had become a jungle, with success depending mainly on the ability to hack through formidable calculations. The "king of invariant theory," Paul Gordan of Erlangen, was notorious for papers consisting almost entirely of equations—in fact, the story goes that he had assistants fill in any words that were necessary. In 1888 Hilbert swept all this away by solving the main problem of invariant theory, in a simple and purely conceptual manner: the *Hilbert basis theorem* showed the existence of the invariants above the quadratic level, without needing to calculate them!

Gordan was at first incredulous and exclaimed, "This is not mathematics, it is theology!" but eventually Hilbert's idea was developed further, to calculate the invariants, and Gordan had to concede that it was mathematics after all. Hilbert, for his part, moved on to conquer other worlds. In fact, this became his modus operandi for most of his career: investigate a topic thoroughly for a few years, turn it upside down, then do something completely different.

Hilbert's triumph in invariant theory secured his position in Königsberg, and in 1892 he married Käthe Jerosch, a very capable woman who acted as secretary and research assistant for many of his works. In particular, she compiled the bibliography for his massive *Zahlbericht* ("Number Report") of 1897, the work in which algebraic number theory came of age. Hilbert was commissioned by the German Union of Mathematicians in 1893 to write a report on algebraic number theory, and the report became a 300-page book (Hilbert (1897)), looking back to quadratic forms and Fermat's last theorem, and forward to *class field theory*, a major topic of the 20th century.

The mathematical public, which had not been ready when Dedekind presented algebraic number theory a few years earlier, now saw the point, and Klein invited Hilbert to assume a chair in mathematics at Göttingen, which he held from 1895 until the end of his life.

After the *Zahlbericht*, Hilbert turned to the foundations of geometry, which we have touched on in Sections 1.6, 2.1, 19.6, and 20.7. Again he scored several triumphs—finally filling the gaps in Euclid, discovering the algebraic meaning of the Pappus and Desargues theorems—but also leaving some unfinished business. Hilbert realized that modeling Euclid's geometry by real-number coordinates is not exactly a proof that geometry is consistent; one still needs to prove that the theory of real numbers is consistent. Hilbert found this far from obvious and made it second on his list of mathematical problems presented in Paris in 1900. Then he dropped the subject in favor of mathematical physics.

However, no one found a consistency proof for the theory of real numbers, and by the 1920s Hilbert felt compelled to return to the subject. *Hilbert's program*, as it became known, called first for a formal language of mathematics, in which the concept of proof itself was mathematically definable, by precise rules for manipulating formulas. This phase of the program was in fact feasible, and was essentially carried out by Whitehead and Russell in their *Principia Mathematica* of 1910. The hard part, however, was proving that the rules of proof could not lead to a contradiction. This is where Hilbert's program stalled, and in 1931 Gödel showed that it could never be completed. His famous *incompleteness theorems* (Chapter 24) showed that such a consistency proof does not exist, and that enlarging the formal language by new axioms only puts the consistency proof further out of reach.

To his credit, Hilbert was among the first to publicize Gödel's work. The first complete proofs of Gödel's theorems are in the book of Hilbert and Bernays (1938). But it was Hilbert's misfortune to end his career, not only with the failure of one of his mathematical dreams, but also with his mathematical community in ruins. The eclipse of Göttingen began in 1933, when the Nazis came to power in Germany and began dismissing Jewish professors. In a few years, most of Germany's mathematical talent had fled, leaving the elderly and frail Hilbert in Göttingen virtually alone. He died on 14 February 1943.

One of the Jewish mathematicians forced to leave Göttingen in 1933 was Emmy Noether (Figure 21.5), who was in many ways a natural successor of Dedekind and Hilbert. Emmy Noether was born in 1882 in Erlangen and died in 1935 in Bryn Mawr, Pennsylvania. She was the oldest of four children of the mathematician Max Noether and of Ida Kaufmann. As a child she loved music, dance, and languages and planned to become a language teacher, qualifying as a teacher of English and French in 1900.

At this time in Germany, women were permitted to study at universities only unofficially, and very few did so, since the permission of the lecturer was also required. However, a few teachers were permitted to attend for

Figure 21.5: Emmy Noether

purposes of "further education," and in 1900 Emmy Noether became one of them, studying mathematics at the University of Erlangen. Here she met the "king of invariants," Paul Gordan, and wrote a thesis under his supervision in 1907. It was on invariant theory, naturally, and Emmy later described it as "crap," but it was not a complete waste of time. Physicists today admire one of her early results, on the invariants of mechanical systems.

In 1910 Gordan retired and there was a reshuffle of positions, leading to the appointment of Ernst Fischer in 1911. Fischer is not well known today, but it seems that Noether's algebraic talent suddenly blossomed through working with him. She dropped the computational approach of Gordan and rapidly mastered the conceptual approach of Dedekind and Hilbert, so much so that Hilbert invited her to Göttingen in 1915. Getting a position was another matter—Hilbert is said to have ridiculed Göttingen's exclusion of women professors by saying "this is a university, not a bathing establishment"—but she was eventually granted an unofficial chair in 1922.

In the 1920s Noether was at the height of her powers, and she found students worthy of her ability. Among them were Emil Artin, who solved two of Hilbert's problems, and B. L. van der Waerden, who brought the ideas of Noether to the world in his *Moderne Algebra* of 1930. Noether herself modestly used to claim that "es steht schon bei Dedekind" ("it's already in Dedekind") and encouraged her students to see for themselves by reading all of Dedekind's supplements. Thus, despite the highly abstract nature of Noether's algebra, her students were made aware of its direct descent from the number theory of Gauss and Dirichlet. In van der Waerden's *Algebra* this connection was unfortunately broken, and many in the next generation of students grew up unaware of it. In recent years there has been a welcome reversal of this trend; in particular, the *Algebra* of Emil Artin's son Michael uses number theory to illustrate the theory of ideals (Artin (1991)).

22

Topology

PREVIEW

In Chapter 15 we saw how Riemann found the topological concept of *genus* to be important in the study of algebraic curves. In the present chapter we will see how topology became a major field of mathematics, with its own methods and problems.

Naturally, topology interacts with geometry, and it is common for topological ideas to be noticed first in geometry. An important example is the *Euler characteristic*, which was originally observed as a characteristic of polyhedra, then later seen to be meaningful for arbitrary closed surfaces. Today, we tend to think that topology comes first, and that it controls what can happen in geometry. For example, the Gauss–Bonnet theorem seems to show that the Euler characteristic controls the value of the *total curvature* of a surface.

Topology also interacts with algebra. In this chapter we focus on the *fundamental group*, a group that describes the ways in which flexible loops can lie in a geometric object. On a sphere, all loops can be shrunk to a point, so the fundamental group is trivial. On the torus, however, there are many closed loops. But they are all combinations of two particular loops, a and b, such that $ab = ba$.

In 1904, Poincaré famously conjectured that a closed three-dimensional space with trivial fundamental group is topologically the same as the three-dimensional sphere. This *Poincaré conjecture* was proved only in 2003, with the help of methods from differential geometry. Thus the interaction between geometry and topology continues.

J. Stillwell, *Mathematics and Its History*, Undergraduate Texts in Mathematics,
DOI 10.1007/978-1-4419-6053-5_22, © Springer Science+Business Media, LLC 2010

22.1   Geometry and Topology

Topology is concerned with those properties that remain invariant under continuous transformations. In the context of Klein's Erlanger Programm (where it receives a brief mention under its old name of *analysis situs*) it is the "geometry" of groups of continuous invertible transformations, or *homeomorphisms*. The "spaces" to which transformations are applied, and indeed the meaning of "continuous," remain somewhat open. When these terms are interpreted in the most general way, as subject only to certain axioms (which we shall not bother to state here), one has *general topology*. The theorems of general topology, important in fields ranging from set theory to analysis, are not very geometric in flavor. *Geometric topology*, which concerns us in this chapter, is obtained when the transformations are ordinary continuous functions on $\mathbb{R}^n$ or on certain subsets of $\mathbb{R}^n$. Examples are the "topological equivalences" between surfaces we spoke about in Section 15.4.

Geometric topology is more recognizably "geometric" than general topology, though the "geometry" is necessarily of a discrete and combinatorial kind. Ordinary geometric quantities—such as length, angle, and curvature—admit continuous variation and hence cannot be invariant under continuous transformations. The kind of quantities that are topologically invariant are such things as the number of "pieces" of a figure or the number of "holes" in it. It turns out, though, that the combinatorial structures of topology can often be reflected by combinatorial structures in ordinary geometry, such as polyhedra and tessellations. In the case of surface topology, this geometric modeling of topological structure is so complete that topology becomes essentially a part of ordinary geometry. "Ordinary" here means geometry with notions of length, angle, and curvature, not necessarily Euclidean geometry. In fact, the natural geometric models of most surfaces are hyperbolic.

It remains to be seen whether topology as a whole will ever be subordinate to ordinary geometry. This seems to be the case in three dimensions, where a "geometrization conjecture" has recently been proved (see Section 22.8). It appears that here, too, hyperbolic geometry is the most important geometry (see Thurston (1997) or Weeks (1985)). In four or more dimensions it would be rash to speculate, though geometric methods have been important in recent breakthroughs (for example Donaldson (1983)). In this chapter we make a virtue of a necessity by confining our discussion mainly

to the topology of surfaces. This is the only area that is sufficiently understandable and relevant when set against the background of the rest of this
book. Fortunately, this area is also rich enough to illustrate some important
topological ideas, while still being mathematically tractable and visual.

We begin the discussion of surface topology at its historical starting
point, the theory of polyhedra.

22.2   Polyhedron Formulas of Descartes and Euler

The first topological property of polyhedra seems to have been discovered
by Descartes around 1630. Descartes's short paper on the subject is lost,
but its contents are known from a copy made by Leibniz in 1676, discovered among Leibniz's papers in 1860 and published in Prouhet (1860). A
detailed study of this paper, including a translation and facsimile of the
Leibniz manuscript, has been published by Federico (1982).

The same property was rediscovered by Euler (1752), and it is now
known as the *Euler characteristic*. If a polyhedron has V vertices, E edges,
and F faces, then its Euler characteristic is $V - E + F$. Euler showed that
this quantity has certain invariance by showing

$$V - E + F = 2$$

for all convex polyhedra, a result now known as the *Euler polyhedron formula*. Descartes already had the same result implicitly in the pair of formulas

$$P = 2F + 2V - 4, \quad P = 2E,$$

where P is the number of what Descartes called "plane angles": corners
of faces determined by pairs of adjacent edges. The relation $P = 2E$ then
follows from the observation that each edge participates in two corners. It
should be stressed that Descartes's "plane angle" has nothing to do with angle measure, and hence is just as topological a concept as Euler's "edges."
Thus Descartes's result belongs to topology just as much as Euler's does,
even though it fails to isolate the concept of Euler characteristic quite as
well. Some rather hairsplitting distinctions have been made between Euler and Descartes in an effort to show that Euler invented topology and
Descartes did not (see Federico (1982) for a review of different opinions).

Actually, neither of these mathematicians understood the polyhedron
formula in a fully topological way. They both used nontopological concepts, such as angle measure, in their proofs, and they did not realize that

"vertices," "edges," and "faces" are meaningful on any surface: edges need not be straight and faces need not be flat. Other early proofs of the Euler polyhedron formula also rely on angle measure and other ordinary geometric quantities. For example, that of Legendre (1794) assumes that the polyhedron can be projected onto the sphere, then uses the Harriot relation between angular excess and area for spherical polygons (Exercises 22.2.1 and 22.2.2).

Probably the first to understand $V - E + F$ purely topologically was Poincaré (1895). In fact, Poincaré generalized the Euler characteristic to n-dimensional figures, but in the case of polyhedra his essential observation was this: a vertex divides an edge into two edges, and an edge divides a face into two faces. It follows that any subdivision of edges or faces of a polyhedron leaves $V - E + F$ unchanged: if a new vertex is introduced on an edge, V and E both increase by 1; if a new edge is introduced across a face, E and F both increase by 1. The reverse processes of amalgamation, where they make sense, likewise leave $V - E + F$ unchanged.

The constancy of $V - E + F$ over, say, the class of convex polyhedra then follows if it can be shown that any polyhedron P_1 in the class can be converted to any other, P_2, by subdivisions and amalgamations. A plausible argument for this, due to Riemann (1851), is to view P_1 and P_2 as subdivisions of the same surface, say a sphere. Assuming that the edges of P_1 and P_2 meet only finitely often, superimposing P_1 on P_2 gives a common subdivision P_3 whose $V - E + F$ value is therefore the same as that of P_1 and P_2. Hence the $V - E + F$ values of P_1 and P_2 are equal. The assumption of only finitely many intersections is hard to justify, however. A different approach, which also yields the value of $V - E + F$ for *nonspherical* surfaces, is explained in the next section.

An engaging recent account of the Euler characteristic and its history is Richeson (2008).

EXERCISES

Here is the proof of the Euler polyhedron formula by Legendre (1794).

22.2.1 Consider the projection of a convex polyhedron onto a sphere, whose faces are therefore spherical polygons. Use the fact that

$$\text{area of a spherical } n\text{-gon} = \text{angle sum} - (n - 2)\pi$$

to conclude that

$$\text{total area} = 4\pi = \left(\sum \text{all angles}\right) - \pi\left(\sum \text{all } n\right) + 2\pi F.$$

22.2.2 Show also that

$$\sum \text{all } n = 2E, \quad \sum \text{all angles} = 2\pi V,$$

whence

$$V - E + F = 2.$$

The invariance of the Euler characteristic gives a simple topological proof that there are only five regular polyhedra. In fact, it shows that only five polyhedra are *topologically regular* in the following sense: for some $m, n > 2$ their "faces" are topological m-gons on a topological sphere, n of which meet at each vertex. We show as follows that $V - E + F = 2$ allows only the pairs

$$(m, n) = (3, 3), (3, 4), (3, 5), (4, 3), (5, 3),$$

corresponding to the known regular polyhedra (Section 2.2).

22.2.3 Given that there are F faces, deduce that $E = mF/2$ and $V = mF/n$.

22.2.4 Apply the formula $V - E + F = 2$ to conclude that $4n/(2m + 2n - mn)$ is a positive integer.

22.2.5 Show that $2m + 2n - mn > 0$, that is, $2\frac{m}{n} + 2 > m$, only for the above pairs (m, n).

22.2.6 Also check that $2m + 2n - mn$ divides $4n$ for these pairs.

22.3   The Classification of Surfaces

Between the 1850s and the 1880s, several different lines of research led to the demand for a topological classification of surfaces. One line, descending from Euler, was the classification of polyhedra. Another was the Riemann surface representation of algebraic curves, coming from Riemann (1851, 1857). Related to this was the problem of classifying symmetry groups of tessellations, considered by Poincaré (1882) and Klein (1882b) (see Section 22.6). Finally, there was the problem of classifying smooth closed surfaces in ordinary space (Möbius (1863)). These different lines of research converged when it was realized that in each case the surface could be subdivided into faces by edges (not necessarily straight, of course) so that it became a generalized polyhedron. The generalized polyhedra are what were traditionally called *closed* surfaces, now described by topologists as *compact and without boundary*.

The subdivision argument for the invariance of the Euler characteristic $V - E + F$ applies to any such polyhedron, not just those homeomorphic to the sphere and not just those with straight edges and flat faces. Various mathematicians, such as Riemann (1851) and Jordan (1866), came to the conclusion that any closed surface is determined, up to homeomorphism, by its Euler characteristic. It also seemed that the different possible Euler characteristics were realized by the "normal form" surfaces seen in Figure 22.1, which were discovered by Möbius (1863). It is certainly plausible that these forms are distinct, topologically, because of their different numbers of "holes." The main part of the proof is to show that any closed surface is homeomorphic to one of them.

The assumptions of Riemann (that the surface is a Riemann surface) and Möbius (that the surface is smoothly embedded in $\mathbb{R}^3$) were a little too special to yield a purely topological proof, and in addition they contained a hidden assumption of *orientability* ("two-sidedness"). A rigorous proof, from an axiomatic definition of generalized polyhedron, was given by Dehn and Heegaard (1907). The closed orientable surfaces indeed turn out to be those pictured in Figure 22.1, but in addition there are *nonorientable* surfaces, which are not homeomorphic to orientable surfaces.

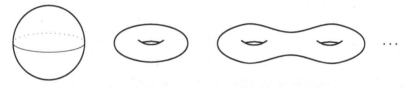

Figure 22.1: Closed orientable surfaces

A nonorientable surface may be defined as one that contains a *Möbius band*, a nonclosed surface discovered independently by Möbius and Listing in 1858 (Figure 22.2).

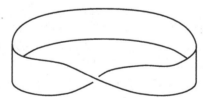

Figure 22.2: The Möbius band

Closed nonorientable surfaces cannot occur as Riemann surfaces, nor can they lie in $\mathbb{R}^3$ without crossing themselves; nevertheless, they include some important surfaces, such as the projective plane (Exercise 8.5.3). The nonorientable surfaces are also determined, up to homeomorphism, by the Euler characteristic.

The Möbius forms of closed orientable surfaces were given standard polyhedral structures by Klein (1882b). These are "minimal" subdivisions with just one face and, except for the sphere, with just one vertex. When the Klein subdivision of a surface is cut along its edges, one obtains a *fundamental polygon*, from which the surface may be reconstructed by identifying like-labeled edges (see Figure 22.3, and also Figures 22.11 and 22.13).

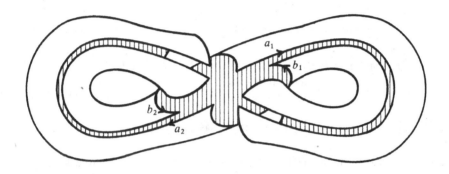

Figure 22.3: Constructing a surface by edge pasting

It is often more convenient to work with the polygon rather than the surface or its polyhedral structure. For example, since Brahana (1921), most proofs of the classification theorem have used polygons rather than polyhedra, "cutting and pasting" them (instead of subdividing and amalgamating) until Klein's fundamental polygons are obtained. The fundamental polygon gives a very easy calculation of the Euler characteristic χ and shows it to be related to the *genus g* (number of "holes") by

$$\chi = 2 - 2g$$

(Exercise 22.3.1). Of course, the genus determines the surface more simply than the Euler characteristic, but we shall see that the Euler characteristic is a better reflection of geometric properties.

EXERCISES

22.3.1 Show that the standard polyhedron for a surface of genus $g \geq 1$ has $V = 1$, $E = 2g$, $F = 1$, whence $\chi = 2 - 2g$.

The standard polygon for the genus g surface has a boundary path of the form $a_1b_1a_1^{-1}b_1^{-1}a_2b_2a_2^{-1}b_2^{-1}\cdots a_gb_ga_g^{-1}b_g^{-1}$, where successive letters denote successive edges and those with exponents -1 have oppositely directed arrows. Edges with the same letter are pasted together, with arrows matching.

22.3.2 Each sequence $a_ib_ia_i^{-1}b_i^{-1}$ is called a *handle*. Justify this term by drawing the surface that results from pasting together the matching edges of the polygon bounded by $a_ib_ia_i^{-1}b_i^{-1}c$. The result should be a "handle-shaped" surface with boundary curve c.

Another fundamental polygon is the "$2n$-gon with opposite edges pasted together," that is, the polygon with boundary of the form $a_1a_2\cdots a_na_1^{-1}a_2^{-1}\cdots a_n^{-1}$.

22.3.3 Show that for both $n = 2$ and $n = 3$ the surface obtained from the polygon $a_1a_2\cdots a_na_1^{-1}a_2^{-1}\cdots a_n^{-1}$ is a torus.

22.3.4 Show that if n is even, the vertices of the polygon $a_1a_2\cdots a_na_1^{-1}a_2^{-1}\cdots a_n^{-1}$ become a single vertex after pasting, and if n is odd they become two. Hence find the Euler characteristic of the surface for any n.

22.4   Descartes and Gauss–Bonnet

The first theorem in the Descartes manuscript is a remarkable statement about the total "curvature" of a convex polyhedron, not at first appearing to have any topological content. It is a spatial analogue of the obvious theorem that the sum of the external angles of a convex polygon is 2π. The latter theorem can be seen intuitively by considering the total turn of a line that is transported around the polygon (Figure 22.4).

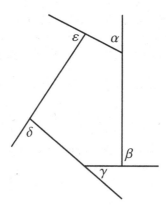

Figure 22.4: Total turn around a polygon

Figure 22.5 shows a different proof, which generalizes to polyhedra.

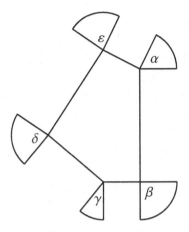

Figure 22.5: Adding the sectors bounded by normals

At each vertex, construct a sector of a unit circle, bounded by normals to the two edges at that vertex. Clearly, the angle of the sector equals the external angle at that vertex. Also, adjacent sides of adjacent sectors are perpendicular to the same edge, hence parallel, so the sectors can be fitted together to form a complete disk, of total angle (circumference) 2π.

To generalize this to polyhedra, define the *exterior solid angle* at each vertex P to be the (area of the) sector of a unit ball bounded by planes normal to the edges at P (Figure 22.6).

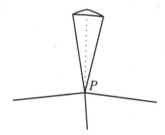

Figure 22.6: The exterior solid angle

As before, adjacent sides of adjacent sectors are parallel; hence the sectors can be fitted together to form a complete ball, of total solid angle (area) 4π. Descartes stated only that the total exterior solid angle is 4π, without even defining exterior solid angle. The foregoing proof is based on the reconstruction by Pólya (1954a).

The theorem about polygons has an analogue for simple closed smooth curves C, namely, $\int_C \kappa \, ds = 2\pi$ (Section 17.2). This leads us to wonder whether the Descartes theorem has an analogue for smooth closed convex surfaces S, say, $\iint_S \kappa_1 \kappa_2 \, dA = 4\pi$, where $\kappa_1 \kappa_2$ is the Gaussian curvature. This is so, and in fact there is a proof like the polyhedron proof using yet another characterization of Gaussian curvature due to Gauss (1827).

If we take a small geodesic polygon $\mathcal{P}$ on the surface S, then the "total curvature" of the portion $\mathcal{P}$ can be represented by an "exterior solid angle" $\mathcal{A}$ bounded by parallels to the normals to S along the sides of $\mathcal{P}$ (Figure 22.7). Gauss showed that the measure of $\mathcal{A}$—the area it cuts out of the unit sphere—is $\iint_{\mathcal{P}} \kappa_1 \kappa_2 \, dA$. But it is also clear, by the parallelism of adjacent sides of adjacent exterior solid angles $\mathcal{A}$, that the $\mathcal{A}$'s corresponding to a partition of S by geodesic polygons $\mathcal{P}$ fit together to form a complete ball. Hence $\iint_S \kappa_1 \kappa_2 \, dA = 4\pi$.

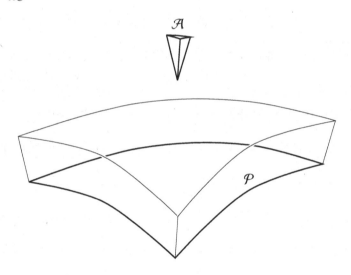

Figure 22.7: The solid angle of total curvature

This is a "global" form of the Gauss–Bonnet theorem. When the Descartes theorem was first published in 1860, the Gauss–Bonnet theorem was already known, and the analogy between the two was noted by Bertrand (1860). Bertrand, however, made the qualification that "the beautiful conception of Gauss could not in any manner be considered as a corollary to that of Descartes." This may be true in a narrow sense; nevertheless, the Descartes and Gauss–Bonnet theorems can be viewed as limiting cases of each other. Gauss–Bonnet $\Rightarrow$ Descartes by concentrating the curvature of a surface at vertices until it becomes a polyhedron, while Descartes $\Rightarrow$ Gauss–Bonnet by increasing the number of vertices of a polyhedron until it becomes a smooth surface. It is interesting, though probably accidental, that Descartes actually uses the word "curvatura" to describe the exterior solid angle.

22.5   Euler Characteristic and Curvature

There is another, more "intrinsic," proof of Descartes's theorem that reveals the fact that total exterior solid angle is really $2\pi \times$ Euler characteristic. In fact, knowledge of the total exterior angle yields a proof of the Euler characteristic polyhedron formula. This seems to have been the way in which Descartes discovered his version of the formula.

The key step is to show that the exterior solid angle at a vertex P is expressible intrinsically as $2\pi - (\alpha_1 + \alpha_2 + \cdots + \alpha_n)$, where $\alpha_1, \alpha_2, \ldots, \alpha_n$ are the face angles that meet at P. These are *not* the angles $\alpha'_1, \alpha'_2, \ldots, \alpha'_n$ between the planes that bound the exterior solid angle, but it turns out (Exercise 22.5.1) that

$$\alpha_i + \alpha'_i = \pi$$

for each i, whence the measure of the exterior solid angle, which comes from $\alpha'_1 + \alpha'_2 + \cdots + \alpha'_n$ by Harriot's theorem (Section 17.6), also comes from $\alpha_1 + \alpha_2 + \cdots + \alpha_n$.

Knowing now that the exterior solid angle at P equals $2\pi - \sum$ face angles at P, we get

$$\text{total exterior solid angle} = 2\pi V - \sum \text{all face angles},$$

where V is the total number of vertices. By grouping the face angles according to the types of faces, we also find (Exercise 22.5.2) that

$$\sum \text{all face angles} = \pi(2E - 2F),$$

whence

$$\text{total exterior solid angle} = 2\pi(V - E + F)$$
$$= 2\pi \times \text{Euler characteristic}.$$

In the case of convex polyhedra, where we already know that total exterior solid angle $= 4\pi$, this gives Euler characteristic $= 2$. More importantly, the derivation is valid for polyhedra of arbitrary Euler characteristic, showing that the total exterior solid angle is really the *same* as the Euler characteristic, up to a constant multiple.

There is a similar intrinsic proof of the Gauss–Bonnet theorem, again valid for arbitrary Euler characteristic, which shows that

$$\text{total curvature} = \iint_S \kappa_1\kappa_2 \, dA = 2\pi \times \text{Euler characteristic}$$

(Exercise 22.5.3). Legendre's (1794) proof of the Euler polyhedron formula is the special case of the argument for constant curvature.

Thus the Euler characteristic regulates the total curvature of a surface. In particular, if the curvature is constant, it must have the same sign as the Euler characteristic. This in turn has implications for the geometry of the surface. As we saw in Section 17.4, surfaces of constant positive curvature have spherical geometry, those of zero curvature have Euclidean geometry, and those of negative curvature have hyperbolic geometry. In the next section we shall see that there is a natural way to impose constant curvature on surfaces of arbitrary Euler characteristic. It will then follow that the natural geometry of a surface is spherical, Euclidean, or hyperbolic according as its Euler characteristic is positive, zero, or negative. Moreover, if the absolute value of the curvature is taken to be 1, then the Gauss–Bonnet theorem gives

$$\text{area} = |2\pi \times \text{Euler characteristic}|.$$

This makes surface topology completely subordinate to geometry, at least for orientable surfaces, because it says that the topology of a surface is completely determined by the sign of its curvature and its area.

These results were implicit in the work of Poincaré and Klein in the 1880s. Perhaps Klein was the first to see clearly how the geometry of a surface determines its topology (see, for example, Klein (1928), p. 264).

EXERCISES

Figure 22.8 shows the region around a vertex P of a polyhedron and the exterior solid angle of P centered at O and bounded by the planes OAB, OBC, OCA perpendicular to the edges through P.

22.5.1 Show that there are right angles where indicated, and hence that

$$\alpha + \alpha' = \pi, \quad \beta + \beta' = \pi, \quad \gamma + \gamma' = \pi.$$

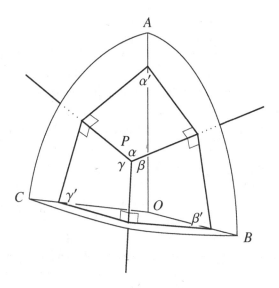

Figure 22.8: The vertex region of a polyhedron

Now to relate face angles to E and F, it helps to write

$$F = F_3 + F_4 + F_5 + \cdots,$$

where $F_3 =$ number of 3-gon faces, $F_4 =$ number of 4-gon faces, and so on.

22.5.2 Show that

$$E = \frac{1}{2}(3F_3 + 4F_4 + 5F_5 + \cdots),$$

and deduce that in an ordinary polyhedron (that is, one with flat faces),

$$\sum \text{ all face angles} = \pi(2E - 2F),$$

using the fact the angle sum of an n-gon is $(n - 2)\pi$.

22.5.3 Prove the global form of the Gauss–Bonnet theorem,

$$\iint_S \kappa_1 \kappa_2 \, dA = 2\pi \times \text{Euler characteristic},$$

by partitioning the closed surface S into geodesic polygons and applying the ordinary form of the Gauss–Bonnet theorem (Section 17.6).

22.6 Surfaces and Planes

In Section 16.5 we noticed that an elliptic function defines a mapping of a plane onto a torus. Such mappings are also interesting in the topological context, where they are called *universal coverings*. In general, a mapping $\varphi : \tilde{S} \to S$ of a surface $\tilde{S}$ onto a surface S is called a *covering* if it is a homeomorphism locally, that is, when restricted to sufficiently small pieces of $\tilde{S}$. The mapping of the plane onto the torus in Section 16.5 is a covering because it is a homeomorphism when restricted to any region smaller than a period parallelogram.

Another interesting example of a covering we have already met is the mapping of the sphere onto the projective plane given by Klein (1874) (Section 8.5). This map sends each pair of antipodal points of the sphere to the same point of the projective plane, and hence is a homeomorphism when restricted to any part of the sphere smaller than a hemisphere.

One more example is Beltrami's (1868a) covering of the pseudosphere by a horocyclic sector (Section 18.4). Topologically, this covering is the same as the covering of a half-cylinder by a half-plane (Figure 22.9). All these coverings are *universal* in the sense that the covering surface $\tilde{S}$ (sphere or plane) can be covered only by $\tilde{S}$ itself.

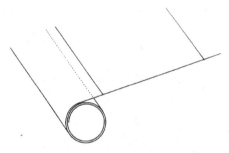

Figure 22.9: Covering a cylinder

An example of a *nonuniversal* covering is the covering of the torus by the cylinder, intuitively like an infinite snake swallowing its own tail (Figure 22.10). This is nonuniversal because the cylinder can in turn be covered by the plane, just as the half-cylinder is covered by the half-plane in Figure 22.9. In fact, by composing the coverings plane $\to$ cylinder $\to$ torus, we recover our first example, the plane $\to$ torus covering.

Since the sphere can be covered only by itself, the first interesting examples of coverings are those of orientable surfaces of genus ≥ 1 (that is, Euler characteristic ≤ 0). All of these surfaces can be covered by planes. Moreover, each nonorientable surface can be doubly covered by an orientable surface in the same way that the projective plane is covered by the

Figure 22.10: Covering a torus

sphere, so the main thing we need to understand is the universal covering
of orientable surfaces of genus ≥ 1 by planes.

The basic idea is due to Schwarz, and it became generally known
through a letter from Klein (1882a) to Poincaré. To construct the universal
covering of a surface S one takes infinitely many copies of a fundamental
polygon F for S and arranges them in the plane so that *adjacent* copies of
F meet in the same way that F meets *itself* on S. For example, the torus T
in Figure 22.11 has the square fundamental polygon F shown, which meets
itself along $\vec{a}$ and $\vec{b}$ in S (where the arrows indicate that edges must agree
in direction as well as label).

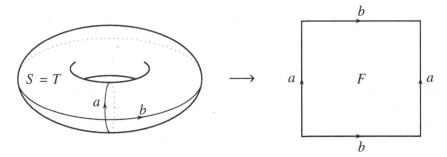

Figure 22.11: The torus and its fundamental polygon

If instead we take infinitely many separate copies of F and join adjacent
copies $\vec{a}$ to $\vec{a}$ and $\vec{b}$ to $\vec{b}$, then we obtain a plane $\tilde{T}$, tessellated as in Figure
22.12. The universal covering $\tilde{T} \to T$ is then defined by mapping each
copy of F in $\tilde{T}$ in the natural way onto the F in T.

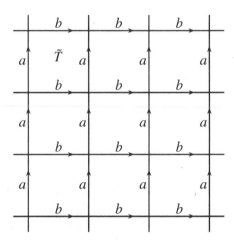

Figure 22.12: Tessellation of the torus cover

The tessellation of Figure 22.12 can of course be realized by squares in the Euclidean plane. We can therefore impose a Euclidean geometry on the torus by defining the distance between (sufficiently close) points on the torus to be the Euclidean distance between appropriate preimage points in the plane. In particular, the "straight lines" (geodesics) on the torus are the images of straight lines in the Euclidean plane. The torus geometry is not quite the geometry of the plane, of course, since there are closed geodesics, such as the images of the line segments a and b. However, it is Euclidean when restricted to sufficiently small regions. For example, the angle sum of each triangle on the torus is π.

For surfaces of genus >1—that is, of negative Euler characteristic—the Gauss–Bonnet theorem predicts negative curvature, and hence the natural covering plane should be hyperbolic. This can also be seen directly from the combinatorial nature of the tessellation on the universal cover. For example, the fundamental polygon F of the surface S of genus 2 is an octagon (Figure 22.13).

In the universal covering, eight of these octagons have to meet at each vertex, since the eight corners of the single F meet on S. Such a tessellation is impossible, by regular octagons, in the Euclidean plane, but it exists in the hyperbolic plane, as Figure 22.14 shows.

In fact, this tessellation is obtained by amalgamating triangles in the Gauss tessellation (Figure 18.15). The tessellations for general genus >1 can similarly be realized geometrically in the hyperbolic plane, and they

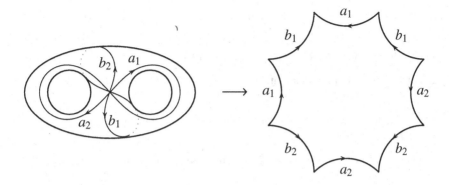

Figure 22.13: Genus-2 surface and its fundamental polygon

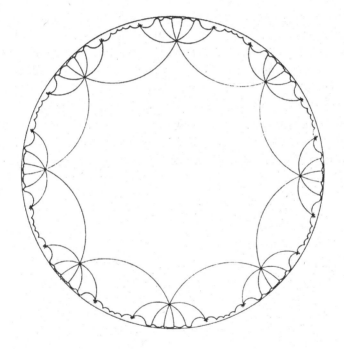

Figure 22.14: Tessellation of the genus-2 covering

were among the hyperbolic tessellations considered by Poincaré (1882) and Klein (1882b). The distance function, hence the curvature and local geometry, can be transported from the covering plane to the surface as we did above for the torus.

When surfaces of genus >1 are realized as surfaces of constant negative curvature, their genus can be read off from their area.

22.6.1 Show that the fundamental polygon for an orientable surface of genus p is a $4p$-gon with angle sum 2π.

22.6.2 Deduce that its Euler characteristic is proportional to its angular defect and hence to its area.

22.6.3 Conclude, using Exercise 22.3.1, that the area determines the genus.

22.7   The Fundamental Group

Another way to explore the meaning of the universal cover $\tilde{S}$ is to use it to plot paths on the surface S. As a point P moves on S, each preimage $\tilde{P}$ of P moves analogously on $\tilde{S}$. The only difference is that as P crosses an edge of the fundamental polygon on S, $\tilde{P}$ crosses from one fundamental polygon to another on $\tilde{S}$. Thus $\tilde{P}$ will not necessarily return to its starting point, even when P does. In fact, we can see that the displacement of $\tilde{P}$ in some way measures the extent to which P winds around the surface S. Figure 22.15 shows an example. As P winds once around the torus, more or less in the direction of $\vec{a}$, $\tilde{P}$ wanders from one end to the other of a segment $\vec{a}$ on $\tilde{S}$.

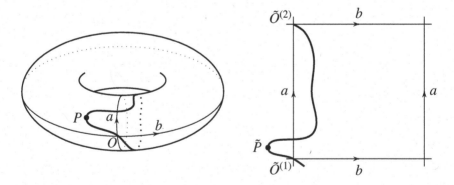

Figure 22.15: Plotting on the covering surface

We say that closed paths p, p' with initial point O on S "wind in the same way," or are *homotopic*, if p can be deformed into p' with O fixed and without leaving the surface. Now if the path p of P is deformed into

p', with O fixed, then the path $\tilde{p}$ of $\tilde{P}$ is deformed into a $\tilde{p}'$ with the same initial and final points, $\tilde{O}^{(1)}$ and $\tilde{O}^{(2)}$, as $\tilde{p}$. Hence each homotopy class corresponds simply to a *displacement* of the universal cover $\tilde{S}$ that moves $\tilde{O}^{(1)}$ to $\tilde{O}^{(2)}$. The different preimages $\tilde{P}$ will of course start at different preimages $\tilde{O}^{(1)}$ of O, but a single displacement of $\tilde{S}$ moves them all to their final positions $\tilde{O}^{(2)}$. Moreover, the displacement moves the whole tessellation of $\tilde{S}$ onto itself: it is a rigid motion of the tessellation.

Thus from the topological notion of homotopic closed paths we arrive once again at ordinary geometry. We also arrive at a group called the *fundamental group* of S. Geometrically, it is the group of motions of $\tilde{S}$ that map the tessellation onto itself (which includes mapping each edge to a like-labeled edge). Topologically, it is the group of homotopy classes of closed paths, with a common initial point O, on S. The product of homotopy classes is defined by successive traversal of representative paths.

The fundamental group was first defined by Poincaré (1895). Poincaré defined it for much more general figures, whose universal covers are not so apparent, so he did not generally view the fundamental group as a covering motion group. However, Poincaré had already studied groups of motions of tessellations (1882). Reconsidering these earlier results topologically in (1904), he arrived at the interpretation just given. This paper was very influential on the later work of Dehn (1912) and Nielsen (1927) and has been indirectly responsible for a recent surge of interest in hyperbolic geometry.

The more general notion of fundamental group in Poincaré (1895) has also been influential outside topology. It turns out, for example, that for any "reasonably described" figure $\mathcal{F}$ it is possible to compute generators and defining relations for the fundamental group of $\mathcal{F}$. The defining relations of a fundamental group can be quite arbitrary (in fact, *completely* arbitrary, as was shown by Dehn (1910) and Seifert and Threlfall (1934), p. 180). So the question arises: can the properties of a group be determined from its defining relations? One would like to know, for example, when two different sets of relations define the same group. The latter question was raised by Tietze (1908) in the first paper to follow up Poincaré's work. Tietze made the remarkable conjecture—which could not even be precisely formulated at the time—that the problem is unsolvable. The *isomorphism problem for groups*, as it came to be known, was indeed shown to be unsolvable by Adyan (1957), in the sense that no algorithm can settle the question for *all* finite sets of defining relations. Adyan's result was based on the theory of algorithms, which will be outlined in Chapter 24.

By combining Adyan's result with some of Tietze (1908) and the result of Seifert and Threlfall mentioned above, Markov (1958) was able to show the unsolvability of the *homeomorphism problem*. This is the problem of deciding, given "reasonably described" figures $\mathcal{F}_1$ and $\mathcal{F}_2$, whether $\mathcal{F}_1$ is homeomorphic to $\mathcal{F}_2$. (A complete proof of the unsolvability of the isomorphism problem and homeomorphism problem may be found in Stillwell (1993), and its history may be found in Stillwell (1982).) Thus Poincaré's construction of the fundamental group led in the end to a quite unexpected conclusion: the basic problem of topology is unsolvable.

EXERCISES

In the following exercises it will be helpful to view the fundamental group as the group of motions of the universal covering plane, diagrammed in the previous section. The diagram shows that any sequence of motions equal to the identity corresponds to a closed path of edges in the diagram.

22.7.1 Explain why the fundamental group of the torus is generated by elements a with defining relation
$$aba^{-1}b^{-1} = 1.$$

22.7.2 Similarly, explain why the fundamental group of the surface of genus 2 is generated by elements a_1, b_1, a_2, b_2 with defining relation
$$a_1 b_1 a_1^{-1} b_1^{-1} a_2 b_2 a_2^{-1} b_2^{-1} = 1.$$

27.7.3 Show that the former group is commutative but the latter is not.

22.8   The Poincaré Conjecture

The homotopy concept, and the associated fundamental group, was just one of Poincaré's contributions to topology. Another was the *homology* concept, which algebraically captures the relationship between topological objects and their boundaries. We can give a glimpse of this relationship in the case of curves on surfaces. Figure 22.16 shows three examples.

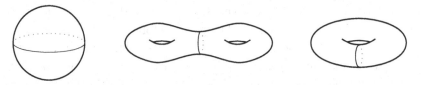

Figure 22.16: Bounding and nonbounding curves

On the left we have a sphere with the equator drawn on it. The equator *bounds* (that is, is the boundary of) both the upper and the lower hemisphere, each of which is topologically a disk. On the right we have a torus and a curve on it that does *not* bound—there is no part of the torus of which this curve is the boundary, as we already observed in Section 16.4.

In the middle we have a surface of genus 2 and a curve on it that also bounds—it bounds both the left half of the surface and the right half. However, this curve does not bound a topological disk so, unlike the equator on the sphere, it cannot be shrunk to a point. Thus the middle example shows that the concept of bounding is coarser than the concept of being homotopic to a point. Still, the concept is refined enough to distinguish the sphere, which we now call $\mathbb{S}^2$, from all other closed surfaces: $\mathbb{S}^2$ *is the only closed surface on which every simple closed curve is bounding.*

There is a 3-dimensional analogue of $\mathbb{S}^2$ called the 3-*sphere* $\mathbb{S}^3$, which may be defined as the set of points in $\mathbb{R}^4$ at unit distance from the origin. $\mathbb{S}^3$ can also be defined, more topologically, as the result of adding a point at infinity to $\mathbb{R}^3$, just as we obtained $\mathbb{S}^2$ from the plane by adding a point at infinity in Section 15.2. $\mathbb{S}^3$ and $\mathbb{R}^3$ are the simplest examples of 3-*dimensional manifolds* (or 3-manifolds for short), which are spaces in which each point has a neighborhood homeomorphic to the interior of a solid ball. $\mathbb{R}^3$ is an "open" 3-manifold, while $\mathbb{S}^3$ is "closed," and one wonders whether $\mathbb{S}^3$ can be distinguished from the other closed 3-manifolds in the same way that $\mathbb{S}^2$ can be distinguished from the other closed surfaces.

In 1900, Poincaré conjectured that $\mathbb{S}^3$ is the only closed 3-manifold in which every closed curve is bounding. He was wrong, because in Poincaré (1904) he discovered a remarkable counterexample. It is now known as a *homology sphere* because it has the "same homology" as $\mathbb{S}^3$ without being homeomorphic to $\mathbb{S}^3$. In Poincaré's homology sphere every simple closed curve bounds a surface but *not* always a topological disk. So Poincaré amended his 1900 conjecture as follows: *if every simple closed curve in a closed connected 3-manifold bounds a disk, then M is homeomorphic to $\mathbb{S}^3$.* This became known as the *Poincaré conjecture*—one of the most famous mathematical problems of the 20th century.

The condition that every simple closed curve bounds a disk is called *simple-connectedness* of *M*. Equivalent statements of this property are:

- Every closed curve in *M* contracts to a point.

- The fundamental group of *M* equals {1}.

The existence of homology spheres shows that three dimensions are more complicated than two, but just *how much* more complicated they are was not immediately clear. Further results on 3-manifolds came with glacial slowness, and they often revealed new complications. Dehn (1910) found infinitely many homology spheres, Alexander (1919) found two non-homeomorphic 3-manifolds with the same fundamental group (but not the group {1}), and Whitehead (1935) found an *open* 3-manifold that is contractible but not homeomorphic to $\mathbb{R}^3$.

In the 1950s and 1960s there was at last some good news, with a series of positive results showing that 3-manifolds are "well behaved" in certain respects. The news did not include a proof of the Poincaré conjecture, however. Instead, progress on the conjecture came in higher dimensions, with a proof by Smale (1961) of the analogous conjecture for $\mathbb{S}^n$ for $n \geq 5$. The analogue states that any closed connected manifold is homeomorphic to a sphere if all the topological spheres in it are contractible. Unfortunately, while three dimensions are harder than two, five are *easier* than three in some respects (topologists say they have "more wiggle room"). So Smale's proof did not throw much light on the classical Poincaré conjecture, or on the analogous conjecture for $\mathbb{S}^4$ either.

The analogue of the Poincaré conjecture for 4-manifolds was finally proved by Freedman (1982). Freedman's proof was a tour de force that simultaneously solved several longstanding problems about 4-manifolds. That his approach worked at all was a surprise to many of his colleagues, and finding a similar approach to the classical Poincaré conjecture seemed out of the question.

Indeed, an entirely new approach to the Poincaré conjecture had already been taking shape in the hands of William Thurston in the late 1970s. Thurston, like Poincaré and Dehn, was interested in *geometric* realizations of manifolds, exemplified by the surfaces of constant curvature that realize all the topological forms of closed surfaces. He conjectured that all 3-manifolds may be realized in a similar, though more complicated, way. Instead of the three 2-dimensional geometries of constant curvature, one has eight "homogeneous" 3-dimensional geometries. And instead of a single geometry for each 3-manifold M one has a "decomposition" of M into finitely many pieces, each carrying one of the eight geometries. Thurston's *geometrization conjecture* states that each closed connected 3-manifold is homeomorphic to one with such a decomposition. The Poincaré conjecture follows from a special case of the geometrization conjecture for manifolds of positive curvature. For more details on the evolution of the Poincaré conjecture up to this point, see Milnor (2003).

Thurston was able to prove many cases of his geometrization conjecture, but geometrization seemed to run out of steam in the early 1980s. This was not entirely disappointing to some topologists, who still hoped for a proof of the Poincaré conjecture by purely topological methods. However, *more* geometry was to come; not less, and *differential* geometry at that. It was not enough to consider manifolds with "homogeneous" geometry; one had to consider manifolds with *arbitrary* smooth geometry, and to let the geometry "flow" towards homogeneity.

The idea of "flowing towards homogeneity" was initiated by Hamilton (1982), and was brought to a triumphant conclusion by Grigory Perelman in 2003, with a proof of Thurston's geometrization conjecture. Perelman had to overcome enormous difficulties, too technical to describe here, but Hamilton's idea and its difficulties can be illustrated with manifolds of lower dimension—curves and surfaces.

A closed 1-dimensional manifold may be realized as a smooth closed curve in the plane, such as that shown in Figure 22.17.

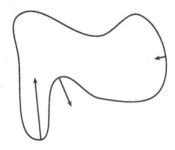

Figure 22.17: Curvature flow of a curve

Assuming that the curve is sufficiently smooth, it has a curvature at each point. We represent the curvature by a directed line, with length proportional to the curvature and direction towards the center of curvature. Now suppose we let the curve "flow" so that each point moves in the direction of the curvature arrow, with speed proportional to the arrow's length. The general tendency is for the curve to shrink to a point, but in the process its *shape* may tend to a limit. It is very plausible, and can be proved, that the shape of the curve tends towards a circle. Thus (not surprisingly) every closed 1-manifold has a geometric realization as a circle.

Now consider a similar process for smooth closed surfaces in $\mathbb{R}^3$. There are some problems. We have to decide which concept of curvature to use, because Gaussian curvature is not the only option. And even if we choose the concept of curvature well, we may not get the outcome we want. Some surfaces homeomorphic to the sphere, such as the one in Figure 22.18, may not flow towards the shape of the standard, constant-curvature, sphere.

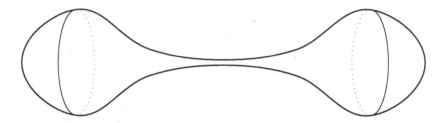

Figure 22.18: A topological sphere that does not flow well

This is because the high curvature of the thin neck causes it to shrink fast in comparison with the low-curvature ends, resulting in a shape whose neck becomes ever-thinner in comparison with the ends. The way out of this situation is drastic but effective. We perform *surgery*, by cutting the neck and smoothly sealing the cuts, as shown in Figure 22.19. Then, if

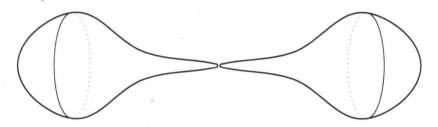

Figure 22.19: The result of surgery

we let the two pieces continue their curvature flow, the shape of each piece tends to that of a standard sphere. With a delicate analysis of the behavior of curvature flow, and the help of surgery, it becomes possible to prove the "2-dimensional Poincaré conjecture": every simply connected closed surface is homeomorphic to $\mathbb{S}^2$. Of course, we can already prove this, more easily, from the topological classification of surfaces. What is important is that the curvature flow idea also works for 3-manifolds, where a classification theorem is not yet available.

The appropriate flow for 3-manifolds is called the *Ricci curvature flow*, introduced by Hamilton (1982). Hamilton was able to show that the Ricci curvature flow works in many cases, but he was stymied by the more complex bad behavior (analogous to the formation of thin necks) that can occur with 3-manifolds. The difficulties were brilliantly overcome by Perelman in 2003. Perelman published his proof only in outline, in three papers posted on the Internet in 2002 and 2003, but experts later found that these papers contained all the ideas necessary to construct a complete proof. See, for example, Morgan and Tian (2007).

Perelman himself, apparently sure that he would be vindicated, published nothing further and seems to have gone into seclusion. In 2006 he was awarded the most prestigious prize in mathematics, the Fields Medal, but declined to accept it. The whole story behind these events remains to be told, but the fascinating *New Yorker* piece of Nasar and Gruber (2006) contains most of what we know so far.

EXERCISES

In the following exercises we study the relation between curves in a manifold that bound a disk (namely those contractible to a point and hence equal to 1 in the fundamental group), and those that bound more general surfaces. We use letters $a, b, \ldots$ to denote closed curves with a fixed initial point, and also the corresponding elements of the fundamental group.

22.8.1 Use Exercise 22.3.2 to show that, for any elements a and b of the fundamental group, there is an element c, equal to $aba^{-1}b^{-1}$ in the fundamental group, that bounds a surface (a "handle"). (Hint: Deform the handle so that its boundary c lies arbitrarily close to the curves a and b.)

22.8.2 Deduce from Exercise 22.8.1 that the elements of the fundamental group that bound surfaces are those that become equal to 1 when the fundamental group is "abelianized," that is, when generators are allowed to commute.

The fundamental group of the Poincaré homology sphere is generated by elements a and b, with defining relations

$$(ab)^2 = a^3 = b^5.$$

This group is called the *binary icosahedral group*. One can now show that this group is nontrivial, but that all elements become equal to 1 when a and b are allowed to commute.

22.8.3 Show that the result of adding the relation $(ab)^2 = 1$ to the relations of the binary icosahedral group is the icosahedral (or dodecahedral) group of Exercise 19.5.5.

22.8.4 Deduce from Exercise 22.8.3 that the binary icosahedral group has at least 60 elements. (Harder: Show that it in fact has 120 elements.)

22.8.5 Show, on the other hand, that adding the relation $ab = ba$ to the binary icosahedral group gives the group $\{1\}$.

22.9   Biographical Notes: Poincaré

Henri Poincaré (Figure 22.20) was born in Nancy in 1854 and died in Paris in 1912. His father, Léon, was a physician and professor of medicine at the University of Nancy, and Henri grew up in a comfortable academic environment. He and his younger sister, Aline, were at first educated by their mother, and Poincaré later traced his mathematical ability to his maternal grandmother. At the age of five he suffered an attack of diphtheria, which weakened his health and excluded him from the more boisterous childhood games. He made up for this by organizing charades and playlets, and he later became a keen dancer. Many photographs of Poincaré and his family may be seen in the centenary volume (1955), which forms the second half of vol. 11 of Poincaré's *Œuvres*.

Figure 22.20: Henri Poincaré

Being excluded from most games, Poincaré had ample time to read and study, and when he began attending school, aged eight, he made rapid progress. His ability first showed in French composition, but by the end of his school career his awesome mathematical talent was also clear. He won first prize in a nationwide mathematics competition and topped the entrance exam to the École Polytechnique in 1873. This, incidentally, was despite the Franco-Prussian War (1870–1871), during which Poincaré's home province of Lorraine bore the brunt of the German invasion. At this time, Poincaré accompanied his father on ambulance rounds, becoming an ardent French patriot as a result. However, he never held German mathematicians responsible for the brutalities of their compatriots. He learned German during the war in order to read the news, and he later put the knowledge to good use in communicating with his German colleagues Fuchs and Klein.

At the École Polytechnique, Poincaré continued to do well, though clumsiness in drawing and experimental work cost him first place. (His marks in drawing, though mediocre, were never zero, despite oft-told tales to that effect. Poincaré's results may be seen in the centenary volume (1955).) Curiously, he planned to become an engineer at this stage and studied at the École des Mines from 1875 to 1879, at the same time writing a doctoral thesis in mathematics. He worked briefly as a mining engineer before becoming an instructor in mathematics at the University of Caen in 1879. It was at Caen that Poincaré made his first important discovery: the occurrence of non-Euclidean geometry in the theory of complex functions. He had been thinking about periodicity with respect to linear fractional transformations, after encountering functions with this property in the work of Lazarus Fuchs. The functions in question arose from differential equations, and Poincaré had been struggling to understand them analytically when he was struck by an unexpected geometric inspiration:

> Just at this time I left Caen, where I was then living, to go on a geological excursion under the auspices of the school of mines. The changes of travel made me forget my mathematical work. Having reached Coutances, we entered an omnibus to go some place or other. At the moment when I put my foot on the step the idea came to me, without anything in my former thoughts seeming to have paved the way for it, that the transformations I had used to define the Fuchsian functions were identical with those of Non-Euclidean geometry.

> Poincaré (1918); translation by Halsted, 1929, p. 387

The discovery of the underlying geometry (and topology, which soon followed) put Fuchsian functions in a completely new light, rather like the illumination of elliptic functions by Riemann's discovery that they belong to the torus. For the next few years Poincaré worked feverishly to develop these ideas, in friendly competition with Klein. There were some reservations about his style—undisciplined and lacking in rigor, though very readable—but his brilliance was not contested. He was appointed to a chair at the University of Paris in 1881 and remained there, winning ever higher honors, until the end of his life. In 1881 he was married to Louise Poulain; they had a son and three daughters.

Poincaré's work on Fuchsian functions led him to topology, as we have seen in Sections 22.6 and 22.7. So did another of his great inventions, the qualitative theory of differential equations. He used this theory, which deals with such questions as the long-term stability of a mechanical system, in his *Les méthodes nouvelles de la mécanique céleste* (1892, 1893, 1899), probably the greatest advance in celestial mechanics since Newton. Poincaré's topological ideas not only breathed new life into complex analysis and mechanics; they amounted to the creation of a major new field, algebraic topology. In papers between 1892 and 1904, Poincaré built up an arsenal of techniques and concepts that were to keep topologists going for the next 30 years. It was not until Hurewicz discovered higher-dimensional analogues of the fundamental group in 1933 that a significant new weapon was added to Poincaré's arsenal.

Poincaré was perhaps the last mathematician to have a general grasp of all branches of mathematics. Like Euler, he wrote fluently and copiously on all parts of mathematics, and in fact he surpassed Euler in his popular writing. He wrote many volumes on science and its philosophy, which were best-sellers in the early part of the 20th century. Poincaré would perhaps have been as prolific as Euler if ill health had not overtaken him in his fifties. In 1911 he took the unusual step of publishing an unfinished paper, on periodic solutions of the three-body problem, believing he might not live to complete the proof. "Poincaré's last theorem" was indeed still open when he died in 1912, but the proof was completed in 1913 by the American mathematician G. D. Birkhoff.

23

Simple Groups

PREVIEW

We saw in Chapter 19 that the group concept came to light when Galois used it to explain why some equations are solvable and some are not. Solving an equation corresponds to "simplifying" a group by forming quotients, so knowing which equations are *not* solvable depends on knowing which groups cannot be "simplified." These are the so-called *simple* groups.

The groups associated with polynomial equations are finite, so one would like to classify the finite simple groups. Galois found one infinite family of such groups—the alternating groups A_n for $n \geq 5$—and three other provocative examples that we now view as the symmetry groups of finite projective lines.

However, classification of the finite simple groups was much harder than could have been foreseen in the 19th century. It turned out to be easier (though still very hard) to classify *continuous* simple groups. This was done by Lie, Killing, and Cartan in the 1880s and 1890s. Each continuous simple group is the symmetry group of a space with hypercomplex coordinates, either from $\mathbb{R}, \mathbb{C}, \mathbb{H}$, or $\mathbb{O}$.

While this classification was in progress, it was noticed that a single continuous simple group can yield infinitely many finite simple groups, obtained by replacing the hypercomplex number system by a finite field. These "finite groups of Lie type" were completely worked out by 1960. Together with the alternating groups and the cyclic groups of prime order, they account for all but finitely many of the finite simple groups.

But identifying all the exceptions—the 26 *sporadic* simple groups— turned out to be the hardest problem of all ...

J. Stillwell, *Mathematics and Its History*, Undergraduate Texts in Mathematics,   495
DOI 10.1007/978-1-4419-6053-5_23, © Springer Science+Business Media, LLC 2010

23.1   Finite Simple Groups and Finite Fields

In Section 19.8 we introduced finite simple groups with the examples of A_5, PSL(2, 5), PSL(2, 7), and PSL(2, 11) discovered by Galois. Our brief account may perhaps have given the impression that Galois and Jordan between them grasped the connections between simple groups, finite fields, and finite geometries. This is hardly possible. These ideas took more than 100 years to unfold, and even the concept of a projective geometry over a finite field did not come fully to light until 1906. In this section we fill in the conceptual parts of the story, up to the discovery of finite geometries, in order to give a better idea of their scope and depth. They center on the concept of a *linear* group, which came to maturity in the book *Linear Groups, with an Exposition of the Galois Field Theory* of Dickson (1901a).

Today, it is easy for us to define a linear group as a group of matrices with entries in a field. Matrices were introduced by Cayley (1855), but they were not commonly used as group elements until the 20th century, perhaps because groups were originally permutation groups, and for a long time this was considered the proper way to represent them. An intermediate stage, typified by Dickson (1901a), was to allow group elements to be linear transformations, defined by linear equations.

The concept of finite field goes back to Galois, as we have seen, and indeed Galois discovered more than the finite fields $\mathbb{F}_p$ mentioned in Section 19.8 and used by Jordan (1870) to define linear groups. Along with the field $\mathbb{F}_p$ whose elements are $0, 1, 2, \ldots, p - 1$ under addition and multiplication mod p, there is field $\mathbb{F}_{p^n}$, for each natural number n, whose elements are polynomials of degree $n - 1$ with coefficients in $\mathbb{F}_p$. For example, the elements of the field $\mathbb{F}_4$ are $0, 1, x$ and $x + 1$ under the obvious mod 2 addition and multiplication, with the additional rule that $x^2 + x + 1 = 0$.

It follows that there are fields with 4, 8, and 9 elements, because $4 = 2^2$, $8 = 2^3$, and $9 = 3^2$. Transformations of the projective lines over these fields give us a new simple group, not noticed by Jordan, and two old ones:

- PSL(2, 4) = PGL(2, 4) has $5 \cdot 4 \cdot 3 = 60$ elements, and it happens to be isomorphic to A_5.

- PSL(2, 9) has $10 \cdot 9 \cdot 8/2 = 360$ elements, and it happens to be isomorphic to A_6.

- PSL(2, 8) = PGL(2, 8) has $9 \cdot 8 \cdot 7 = 504$ elements, and it is a new simple group, discovered by Cole (1893).

Cole's construction of a simple group from a field with a nonprime number of elements sparked the investigation of Moore (1893), showing that every finite field is isomorphic to one of the Galois fields $\mathbb{F}_{p^n}$, and that all the groups $PSL(2, p^n)$ are simple when $p > 3$ and $n > 1$. In fact, all the groups $PSL(2, p^n)$ are simple when $p^n > 3$, and indeed $PSL(m, q)$ is simple for all $m, q \geq 2$ except $(m, q) = (2, 2)$ and $(2, 3)$. This was proved by Dickson (1901a), using a definition of $PSL(m, q)$ as a certain group of linear transformations in m variables. Today, we define $PSL(m, q)$ as the group of $m \times m$ matrices, with entries in $\mathbb{F}_q$, and determinant 1, quotiented by the subgroup consisting of the identity matrix and its negative.

At this point in history, linear groups were better understood than the spaces they "transform." Geometry caught up in 1905 when Veblen defined the m-dimensional projective space over the field $\mathbb{F}_q$ (see Veblen and Bussey (1906)). And projective geometry was not the only kind of geometry that could be "finitized" by replacing the underlying number line by a finite field. Groups like the "rotation group of $\mathbb{R}^n$" were also found to have finite counterparts, which typically were simple groups. It seemed then, that further progress in the investigation of finite simple groups depended on a better understanding of *continuous groups*, such as groups of rotations. We take up this viewpoint in Section 23.3. It turns out to be amazingly fruitful, but it does not account for *all* finite simple groups.

Even before continuous groups were seen to yield many finite groups, five mysterious finite simple groups emerged from nowhere in the 1860s. They are now known as the *Mathieu groups*, after their discoverer Émile Mathieu. They do *not* arise as finite-field analogues of continuous groups. 100 years were to elapse before their exceptional nature was realized.

EXERCISES

We investigate $\mathbb{F}_4$ by naming its elements $0, 1, x, x + 1$ and finding their addition and multiplication tables. These enable us to determine the basic linear fractional functions $y \mapsto y + l$, $y \mapsto ky$ for $k \neq 0$, and $y \mapsto 1/y$ on the projective line $\mathbb{F}_4 \cup \{\infty\}$.

23.1.1 Check that the elements of $\mathbb{F}_4$ have the following addition and multiplication tables (omitting the obvious results for multiplication by zero):

+	0	1	x	$x + 1$
0	0	1	x	$x + 1$
1	1	0	$x + 1$	x
x	x	$x + 1$	0	1
$x + 1$	$x + 1$	x	1	0

×	1	x	$x + 1$
1	1	x	$x + 1$
x	x	$x + 1$	1
$x + 1$	$x + 1$	1	x

23.1.2 With the help of these tables, or otherwise, show that the functions $y \mapsto y + 1$, $y \mapsto ky$ for $k \neq 0$, and $y \mapsto 1/y$ on $\mathbb{F}_4 \cup \{\infty\}$ are all even permutations.

23.1.3 Deduce from Exercise 23.1.2 that $PGL(2, 4) = PSL(2, 4)$ has $5 \cdot 4 \cdot 3 = 60$ elements. Why does this imply that it is isomorphic to A_5?

In the following exercises it may be helpful to look again at Section 19.8 for the geometric meaning of $PGL(2, 7)$ and $PGL(2, 11)$.

23.1.4 Find odd permutations in $PGL(2, 7)$ and $PGL(2, 11)$.

23.1.5 Deduce that the subgroup of even permutations in $PGL(2, 7)$ has 168 elements and that the subgroup of even permutations in $PGL(2, 11)$ has 660 elements.

23.1.6 Show that no symmetric groups have size 336 or 1320.

23.2   The Mathieu Groups

Back in the middle of the 19th century, when all groups were viewed as permutation groups, a burning question was how "transitive" a group G may be. If the elements of G permute a certain set S, then G is called 1-*transitive* if any member of S may be sent to any other member of S by a permutation in G, 2-*transitive* if any ordered pair of members of S may be sent to any other ordered pair of members of S by a member of G, and so on.

The symmetric group S_n, whose members are all the permutations of $\{1, 2, \ldots, n\}$, is k-transitive for each $k \leq n$. This is clear, because any k-tuple $(a_1, a_2, \ldots, a_k)$ of numbers $\leq n$ may be sent to any other such k-tuple, $(b_1, b_2, \ldots, b_k)$, by a permutation σ of $\{1, 2, \ldots, n\}$ such that

$$\sigma(a_1) = b_1, \quad \sigma(a_2) = b_2, \quad \ldots, \quad \sigma(a_k) = b_k$$

The alternating group A_n is also k-transitive, for *odd* numbers $k \leq n$, as may be easily proved (exercise).

But except for these obvious examples, highly transitive groups are hard to find. The best result in this direction was discovered by Mathieu (1861, 1873), who found four permutation groups that are 4- or 5-transitive, and a related group that is 3-transitive. Remarkably, the Mathieu groups are also simple, which earns them their place in this chapter. They were the first simple groups to be discovered outside the infinite families of alternating and projective groups.

The five Mathieu groups are called $M_{11}, M_{12}, M_{22}, M_{23}$, and M_{24}—the subscript denoting the number of objects being permuted. The transitivity and *order* (number of elements) of each is given in the following table.

Group	Transitivity	Order
M_{11}	4	$11 \cdot 10 \cdot 9 \cdot 8$
M_{12}	5	$12 \cdot 11 \cdot 10 \cdot 9 \cdot 8$
M_{22}	3	$22 \cdot 21 \cdot 20 \cdot 16 \cdot 3$
M_{23}	4	$23 \cdot 22 \cdot 21 \cdot 20 \cdot 16 \cdot 3$
M_{24}	5	$24 \cdot 23 \cdot 22 \cdot 21 \cdot 20 \cdot 16 \cdot 3$

We now know that the Mathieu groups M_{11}, M_{12}, M_{23} and M_{24} are the *only* 4- and 5-transitive finite groups, other than S_n and A_n. This became known for certain only after all the finite simple groups had been found, in the 1980s. However, the existence of these extreme objects implies the existence of extreme objects in other parts of mathematics, some of which were observed independently. Perhaps the most spectacular was a manifestation of M_{23} and M_{24} known as the *Golay code*.

Coding theory developed in the 1940s, in response to mathematical problems of "communication in the presence of noise." Most communication suffers from errors due to noise, so the question arises, what is the best way to encode messages so that errors can be detected and corrected?

Typically, a message is broken into "characters," which are sequences of 0s and 1s (*binary* sequences) of a certain fixed length k. The sequences chosen to be characters make up the *code*. It is crucial that not all sequences of length k belong to the code, so that a (not too large) number of erroneous digits will produce a k-digit sequence *not* in the code, thus revealing that an error has occurred. Moreover, if any two characters in the code differ in, say, d or more digits then we can *correct* a k-digit sequence σ with fewer than $d/2$ errors by replacing it by the (unique) sequence τ in the code that differs for σ in fewer than $d/2$ digits.

Obviously, one can detect and correct errors by sending each character twice, or three times, but this greatly increases the length of the message. The goal of coding theory is to attain the maximum amount of error correction with the minimum increase in the length of the message. That is, for a given d, and a given number of characters, one wants k as small as possible. For example, for $d = 3$ it is known that $k = 7$ is the minimum length one can use to get 16 characters, and this is achieved by the following code:

```
0000000    0100101    1000110    1100011
0001111    0101010    1001001    1101100
0010011    0110110    1010101    1110000
0011101    0111001    1011010    1111111
```

This code is due to Hamming (1950), and it is known as the *Hamming* (7, 4) *code*. An even more remarkable code is the amazing (23, 12) code of Golay (1949). The *Golay code* consists of $2^{12} = 4096$ binary sequences of length 23, any two of which differ in at least seven digits (so that three erroneous digits can be corrected).

If we view 0 and 1 as elements of the field $\mathbb{F}_2$, then the Golay code is a highly symmetric set of 4096 points in the 23-dimensional space $\mathbb{F}_2^{23}$. The symmetries of the code can be realized by a group of linear transformations of the space $\mathbb{F}_2^{23}$, and this symmetry group turns out to none other than M_{23}. The group M_{24} turns up nearby, as the symmetry group of a related subset of $\mathbb{F}_2^{24}$, the so-called *extended Golay code* consisting of 4096 binary sequences of length 24, any two of which differ in at least seven digits. These discoveries are due to Paige (1957) and Assmus and Mattson (1966).

The connection with coding theory led to renewed interest in the Mathieu groups, culminating in the discovery of new finite simple groups in the 1960s, as we will see. But first it was necessary to attain a better understanding of the "old" simple groups, and their relation to the so-called "continuous" groups.

EXERCISES

Transitivity is an important feature in projective geometry, where any three points may be projected to any three points on a projective line, as Figure 23.1 suggests. It follows that the group of linear fractional transformations of $\mathbb{RP}^1$ is 3-transitive.

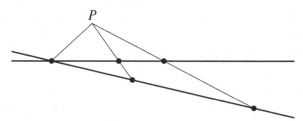

Figure 23.1: Projecting three points to three points

23.2.1 Verify this result algebraically, and explain why the argument is valid for the finite projective groups PGL(2, q).

23.2.2 Deduce from Exercise 23.2.1 that the subgroup PSL$(2, q)$ of even permutations in PGL$(2, q)$ is 3-transitive.

The first two Mathieu groups are called *sharply transitive*, because of the following property.

23.2.3 Show that M_{11} has the smallest possible size for a 4-transitive group that permutes 11 objects, and that M_{12} has the smallest possible size for a 5-transitive group that permutes 12 objects.

The Hamming $(7, 4)$ code and the Golay $(23, 12)$ code are known as *perfect* codes, for the following reasons. The "one-error-correcting" property of the Hamming code means that each element of the 7-dimensional space $\mathbb{F}_2^7$ *not* in the code differs from a unique code member in one or more digits. What makes the Hamming code perfect is that, in fact, each element of $\mathbb{F}_2^7$ not in the code differs from a unique code member in *exactly* one digit. To put this another way: if we define the 1-*neighborhood* of a code member τ to consist of those σ in $\mathbb{F}_2^7$ that differ from τ in at most one digit, then the 16 1-neighborhoods of code members fill the space $\mathbb{F}_2^7$ without overlapping.

23.2.4 Show that the 1-neighborhood of any member of the Hamming code has eight elements, and deduce that the Hamming code is perfect.

We similarly define the 3-neighborhood of a member τ of the Golay $(23, 12)$ code to consist of those σ in $\mathbb{F}_2^{23}$ that differ from τ in at most three digits.

23.2.5 Show that the 3-neighborhood of any member of the Golay $(23, 12)$ code has 2^{11} elements, and deduce that the Golay $(23, 12)$ code is perfect.

23.3 Continuous Groups

The theory of continuous groups was created by the Norwegian mathematician Sophus Lie in the 1870s. Initially, his goal was to develop a theory of differential equations like the Galois theory of polynomial equations. He saw that each differential equation has a group, analogous to the Galois group but "continuous" rather than finite, and that "simple" groups present an obstacle to solvability. Thus his attention quickly shifted to the problem of classifying continuous groups and (particularly) identifying the simple groups among them.

The definition of a "continuous group"—or what we now call a *Lie group*—is somewhat subtle, as is the definition of "simple" for these groups. Here we are content merely to give a few examples, and to prove simplicity of one of them. For a more thorough, but still elementary, account see Stillwell (2008).

The most easily understood example of a continuous group is the number line $\mathbb{R}$, under the operation of addition. This group is "continuous" in the sense that the group operation $x, y \mapsto x + y$, and also the group inverse operation $x \mapsto -x$, is a continuous function. A related example is the unit circle

$$\mathbb{S}^1 = \{z : |z| = 1\}$$

in the plane of complex numbers, under the (obviously continuous) operation of multiplication of complex numbers. $\mathbb{S}^1$ is also called SO(2), the first in a family called the *special orthogonal*, or *rotation*, groups. We can interpret a member z of SO(2) as a *rotation of the plane*, because

$$z = \cos \theta + i \sin \theta, \quad \text{for some } \theta,$$

and multiplying each complex number by z rotates the plane $\mathbb{C}$ about O through angle θ. Thus the group operation in SO(2) can also be viewed as addition of angles, which is another way to see that SO(2) is continuous.

Both $\mathbb{R}$ and SO(2) are abelian groups, so they are not very interesting.

The first really interesting continuous group is SO(3), the rotation group of the three-dimensional space $\mathbb{R}^3$. If we take a rotation r of $\mathbb{R}^3$ to be given by an *axis* $\mathcal{A}$ through O and a turn of angle θ about $\mathcal{A}$, then it is not even obvious that the space rotations form a group. Given a rotation r with axis $\mathcal{A}$ and angle θ, and a rotation s with axis $\mathcal{B}$ and angle φ, can we be sure that the combination sr even has a well-defined axis C and angle χ? The answer (yes) was apparently first found by Euler (1776), but we can now find this answer much more easily. The trick is to *view each rotation as a product of two reflections*, as shown in Figure 23.2.

The left picture in the figure shows a pair of lines in the plane, $\mathcal{L}$ and $\mathcal{M}$, meeting at O at angle $\theta/2$. If a point X is reflected in $\mathcal{L}$ (to X'), then in $\mathcal{M}$ (to X''), the angle between X and X'' is clearly θ . More generally, it is clear that a rotation about any point Y through angle θ can be realized by successive reflections in any two lines through Y meeting at angle $\theta/2$ (measured in the appropriate sense). The same is true for rotation of a sphere, and hence of $\mathbb{R}^3$, about any axis. To rotate through angle θ about the axis YY' (right picture) it suffices to reflect in any two great circles through Y and Y' that meet at angle $\theta/2$. Equivalently, one reflects the sphere in any two planes that meet along the line YY' at angle $\theta/2$.

Now suppose that we want to find the result of performing rotation r of the sphere, with axis through P and angle θ, then rotation s with axis

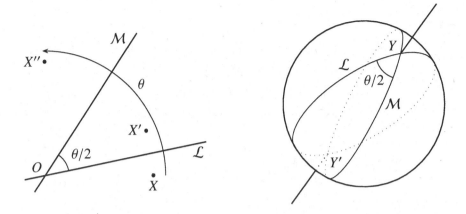

Figure 23.2: Rotation via a pair of reflections

through Q and angle φ. Making use of our freedom to choose the great circles of reflection, we realize r by the pair of reflections in the great circles
$\mathcal{L}$ and $\mathcal{M}$ through P that are angle $\theta/2$ apart, *where $\mathcal{M}$ passes through P
and Q* (Figure 23.3).

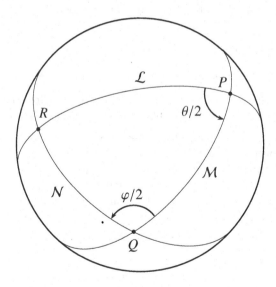

Figure 23.3: Finding the product of rotations

Then we realize s by the pair of reflections in $\mathcal{M}$ and $\mathcal{N}$ through Q that
are angle $\varphi/2$ apart. It follows, since successive reflections in $\mathcal{M}$ cancel,

that

$$sr = \text{reflection in } \mathcal{L} \text{ then reflection in } \mathcal{M} \text{ then}$$
$$\text{reflection in } \mathcal{M} \text{ then reflection in } \mathcal{N}$$
$$= \text{reflection in } \mathcal{L} \text{ then reflection in } \mathcal{N}$$
$$= \text{rotation about axis } RR' \text{ through angle } \chi,$$

where R is the third vertex, and $\chi/2$ is the third angle, in the spherical triangle formed by the great circles $\mathcal{L}$, $\mathcal{M}$, and $\mathcal{N}$.

Thus the product of rotations is a rotation. As always, this "product" operation is associative, because it is the product of functions. It is also clear that the inverse of a rotation is a rotation (same axis, negative of the angle), so the rotations form a group under the product operation. Finally, it is intuitively clear that the product and inverse depend continuously on the position of the axis and the angle of rotation. So this group SO(3) is continuous. In the next section we will see that continuity is crucial in proving that SO(3) is a simple group.

EXERCISES

Another important Lie group, closely related to SO(3), is the group of unit quaternions

$$q = a + b\mathbf{i} + c\mathbf{j} + d\mathbf{k}, \quad \text{where} \quad a^2 + b^2 + c^2 + d^2 = 1,$$

under the operation of quaternion multiplication.

23.3.1 Use the properties of quaternion norm and inverse, from the exercises to Section 20.5, to show that the unit quaternions form a continuous group.

As a geometric object, the group of unit quaternions is known as the 3-*sphere* $\mathbb{S}^3$, because it consists of the points at unit distance from O in $\mathbb{R}^4$. As a group, it is known as SU(2), where SU stands for "special unitary" and the 2 is there because each quaternion may be viewed as a 2×2 complex matrix, as we saw in Section 20.5. There we saw a rather roundabout connection between quaternions and rotations of the sphere. Here is a more direct one, discovered by Cayley (1845a).

A unit quaternion q effects a *rotation* of the 3-dimensional space $\mathbb{R}\mathbf{i} + \mathbb{R}\mathbf{j} + \mathbb{R}\mathbf{k}$ of pure imaginary quaternions

$$p = b\mathbf{i} + c\mathbf{j} + d\mathbf{k}$$

by sending each such p to $q^{-1}pq$. This fact can be verified as follows.

23.3.2 Show that each unit quaternion may be written uniquely in the form $q = \cos\theta + u\sin\theta$, for some angle θ and some pure imaginary quaternion u of unit length.

23.3.3 Show that, in the notation of Exercise 23.3.2, $q^{-1} = \cos\theta - u\sin\theta$. Deduce that $q^{-1}aq = a$ for any real a, and $q^{-1}uq = u$.

23.3.4 Using the multiplicative norm property $|q_1 q_2| = |q_1||q_2|$ from the exercises in Section 20.5, and the fact that $|r - s|$ is the distance between any two quaternions r and s, show that multiplication by any unit quaternion q preserves distance.

It follows that $p \mapsto q^{-1}pq$ maps the 3-dimensional space $\mathbb{R}\mathbf{i} + \mathbb{R}\mathbf{j} + \mathbb{R}\mathbf{k}$ onto itself and fixes the line of real multiples of u. This map is in fact a rotation of $\mathbb{R}\mathbf{i} + \mathbb{R}\mathbf{j} + \mathbb{R}\mathbf{k}$ with axis u. We can verify this fact, and at the same time find the angle of rotation, by observing what the map $p \mapsto q^{-1}pq$ does to a pair of perpendicular unit vectors v, w in $\mathbb{R}\mathbf{i} + \mathbb{R}\mathbf{j} + \mathbb{R}\mathbf{k}$ that are also perpendicular to u.

23.3.5 Explain, using Exercise 20.5.4, why we can assume that $uv = w$, $vw = u$, $wu = v$, $uv = -vu$, $vw = -wv$, and $wu = -uw$.

23.3.6 Verify that $q^{-1}vq = v\cos 2\theta - w\sin 2\theta$ and $q^{-1}wq = v\sin 2\theta + w\cos 2\theta$, and deduce that $p \mapsto q^{-1}pq$ is a rotation through angle 2θ.

Thus the rotation with axis u and angle φ may be effected by the map $p \mapsto q^{-1}pq$, where $q = \cos\frac{\varphi}{2} + u\sin\frac{\varphi}{2}$. Also, to follow it by the rotation about axis u' with angle φ' we follow the map by $p \mapsto q'^{-1}pq'$, where $q' = \cos\frac{\varphi'}{2} + u'\sin\frac{\varphi'}{2}$. The result is the map $p \mapsto (qq')^{-1}p(qq')$, so *the product of rotations corresponds to the product of quaternions.*

It seems that SU(2) is pretty much the same group as SO(3). But not quite. *Two* unit quaternions correspond to each space rotation. If q effects some rotation, so does $-q$, because $(-q)^{-1}p(-q) = q^{-1}pq$. Thus each element of SO(3) actually corresponds to an *antipodal point pair* $\pm q$ on the 3-sphere of unit quaternions. Sound familiar? If you recall from Section 8.5 that the "points" of the projective plane $\mathbb{RP}^2$ are antipodal point pairs on the ordinary sphere, then it will be clear that antipodal point pairs $\pm q$ on the 3-sphere should be "points" of the *projective space $\mathbb{RP}^3$*. Thus SO(3), as a geometric object, is none other than $\mathbb{RP}^3$.

23.4   Simplicity of SO(3)

To prove that SO(3) is simple we consider a normal subgroup $H \neq \{1\}$, and aim to show that $H = $ SO(3). Since H is normal, $gH = Hg$, and hence $gHg^{-1} = H$, for each g in SO(3). In other words, ghg^{-1} is in H for each g in SO(3) and each h in H. This enables us to build many elements of H from one nontrivial element h, and in fact we can build all elements of SO(3). We build them in three steps, starting with a specific h with axis $\mathcal{A}$ and nonzero angle θ:

Step 1. *H includes the rotation through angle θ about any axis ℬ.*

To see why, let g be a rotation that moves axis $\mathcal{A}$ to axis $\mathcal{B}$. Then ghg^{-1} is the rotation through angle θ about axis $\mathcal{B}$ because

- g^{-1} moves axis $\mathcal{B}$ to the position of axis $\mathcal{A}$.
- h rotates $\mathbb{R}^3$ about axis $\mathcal{A}$ through angle θ.
- g moves the axis at position $\mathcal{A}$ back to its original position $\mathcal{B}$.

Step 2. *H includes rotations through all angles in an interval between some α and β, α < β.*

As we know from the previous section, the product of a rotation r about axis PP' through angle θ and a rotation s about axis QQ' through angle θ is a rotation about axis RR' through angle χ, where R and $\chi/2$ are as shown in Figure 23.4.

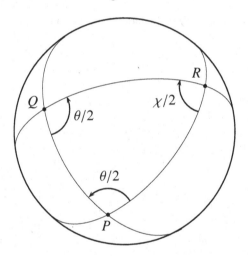

Figure 23.4: Angle of the product rotation

Now suppose that P is fixed and that Q is allowed to vary continuously along a fixed great circle through P. When Q is near P, so is R; hence the triangle PQR is almost Euclidean and its angle sum is close to π. It follows that $\chi/2$ is close to $\pi - \theta$. As P moves farther away, the spherical triangle PQR becomes larger, hence so does its angle sum by Section 17.6, so $\chi/2$ becomes larger. Since $\chi/2$ varies continuously with the position of Q, it necessarily takes all values in an interval between some α and β, where $\alpha < \beta$.

Step 3. *H includes rotations through any angle.*

Since $\alpha < \beta$, the interval between α and β includes a subinterval of the form $[2m\pi/n, 2(m+1)\pi/n]$, for some integers m and n. Thus it follows from Step 2 and Step 1 that H includes all rotations, about some fixed axis $\mathcal{B}$, with angles between $2m\pi/n$ and $2(m+1)\pi/n$. Of course, H also includes all products of its members, and hence all their nth powers—which multiply angles by n. But if we multiply the angles between $2m\pi/n$ and $2(m+1)\pi/n$ by n we get *all* angles, as required.

Applying Step 1 again, we get rotations with all angles and all axes, so H includes all elements of SO(3), as claimed. □

Lie observed the simplicity of many Lie groups, including SO(3), but his concepts of "group" and "simplicity" were somewhat different from ours. In his view, a "group" included "infinitesimal elements," and he used these to determine simplicity. Today, we call the "infinitesimal elements" of a Lie group its *tangent vectors at the identity*, and we build from them a separate algebraic structure called the *Lie algebra* of the Lie group. A Lie algebra has a "product" operation, called the *Lie bracket*, that is quite different from a group operation; for example, the Lie bracket is not associative.

Nevertheless, it is a good idea to look at Lie algebras. There is a natural concept of simplicity for Lie algebras, such that a simple Lie group has a simple Lie algebra, and testing simplicity is somewhat easier for algebras than for groups. The downside, if there is one, is that a simple Lie algebra may *not* come from a simple Lie group. For example, the group SU(2) from the previous exercise set has the same Lie algebra as SO(3), so the Lie algebra of SU(2) is simple. However, the group SU(2) is *not* simple; it has a normal subgroup with the elements 1 and -1. The problem is that the Lie algebra cannot "detect" group elements that are far from the identity element, so it can miss a normal subgroup whose nonidentity members are all far away (which is the case for SU(2)). This is not necessarily a bad thing, and in fact many authors *define* a Lie group to be simple if its Lie algebra is simple.

EXERCISES

For an easily visualized Lie group, such as the circle SO(2) = $\{z : |z| = 1\}$, the tangent space at the identity is also easily visualized, in this case as the vertical line through 1. To get at the tangent space for a higher-dimensional Lie group, such as

SO(3) or SU(2), we need to view tangent vectors from "inside" the group. We do so by taking a tangent vector to be the *velocity* vector of a moving point as it passes through the identity element 1. In the case of SO(2), we can see that the velocity interpretation gives the same result as the geometric tangent interpretation—any point moving around the circle has velocity in the tangential direction.

For groups such as SU(2), the velocity interpretation allows us to *calculate* velocity vectors by differentiation. Here is how. We suppose that $q(t)$ gives the position of a point in SU(2) as a (differentiable) function of t, with $q(0) = 1$ for convenience. Thus the velocity of the moving point as it passes through 1 is $q'(0)$.

23.4.1 Explain why the condition for $q(t)$ to lie in SU(2) is $q(t)\overline{q(t)} = 1$.

23.4.2 By differentiating the equation $q(t)\overline{q(t)} = 1$, then setting $t = 0$, deduce that each tangent vector $q'(0)$ satisfies

$$q'(0) + \overline{q'(0)} = 0,$$

which implies that $q'(0)$ is a pure imaginary quaternion (why?).

Conversely, we can show that each pure imaginary quaternion is a tangent vector to SU(2).

23.4.3 By choosing p so that $q(t) = \cos \theta t + p \sin \theta t$ is a path in SU(2), show that each pure imaginary quaternion is a tangent vector to SU(2) at the identity.

Thus the tangent space of SU(2) at the identity is the space $\mathbb{R}\mathbf{i} + \mathbb{R}\mathbf{j} + \mathbb{R}\mathbf{k}$ of pure imaginary quaternions. So far, we have a nice parallel with SO(2), whose tangent space at the identity is a vertical line, that is, a line in the imaginary direction. However, the vectors in a line have very little algebraic potential—they can only be added and multiplied by real numbers. The vectors in the tangent space to SU(2), on the other hand, also have an interesting *Lie bracket* operation that reflects the *conjugation* operation $u, v \mapsto uvu^{-1}$ in the group SU(2).

To see how conjugation in SU(2) plays out in the tangent space, we consider two paths, $u(s)$ and $v(t)$, through 1 in SU(2), with $u(0) = v(0) = 1$. By differentiating the path

$$w_s(t) = u(s)v(t)u(s)^{-1}$$

with respect to t we get a path

$$w'_s(0) = u(s)Vu(s)^{-1} = x(s)$$

in the tangent space $\mathbb{R}\mathbf{i} + \mathbb{R}\mathbf{j} + \mathbb{R}\mathbf{k}$, where $V = v'(0)$ is the tangent vector to $v(t)$ at the identity.

23.4.4 Explain why the tangent vector $x'(0)$ of the path $x(s)$ is also a member of $\mathbb{R}\mathbf{i} + \mathbb{R}\mathbf{j} + \mathbb{R}\mathbf{k}$, and show by differentiation that

$$x'(0) = UV - VU,$$

where $U = u'(0)$ is the tangent vector to $u(s)$ at the identity.

Thus we have an operation $U, V \mapsto UV - VU$ on the tangent space that reflects conjugation in the group. $UV - VU$ is called the *Lie bracket* of U and V and is often written $[U, V]$.

23.4.5 Calculate $[U, V]$ for $U, V = \mathbf{i}, \mathbf{j}, \mathbf{k}$ and hence show directly that $[U, V]$ is in $\mathbb{R}\mathbf{i} + \mathbb{R}\mathbf{j} + \mathbb{R}\mathbf{k}$ for any U and V in $\mathbb{R}\mathbf{i} + \mathbb{R}\mathbf{j} + \mathbb{R}\mathbf{k}$.

23.4.6 Show that, if $\mathbf{i}' = \mathbf{i}/2$, $\mathbf{j}' = \mathbf{j}/2$, and $\mathbf{k}' = \mathbf{k}/2$, then the Lie bracket operation on $\mathbf{i}', \mathbf{j}', \mathbf{k}'$ satisfies exactly the same relations as the vector product operation on $\mathbf{i}, \mathbf{j}, \mathbf{k}$.

Thus the *Lie algebra* of SU(2), which is what we call the tangent space of SU(2) under its Lie bracket operation, is essentially the same as $\mathbb{R}^3$ with the vector product operation. It is nice to discover that this Lie algebra is something we already know.

23.4.7 Why do SO(3) and SU(2) have the same Lie algebra?

23.5 Simple Lie Groups and Lie Algebras

The rotation group SO(3) is the prototype for many other simple Lie groups, obtained by generalizing the concept of "rotation" in two ways. We can generalize from rotations of $\mathbb{R}^3$ to rotations of $\mathbb{R}^n$, and we can replace $\mathbb{R}$ by $\mathbb{C}$ or $\mathbb{H}$.

A *rotation of* $\mathbb{R}^n$ is defined to be a linear transformation of $\mathbb{R}^n$ that preserves length and orientation. If we denote a linear transformation of $\mathbb{R}^n$ by its matrix A, then it turns out that

$$A \text{ preserves length} \Leftrightarrow AA^{\mathrm{T}} = \mathbf{1},$$

where A^{T} denotes the transpose of A and $\mathbf{1}$ denotes the identity matrix. It follows, by taking the determinant of this equation, that $\det A = \pm 1$. Such matrices are called *orthogonal*, and those with determinant 1 are called *special orthogonal*. The latter are the matrices that preserve orientation, whence the name *special orthogonal group* SO(n) for the rotations of $\mathbb{R}^n$.

The Lie algebra of SO(n) is denoted by $\mathfrak{so}(n)$, and its members are matrices of the form $A'(0)$, where $A(t)$ is a smooth path of matrices through the identity in SO(n). It follows from the definition of SO(n) that

$$A(t)A(t)^{\mathrm{T}} = \mathbf{1}.$$

Differentiating this equation with respect to t and using $A(0) = \mathbf{1}$, one finds (exercise) that

$$A'(0) + A'(0)^{\mathrm{T}} = \mathbf{0},$$

where **0** is the zero matrix. This means that each matrix $A'(0)$ in $\mathfrak{so}(n)$ is *skew-symmetric*, that is, the entry in row i and column j is the negative of the entry in row j and column i. In particular, all the diagonal entries are zero.

Thus the Lie algebra $\mathfrak{so}(n)$ consists of $n \times n$ skew-symmetric matrices (in fact all of them, as we show for the case $n = 3$ in the exercises). The Lie bracket operation that reflects conjugation in $SO(n)$ is $[U, V] = UV - VU$, by the same argument that we used for $SU(2)$ in the previous section— except with matrices U, V instead of quaternions.

All of the Lie algebras $\mathfrak{so}(n)$, for $n \geq 3$, are simple, with the curious exception of $\mathfrak{so}(4)$. These results were discovered by Lie in the 1880s, though in the different language of "infinitesimal" transformations that we mentioned in the previous section.

Simple Lie algebras reflect the simplicity of the corresponding groups, or nearly so. $SO(4)$ is definitely *not* simple; $SO(3), SO(5), SO(7), \ldots$ are simple, while $SO(6), SO(8), SO(10), \ldots$ are *almost* simple in the sense that the only normal subgroup of each is $\{\mathbf{1}, -\mathbf{1}\}$.

When $\mathbb{R}$ is replaced by $\mathbb{C}$ there is an analogous group of "rotations" of $\mathbb{C}^n$ called the *special unitary group* $SU(n)$. Finally, there is an analogous group of "rotations" of $\mathbb{H}^n$ called the *symplectic group* $Sp(n)$. The Lie algebras of these generalized rotation groups are denoted by $\mathfrak{su}(n)$ and $\mathfrak{sp}(n)$ respectively. Lie also found them to be simple, so the corresponding groups $SU(n)$ and $Sp(n)$ are "almost simple." For each of these groups, the largest normal subgroup is finite, so its nonidentity elements are not near the identity, and hence they are not detected by the Lie algebra.

It is a remarkable fact that the simple Lie algebras found by Lie include *all but five* of the simple Lie algebras in existence. The five exceptions were discovered by Killing (1888) and Cartan (1894). (Cartan showed that two Lie algebras that Killing believed to be distinct are actually identical, and he filled some gaps in Killing's proofs.) These *exceptional Lie algebras*, and the corresponding groups, had not been suspected by Lie, and Killing's discovery of them is considered by many to be one of the greatest mathematical achievements of all time. (See, for example, the paper of Coleman (1989).) Certainly, it was one of the greatest mathematical achievements of the 19th century.

The five exceptional Lie algebras are called G_2, F_4, E_6, E_7, and E_8, and these names are also given, somewhat loosely, to the corresponding groups. The algebras G_2, F_4, E_6, E_7 and E_8 are of dimension 14, 52, 78, 133, and

248 respectively. In an extraordinarily apt way they extend Lie's classical families of Lie algebras arising from spaces with $\mathbb{R}$, $\mathbb{C}$, and $\mathbb{H}$ coordinates, because they arise from the octonions $\mathbb{O}$. The classical families are infinite because the classical spaces can have arbitrary dimension n but, as was mentioned in Chapter 20, $\mathbb{O}$ is different because it does not support a projective space of dimension $n \geq 3$. Indeed, all five exceptional Lie algebras and their groups are connected to $\mathbb{O}$ and the octonion projective plane $\mathbb{OP}^2$.

The first to point out a connection between exceptional Lie groups and $\mathbb{O}$ was Cartan (1908), who observed that the 14-dimensional group G_2 is the group of *automorphisms* of $\mathbb{O}$. Automorphisms are the invertible mappings φ of $\mathbb{O}$ onto itself such that

$$\varphi(u + v) = \varphi(u) + \varphi(v) \quad \text{and} \quad \varphi(uv) = \varphi(u)\varphi(v) \quad \text{for all } u, v \text{ in } \mathbb{O}.$$

(By way of precedent for this result, it is worth mentioning that SO(3) is the automorphism group of $\mathbb{H}$. Each automorphism of $\mathbb{H}$ is in fact a rotation of the space of pure imaginary quaternions.) The octonion projective plane $\mathbb{OP}^2$ has two natural groups of transformations: the group of transformations that preserve length, which happens to be F_4; and the group of transformations that preserve straight lines, which happens to be E_6. The groups E_7 and E_8 also arise from $\mathbb{OP}^2$, but in a way that is too complicated to describe here. An almost magical web of relationships between $\mathbb{O}$ and the exceptional Lie groups was worked out by Freudenthal (1951) and Tits (1956).

EXERCISES

As an example of matrix differentiation, consider the following path of matrices in SO(3):

$$A(t) = \begin{pmatrix} \cos t & -\sin t & 0 \\ \sin t & \cos t & 0 \\ 0 & 0 & 1 \end{pmatrix}.$$

23.5.1 Describe the space rotation represented by $A(t)$.

23.5.2 By computing $A'(0)$, show that

$$I = \begin{pmatrix} 0 & -1 & 0 \\ 1 & 0 & 0 \\ 0 & 0 & 0 \end{pmatrix} \quad \text{is in } \mathfrak{so}(3).$$

By finding other suitable matrices in SO(3), show also that $\mathfrak{so}(3)$ includes the matrices

$$J = \begin{pmatrix} 0 & 0 & -1 \\ 0 & 0 & 0 \\ 1 & 0 & 0 \end{pmatrix} \quad \text{and} \quad K = \begin{pmatrix} 0 & 0 & 0 \\ 0 & 0 & -1 \\ 0 & 1 & 0 \end{pmatrix}.$$

Now we apply a basic rule of differentiation, the product rule, which works even when products are noncommutative.

23.5.3 By imitating the usual proof of the product rule, show that

$$\frac{d}{dt}A(t)B(t) = A'(t)B(t) + A(t)B'(t).$$

23.5.4 Deduce from Exercise 23.5.3 that $A'(0)+A'(0)^{\mathrm{T}} = 0$ for a path $A(t)$ through the identity 1 in SO(3) with $A(0) = 1$.

23.5.5 Suppose that $A(t)$ and $B(t)$ are paths through the identity 1 in SO(3), with $A(0) = B(0) = 1$. Show, by differentiating $A(t)B(t)$ and $A(rt)$, that $\mathfrak{so}(3)$ is closed under sums and multiples by real numbers r.

It follows from Exercises 23.5.2 and 23.5.5 that $\mathfrak{so}(3)$ includes *all* the real skew-symmetric matrices

$$xI + yJ + zK = \begin{pmatrix} 0 & -x & -y \\ x & 0 & -z \\ y & z & 0 \end{pmatrix}.$$

Now we investigate the Lie bracket, $[U, V] = UV - VU$, of such matrices.

23.5.6 Verify that $[I, J] = K$, $[J, K] = I$, $[K, I] = J$, and hence explain why the matrices $xI + yJ + zK$, under the Lie bracket operation, behave the same as the vectors $x\mathbf{i} + y\mathbf{j} + z\mathbf{k}$ under the vector product.

This confirms the isomorphism between $\mathfrak{so}(3)$ and the vector product algebra, previously noticed in Exercise 23.4.7 via the isomorphism between $\mathfrak{su}(2)$ and $\mathfrak{so}(3)$. Now we use properties of the vector product to confirm that $\mathfrak{so}(3)$ is simple. But first, what should it mean for a Lie algebra to be "simple"?

It follows from the parallel between conjugation in a Lie group G and the Lie bracket in its Lie algebra $\mathfrak{g}$ that a normal subgroup H of G (which is closed under conjugation by all elements of G) corresponds to a subspace $\mathfrak{h}$ of $\mathfrak{g}$ that is closed under Lie brackets with all elements of $\mathfrak{g}$. Such a subspace is called an *ideal*, because it is roughly analogous to an ideal in a ring (Section 21.4). Continuing the parallel, we call a Lie algebra *simple* if its only ideals are {0} and itself.

It follows that a simple Lie group G always has a simple Lie algebra $\mathfrak{g}$. However, it is often easier to prove the simplicity of $\mathfrak{g}$ directly. This is the case for $\mathfrak{so}(3)$, when we view it as the vector product algebra.

23.5.7 Suppose that $\mathfrak{I}$ is an ideal of the vector product algebra $\mathbb{R}\mathbf{i} + \mathbb{R}\mathbf{j} + \mathbb{R}\mathbf{k}$ that includes a nonzero element $x\mathbf{i} + y\mathbf{j} + z\mathbf{k}$. By taking vector products with suitable elements of $\mathbb{R}\mathbf{i} + \mathbb{R}\mathbf{j} + \mathbb{R}\mathbf{k}$, show that $\mathfrak{I}$ also includes an element $r\mathbf{i}$ for some real $r \neq 0$.

23.5.8 Deduce from Exercises 23.5.5 and 23.5.7 that $\mathfrak{I} = \mathbb{R}\mathbf{i} + \mathbb{R}\mathbf{j} + \mathbb{R}\mathbf{k}$, so $\mathfrak{so}(3)$ is simple.

23.6 Finite Simple Groups Revisited

The classification of simple Lie algebras, by Lie, Killing, and Cartan, breathed new life into the search for finite simple groups. The idea of inserting finite fields in place of the fields, $\mathbb{R}$ and $\mathbb{C}$, previously known for projective groups, could now be applied to many other continuous groups. Dickson (1901a,b) found infinite families of new simple groups corresponding to the infinite families found by Lie, and also an infinite family corresponding to the exceptional group G_2. Four years later he also found an infinite family corresponding to E_6. Then there was a hiatus for nearly 50 years, before the story of these *finite groups of Lie type* was finally wrapped up in the 1950s, mostly through the work of Chevalley (1955).

The general picture of the finite simple groups at that time was quite similar to that of the simple Lie algebras, only larger. Most of the finite simple groups fell into infinite families: the cyclic groups of prime order, the alternating groups A_n for $n \geq 5$, and an infinite family for each simple Lie algebra (with members corresponding to finite fields). Since each exceptional Lie algebra spawns an infinite family of finite simple groups, any finite simple groups outside the infinite families are even *more* exceptional than the exceptional Lie algebras. They are called *sporadic* simple groups, following Burnside (1911), who referred to the Mathieu groups by that name. As of 1960, the Mathieu groups remained the only known sporadic groups, having stubbornly resisted classification for almost 100 years.

It turned out that 21 more sporadic groups remained to be discovered, and it took heroic efforts to bring them to light and to prove that no other finite simple groups exist.

Mathematicians began to realize the difficulty of existence questions for simple groups with the publication of Feit and Thompson (1963). Feit and Thompson gave a negative answer to a question raised by Burnside in 1911: is there a nonabelian simple group of odd order? Burnside's question was a natural one to ask, because all the known nonabelian simple groups had even order. (As we have seen, the first few have orders 60, 168, 360, 504, 660, ...) But, to answer it, Feit and Thompson had to devise a massively intimidating argument that filled 255 pages. This length of proof was unprecedented in group theory—and perhaps anywhere in mathematics— but it was only the beginning. Before the classification of finite simple groups was over, the experts had to digest proofs of over 1000 pages.

And alongside theorems like those of Feit and Thompson—asserting that broad classes of groups do *not* include simple groups—were equally grueling searches for the tiny places where simple groups were still allowed to exist, like flowers growing through cracks in the pavement.

The first finite simple groups to be discovered in nearly 100 years were found by Janko in 1965. Janko was trying to prove that there are *no* simple groups in a certain class, but his attempt ran into an obstacle that he eventually recognized as a new simple group—one of order 175,560—now called J_1. It was the first, and smallest, of four simple groups eventually discovered by Janko, and now called J_1, J_2, J_3, J_4.

Another remarkable family of simple groups was discovered by Conway in 1967. Conway's starting point was in the neighborhood of the Mathieu groups M_{23} and M_{24}, and the Golay codes. As we mentioned in Section 23.2, the extended Golay code is a highly symmetric subset of the space $\mathbb{F}_4^{24}$ whose points are 24-tuples with coordinates 0 or 1. The symmetry group of this set is M_{24}. We can also use M_{24} to construct a highly symmetric set of points with integer coordinates in the "ordinary" 24-dimensional space $\mathbb{R}^{24}$. This was done by Leech (1967), and his highly symmetric set is now called the *Leech lattice*. (A "lattice" in this context means a set of integer points such that the vector sum of any two points is also in the set.)

Leech tried to interest group theorists in the symmetries of his lattice, but with little success. Conway was not a group theorist, but he decided to set aside a little time on the Leech lattice each week, because his other work was not going anywhere. The very first night he sat down to work on the problem he made dramatic progress: he found the number

$$2^{22} \cdot 3^9 \cdot 5^4 \cdot 7^2 \cdot 11 \cdot 13 \cdot 23 = 8,315,553,613,086,720,000,$$

which he believed to be the order (or, possibly, twice the order) of a new simple group. With some trepidation, he phoned Thompson and told him the number, because Thompson's mastery of group theory was such that he could recognize a potential simple group from its order alone. Sure enough, Thompson phoned back in 20 minutes, confirming that Conway needed to halve the number to get a simple group, and adding that there were two other potential simple groups in its vicinity. Six hours later, Conway found the first group, and went to bed. Within a week, he had found the other two groups that Thompson had predicted.

The Leech lattice seems to be some kind of center of attraction in the universe of sporadic simple groups, with no fewer than 12 groups in its gravitational field—the five Mathieu groups, the three Conway groups, and four others. It reinforces the impression that extreme mathematical objects tend to cluster together. M_{24} is an extremely transitive group, the Golay code is an extremely efficient code, and the Leech lattice is an extremely dense *sphere-packing* in $\mathbb{R}^{24}$. That is, if one makes each point of the Leech lattice the center of a unit sphere, then the spheres just touch, and they fill $\mathbb{R}^{24}$ as densely as any lattice arrangement of spheres possibly can. The Leech lattice was long suspected to have this property, but it was proved only recently, by Cohn and Kumar (2004).

There are two excellent books that tell more of the Leech lattice story, from both the mathematical and human points of view: Thompson (1983) and Ronan (2006). The latter also takes the story of finite simple groups to its climax, with the discovery of the so-called *monster*, whose almost unbelievable properties are discussed in the next section.

23.7   The Monster

The largest sporadic group, and the last confirmed to exist, is a group now known as the Monster. Its existence was first suspected by Gerd Fischer in 1973, but the Monster lived in mathematical limbo until the 1980s, when enough became known about it to enable its explicit construction. The first important fact was its order, found in 1974 by Conway to be

$$2^{46} \cdot 3^{20} \cdot 5^9 \cdot 7^6 \cdot 11^2 \cdot 13^3 \cdot 17 \cdot 19 \cdot 23 \cdot 29 \cdot 31 \cdot 41 \cdot 47 \cdot 59 \cdot 71,$$

a number that is approximately 10^{54}. The huge size of the Monster was one thing that made it difficult to approach. If one attempts to view the Monster as a symmetric object in some space $\mathbb{R}^n$, for example, then the least dimension that works is $n = 196883$. This number, incidentally, is the product $47 \cdot 59 \cdot 71$ of the last three primes in the factorization of its order.

When the Monster was finally constructed, by Robert Griess in 1980 (published in Griess (1982)), there was a great deal of relief, but also a huge accumulation of unfinished business. On the one hand, it remained to be proved that the list of finite simple groups was now complete. Thousands more pages had to be written before (around 2004) group theorists became convinced that the job was done, and we do not know whether there will *ever* be a really accessible proof. On the other hand, much of the information gathered about the Monster only added to its mystery.

The center of the mystery is the number 196883, because its successor, 196884, was a number already known in mathematics, but in a branch with no apparent connection to the Monster: the theory of the modular function. We have seen several appearances of the modular function already in this book: solving the quintic equation in Section 6.7, in the theory of elliptic functions in Section 12.6, as a function with non-Euclidean periodicity in Section 16.6, and in the study of quadratic integers in Section 21.6. However, it is relatively easy to tie these appearances of the modular function together, as may be seen in the book of McKean and Moll (1997).

The connection between the 196883 of the Monster and the 196884 of the modular function, first noticed by John McKay in 1978, looks more like a bizarre coincidence. After all, many 6-digit numbers turn up in mathematics; it is not out of the question for two of them to be close together. But let us see exactly how 196884 turns up in the modular function $j(\tau)$. The function $j(\tau)$ repeats its value when we replace τ by $\tau + 1$, as we have seen in the exercises to Section 16.5. Thus $j(\tau)$ has *period* 1. Because of this, j may be expanded in Fourier series, which it is appropriate to write as a series in powers of

$$\cos 2\pi\tau + i \sin 2\pi\tau = e^{2i\pi\tau} = q.$$

When this is done, it turns out that

$$j(\tau) = q^{-1} + 744 + 196884q + 21493760q^2 + \cdots.$$

The first such expression for j is due to Hermite (1859) but, ironically, there is an error and the published expression is

$$j(\tau) = q^{-1} + 744 + 196880q + \cdots.$$

If this error had gone undetected, perhaps the connection with the Monster would not have been noticed!

What puts McKay's observation beyond coincidence is the *totality* of coefficients in the series for $j(\tau)$. There is a series of numbers describing the Monster, called its *character degrees*, which goes 1, 196883, 21296876, 842609326, ..., and the *sum of the first $n + 1$ character degrees equals the coefficient of q^n in the series for $j(\tau)$*, for as many terms as anyone cares to calculate. Something must explain the agreement between these two series of numbers, but what? Conway called the unknown theory "monstrous moonshine," drawing on both British and American senses of the word "moonshine": the subject was dimly lit, perhaps illusory, and based on illegally distilled information.

McKay also observed a connection between the modular function and the exceptional Lie group E_8, and others noticed a connection with the Leech lattice. Moonshine seems to illuminate many extreme objects.

We now have rational explanations for most of the connections brought to light by moonshine, using some beautiful new mathematics. A wide-ranging account of this new mathematics may be found in Gannon (2006). Most surprisingly, one of its key ingredients comes from physics, and indeed from the highly speculative part of physics known as *string theory*. This was discovered by Borcherds (1994). String theory is perfectly sound mathematics, but it is not yet acceptable physics, because no physical traces of the *mathematical* objects called "strings" have yet been observed.

This is a curious state of affairs, but perhaps not unprecedented. Consider Kepler's model of the solar system using nested polyhedra (Section 2.2). The model turns out to be physically false, but there is still nothing wrong with the regular polyhedra. Indeed, they are useful in other parts of physics, such as the theory of crystals. Some mathematical physicists hold out hope that some day even the Monster may be found to play some role in the physical universe.

EXERCISES

23.7.1 Check that the sum of the first three character degrees of the Monster equals the coefficient of q^2 in the expansion of the modular function.

The expansion of the modular function led Hermite (1859) to a strange numerical result:

$$e^{\pi \sqrt{163}} = 262537412640768744$$

(an integer!) correct to 12 decimal places. In Hermite's time, it would have been very hard work to show that $e^{\pi \sqrt{163}}$ is *not* an integer by brute calculation. But Hermite knew that $e^{\pi \sqrt{163}}$ must differ from an integer by a very small amount, thanks to a remarkable result discovered by Kronecker (1857): *if the integers in* $\mathbb{Q}[\sqrt{-N}]$ *have unique prime factorization, then* $j(\sqrt{-N})$ *is an ordinary integer.* Kronecker's result stems from the connection between the modular function and lattice shapes, mentioned briefly in Section 21.6.

The largest value of N for which the integers in $\mathbb{Q}[\sqrt{-N}]$ have unique prime factorization happens to be $N = 163$. This instance of unique prime factorization was known to Hermite, so he knew that

$$j\left(\text{any integer in } \mathbb{Q}[\sqrt{-163}]\right) = \text{an ordinary integer.}$$

23.7.2 Show that $\tau = (1 - \sqrt{-163})/2$ is an integer of $\mathbb{Q}[\sqrt{-163}]$ by finding the quadratic equation that τ satisfies.

23.7.3 By substituting $\tau = (1 - \sqrt{-163})/2$ in the series for $j(\tau)$ in powers of $q = e^{2\pi\tau}$, show that $e^{\pi\sqrt{163}} = \text{integer} - \text{tiny number}$.

23.7.4 Verify that the integer in Exercise 23.7.3 equals $(640320)^3 + 744$.

23.8   Biographical Notes: Lie, Killing, and Cartan

Sophus Lie was born on December 17, 1842, in the small farming community of Norfjordeid in Norway. His father, Johan Herman Lie, was a Lutheran pastor and former teacher, his mother the former Mette Maren Stabell from Trondheim. Sophus was the youngest of their six children. Johan taught his children music, languages, history, and geography, but it appears that Sophus first learned mathematics from his Aunt Edle.

At the end of 1850 Johan received a posting to Moss, a port in southeastern Norway near Oslo (then known as Christiania). During the family's move to Moss they were lucky enough to observe a total eclipse of the sun, on July 28, 1851. Sophus went to Nissen's Latin School in Christiania in 1857, to prepare for university, and he entered the University of Christiania in 1859. Figure 23.5 shows him during his student days.

Figure 23.5: Sophus Lie

Lie enjoyed his time as a student, becoming a keen gymnast, hiker, and climber, and an active member of the Science Students Association. He does not seem to have had a special interest in mathematics, but in 1862 he was taught by the mathematician Ludvig Sylow, who planted some seeds of Lie's future mathematical development. Sylow made pioneering contributions to group theory and taught one of the first courses in Galois theory, in which Lie was in an audience of just two or three students. Lie was more interested in differential geometry at this time—he gave a talk on evolutes to the Science Students Association in 1863—but the idea that solvable equations have "solvable groups" lodged in the back of his mind.

As Lie was winding up his science studies in 1864, Norway was in a period of nationalist turmoil. Relations with Sweden and Denmark were tense, and many students enrolled in officer training school. Lie himself considered a military career, but eventually dropped the idea because of his weak eyesight. He experienced a period of depression in 1866, but picked up enough in 1867 to give some public lectures on astronomy. Then in 1868 he reached a turning point, beginning a serious study of geometry with a growing conviction that group theory was going to be important.

In 1869 Lie traveled to Germany and France to further his studies of geometry. Around October 24 he crossed paths with the 20-year-old Felix Klein in Berlin. Lie and Klein had a convergence of interests: first in the work of Klein's teacher, Julius Plücker, and second in the field they were both just becoming aware of—group theory.

In 1870 Lie traveled to Paris, where he was joined by Klein. They met the geometer Gaston Darboux, whose work on the differential geometry of surfaces was a great inspiration to Lie, and Camille Jordan, a mathematician of many talents who at that time had just completed his great book on Galois theory and the associated theory of finite groups, Jordan (1870). This experience confirmed, for both of them, that their future lay in group theory. However, they took different paths; Lie into the theory of continuous groups and Klein into the theory of discrete groups.

In July 1870 the Franco-Prussian war broke out, and Klein had to leave immediately for Germany. Lie stayed a little longer, since Norway was neutral in the war, but in August he was mistaken for a German spy and arrested. The mathematical symbols in his notes were thought to be a secret code. He spent four weeks in prison before he was able to convince the authorities that he was indeed a mathematician, partly due to the intervention of Darboux.

After this close shave, Lie quickly left France but continued to travel in Italy and Germany, including a reunion with Klein in Düsseldorf. He returned to Christiania in December, becoming something of a celebrity because of his adventures in France. In 1871 he was awarded a doctorate from the University of Christiania and in 1872 he was appointed professor of mathematics, a position he held until 1886. During this period, Lie laid the foundations of his theory of continuous groups, single-handedly discovering the main theorems of the subject. He also found time to co-edit Abel's collected works (with Sylow) and to marry Anna Birch and father three children.

Nevertheless, Lie was isolated in Norway, and disappointed by the little recognition that he received. This was partly due to the novelty of his work, but also due to lack of clarity on Lie's part. In 1884, Klein had the happy idea of sending his recently graduated student Friedrich Engel from Leipzig to Christiania as Lie's assistant. Engel proved to be the ideal editor for Lie, and together they produced three massive volumes on the theory of "transformation groups" between 1888 and 1893. In the meantime, Klein left Leipzig for Göttingen in 1886 and Lie was appointed to replace him, so his fruitful collaboration with Engel continued in Leipzig—for a while.

Lie never became really comfortable with the German language and culture. His workload was heavy and he missed the landscape of Norway. In 1889–1890 he suffered another bout of depression and was hospitalized for seven months. Around this time his relationships with Engel and Klein broke down, and they never completely recovered. In the preface to the final volume of his book on transformation groups Lie wrote:

> I am no pupil of Klein's, nor is the reverse the case, although
> it might be closer to the truth.

Whatever Klein may have felt about this slight, it did not stop him from recommending Lie for the inaugural Lobachevsky Prize, offered by the Kazan Scientific Society in 1897. Lie won, and was thus able to return to Christiania in triumph. However, his health was failing, so he had little time to enjoy his homecoming. He died of pernicious anemia in 1899.

For more details of Lie's life, particularly his life outside mathematics, see the excellent biography of Stubhaug (2002).

Wilhelm Killing was born on May 10, 1847 in the west German town of Burbach, where his father was a legal clerk and his mother Katharina was the daughter of a pharmacist. Until the late 1880s his career and Lie's ran on almost parallel tracks—neither being aware of the other's work—until Killing's great discovery of the exceptional Lie algebras.

Killing spent his early life in a series of towns in the Westphalia region of Germany near his birthplace. He attended high school in Brilon in 1860 and commenced university studies in Münster in 1865. However, the level of instruction in Münster was then quite low and Killing was largely self-taught. Like Lie, he was particularly impressed by the works of Plücker. Seeking a better education, he moved to Berlin in 1867, where Weierstrass and Kummer were at the height of their powers. Under the direction of Weierstrass, Killing completed a thesis on families of second-degree surfaces in 1872.

In Berlin he also met Anna Commer, the daughter of a professor at the Academy of Art. They were married in 1875 and eventually had four sons and two daughters.

Killing was full of admiration for Weierstrass, but this did not mean that he wanted to become a research mathematician. Weierstrass himself spent many years as a high school teacher, and Killing did likewise. He interrupted his studies in 1870–1871 to teach in a high school in the town of Rüthen, where his father had become mayor. The school was on the point of collapse due to lack of teachers, and Killing spent up to 36 hours per week in the classroom, teaching or tutoring in all subjects. This delay of one year was a stroke of luck, however, because he returned to Berlin just as Weierstrass was turning his attention to non-Euclidean geometry. Killing became fascinated by spaces of constant curvature, and started to do research on them even as he left Berlin to become a high school mathematics teacher. In 1879 he was back teaching in the very school in Brilon where he had been a student.

Killing's spare-time research enabled him to publish a couple of papers, and this led him to a new job. On the recommendation of Weierstrass, he was appointed to the Lyceum Hosianum—a school for the training of Catholic priests—in the remote East Prussian town of Braunsberg in 1880. In this unlikely setting, out of touch with other mathematicians, Killing gradually uncovered a strange new world. Non-Euclidean geometry led him to ask what forms of space could support the concepts of geometry, which led him in turn to groups of transformations and their tangent spaces. Independently of Lie, he had discovered what we now call Lie algebras.

Like Lie, Killing discovered that the generalized rotation groups have simple Lie algebras. But unlike Lie, he noticed that the rotation groups do not account for all possible simple algebras. The two mathematicians became aware of each other in 1884, through Felix Klein. Lie was dismissive of Killing's work, though Engel praised it, and Engel's encouragement probably helped Killing to press on to his discovery of the five exceptional algebras. For a blow-by-blow account of this episode, and more details of Killing's life, see Hawkins (2000), Chapters 4 and 5. Figure 23.6 shows Killing in 1889, around the time that he published his discoveries.

Figure 23.6: Wilhelm Killing

Killing returned to Münster in 1892 as professor of mathematics, and later became rector of the university. He spent the rest of his life in teaching, administration, and charitable work. Fittingly, his research was recognized by the second Lobachevsky Prize, in 1900. He died on February 11, 1923.

Élie Cartan was born on April 9, 1869, in Dolomieu in eastern France, near Lyon. He was the son of Joseph Cartan, the village blacksmith, and Anne Cottaz. While in primary school he was noticed by Antonin Dubost, a school inspector (and later a prominent politician). Dubost obtained funds for Élie to study at a high school in Grenoble and later at the École Normale Supérieure, which he entered in 1888. In 1894 he obtained his doctorate from the École Normale with one of the most famous theses of all time, *Sur la structure des groupes simples finis et continus.*

In his thesis, Cartan gave the first complete proofs of Killings's results on the exceptional Lie algebras, correcting some errors made by Killing. In particular, it was really Cartan who found that there are five exceptions—Killing had listed six, but he failed to notice that two of the algebras on his list were isomorphic. Because Cartan was the first to produce a complete and correct proof of Killing's result, many later authors ignored Killing's work and attributed some of his key ideas to Cartan. Cartan himself gave Killing full credit, but Killing's eclipse was perhaps unavoidable, as Cartan went from strength to strength.

In 1903 he was appointed professor at the University of Nancy and in 1904 he married Marie-Louise Brianconi. They eventually had three sons and a daughter. Two of the sons died tragically in their youth, but the oldest son, Henri, became a prominent mathematician and died as recently as 2008, at the age of 104. Thus only two mathematical generations separate us from the world of mathematics in 1869, before the discovery of Lie groups. Many of the changes that have taken place since 1869 are due to the Cartans, father and son.

Élie Cartan moved to Paris in 1909, and began a long career at the Sorbonne; first as a lecturer, then in 1912 as professor of differential and integral calculus, in 1920 as professor of rational mechanics, and in 1924 as professor of higher geometry. He retired in 1940 and died on May 6, 1951. Figure 23.7 is a picture of him taken around 1930.

Figure 23.7: Élie Cartan

As his accumulation of chairs suggests, Cartan brought the theory of Lie groups into full bloom, uniting it with differential geometry and mathematical physics. Cartan also brought to light connections with hypercomplex numbers and topology that were not dreamed of in the 19th century. Nice examples are the Lie groups that occur in the complex numbers and quaternions, because multiplication of complex numbers and quaternions are continuous group operations.

As mentioned in the exercises to Section 23.4, the complex numbers of absolute value 1 are closed under multiplication and they form a Lie group homeomorphic to the circle, or "1-sphere," $\mathbb{S}^1$. Similarly, the quaternions of absolute value 1 form a Lie group homeomorphic to the 3-sphere $\mathbb{S}^3$. So $\mathbb{S}^1$ and $\mathbb{S}^3$, which are well known as geometric or topological objects, can also be viewed as Lie groups. The surprise is that they are *exceptional* in this respect. Using methods of algebraic topology, Cartan (1936) proved the remarkable result that $\mathbb{S}^1$ and $\mathbb{S}^3$ are the *only* spheres with a continuous group structure.

Élie Cartan's fame in Lie theory is matched by his fame in differential geometry and mathematical physics, where he introduced not only Lie groups, but also ideas such as differential forms, moving frames, and spinors. For more information on his life, and particularly his work, see Akivis and Rosenfeld (1993).

24

Sets, Logic, and Computation

PREVIEW

In the 19th century, perennial concerns about the role of infinity in mathematics were finally addressed by the development of *set theory* and *formal logic*. Set theory was proposed as a mathematical theory of infinity and formal logic was proposed as a mathematical theory of proof (partly to avoid the paradoxes that seem to arise when reasoning about infinity).

In this chapter we discuss these two developments, whose interaction led to mind-bending consequences in the 20th century. Both set theory and logic throw completely new light on the question, "What is mathematics?" But they turn out to be double-edged swords.

- Set theory brings remarkable clarity to the concept of infinity, but it shows infinity to be unexpectedly complicated—in fact, more complicated than set theory itself can describe.

- Formal logic encompasses all known methods of proof, but at the same time it shows these methods to be *incomplete*. In particular, any reasonably strong system of logic cannot prove its own consistency.

- Formal logic is the origin of the concept of *computability*, which gives a rigorous definition of an *algorithmically solvable problem*. However, some important problems turn out to be *unsolvable*.

It might be thought that the limits of formal proof are too remote to be of interest to ordinary mathematicians. But in the next chapter we will show how these limits are now being reached in one of the most down-to-earth fields of mathematics: combinatorics.

J. Stillwell, *Mathematics and Its History*, Undergraduate Texts in Mathematics,   525
DOI 10.1007/978-1-4419-6053-5_24, © Springer Science+Business Media, LLC 2010

24.1   Sets

Sets became established in mathematics in the late 19th century as a result of attempts to answer certain questions about the real numbers. Our intuition of the real numbers—that they form a "line" without gaps—is a mystery that mathematicians have struggled to explain since ancient times. It underlies the concept of "motion" that Zeno tried to challenge with his paradoxes; it resurfaced with calculus in the 17th century; and it intruded into algebra when Gauss used the intermediate value theorem in his 1816 proof of the fundamental theorem of algebra. As we mentioned in Section 14.6, Bolzano (1817) realized that the intermediate value theorem demands a proof, but he did not have a concept of real number on which a proof could be soundly based.

Bolzano did, however, realize the need for a *completeness* property of $\mathbb{R}$ that expresses the absence of gaps. He identified the *least upper bound property*, that every bounded set of real numbers has a least upper bound, and the equivalent *nested interval property*, that if

$$a_0 < a_1 < a_2 < \cdots < b_2 < b_1 < b_0$$

then there is a number x such that

$$a_0 < a_1 < a_2 < \cdots \leq x \leq \cdots < b_2 < b_1 < b_0.$$

To prove such properties, we have to answer the question, what *is* a real number? Several equivalent answers were given around 1870, all involving infinite sets or sequences. The simplest was that of Dedekind (1872), who defined a real number to be a partition (or "cut") of the rational numbers into two sets, L and U, such that each member of L is less than all members of U. If one has a preconceived notion of real number, such as a point x on a line, then L and U are uniquely determined by x as the sets of rational points to left and right of it, respectively. Thus if x is preconceived, then L and U are no more than auxiliary concepts that enable x to be handled in terms of rationals, as Eudoxus did (Section 4.2). Dedekind's breakthrough was to realize that no preconceived x was necessary: x could be *defined* as the pair (L, U). Thus the concept of sets of rationals was a basis for the concept of real number.

Dedekind cuts give a precise model for the continuous number line $\mathbb{R}$, since they fill all the gaps in the rationals. Indeed, wherever there is a gap in the rationals, the real number that fills it is essentially the gap itself: the pair of sets L, U to left and right of it. Other formulations of this completeness property of $\mathbb{R}$ are also easy consequences of Dedekind's definition. For example, each bounded set of reals (L_i, U_i) has a least upper bound (L, U): L is simply the union of the sets L_i.

Dedekind seemed to have settled the ancient problem of explaining the continuous in terms of the discrete, but in penetrating as far as he did, he also uncovered deeper problems. The central problem is that the completeness of $\mathbb{R}$ entails its *uncountability*, a phenomenon discovered by Cantor (1874). The *countable* sets are those that can be put in one-to-one correspondence with $\mathbb{N} = \{0, 1, 2, \ldots\}$, and they include the set of rationals and the set of algebraic numbers, as Cantor also discovered. But if $\mathbb{R}$ is countable, this means that all reals can be included in a sequence $x_0, x_1, x_2, \ldots$. Cantor (1874) showed that this is impossible by selecting from each sequence $\{x_m\}$ of distinct reals a subsequence $a_0, b_0, a_1, b_1, a_2, b_2, \ldots$, such that

$$a_0 < a_1 < a_2 < \cdots < b_2 < b_1 < b_0$$

and with each x_m *outside* one of the nested intervals $(a_0, b_0) \supset (a_1, b_1) \supset (a_2, b_2) \supset \cdots$. It follows that any common element of all the (a_n, b_n) is a real $x \neq$ each x_m. A common element obviously exists if the sequence of intervals is finite, and if the sequence is infinite, it exists by completeness, as the least upper bound of the a_n. The common element x is a "gap" in the given sequence $\{x_m\}$.

EXERCISES

Cantor's 1874 proof of the uncountability of $\mathbb{R}$ is based on the following construction. Given a sequence $x_0, x_1, x_2, \ldots$ of distinct reals, he found a "gap" in them by picking out $a_0, b_0, a_1, b_1, \ldots$ as follows:

$$a_0 = x_0,$$
$$b_0 = \text{first } x_m \text{ with } a_0 < x_m,$$
$$a_1 = \text{first } x_m \text{ after } b_0 \text{ with } a_0 < x_m < b_0,$$
$$b_1 = \text{first } x_m \text{ after } a_1 \text{ with } a_1 < x_m < b_0,$$
$$a_2 = \text{first } x_m \text{ after } b_1 \text{ with } a_1 < x_m < b_1.$$

$$\vdots$$

24.1.1 Explain why the sequence $a_0, b_0, a_1, b_1, a_2, b_2, \ldots$ has the "gap" property described above: each x_m is outside one of the nested intervals $(a_0, b_0) \supset (a_1, b_1) \supset (a_2, b_2) \supset \cdots$.

We now explore how far we can enlarge the set of natural numbers and still have a countable set.

24.1.2 Give a rule for continuing the sequence

$$\frac{1}{1}, \frac{2}{1}, \frac{1}{2}, \frac{3}{1}, \frac{2}{2}, \frac{1}{3}, \frac{4}{1}, \frac{3}{2}, \cdots$$

so as to include all positive rationals.

24.1.3 How can one then conclude that the set of all rationals is countable?

24.1.4 The words on a fixed finite alphabet can be enumerated by listing first the one-letter words, then the two-letter words, and so on. Use this observation to show that the set of polynomial equations with integer coefficients is countable and hence that the set of algebraic numbers is countable.

Cantor used the latter result to prove the existence of transcendental numbers. Namely, let $\{x_m\}$ be the sequence of algebraic numbers; we know that these are not all the real numbers, so any other real number is transcendental.

24.2   Ordinals

The uncountability of $\mathbb{R}$ has been a great challenge to set theorists and logicians ever since its discovery. The most successful response to this challenge has been the theory of *ordinal numbers*, which grew out of Cantor's (1872) investigation of Fourier series (see Section 13.6). The existence of a Fourier series for a function f depends largely on the structure of the set of discontinuities of f, and thus leads to the problem of analyzing the complexity of point sets. Cantor measured complexity by the number of iterations of the prime operation (') of taking the limit points of a set. For example, if $S = \{0, 1/2, 3/4, 7/8, \ldots, 1\}$, then the prime operation can be applied once, and $S' = \{1\}$. It can happen that S' itself has limit points, so that S'' also exists. In fact, one can find a set S for which $S', S'', \ldots, S^{(n)}, \ldots$ exist for all finite n, so one can envisage iterating the prime operation an infinite number of times. In the case where all the $S^{(n)}$ exist, Cantor (1880) took their intersection, thereby defining

$$S^{\infty} = \cap_{n=1,2,3,\ldots} S^{(n)}.$$

He viewed ∞ as the first infinite ordinal number. To avoid confusion with higher infinite numbers soon to appear, I shall use the modern notation ω for the first infinite ordinal.

Having made the leap to ω, it is easy to go further: $\left(S^{(\omega)}\right)' = S^{(\omega+1)}$, $\left(S^{(\omega+1)}\right)' = S^{(\omega+2)}, \ldots$, and the intersection of this new infinite sequence is $S^{\omega \cdot 2}$, where $\omega \cdot 2$ is the first infinite number after ω, $\omega + 1$, $\omega + 2, \ldots$. After $\omega \cdot 2$, one has

$$\omega \cdot 2 + 1, \quad \omega \cdot 2 + 2, \quad \ldots, \quad \omega \cdot 3, \quad \ldots, \quad \omega \cdot 4, \quad \ldots, \quad \ldots, \quad \omega \cdot \omega, \quad \ldots.$$

All these can actually be realized as numbers of iterations of the prime operation on sets of reals. We can also investigate the ordinal numbers independently of this realization, as an extension of the concept of natural number.

Cantor (1883) viewed the ordinals as the result of two operations:

(i) Successor, which for each ordinal α gives the next ordinal, $\alpha + 1$.

(ii) Least upper bound, which for each set $\{\alpha_i\}$ of ordinals gives the least ordinal $\geq$ each α_i.

The most elegant formalization of these notions was given by von Neumann (1923). The empty set $\emptyset$ (not considered by Cantor) is taken to be the ordinal 0, the successor of α is $\alpha \cup \{\alpha\}$, and the least upper bound of $\{\alpha_i\}$ is simply the union of the α_i. Thus

$$0 = \emptyset,$$
$$1 = \{0\},$$
$$2 = \{0, 1\},$$
$$\cdots$$
$$\omega = \{0, 1, 2, \ldots, n, \ldots\},$$
$$\omega + 1 = \{0, 1, 2, \ldots, n, \ldots, \omega\},$$

and so on. The natural ordering of the ordinals is then given by set membership, $\in$, and, in particular, the members of an ordinal α are all ordinals smaller than α.

Cantor's principle (ii) generates ordinals of breathtaking size, since it gives the power to transcend any set of ordinals already defined. In particular, an ordinal of uncountable size is on the horizon as soon as one thinks of the concept of countable ordinal, as Cantor did (1883). He defined an ordinal α to be countable (or, as he later put it, of *cardinality* or cardinal number $\aleph_0$) if α could be put in one-to-one correspondence with $\mathbb{N}$. For example,

$$\omega \cdot 2 = \{0, 1, 2 \ldots, \omega, \omega + 1, \omega + 2, \ldots\}$$

is countable because of its obvious correspondence with

$$\mathbb{N} = \{0, 2, 4, \ldots, 1, 3, 5, \ldots\}.$$

The least upper bound of the countable ordinals is the least *uncountable* ordinal, ω_1. Sets in one-to-one correspondence with ω_1 are of the next cardinality, $\aleph_1$. Ordinals of cardinality $\aleph_1$ have a least upper bound ω_2 of cardinality $\aleph_2$, and so on.

Having found this orderly way of generating successive uncountable cardinals, Cantor reconsidered the uncountable set $\mathbb{R}$. Although no method of generating members of $\mathbb{R}$ in the manner of ordinals was apparent, Cantor conjectured that the cardinality of $\mathbb{R}$ was $\aleph_1$. This conjecture has since become known as the *continuum hypothesis*. By 1900 it was recognized as the outstanding open problem of set theory, and Hilbert (1900a) made it number one on the famous list of problems he presented to the mathematical community. There have been two outstanding results on the continuum problem since 1900, but they seem to make it less likely that we will ever know whether the continuum hypothesis is correct. Gödel (1938) showed that the continuum hypothesis is *consistent* with standard axioms for set theory, but Cohen (1963) showed that its negation is also consistent. Thus the continuum hypothesis is independent of standard set theory, in the same way that the parallel postulate is independent of Euclid's other postulates. Whether this means that the notion of "set" is open to different natural interpretations, like the notion of "straight line," is not yet clear.

EXERCISES

For each countable ordinal α there is a set of rationals in [0,1] with *order type* α. For example, the set $\{0, 1/2, 3/4, 7/8, \ldots\}$ has order type ω.

24.2.1 Give an example of a set of rationals in [0,1] with order type $\omega \cdot 2$.

24.2.2 Give an example of a set of rationals in [0,1] with order type $\omega \cdot \omega$.

24.2.3 Given sets of rationals in [0,1] with order types $\alpha_1, \alpha_2, \alpha_3, \ldots$, explain how to obtain a set of rationals in [0,1] with order type at least as large as the least upper bound of $\{\alpha_1, \alpha_2, \alpha_3, \ldots\}$.

24.2.4 Explain why there is a set of rationals in [0,1], with order type α, for each countable ordinal α.

24.3 Measure

The reason for investigating sets of discontinuities in the theory of Fourier series was the discovery of Fourier (1822) that these series depend on integrals. Assuming that

$$f(x) = \frac{1}{2}a_0 + \sum_{n=1}^{\infty} (a_n \cos n\pi x + b_n \sin n\pi x),$$

Fourier derived the formulas

$$a_n = \int_{-1}^{1} f(x) \cos n\pi x \, dx, \quad b_n = \int_{-1}^{1} f(x) \sin n\pi x \, dx.$$

Thus the existence of the series depends on the existence of the integrals for a_n and b_n, and this in turn depends on how discontinuous f is. It was known (though not rigorously proved) that every continuous function has an integral, so the next question was how the integral should, or could, be defined for discontinuous functions. The first precise answer was the Riemann (1854a) integral concept, familiar to all calculus students, and based on approximating the integrand by step functions. Any function with a finite number of discontinuities has a Riemann integral, and indeed so have certain functions with infinitely many discontinuities, but not all. The classic function for which the Riemann integral does not exist is the function of Dirichlet (1829):

$$f(x) = \begin{cases} 1 & \text{if } x \text{ is rational,} \\ 0 & \text{if } x \text{ is irrational.} \end{cases}$$

Eventually a more general integral, the Lebesgue integral, was introduced to cope with such functions, but not until the focus of attention had shifted from the problem of integration to the more fundamental problem of *measure*. Measure generalizes the concept of length (on the line $\mathbb{R}$), area (in the plane $\mathbb{R}^2$), and so on, to quite general point sets. Since an integral can be viewed as the area under a graph, its dependence on the concept of measure is clear, though it was not immediately realized that the measure of sets on the line had to be clarified first.

The need for clarification arose from the discovery of Harnack (1885) that any countable subset $\{x_0, x_1, x_2, \ldots\}$ of $\mathbb{R}$ could be covered by a collection of intervals of arbitrarily small total length. Namely, cover x_0 by an interval of length $\varepsilon/2$, x_1 by an interval of length $\varepsilon/4$, x_2 by an interval of length $\varepsilon/8$, ..., so that the total length of intervals used is $\leq \varepsilon$. (This is another proof, by the way, that $\mathbb{R}$ is *not* a countable set.) This seemed to show that countable sets were "small"—of *measure zero*, as we now say— but mathematicians were reluctant to say this of dense countable sets, like the rationals. The first response, by Jordan (1892), was to define measure analogously to the Riemann integral, using finite unions of intervals to approximate subsets of $\mathbb{R}$. Under this definition, "sparse" countable sets like $\{0, 1/2, 3/4, 7/8, \ldots\}$ did have measure zero, but dense sets like the rationals were not measurable at all.

The first to take the hint from Harnack's result that countable unions of intervals should be used to measure subsets of $\mathbb{R}$ was Borel (1898). He defined the measure of any interval to be its length, and he extended measurability to more and more complicated sets by *complementation* and *countable disjoint unions*. That is, if a set S contained in an interval I has measure $\mu(S)$, then

$$\mu(I - S) = \mu(I) - \mu(S),$$

and if S is a disjoint union of sets S_n with measures $\mu(S_n)$, then

$$\mu(S) = \sum_{n=1}^{\infty} \mu(S_n).$$

The sets that can be formed from intervals by complementation and countable unions are now called *Borel sets*. Borel's idea was pushed to its logical conclusion by Lebesgue (1902), who assigned measure zero to any subset of a Borel set of measure zero. Since not all such sets are Borel, this extended measurability to a larger class of sets: those that differ from Borel's by sets of measure zero. It can be proved that the class of Lebesgue measurable sets has the same cardinality as the class of all subsets of $\mathbb{R}$. But whether the measurable sets *are* all subsets of $\mathbb{R}$ is an interesting question to which we shall return shortly.

The distinctive property of Borel–Lebesgue measure is *countable additivity*: if $S_0, S_1, S_2, \ldots$ are disjoint measurable sets, then

$$\mu(S_0 \cup S_1 \cup S_2 \cup \cdots) = \mu(S_0) + \mu(S_1) + \mu(S_2) + \cdots.$$

This follows easily from Borel's definition of measure for countable disjoint unions, because any countable union can be reassembled as a countable disjoint union.

Lebesgue showed that countable additivity gives a concept of integral that is better behaved with respect to limits than the Riemann integral. For example, one has the *monotone convergence property*: if $f_0, f_1, f_2, \ldots$ is an increasing sequence of positive integrable functions, and $f_n \to f$ as $n \to \infty$, then $\int f_n \, dx \to \int f \, dx$ for the Lebesgue integral, whereas this is not generally true for the Riemann integral (see Exercise 24.3.1).

Another motivation for countable additivity that Borel pointed out was the theory of probability. If an "event" E is formalized as a set S of points ("favorable outcomes"), then the probability of E can be defined as the measure of S. Some quite natural events turn out to be countable unions; hence it is necessary for probability measure to be countably additive. In informal probability theory, countable additivity was assumed as far back as 1690, when Jakob Bernoulli answered the following question he had posed in 1685:

> *A* and *B* play with a die, the one that throws an ace first being declared the winner. *A* throws once, then *B* throws once also. *A* then throws twice, and *B* does the same, and so on, until one wins. What is the ratio of their chances of success?

To solve this problem, Jakob Bernoulli (1690) decomposed the event of a win for *A* (or *B*) into the subevents of a win at *A*'s (*B*'s) first, second, third, ..., turn and summed the probabilities of these countably many subevents. Formal probability theory, which was created by Kolmogorov (1933), bases all such arguments on the theory of countably additive measures.

It could be said that set theory paved the way for measure theory by showing the uncountability of $\mathbb{R}$, thus enabling countable subsets of $\mathbb{R}$ to be regarded as "small." On the other hand, measure theory itself shows the uncountability of $\mathbb{R}$ (by Harnack's result), and in fact measure theory's assessment of the "smallness" of countable sets greatly influenced the later development of set theory.

"Measure theoretically desirable" axioms, such as the measurability of all subsets of $\mathbb{R}$, turned out to conflict with "set theoretically desirable" axioms such as the continuum hypothesis, and efforts to resolve the conflict brought more fundamental questions about sets to light. These questions do not reduce to clear-cut alternatives—the way geometric questions reduce to alternative parallel axioms, for example—but they do seem to gravitate toward the so-called *choice* and *large cardinal* axioms, discussed in the next section.

24.3.1 Show that a function f_n that is zero at all but n points has Riemann integral zero over any interval and that the non-Riemann integrable function of Dirichlet is a limit as $n \to \infty$ of such functions f_n.

The complexity of Borel sets may be roughly measured by the number of countable unions and complements needed to define them. Here are a few of the simpler ones.

24.3.2 Show that a single point is the complement of a countable union of intervals and hence that any countable set is a Borel set.

24.3.3 Deduce that the set of irrational numbers is a Borel set.

24.3.4 What is the measure of the set of irrationals between 0 and 1?

24.4   Axiom of Choice and Large Cardinals

The usual axiom of choice states that for any set S (of nonempty sets) there is a *choice function* f such that $f(x) \in x$ for each $x \in S$. (Thus f "chooses" an element from each set x in S.) The axiom seems so plausible that early set theorists used it almost unconsciously, and it first attracted attention in Zermelo's (1904) proof that any set S could be *well ordered* (that is, put in one-to-one correspondence with an ordinal). This looked like progress toward the continuum hypothesis. But Zermelo's proof gave no more than the existence of a well-ordering of S, given a choice function for the set of subsets of S. There was still no sign of an explicit well-ordering of $\mathbb{R}$. And of course if one doubted the existence of a well-ordering of $\mathbb{R}$, this threw doubt on the axiom of choice. Further doubts were raised when the axiom of choice was found to have incredible consequences in measure theory.

The first of these, discovered by Vitali (1905), was that the circle can be decomposed into countably many disjoint congruent sets. Since congruent sets have the same Lebesgue measure, it easily follows that the sets in question are not Lebesgue measurable (by countable additivity; see Exercises 23.4.2–23.4.4).

Even more paradoxical decompositions were given by Hausdorff (1914) (for the sphere) and Banach and Tarski (1924) (for the ball). The Banach–Tarski theorem states that the unit ball can be decomposed into finitely many sets that, when rigidly moved in space, form *two* unit balls! This shows that not all subsets of the ball are measurable, even if one asks only for finite, rather than countable, additivity. For an excellent discussion of the paradoxical decompositions and their connections with other parts of mathematics, see Wagon (1985).

The measure-theoretic consequences of the paradoxical decompositions follow from the geometrically natural assumption that congruent sets have the same measure. If one drops this assumption and asks only for countable additivity and nontriviality (that is, not all subsets have measure zero), then the conflict with the axiom of choice seems to disappear. No contradiction has yet been derived from these assumptions, but Ulam (1930) showed that any set possessing such a measure must be extraordinarily large—as large, in fact, as a model of set theory itself, and in particular larger than the cardinals $\aleph_1, \aleph_2, \ldots, \aleph_\omega, \ldots$. Thus if $\mathbb{R}$ has a nontrivial countably additive measure, then $\mathbb{R}$ must be far larger than $\aleph_1$, and we still have a conflict with the continuum hypothesis. (For more on the "largeness" of models, see Section 23.8.)

An even more desirable axiom than measurability would be Lebesgue measurability of all subsets of $\mathbb{R}$. This conflicts with the axiom of choice, by Vitali's theorem, but it was nevertheless shown to be consistent with set theory by Solovay (1970), assuming the existence of a large cardinal. Shelah (1984) showed that the large cardinal assumption is necessary.

Thus measurability of all subsets of $\mathbb{R}$ is intimately connected with the existence of sets large enough to model the whole of set theory. This mind-boggling concept seems to be the answer to many fundamental questions. We shall find ourselves drawn to it again in the next sections when we explore the influence of set theory on logic. Meanwhile, for those who would like a more detailed account of the development of set theory, and the contentious axioms in particular, we refer to van Dalen and Monna (1972). For recent developments in the theory of large cardinals, which some believe will throw new light on the continuum hypothesis, see Kanamori (1994) and Woodin (1999).

EXERCISES

The axiom of choice turns up even in elementary analysis, when one attempts to formalize the idea of a continuous function. A natural definition in terms of infinite sequences is equivalent to the standard ε-δ definition only if we assume the axiom of choice.

Call f *sequentially continuous* at $x = a$ if, for any sequence $\{a_n\}$ such that $a_n \to a$, we have $f(a_n) \to f(a)$.

24.4.1 Show, assuming the axiom of choice, that if f is not continuous at a then

f is not sequentially continuous at a. (It is a consequence of Cohen (1963) that this result *cannot* be proved without the axiom of choice.)

Vitali's decomposition of the circle is created as follows. For each θ between 0 and 2π let $S(\theta)$ be the set of points on the unit circle whose angle differs from θ by a rational multiple of 2π. Thus $S(\theta) = S(\phi)$ if $\theta - \phi = 2\pi \times$ a rational, and $S(\theta) \cap S(\phi) = \emptyset$ otherwise.

24.4.2 Let S be a set (existing by virtue of the axiom of choice) that contains exactly one element from each distinct $S(\theta)$ and let

$$S + 2\pi r = \{\theta + 2\pi r : \theta \in S\} \quad \text{for each rational } r.$$

(Thus $S + 2\pi r$ is S rotated through the rational multiple $2\pi r$ of 2π.) Show that any two of the sets $S + 2\pi r$ are either identical or disjoint.

24.4.3 Show that the circle is a countable union of sets $S + 2\pi r$.

24.4.4 Show that both assumptions $\mu(S) = 0$ and $\mu(S) > 0$ lead to contradictions, and hence conclude that S is nonmeasurable.

24.5   The Diagonal Argument

The uncountability of $\mathbb{R}$ was shown again in a strikingly simple way by Cantor (1891). His argument applies most directly to the set $2^{\mathbb{N}}$ of all subsets of $\mathbb{N}$, but there are variants that work similarly on the set $\mathbb{N}^{\mathbb{N}}$ of integer functions and on $\mathbb{R}$ (which can be identified with a set of integer functions in various ways). To show that there are uncountably many subsets of $\mathbb{N}$ one shows that any countable collection $S_0, S_1, S_2, \ldots$ of sets $S_n \subseteq \mathbb{N}$ is incomplete, by constructing a new set S, different from each S_n. S is the so-called *diagonal set* $\{n : n \notin S_n\}$, which obviously differs from S_n with respect to the number n. Q.E.D.

The "diagonal" nature of S can be seen by visualizing a table of 0's and 1's in which

$$m\text{th entry in } n\text{th row} = \begin{cases} 0 & \text{if } m \notin S_n, \\ 1 & \text{if } m \in S_n. \end{cases}$$

In other words, the nth row consists of the values of the characteristic function of S_n. The characteristic function of S is simply the diagonal of the table, with all values reversed. A sequence $x_0, x_1, x_2, \ldots$ of real numbers can be diagonalized similarly by forming the table whose nth row consists of the decimal digits of x_n. A suitable way to "reverse" the digits on the diagonal is to change any 1 to a 2 and any other digit to a 1. (The resulting sequence of 1's and 2's, after a decimal point, then defines a real number x whose decimal expansion is unique. Hence x is not just different from each x_n in its decimal expansion but is definitely a different number.)

More generally, for any table of rows of integers, that is, any sequence of integer functions f_n, one can construct an integer function f unequal to each f_n by changing the values along the diagonal of the table. The diagonal argument was in fact first given in this context, by du Bois-Reymond (1875), in order to construct an f with a greater rate of growth than all functions in a sequence $f_0, f_1, f_2, \ldots$ (Exercise 24.5.1). With hindsight, one can even see a diagonal construction in Cantor's first (1874) argument for the uncountability of $\mathbb{R}$ (Exercise 24.5.2).

The diagonal argument is important in set theory because it readily generalizes to show that every set has more subsets than elements (Exercise 24.5.3), and hence that there is no largest set. What was not noticed at first is that the diagonal argument also has consequences at a more concrete level. This is because the diagonal of a table is *computable* if the table as a whole is computable. Hence the argument does not merely show how to add a new function f to a list $f_0, f_1, f_2, \ldots$—it shows how to add a new computable function to a computable list. In other words, it is *impossible to compute a list of all computable functions*. And of course the same goes for lists of computable real numbers. This remarkable result went unnoticed in the early days of the diagonal argument because computability was not then regarded as an interesting concept, or indeed as a mathematical concept at all. The controversies over the axiom of choice, however, helped to sharpen awareness of the difference between constructive and nonconstructive functions. In the 1920s logicians began to investigate the concept of computability more seriously, and by a "kind of miracle," as Gödel (1946) later expressed it, computability turned out to be a mathematically precise notion.

EXERCISES

The diagonal construction is quite a natural way to construct a function or real number "larger" than the members of a given countable set.

24.5.1 Given integer functions $f_0, f_1, f_2, \ldots$, define an integer function f such that $f(m)/f_n(m) \to \infty$ as $m \to \infty$, for each n. *Hint*: Arrange that $f(m) \geq n f_n(m)$ for all $m \geq n$.

24.5.2 Show that if $a_0 < a_1 < a_2 < \cdots$ is a bounded sequence of real numbers, then $a = $ least upper bound of $\{a_0, a_1, a_2, \ldots\}$ is a "diagonal number" of the sequence in the following sense. There are integers $k_0 < k_1 < k_2 < \cdots$ such that the decimal digits of a exceed those of a_n after the k_nth place.

The last exercise applies the diagonal construction to *any* set I, to show that I has more subsets than members.

24.5.3 Let I be any set, and let $\{S_i\}$ be a collection of subsets of I in one-to-one correspondence with the elements i of I. Show that the natural "diagonal" set S of this collection is a subset of I unequal to each S_i.

24.6 Computability

The notion of computability was first formalized by Turing (1936) and Post (1936), who arrived independently at a definition of computing machine, now called a *Turing machine*. A Turing machine M is given by two finite sets, $\{q_0, q_1, \ldots, q_m\}$ of *internal states* and $\{s_0, s_1, \ldots, s_n\}$ of *symbols*, and a *transition function* T that formalizes the behavior of M for pairs (q_i, s_j). The machine M is visualized as having an infinite tape, divided into squares, each of which can carry one of the symbols s_j. (For most purposes, M is assumed to start on a tape with all but finitely many squares blank: s_0 is taken to denote the blank symbol.) Depending on its internal state q_i, M will make a *transition*: changing s_j to s_k, then moving one square right or left and going into a new state q_l. Thus the transition function is given by finitely many equations

$$T(q_i, s_j) = (m, s_k, q_l),$$

where $m = \pm 1$ indicates a move to right or left.

To use M to compute a function $f : \mathbb{N} \to \mathbb{N}$, some convention must be adopted for inputs (arguments of f) and outputs (values of f). The simplest is seen in Figure 24.1. M starts in state q_0 on the leftmost 1 of a block of n 1s, on an otherwise blank tape, and halts on the leftmost 1 of a block of $f(n)$ 1s, on an otherwise blank tape. M halts by virtue of entering a *halting state*, that is, a state q_h for which M has no transition from the pair $(q_h, 1)$. A *computable function* f is one that can be represented in this way by a Turing machine M.

It follows that there are only countably many computable functions $f :$ $\mathbb{N} \to \mathbb{N}$, since there are only countably many Turing machines. In fact, we can compute a list of all Turing machines by first listing the finitely many machines with one transition, then those with two transitions, and so forth.

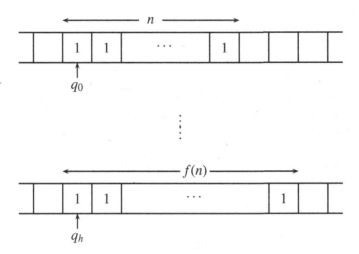

Figure 24.1: Computing a function by Turing machine

This may seem to contradict the discovery from the previous section that a list of all computable functions cannot be computed, but, as Turing (1936) realized, it does not. The catch is that not all machines define functions, and *it is impossible to pick out those that do*. Of course, it is possible to rule out any machine that halts in a situation unlike that in Figure 24.1; the difficulty is in knowing whether halting is going to occur. It is precisely this difficulty that prevents computation of the diagonal function.

If it could be decided, for each machine M and each input, whether M eventually halts, then we could find the first machine to halt on input 1, the next after that to halt on input 2, the next after that to halt on input 3, and so on. By changing the corresponding outputs according to some rule (say, adding 1 if the output is a number, and taking the value 1 otherwise), we could compute a function different from each computable function.

This contradiction shows that the problem of deciding, given a machine and an input, whether halting eventually occurs, is *unsolvable*. This problem is called the *halting problem* and its unsolvability means that no Turing machine can solve it. That is, if the questions "Does M on input n eventually halt?" are written in some fixed finite alphabet, then there is no machine that, given these questions as inputs, will give their answers as outputs. The point is that, as far as we know, all possible rules or algorithms for answering infinite sets of questions can be realized by Turing machines. This is the "kind of miracle" referred to by Gödel (1946).

Now that computers are everywhere, it is taken for granted that "computability" has a precise, absolute meaning—synonymous with Turing machine computability. It is even a familiar fact that all computations can be done on a single, sufficiently powerful machine; this corresponds to the discovery of Turing (1936) of a *universal Turing machine*. However, these claims were surprising in the 1930s, particularly to Gödel, who had shown (1931) that the related notion of "provability" is *not* absolute. This will be discussed further in the next section. Briefly, the reason for the difference is that new computable functions cannot be created by diagonalization, whereas new theorems can.

The halting problem was of no obvious mathematical significance in 1936, but it seemed no more difficult than other unsolved algorithmic problems in mathematics. Thus for the first time it was reasonable to suspect that some ordinary mathematical problems were unsolvable. Moreover, if it could be shown that a solution of a particular problem P implied a solution of the halting problem, then the unsolvability of P would be rigorously established. This method was used to demonstrate the unsolvability of some problems in formal logic by Turing (1936) and Church (1936). Church (1938) also put forward a strong candidate for unsolvability in ordinary mathematics: the word problem for groups.

This is the problem of deciding, given a finite set of defining relations for a group G (Section 19.7) and a word w, whether $w = 1$ in G. There is more than a superficial analogy between the word problem and the halting problem. The group G corresponds to a machine M, words in G correspond to expression on M's tape, and $w = 1$ corresponds to halting. The defining relations of G roughly correspond to the transition function of M, but unfortunately there is no machine equivalent of the cancellation of inverses in G. This creates fierce technical difficulties, but they were overcome by Novikov (1955). He succeeded in establishing the validity of the analogy and hence the unsolvability of the word problem. This led to unsolvability results for a host of significant mathematical problems, among them the homeomorphism problem mentioned in Section 22.7. (The reference given there, Stillwell (1993), also includes a proof of the unsolvability of the word problem.)

A profound reworking of Novikov's ideas, by Higman (1961), shows that computability is a mathematically natural concept in the context of groups. Higman showed that a finitely generated group H has a computable set of defining relations if and only if H is a subgroup of a finitely generated group F with a finite set of defining relations. Thus "computation" is the same as "generation" in a group that is "finitely defined" by generators and relations.

EXERCISES

Turing (1936) actually discovered the unsolvability of the halting problem by considering computable *real numbers* and applying the diagonal argument to them. The argument is similar to the one above using computable functions, but a little messier. Define a real number x to be *computable* if there is a Turing machine M that represents x in the following manner.

- Starting on a blank tape, M prints the decimal digits of x on successive squares of tape, eventually filling each square to the right of the square initially scanned (if necessary, printing all 0s beyond a certain point).

- The squares to the left may be used, and reused, for preliminary computation, but squares to the right, once written, may not be rewritten.

24.6.1 Show that there is no algorithm for recognizing the Turing machines that define real numbers in this way, since such an algorithm would give a way to compute a number different from every computable number.

24.6.2 Explain informally how each Turing machine M may be converted to a machine M' such that M defines a computable number if and only if M' does not halt.

24.6.3 Hence prove that no Turing machine can solve the halting problem.

24.7   Logic and Gödel's Theorem

Since the time of Leibniz, and perhaps earlier, attempts have been made to mechanize mathematical reasoning. Little success was achieved until the late 19th century, when the subject matter of mathematics was clarified by defining all mathematical objects in terms of sets. The reduction of the many concepts of number, space, function, and the like, to the single concept of set brought with it a corresponding reduction in the number of axioms that seemed to be necessary for mathematics. At about the same time, investigation of the principles of logic by Boole (1847), and particularly Frege (1879), led to a system of rules by which all logical consequences of a given set of axioms could be inferred. These two lines of investigation together offered the possibility of a complete, rigorous, and, in principle, *mechanical* system for the derivation of all mathematics.

The most thorough attempt to realize this possibility was the massive *Principia Mathematica* of Whitehead and Russell (1910). *Principia* used axioms of set theory, together with a small collection of rules of inference, to derive a substantial part of ordinary mathematics in a completely formal language. The purpose of the formal language was to avoid the vagueness and ambiguity of natural language, so that proofs could be checked mechanically. Mechanical proof-checking was not then regarded as a goal in itself but rather as a guarantee of rigor. When Whitehead and Russell began writing their *Principia* in 1900, they believed that they were about to reach the 19th-century goal of a complete and absolutely rigorous mathematical system. They did not realize that the rigor of their system—the possibility of checking proofs mechanically—was in fact *incompatible* with completeness. Gödel (1931) showed that there are true sentences that can be expressed in the language of *Principia Mathematica* but that do not follow from its axioms. (Unless *Principia* is inconsistent, in which case all sentences follow from its axioms. The assumption of consistency is actually a weighty one, as we shall see by the end of this section.)

Gödel's theorem created a sensation when it first appeared. Not only did it shatter previous conceptions of mathematics and logic, but its proof was of a new and bewildering kind. Gödel exploited the mechanical nature of proof in *Principia* to define the relation "the nth sentence of *Principia* is provable" within the language of *Principia* itself. Using this, he was able to concoct a sentence that says, in effect, "This sentence is not provable." The Gödel sentence, if true, is therefore not provable. And if false, it is provable, and so *Principia* proves a false sentence. Either way, provability in *Principia* is not the same as truth.

Gödel's proof was very difficult for his contemporaries to understand. Combined with the novelty of treating sentences as mathematical objects was the near inconsistency of a sentence that expresses its own unprovability (a sentence that says, "This sentence is not true" *is* inconsistent). Post (1944) presented Gödel's theorem in a less paradoxical way by deriving it from the classical diagonal argument. The key to Post's approach is the concept of a *recursively enumerable set*. A set W is called recursively enumerable if a list of its members can be computed, say by a Turing machine that prints them on its tape. (Of course if W is infinite, the computation lasts forever.) The paradigm of a recursively enumerable set is the set of theorems of a formal system, such as *Principia Mathematica*. For such a system one can compute a list of all sentences, a list of all finite sequences of sentences, and, by picking out those sequences that are proofs, a list of all theorems—since a theorem is simply the last line of a proof.

Post's idea was to look at the theorems about recursively enumerable sets proved in a given system Σ and to compute a "diagonal sentence" from them. Since recursively enumerable sets are associated with Turing machines, it is possible to enumerate the recursively enumerable subsets of $\mathbb{N}$ as $W_0, W_1, W_2, \ldots$ by letting W_n be the set of numbers output by the nth machine, under some reasonable convention. (Incidentally, there is no problem of picking out suitable machines, as there is for computable functions, since we do not mind if W_n is empty.) The diagonal set

$$D = \{n : n \notin W_n\},$$

being unequal to each W_n, is of course not recursively enumerable, but the following set is:

$$\Pr(D) = \{n : \Sigma \text{ proves } "n \notin W_n"\}.$$

This "provable part" of D is recursively enumerable because we can list the theorems of Σ and select those of the form "$n \notin W_n$." We have $\Pr(D) \subseteq D$, assuming that Σ proves only correct sentences, but $\Pr(D) \neq D$ since $\Pr(D)$ is recursively enumerable and D is not. This shows immediately that there is an n_0 in D that is not in $\Pr(D)$, that is, an $n_0 \notin W_{n_0}$ for which "$n_0 \notin W_{n_0}$" is not provable.

Better still, a specific n_0 with this property is the index of the recursively enumerable set $\Pr(D)$. If $W_{n_0} = \Pr(D)$, then $n_0 \in W_{n_0}$ is equivalent to $n_0 \in \Pr(D)$, which means that "$n_0 \notin W_{n_0}$" is provable. But then it is true that $n_0 \notin W_{n_0}$, assuming that Σ proves only correct sentences, and we have a contradiction. Thus $n_0 \notin W_{n_0}$. This in turn is equivalent to $n_0 \notin \Pr(D)$, which means "$n_0 \notin W_{n_0}$" is not provable. (Notice, incidentally, that the last part of this argument reveals "$n_0 \notin W_{n_0}$" to be a sentence that expresses its own unprovability.)

It seems that Post was aware of this approach to Gödel's theorem in the 1920s, before Gödel's own proof appeared. However, Post's more general view of incompleteness as a property of arbitrary recursively enumerable systems held him up until he was satisfied that computability was a mathematically definable concept. In December 1925 Post formulated a plan for proving *Principia Mathematica* incomplete but, as he later wrote, "The plan, however, included prior calisthenics at other mathematical and logical work, and did not count on the appearance of a Gödel!" (Post (1941), p. 418).

Gödel's theorem comes from reflection on the nature of proofs in ordinary mathematics. An even more devastating theorem, known as Gödel's second theorem, comes from reflection on the proof of Gödel's theorem itself. The latter proof, unusual though it is, can in fact be expressed in ordinary mathematical language.

We described Post's proof of Gödel's theorem in an informal language of Turing machines, but, with some effort, it can be expressed in a small language for number theory called *Peano arithmetic* (PA). Indeed, this "arithmetization of syntax" was one of Gödel's most important ideas. By expressing his proof in PA, he showed the incompleteness of classical mathematics. PA is a language of addition and multiplication on $\mathbb{N}$, with basic logic and mathematical induction as the proof machinery. Turing machines can be discussed in PA by interpreting sequences of symbols on the tape as numerals, so that the changes they undergo in the course of a computation become operations on numbers. Under this interpretation, "$n_0 \notin W_{n_0}$" and "Σ does not prove '$n_0 \notin W_0$'" become sentences of PA.

At this point it is important to recall the hypothesis about Σ used in the proof of Gödel's theorem: Σ proves only correct sentences. This assumption cannot be dropped (since one incorrect theorem usually allows *all* sentences to be proved), but it can be weakened to the assumption that Σ does not prove the sentence "$0 = 1$." Since the latter assumption says that a certain element (the number of the sentence "$0 = 1$") does not belong to a certain recursively enumerable set (the set of theorems of Σ), it can be expressed as a sentence of PA, $\mathrm{Con}(\Sigma)$. In particular, PA expresses its own consistency by the sentence $\mathrm{Con}(\mathrm{PA})$. With these modifications, Gödel's theorem for $\Sigma = \mathrm{PA}$ becomes the following sentence of PA:

$$\mathrm{Con}(\mathrm{PA}) \Rightarrow \text{PA does not prove "}n_0 \notin W_{n_0}\text{."}$$

As we have seen, the sentence "$n_0 \notin W_{n_0}$" is equivalent to its own unprovability, so an equivalent sentence is simply

$$\mathrm{Con}(\mathrm{PA}) \Rightarrow n_0 \notin W_{n_0}.$$

Now Gödel noticed that his proof could be carried out in PA. (The rather laborious verification was carried out by Hilbert and Bernays (1936)). Consequently, if $\mathrm{Con}(\mathrm{PA})$ can be proved in PA, then so can "$n_0 \notin W_{n_0}$," by basic logic. But if PA is consistent, "$n_0 \notin W_{n_0}$" *cannot* be proved in it, by Gödel's theorem, hence neither can $\mathrm{Con}(\mathrm{PA})$. (Gödel of course had a different unprovable sentence, but it was similarly implied by $\mathrm{Con}(\mathrm{PA})$, and equivalent to its own unprovability.)

Thus the assertion Con(PA) that the axioms of PA are consistent is in some way stronger than the axioms themselves. Similarly, if Σ is any system that includes PA (such as *Principia Mathematica* and other systems of set theory), then Con(Σ) cannot be proved in Σ, if Σ is consistent. This is Gödel's second theorem.

EXERCISES

It is instructive to spell out why the sentence "$n_0 \notin W_{n_0}$" expresses its own unprovability, if this is not already obvious.

24.7.1 Fill in the gap so as to establish a chain of equivalences:

$$n_0 \notin W_{n_0} \Leftrightarrow \cdots \Leftrightarrow \Sigma \text{ does not prove } "n_0 \notin W_{n_0}".$$

A remarkable new form of Gödel's theorem was discovered by Chaitin (1970). Like Gödel's own version, it is most easily explained in terms of computation. Let us call a finite sequence σ of 0s and 1s *computationally random* if it cannot be produced (from a blank tape) by a Turing machine whose description is shorter than σ. To compare lengths fairly we assume that Turing machines are themselves written as sequences of 0s and 1s. (This makes the definition of "computationally random" dependent on the way we encode Turing machines, but never mind—the proof of Chaitin's theorem assumes only that the method of encoding is computable.)

24.7.2 Give an informal argument to explain why the sequence of 10^{100} consecutive 0s is *not* computationally random.

24.7.3 Show that at most $2^n - 1$ Turing machines have descriptions of length less than n.

24.7.4 Deduce from Exercise 24.7.3 that there are infinitely many computationally random sequences.

Despite the prevalence of computationally random sequences, they are very hard to find. Chaitin's incompleteness theorem states: *any sound formal system proves only finitely many theorems of the form "σ is computationally random."*

To prove Chaitin's theorem suppose, on the contrary, that there is a formal system, and hence a Turing machine M, that generates infinitely many theorems of the form "σ is computationally random," and no false statements of this form. Suppose, for example, that M has length 10^6.

24.7.5 Explain informally how to convert M to a machine M' that finds the first theorem of the form "σ is computationally random" output by M, where σ has at least 10^{100} digits.

24.7.6 Also explain informally why the length of M' is less than 10^{100}.

24.7.7 Deduce from Exercise 24.7.6 that we have a contradiction; hence M does not exist.

24.8   Provability and Truth

The previous section stressed that Gödel's theorem is a statement of alternatives: a formal system Σ either fails to prove a true sentence or else proves a false one. Gödel's second theorem identifies a sentence, $\mathrm{Con}(\Sigma)$, which is either true and unprovable or false and provable, but the proof does not say which alternative actually holds for a particular Σ, such as PA or *Principia Mathematica*. How could it, without violating Gödel's theorem itself? Unless Σ actually is *inconsistent*, there can be no formal proof that $\mathrm{Con}(\Sigma)$ is true!

Nevertheless, Gödel's theorem tells us that we have nothing to lose by adding $\mathrm{Con}(\Sigma)$ to the system Σ. If Σ is inconsistent, then it is already worthless, and we are no worse off for having added $\mathrm{Con}(\Sigma)$. And if Σ is consistent, we actually gain, because $\mathrm{Con}(\Sigma)$ is a new mathematical truth not provable from Σ alone. In this case, Gödel's theorem gives a way to transcend any given formal system. Knowing that $\mathrm{Con}(\Sigma)$ is beyond the scope of Σ (if Σ is consistent) is of practical value to mathematicians, for it means there is no point trying to prove any sentence that implies $\mathrm{Con}(\Sigma)$. If one wants to use such a sentence, it should be taken as a new axiom.

Sentences of mathematical interest actually arise in this way, most simply in set theory, where consistency is implied by the existence of a "large set." The usual axioms of set theory (called the Zermelo–Fraenkel, or ZF, axioms) say roughly that

(i) $\mathbb{N}$ is a set.

(ii) Further sets result from certain operations, the most important of which are *power* (taking all subsets of a set) and *replacement* (taking the range of a function whose domain is a set).

Because of this, the axioms of ZF can be modeled by any set that contains $\mathbb{N}$ and is closed under power and replacement. Such a set has to be very large—larger than any set whose existence can be proved in ZF—but if it exists then ZF must be consistent, since two contradictory sentences cannot be true of an actually existing object. Thus the existence of a set that is large in the above sense implies $\mathrm{Con}(\mathrm{ZF})$.

If ZF is consistent, then $\mathrm{ZF} + \mathrm{Con}(\mathrm{ZF})$ is also consistent, but an even larger set is required to satisfy the enlarged axiom system. These large-set existence axioms are called *axioms of infinity*. Since they imply $\mathrm{Con}(\mathrm{ZF})$, they cannot be proved in ZF. In particular, one cannot prove the existence of a nontrivial measure on all subsets of $\mathbb{R}$ since, as mentioned in Section 24.4, this implies the existence of a large set. In fact, the existence of a nontrivial measure on $\mathbb{R}$ is an axiom of infinity far stronger than those previously mentioned. Gödel (1946) made the interesting speculation that any true but unprovable proposition is a consequence of some axiom of infinity.

More recently, some "largeness" properties in number theory have been found to imply Con(PA). The first of these was found by Paris and Harrington (1977), using a modification of a combinatorial theorem of Ramsey (1929). Paris and Harrington found a sentence σ that says that for each $n \in \mathbb{N}$ there is an m such that sets of size $\geq m$ have a certain combinatorial property $C(n)$. They showed that σ follows from Ramsey's theorem on infinite sets (see Section 25.7), but that the function

$$f(n) = \text{least } m \text{ such that sets of size } m \text{ have property } C(n)$$

grows faster than any computable function whose existence can be proved in PA. Thus σ in some sense asserts the existence of a "large" function. The property $C(n)$ is such that one can decide whether a finite set has it or not; hence σ implies (very simply, and certainly in PA) that f is computable. This shows immediately that σ cannot be proved in PA, but Paris and Harrington in fact proved the stronger result that σ implies Con(PA).

Gödel's theorem shows that something is missing in the purely formal view of mathematics, and the axioms of infinity show that the missing elements may be mathematically interesting and important. Despite this, the official view still seems to be that mathematics consists in the formal deduction of theorems from fixed axioms. As early as 1941 Post protested against this view:

> It is to the writer's continuing amazement that ten years after Gödel's remarkable achievement current views on the nature of mathematics are thereby affected only to the point of seeing the need of many formal systems, instead of a universal one. Rather has it seemed to us to be inevitable that these developments will result in a reversal of the entire axiomatic trend of the late 19th and early 20th centuries, with a return to meaning and truth.

> Post (1941), p. 345

I believe that what Post was saying was this: Before Gödel, the goal of mathematical logic had been to distill all mathematics into a set of axioms.

It was expected that all of number theory, for example, could be recovered by formal deduction from PA, that is, *by forgetting that the axioms of PA had any meaning*. Gödel showed that this was not so, and in particular that the sentence Con(PA), which expresses consistency, could not be so recovered. But it is precisely by knowing the *meaning* of the PA axioms that one knows they are consistent: contradictory sentences cannot hold in the actual structure of $\mathbb{N}$ with $+$ and $\times$. Thus it is the ability to see meaning in PA that enables us to see the truth of Con(PA) and hence to transcend the power of formal proof.

Exercises

An argument for the unprovability of "large" sets that does not assume the unprovability of consistency was discovered by Zermelo in 1928 (Zermelo's announcement is mentioned in Baer (1928)). Since this was before Gödel's own work, it seems fair to call this *Zermelo's incompleteness theorem*. It states that, *if "large" sets exist, then this fact is not provable in ZF.*

To pave the way for Zermelo's argument, we need to explain how ordinals measure the "complexity level"—called the *rank*—of sets. The simplest set is the empty set 0, which is assigned rank 0. For each ordinal α, the sets of rank $\leq \alpha + 1$ are those of rank $\leq \alpha$, together with all subsets of the set of sets of rank $\leq \alpha$.

24.8.1 Show that $1 = \{0\}$ has rank 1, and more generally that $n + 1 = \{0, 1, \ldots, n\}$ has rank $n + 1$.

If λ is an ordinal *not* of the form $\alpha + 1$, the sets of rank $\leq \lambda$ are those of rank $\alpha < \lambda$, together with all subsets of the set of sets of rank $< \lambda$.

24.8.2 Show that the ordinal $\omega = \{0, 1, 2, \ldots\}$ has rank ω.

24.8.3 More generally, show that any ordinal α has rank α.

It is essentially an axiom of ZF (the *axiom of foundation*) that every set has a rank.

An ordinal λ is called *inaccessible* if the sets of rank $< \lambda$ are closed under the power and replacement operations. Thus, if an inaccessible λ exists, the sets of rank $< \lambda$ form a model of ZF. Also, if inaccessible ordinals exist, there is a *least* inaccessible, μ.

24.8.4 Show that the sets of rank $< \mu$ are a model of ZF plus the sentence "there is no inaccessible ordinal."

24.8.5 Deduce from Exercise 24.8.4 that, if inaccessible ordinals exist, this fact is not provable in ZF.

24.9   Biographical Notes: Gödel

Kurt Gödel (Figure 24.2) was born in 1906 in Brünn, Moravia (now Brno, Czech Republic) and died in Princeton in 1978. He was the second son of Rudolf Gödel, the manager of a textile firm, and Marianne Handschuh. Both his parents were members of the substantial German-speaking minority of the region, and his mother had received some of her education at the French school in Brünn. Her influence seems to have been dominant in Kurt's upbringing, at least in the matter of church and school. He attended Lutheran institutions and was unsympathetic to the Catholic church, to which his father nominally belonged.

Figure 24.2: Kurt Gödel

Gödel had a generally happy childhood and was noted for his curiosity, being known to his family as *Herr Warum* (Mr Why). The family was fortunate that Brünn was relatively untouched by World War I, and even after the war the absorption of Moravia into the new nation of Czechoslovakia had little effect on the Gödel family. The most disturbing event of Gödel's childhood was an attack of rheumatic fever at the age of six or seven, followed by his learning, at the age of eight, that rheumatic fever can damage the heart. To the end of his life he was convinced that he had a weak heart and, when doctors found no evidence of this, he developed a distrust of the medical profession as well. This led to a brush with death in the 1940s, when he left a duodenal ulcer untreated, and he became obsessively cautious and prone to depression,

After completing secondary school, Gödel moved to Vienna (his father's birthplace) to enter university. He was at first undecided between mathematics and physics but opted for mathematics after hearing a brilliant cycle of lectures by the number theorist Fürtwängler. He was introduced to logic and set theory by Hans Hahn, who was interested in point set problems in the theory of real functions. Hahn got Gödel involved in the famous Vienna Circle of philosophers in 1926–1928 and later became his thesis supervisor. The Vienna Circle aimed to put science and philosophy on a rigorous basis by means of formal logic and no doubt had a strong influence on Gödel's work. However, his incompleteness theorem was obviously a blow to the Vienna Circle, just as it was to formalists in mathematics. In fact, Gödel began to drift away from the Vienna Circle long before he discovered his theorem, since his philosophical position tended toward the diametric opposite of theirs. The Vienna Circle based its philosophy on strictly material data, whereas Gödel was metaphysical to the point of being interested in ghosts and demons (see for example Kreisel (1980), p. 155).

In 1927 Gödel met his future wife, Adele Porkert, a dancer at a nightclub in Vienna. His parents objected to her, on the grounds that she was six years older than Gödel and had been married before, and the couple did not marry until 1938. The marriage endured, and friends noted how much warmer Gödel became in her company. They had no children, and Adele was probably the only person in Gödel's life who could bring him down to earth occasionally.

Gödel became an Austrian citizen in 1929 and rapidly rose to fame after the publication of the incompleteness theorem in 1931. He was invited to the United States and made three visits to the Institute for Advanced Study in Princeton. In between, however, he suffered bouts of depression and spent some time in mental hospitals. In 1938 Hitler annexed Austria and the atmosphere became increasingly oppressive, though Gödel does not seem to have been perceptive about the menace of Nazism. He blamed the situation on Austrian "sloppiness" and decided to leave only when he was judged fit for military service—an obviously incompetent judgment in his opinion.

During this tense period of his life (1937–1940), Gödel tackled the main problems of set theory and proved the consistency of the axiom of choice and the continuum hypothesis. Thus he arrived at Princeton in 1940 on a second wave of fame. He settled into a position at the Institute for Advanced Study, where he was to stay for the rest of his life. In the early 1940s he continued to work hard on set theory. In 1942 he found a proof of the independence of the axiom of choice but left his work unpublished when he found he was unable to do the same for the continuum hypothesis (namely, to show that if set theory is consistent, one can consistently assume that the axiom of choice is true but the continuum hypothesis is false). These are the results, of course, that were eventually obtained by Cohen (1963).

From 1943 onward, Gödel devoted himself mainly to philosophy. Indeed, Kreisel (1980), p. 150, argues that *all* of Gödel's discoveries stemmed from his philosophical acuteness—allied with the appropriate, but generally elementary, mathematical techniques. The incompleteness theorem, for example, comes from observing the difference between provability and truth. Gödel (1949) made an unexpected foray into another area of mathematics of philosophical interest, the theory of relativity. He showed that there are solutions of Einstein's equations that contain closed timelike lines, theoretically allowing the possibility of time travel. Gödel later calculated that the amount of energy required to travel into one's own past was prohibitively large, but the feasibility of signals to and from the past remained open. Indeed, he seems to have believed that this was a possible basis for the existence of ghosts (Kreisel (1980), p. 155).

Gödel was understandably reticent about expressing such opinions publicly. Even in the case of the incompleteness theorem, whose implications for the question of minds versus machines were widely debated, he did not publish his opinions. His private view, that the mind is more powerful than a machine, may, however, have been important in enabling him to foresee the incompleteness theorem in the first place. Indeed, it may not be too much to say that Gödel's receptiveness to scientifically unconventional ideas paved the way for his unconventional theorems.

25

Combinatorics

PREVIEW

In this final chapter we look at another field that came to maturity in the 20th century: combinatorics. Like number theory before the 19th century, combinatorics before the 20th century was thought to be an elementary topic without much unity or depth. We now realize that, like number theory, combinatorics is infinitely deep and linked to all parts of mathematics. Here we emphasize the parts that link nicely to topics from earlier chapters, but without completely sacrificing the distinctive features of the subject.

Combinatorics is often called "finite mathematics" because it studies finite objects. But there are infinitely many finite objects, and it is sometimes convenient to reason about all members of an infinite collection at once. In fact, combinatorics pioneered this idea with the use of *generating functions* (already seen in Section 10.6).

Other important infinite principles in combinatorics are the *infinite pigeonhole principle* and the *Kőnig infinity lemma*. We illustrate these first by some classical proofs in number theory and analysis, then in the 20th-century fields of *graph theory* and *Ramsey theory*. Ramsey theory leads us to a proof of the Paris–Harrington theorem, mentioned in Section 24.8 as a theorem that *cannot* be proved in the strictly finite reasoning of PA.

Infinite reasoning is likewise essential for graph theory. The field had its origins in topology, and it is still relevant there, but it has expanded extraordinarily far in other directions. Graph theory today is exploring the boundaries of finite provability first exposed by Gödel's incompleteness theorem.

J. Stillwell, *Mathematics and Its History*, Undergraduate Texts in Mathematics,   553
DOI 10.1007/978-1-4419-6053-5_25, © Springer Science+Business Media, LLC 2010

25.1 What Is Combinatorics?

Combinatorics is a large and rapidly-growing area of mathematics with a long history. But, until recently, it consisted of isolated fragments without a sense of unity. We have seen some of these fragments in earlier chapters:

- Permutations and combinations. As we saw in Section 11.1, the binomial coefficients $\binom{n}{k}$ and Pascal's triangle were used in medieval Chinese algebra. Independently, Levi ben Gershon interpreted $\binom{n}{k}$ as the number of combinations of n things, taken k at a time, and used this interpretation to show that

$$\binom{n}{k} = \frac{n!}{(n-k)!k!}.$$

  Today we would say that he interpreted $\binom{n}{k}$ *combinatorially*. This interpretation shows, among other things, that $\frac{n!}{(n-k)!k!}$ is always an integer—a result that is not obvious on arithmetic grounds. In fact, Gauss's *Disquisitiones*, Section 127, contains the first "direct" proof of this result using the concepts of prime factorization and divisibility. Gauss's proof is considerably longer than the combinatorial proof.

- Generating functions. The *algebraic* interpretation of $\binom{n}{k}$, as the coefficient of x^k in $(1+x)^n$, is convenient for proving other properties of the binomial coefficients; for example, the Pascal's triangle property

$$\binom{n+1}{k} = \binom{n}{k-1} + \binom{n}{k}.$$

  A function that packages a sequence of numbers as coefficients of powers of x—the way that $\binom{n}{0}, \binom{n}{1}, \ldots, \binom{n}{n}$ are packaged by

$$(1+x)^n = \binom{n}{0} + \binom{n}{1}x + \cdots + \binom{n}{n}x^n$$

  —is called a *generating function*. As we saw in Section 10.6, a generating function can also package an infinite sequence of numbers in a compact way. In particular, the Fibonacci sequence $F_0, F_1, F_2, \ldots$ is packaged by the function

$$\frac{x}{1-x-x^2} = F_0 + F_1x + F_2x^2 + \cdots .$$

This leads to the surprising formula

$$F_n = \frac{1}{\sqrt{5}}\left[\left(\frac{1+\sqrt{5}}{2}\right)^n - \left(\frac{1-\sqrt{5}}{2}\right)^n\right].$$

- The Euler polyhedron formula, discussed in Section 22.2, reveals a combinatorial property of polyhedra. Regardless of the shape or size of the polyhedron, the *numbers* V, E, and F of its vertices, edges, and faces must satisfy $V - E + F = 2$.

What these fragments have in common is a focus on *finite, discrete* aspects of mathematical objects that can be *counted* by natural numbers. For this reason, combinatorics is also known as *finite mathematics, discrete mathematics*, or simply "counting." If one were to attempt a formal axiomatic definition, one could say that combinatorics is the theory of finite sets. There is a standard set of axioms for finite set theory, namely the ZF axioms mentioned in Section 24.8, minus the axiom asserting the existence of an infinite set. We call this axiom set ZT − Infinity for short.

In practice, combinatorialists do not prove theorems in ZF − Infinity, since this would be unbearably tedious. Nevertheless, there are some insights to be gained from this point of view. For one thing, it reveals that combinatorics is equivalent, in a certain sense, to elementary number theory. This follows from the work of Ackermann (1937), who pointed out that number theory and finite set theory "contain" each other.

Finite set theory contains number theory because the natural numbers can be defined as certain finite sets, as we saw in Section 24.2. Namely, 0 is the empty set and

$$1 = \{0\}, \quad 2 = \{0, 1\}, \quad \ldots, \quad n + 1 = \{0, 1, 2, \ldots, n\}, \quad \ldots.$$

The axioms of finite set theory also allow one to prove the *induction* principle, which, as Dedekind and Peano discovered, is the basis for elementary number theory.

Conversely, elementary number theory "contains" finite set theory, but in a more subtle sense. Finite sets can be *encoded* by natural numbers, and operations on finite sets can be encoded by operations on natural numbers (in fact, by operations definable from addition and multiplication). This was discovered by Gödel, as part of the machinery of his incompleteness proof (Section 24.7). It follows that every theorem of combinatorics is encoded by a theorem of number theory, so combinatorics and number theory are essentially equivalent.

Thus, if combinatorics is defined to be finite set theory, we can say that combinatorics has exactly the same depth and degree of difficulty as elementary number theory—which of course is plenty. In fact, this definition raises the specter of Gödel incompleteness in combinatorics. Just as there are sentences of elementary number theory that cannot be proved from the axioms of elementary number theory (the "Peano axioms"), there are sentences of finite set theory that cannot be proved from the axioms of finite set theory. What is remarkable is that such sentences actually arise more naturally in combinatorics than they do in number theory. We will see examples in Section 25.8.

For these reasons combinatorics cannot be entirely "finite" mathematics. The evolution of combinatorics since the 19th century bears this out. Assumptions about infinite sets have been used increasingly often in combinatorics (as they have been in number theory for a longer period)—often for convenience, but sometimes out of logical necessity. We study the evolution from finite to infinite in the sections that follow.

EXERCISES

A spectacular example of a generating function is the one for the so-called *Catalan numbers* C_n, which count the valid strings of n pairs of parentheses. The valid strings may be defined inductively as follows:

- The empty string is a valid string.

- If a and b are valid strings, possibly empty, then so are $a(b)$ and $(a)b$.

25.1.1 Confirm that $C_0 = 1, C_1 = 1, C_2 = 2$, and $C_3 = 5$ by enumerating the possible valid strings.

25.1.2 Explain why $C_{n+1} = C_0C_n + C_1C_{n-1} + C_2C_{n-2} + \cdots + C_nC_0$.

Now let
$$C(x) = C_0 + C_1x + C_2x^2 + \cdots$$
be the generating function of the Catalan numbers.

25.1.3 Compute $C(x)^2$, and hence show that

$$\text{coefficient of } x^n = C_0C_n + C_1C_{n-1} + C_2C_{n-2} + \cdots + C_nC_0.$$

25.1.4 Deduce from Exercise 25.1.3 that $C(x)$ satisfies the equation

$$1 + xC(x)^2 = C(x),$$

so that
$$C(x) = \frac{1 \pm (1 - 4x)^{1/2}}{2x}.$$

25.1.5 Expanding $(1 - 4x)^{1/2}$ by the binomial theorem, deduce from Exercise 25.1.4 that
$$C_n = \frac{1 \cdot 3 \cdot 5 \cdots (2n - 1)}{(n + 1)!} 2^n.$$

25.1.6 Show also that
$$C_n = \frac{1}{n + 1} \binom{2n}{n}.$$

25.2   The Pigeonhole Principle

An obvious property of finite sets is that if n objects ("pigeons") are distributed among fewer than n sets ("pigeonholes") then at least one set contains more than one object. This property is called the *pigeonhole principle*, and it was first used to prove serious theorems by Dirichlet around 1840 (see Dedekind's Supplement VIII to Dirichlet (1863)). One of them is the following theorem of number theory.

Dirichlet's approximation theorem. *For any irrational number α and any integer $Q > 1$ there are positive integers p and q with $0 < q < Q$ and $|q\alpha - p| \leq 1/Q$.*

For example, suppose we choose $Q = 100$ and $\alpha = \pi$. According to the approximation theorem, there are an integer $q < 100$ and an integer p such that
$$|q\pi - p| < 1/1000.$$

Indeed, this happens for the rational approximation $\frac{p}{q} = \frac{355}{113}$ to π that was discovered by the Chinese mathematician Zu Chongzhi (429–500 CE). The difference between π and $\frac{355}{113}$ is less than 1000000, so we actually have
$$|113\pi - 355| < \frac{1}{10000}.$$

In this case we do even better than the theorem leads us to expect!

To prove the theorem we consider the $Q + 1$ numbers
$$0, \quad 1, \quad \alpha - p_1, \quad 2\alpha - p_2, \quad \ldots, \quad (Q - 1)\alpha - p_{Q-1},$$

where $p_1, p_2, \ldots, p_{Q-1}$ are integers chosen so that all the numbers lie between 0 and 1. If we now divide the interval from 0 to 1 into subintervals of length $1/Q$, then we have Q subintervals containing $Q + 1$ numbers. It follows, by the pigeonhole principle, that at least two numbers are in the same subinterval, and hence their distance apart is $\leq 1/Q$. Since the difference between these numbers is necessarily of the form $q\alpha - p$, where p and $q \leq Q - 1$ are integers, we have

$$|q\alpha - p| \leq 1/Q, \quad \text{where} \quad 0 < q < Q. \qquad \square$$

Pell's equation revisited

Dirichlet used his approximation theorem to prove that the Pell equation $x^2 - Dy^2 = 1$ (Section 3.4) always has an integer solution. In his proof he also made crucial use of the *infinite pigeonhole principle*, which states that distributing infinitely many objects among finitely many sets results in at least one infinite set.

Given a nonsquare integer D, Dirichlet finds integers x and y such that $x^2 - Dy^2 = 1$ by thinning out the infinite set of all integer pairs p, q.

His first step is to find infinitely many such pairs for which

$$p^2 - Dq^2 = (p - q\sqrt{D})(p + q\sqrt{D}) \leq 3\sqrt{D}.$$

This is done by letting $Q \to \infty$ in the Dirichlet approximation theorem with $\alpha = \sqrt{D}$, which gives infinitely many pairs p, q such that

$$|p - q\sqrt{D}| \leq 1/q.$$

Then, since

$$|p + q\sqrt{D}| = |p - q\sqrt{D} + 2q\sqrt{D}| \leq |p - q\sqrt{D}| + |2q\sqrt{D}|,$$

we have

$$|p + q\sqrt{D}| \leq 3q\sqrt{D}.$$

And therefore

$$p^2 - Dq^2 = (p - q\sqrt{D})(p + q\sqrt{D}) \leq \frac{1}{q} \cdot 3q\sqrt{D} = 3\sqrt{D}.$$

His second step is to apply the infinite pigeonhole principle:

- Since there are only finitely many integers less than $\sqrt{D}$, $p^2 - Dq^2$ must have the same value N for infinitely many pairs p, q.

- Among these pairs p, q, infinitely many contain integers p leaving the same remainder, A, on division by N.

- Among the pairs p, q for which p leaves remainder A, infinitely many contain integers q leaving the same remainder, B, on division by N.

Thus at the end of the second step we have an integer N and an infinite set of pairs p, q such that

- For each pair, $p^2 - Dq^2 = N$.

- All p leave the same remainder, A, on division by N.

- All q leave the same remainder, B, on division by N.

Now take two pairs, p_1, q_1 and p_2, q_2, from this set, so the corresponding numbers, $p_1 - q_1 \sqrt{D}$ and $p_2 - q_2 \sqrt{D}$, are distinct. It follows that the number

$$x - y\sqrt{D} = \frac{p_1 - q_1 \sqrt{D}}{p_2 - q_2 \sqrt{D}} \qquad (*)$$

has both x and y nonzero. It also follows that

$$x + y\sqrt{D} = \frac{p_1 + q_1 \sqrt{D}}{p_2 + q_2 \sqrt{D}},$$

and hence (multiplying the last two equations)

$$x^2 - Dy^2 = \frac{p_1^2 - Dq_1^2}{p_2^2 - Dq_2^2} = \frac{N}{N} = 1.$$

Thus we have a nontrivial solution of the Pell equation, provided that x and y are integers. This last step is a routine calculation. First, we calculate from $(*)$ that

$$x = \frac{p_1 p_2 - q_1 q_2 D}{N} \quad \text{and} \quad y = \frac{q_1 p_2 - q_2 p_1}{N}.$$

We also know, since all p leave remainder A and all q leave remainder B, that

$$p_1 = a_1 N + A, \quad p_2 = a_2 N + A, \quad q_1 = b_1 N + B, \quad q_2 = b_2 N + B \quad (**)$$

for some integers a_1, a_2, b_1, b_2. So it remains to substitute the expressions (**) in $p_1 p_2 - q_1 q_2 D$ and $q_1 p_2 - q_2 p_1$ and see whether the results are divisible by N. This last step is easy (given some guidance on how to handle the term $A^2 - B^2 D$), so we leave it to the exercises.

EXERCISES

25.2.1 Show that

$$\frac{p_1 - q_1 \sqrt{D}}{p_2 - q_2 \sqrt{D}} = \frac{p_1 p_2 - q_1 q_2 D}{N} - \frac{q_1 p_2 - q_2 p_1}{N} \sqrt{D}.$$

25.2.2 Show that N divides $q_1 p_2 - q_2 p_1$ by substituting the expressions (**).

25.2.3 Given that $p_1^2 - q_1^2 D = N$, show, by substituting the expressions (**), that N divides $A^2 - B^2 D$.

25.2.4 Show, by substituting the expressions (**) and using Exercise 25.2.3, that N divides $p_1 p_2 - q_1 q_2 D$.

25.3 Analysis and Combinatorics

The infinite pigeonhole principle made another important appearance in 19th-century mathematics, in the following theorem about real numbers. A theorem like this was first proved by Bolzano (1817), in the course of his attempt to prove the intermediate value theorem. However, the theorem was not appreciated until Weierstrass took it up again in the 1860s, as part of a comprehensive theory of real numbers, limits, and continuity.

Bolzano–Weierstrass theorem. *If S is an infinite set of points between 0 and 1, then there is a point X, every neighborhood of which contains points of S other than itself.*

A *neighborhood* of X consists of all the points within distance ε of X, where $\varepsilon > 0$. We call a point X a *limit point* of S if every neighborhood of X contains infinitely many points of S. Thus the Bolzano–Weierstrass theorem asserts that any infinite set of points in the unit interval has a limit point. This makes it a theorem of analysis but, as we will see in Section 25.6, the Bolzano–Weierstrass theorem is not out of place in a chapter on combinatorics.

To prove the theorem we apply the infinite pigeonhole principle infinitely many times.

We start by by dividing the unit interval $[0, 1]$ into two halves, $[0, 1/2]$ and $[1/2, 1]$. Since the infinite set S is distributed between these two halves, at least one of them contains infinitely many members of S. Pick such a half (say, the leftmost that contains infinitely many members of S), and call it I_1. We similarly find a half of I_1, call it I_2, that contains infinitely many points of S, and so on.

The result of this process, illustrated in Figure 25.1, is an infinite sequence $I_1, I_2, I_3, \ldots$ of subintervals of $[0, 1]$, each of which contains infinitely many members of S. Also, each interval I_{n+1} is a half of I_n, so there is a single point X common to all of $I_1, I_2, I_3, \ldots$, by the nested interval property mentioned in Section 24.1.

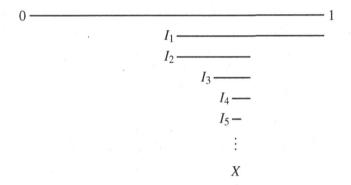

Figure 25.1: Constructing the limit point X

Clearly, each neighborhood of X contains one of the intervals I_n, and hence infinitely many points of S, so X is a limit point of S.   □

An interesting consequence of the Bolzano–Weierstrass theorem is the following:

Monotonic subsequence theorem. *Any infinite sequence $x_1, x_2, x_3, \ldots$ of real numbers contains an infinite monotonic subsequence.*

We call a subsequence $y_1, y_2, y_3, \ldots$ *monotonic* if $y_1 \leq y_2 \leq y_3 \leq \cdots$ or $y_1 \geq y_2 \geq y_3 \geq \cdots$. For example, the sequence

$$0, 1/2, 1/3, 2/3, 1/4, 3/4, 1/5, 4/5, \ldots$$

contains the monotonic subsequence $0, 1/2, 2/3, 3/4, 4/5, \ldots$.

To prove the theorem we suppose that the sequence $x_1, x_2, x_3, \ldots$ contains infinitely many *different* numbers. (If not, then the sequence must contain a constant sequence, which is trivially monotonic.) If the sequence is unbounded, we can find an infinite monotonic subsequence by continually choosing members further in the unbounded direction. If the sequence is bounded, then the set S of its members has a limit point X, by the Bolzano–Weierstrass theorem.

There are now two possibilities: S has infinitely many members greater than X, or S has infinitely many members less than X. If S has infinitely many members greater than X, let y_1 be any one of them. Since X is a limit point of S, there are infinitely many members of S between X and y_1. So, by looking along the sequence $x_1, x_2, x_3, \ldots$ beyond y_1 we will eventually find a y_2 between X and y_1, then a y_3 between X and y_2, and so on. This gives an infinite monotonic subsequence $y_1 > y_2 > y_3 > \cdots$ above X.

Similarly, if there are infinitely many members of S less than X, we only have to pick one of them, y_1, and an infinite monotonic subsequence $y_1 < y_2 < y_3 < \cdots$ below X will follow.   □

The monotonic subsequence theorem can be proved more "combinatorially," without appealing to limit points, as we will see in Section 25.7. However, the limit point X makes for an easier proof, by providing a "target" for the monotonic subsequence to approach, like a heat-seeking missile. (If there is more than one limit point, all the better: there is more than one target for the missile.) This is the first intimation that analysis—our intuition about continuous structures like the line—can guide us to results in discrete mathematics. In Section 25.7 we will see how this intuition can even guide us to results about *finite* sets.

But first, it is high time we looked at some typical concepts and results of finite combinatorics.

EXERCISES

25.3.1 Prove the monotonic sequence theorem directly for *integer* sequences.

25.3.2 Prove the following two-dimensional version of Bolzano–Weierstrass: any infinite set of points in the unit square has a limit point.

25.3.3 Show that a convergent sequence of distinct points $P_1, P_2, P_3, \ldots$ in the plane contains a convergent subsequence of points $Q_i = (x_i, y_i)$ that is "monotonic" in the sense that $x_1, x_2, x_3, \ldots$ is monotonic and $y_1, y_2, y_3, \ldots$ is monotonic.

25.4   Graph Theory

In the first three sections of this chapter I have placed combinatorics against a classical background of number theory, geometry, and analysis. My purpose was to show that combinatorics has deep roots, so presumably we can expect big things of the subject. All this is true, but it is also true that combinatorics is the most naive branch of mathematics, accessible with almost no background. The most naive branch of combinatorics is *graph theory*, a subject that is visual and easily grasped, yet rich in connections with other parts of mathematics. In this section we illustrate these connections with the example of the Euler polyhedron formula, previously discussed in Section 22.2.

First, what are the "graphs" studied in graph theory? They are *not* graphs of functions as studied in calculus and analytic geometry. They are (usually finite) structures consisting of *vertices* and *edges*. As in geometry, we can think of vertices as points (but they are denoted by thick dots in diagrams) and of edges as arcs connecting pairs of distinct vertices. The positions of the vertices and the shapes of the edges are irrelevant: the graph is completely specified by saying *which* vertices are connected by edges. A common convention is that *at most one* edge connects a given pair of vertices, so a graph is essentially just a pair of sets: a set of objects called vertices, and a set of pairs of distinct vertices (the set of edges).

This is an abstract definition of graph, but usually we just draw pictures, such as the four graphs shown in Figure 25.2.

Figure 25.2: Four graphs

To be precise, there are four *connected* graphs in Figure 25.2, a graph being called "connected" when it contains a "path" between any two of its vertices. A *path* is a sequence of distinct edges, each of which has a vertex in common with the next. It is also true that Figure 25.2 can be viewed as a picture of one disconnected graph with four "connected components." However, we usually confine our attention to connected graphs.

Trees

A *tree* is a connected graph containing no closed paths, where a path is called *closed* if its final vertex equals its initial vertex. Thus the first and last graphs in Figure 25.2 are trees. (Moreover, the edges in the last graph in Figure 25.2 form a path, and those in the second graph in Figure 25.2 form a closed path.) Another tree is shown in Figure 25.3.

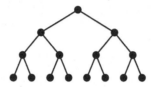

Figure 25.3: A tree

If we let V denote the number of vertices, and E the number of edges, of a tree, then we have the following:

"Euler formula" for trees. *For any tree, $V - E = 1$.*

We prove this by induction on V. If $V = 1$ we have a tree with one vertex, and hence no edges, so $V - E = 1$. Now assume that the theorem is true for trees with $\leq k$ vertices, and suppose we are given a tree T with $k + 1$ vertices. In T we can find a vertex v that is the endpoint of only one edge e, by following any path as far as possible. The path must reach an "end" vertex v because T has no closed paths. Then if we remove both v and e we still have tree, T', with $V' = k$ vertices and hence $E' = k - 1$ edges by induction. But then T has k edges, as required.   □

Plane graphs

We call the formula $V - E = 1$ for trees an "Euler formula" because it is a natural precursor to the Euler polyhedron formula. In fact, it leads to a generalization of the polyhedron formula for graphs drawn in the plane.

A graph is called a *plane graph* if its vertices are points in the plane and its edges are arcs in the plane that meet only at their endpoints. The plane graphs include graphs of all convex polyhedra, because any convex polyhedron can be projected one-to-one onto the plane (for example, by first projecting the polyhedron onto a sphere around it, then projecting the sphere stereographically onto the plane). As examples, consider the plane graphs of the tetrahedron, cube, and octahedron shown in Figure 25.4.

Figure 25.4: Plane graphs of polyhedra

A plane graph G has, in addition to vertices and edges, *faces*. The faces of G are the regions into which the plane is cut by the edges of G (think of actually cutting the plane along the edges of G). One sees, for example, that the cube graph has six faces, as it should. The first proof of the Euler polyhedron formula using plane graphs was given by Cauchy (1813b). Cauchy considered only polyhedral graphs, but his idea extends naturally to all plane graphs, and is considerably simplified thereby.

Any tree can drawn as a plane graph, necessarily with one face, because there are no closed paths to create separate regions in the plane. Thus plane graphs of trees satisfy $V - E + F = 2$, which is none other than the Euler polyhedron formula. In fact we have:

Euler plane graph formula. *For any connected plane graph G with V vertices, E edges, and F faces,*

$$V - E + F = 2.$$

This can be proved by induction on the number of closed paths in G. If there are none, then G is a tree and we have $V - E + F = 2$ as explained above.

If G contains a closed path p, consider any edge e in p. The graph G' that results from G by removing e is connected, because any vertices previously connected via e are still connected (the "long way round") via what remains of p. Also, G' has fewer closed paths, so the numbers V', E', and F' of its vertices, edges, and faces satisfy $V' - E' + F' = 2$, by induction.

For G itself, $V = V'$, $E = E' + 1$ (because G has the additional edge e), and $F = F' + 1$ (because G has a face on either side of the edge e, and these two faces become one when e is removed). Thus $V - E + F = 2$ also, as required to complete the induction. □

It must be confessed that the penultimate step of this proof is not as easy as it looks. It *seems* obvious that a closed path p in the plane has an "inside" and an "outside"—and hence there is one face on the "in" side of e and another on the "out" side of e—but this is actually a subtle theorem of topology known as the *Jordan curve theorem*. Jordan (1887) recognized that the theorem requires proof, but his proof was incorrect! The first proof that meets modern standards of rigor was given by Veblen (1905).

The Jordan curve theorem is hard because it concerns arbitrary curves, which can be infinitely complicated. These complications do not really belong to combinatorics, so we have glossed over them in the proof above. A complete proof of the Euler formula nevertheless has to do *something* about closed paths in the plane, such as proving the Jordan curve theorem for polygons. This is easier, and the restriction to polygons can be justified by assuming that all plane graphs have straight edges (a valid assumption, by a theorem of Wagner (1936)).

EXERCISES

A useful concept in graph theory is that of the *degree of a vertex V*, which is the number of edges that contain V.

25.4.1 Show that the sum of the degrees equals twice the number of edges, and hence that the number of vertices of odd degree is even.

With the help of this simple observation we prove a rather surprising result known as *Sperner's lemma*, due to Sperner (1928). The lemma concerns subdivisions of the triangle into subtriangles, such as the one in Figure 25.5.

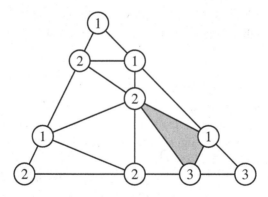

Figure 25.5: Division of a triangle into subtriangles with colored vertices

The vertices of the subdivision are "colored" by labeling them 1, 2, or 3. The coloring is arbitrary for vertices inside the triangle, but vertices on edges of the triangle obey the following rules:

- The vertices V_1, V_2, V_3 are labeled 1, 2, 3 respectively.

- Vertices on V_1V_2 are labeled 1 or 2.

- Vertices on V_2V_3 are labeled 2 or 3.

- Vertices on V_3V_1 are labeled 3 or 1.

The claim of Sperner's lemma is that *at least one subtriangle has vertices of all three colors.* In the example of Figure 25.5 this is the shaded triangle.

To prove Sperner's lemma we construct a graph with the following vertices and edges.

- A vertex inside each subtriangle, and also a vertex in the region outside the triangle $V_1V_2V_3$.

- An edge connecting any two of the vertices u, v just described, provided that the regions containing u, v meet along an edge e whose vertices are labeled 1 and 2 (in which case the connecting edge crosses e).

25.4.2 Explain why the edges from the vertex in the region outside $V_1V_2V_3$ cross the line V_1V_2, and hence show that the degree of this vertex is odd.

25.4.3 Explain why, for any other vertex u, the degree is

- 0 if u lies in a subtriangle whose vertices lack one of the labels 1, 2,

- 1 if u lies in a subtriangle whose vertices have all the labels 1, 2, 3,

- 2 if u lies in a subtriangle whose vertices have the labels 1, 2 only.

25.4.4 Deduce from Exercises 25.4.1 and 25.4.3 that there is an odd (and hence *nonzero!*) number of subtriangles whose vertices have all the labels 1, 2, 3.

25.5 Nonplanar Graphs

As we emphasized at the beginning of Section 25.4, a graph is really an abstract structure (a set of objects called vertices, and a set of pairs of distinct vertices called edges) with many concrete realizations, such as drawings in the plane. Naturally, we prefer realizations that are as simple as possible, such as plane drawings in which edges do not cross.

However, not every graph can be drawn in the plane without edges crossing. Two notorious examples are the graphs K_5 and $K_{3,3}$ shown in Figure 25.6.

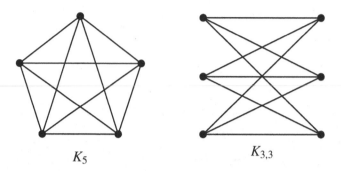

K_5 $K_{3,3}$

Figure 25.6: Two nonplanar graphs

Such graphs are called *nonplanar*. K_5 and $K_{3,3}$ are sometimes said to be given the letter K in honor of the Polish mathematician Kasimierz Kuratowski. Kuratowski (1930) proved that *any* nonplanar graph "contains" (in a sense we explain below) a copy of K_5 or $K_{3,3}$. K_5 is also called the *complete graph on five vertices*. It is one of an infinite family of graphs K_n, each of which contains an edge between any two of its n vertices. $K_{3,3}$ is one of an infinite family of graphs $K_{m,n}$, each of which has $m + n$ vertices and an edge from each of the first m vertices to each of the last n vertices.

It is quite hard to prove that each nonplanar graph "contains" (in a suitable sense) a K_5 or a $K_{3,3}$. However, we can immediately prove that K_5 and $K_{3,3}$ themselves are nonplanar, using the Euler plane graph formula.

Nonplanarity of K_5. Suppose, for the sake of contradiction, that K_5 can be realized as a plane graph. We can see that $V = 5$ and $E = 10$ for K_5, so if F is the number of faces in its plane realization we have

$$5 - 10 + F = 2, \quad \text{by the Euler plane graph formula.}$$

Thus $F = 7$.

Now, each face has at least three edges, because a face with two edges has its two vertices joined by different edges, which does not happen in K_5. Therefore, the total number of edges, E, satisfies

$$E \geq \frac{7 \times 3}{2}.$$

(We have to divide by 2 because each edge belongs to two faces, and hence is counted twice.) This contradicts the actual number $E = 10$, so K_5 cannot be realized as a plane graph. □

Nonplanarity of $K_{3,3}$. Again suppose, for the sake of contradiction, that $K_{3,3}$ can be realized as a plane graph. We can see that $V = 6$ and $E = 9$ for $K_{3,3}$, so if F is the number of faces in its plane realization we have

$$6 - 9 + F = 2, \quad \text{by the Euler plane graph formula.}$$

Thus $F = 5$.

Now in $K_{3,3}$ there are no triangles, because any path of three edges starts and ends in opposite sets of vertices, hence is not closed. Therefore, each face in the plane realization of $K_{3,3}$ has at least four edges, so the total number of edges satisfies

$$E \geq \frac{5 \times 4}{2}$$

(dividing by 2, as before, because each edge belongs to two faces). This contradicts the actual number $E = 9$, so $K_{3,3}$ cannot be realized as plane graph. □

Since K_5 is nonplanar, so is any *subdivided K_5*, where subdivision is the process of replacing certain edges of K_5 by paths. An example is the graph shown in Figure 25.7, in which the bottom edge has been replaced by a 3-edge path, and the edge parallel to it by a 2-edge path. Intuitively, we "subdivide" edges by inserting extra vertices in them. If a subdivided K_5 has a plane realization, then so has K_5, simply by erasing the extra vertices, hence each subdivided K_5 is nonplanar.

Figure 25.7: A subdivided K_5

Similarly, the nonplanarity of $K_{3,3}$ implies the nonplanarity of any subdivided $K_{3,3}$.

These two results on the nonplanarity of subdivided graphs make up the "easy direction" of Kuratowski's theorem.

Kuratowski's theorem. *A graph is nonplanar if and only if it contains a (possibly subdivided) K_5 or $K_{3,3}$.*

The "hard direction" is to prove that any nonplanar graph contains a (possibly subdivided) K_5 or $K_{3,3}$. We do not attempt to explain the proof. However, the exercises give enough examples to show that it is not easy to find the K_5 or $K_{3,3}$ even in quite small nonplanar graphs. Nevertheless, Kuratowski's theorem gives a guaranteed test for nonplanarity, because we can test whether or not a finite graph contains a subdivided K_5 or $K_{3,3}$ by exhaustive search.

EXERCISES

Another famous nonplanar graph is the *Petersen graph* shown in Figure 25.8.

Figure 25.8: The Petersen graph

25.5.1 Use the Euler planar graph theorem to prove that the Petersen graph is nonplanar. (Hint: You may need to assume that the graph contains no quadrilaterals.)

25.5.2 Show that the Peterson graph "contains" a subdivided $K_{3,3}$, thus giving a different proof that the Petersen graph is nonplanar.

25.5.3 Show that the "twisted cube graph" shown in Figure 25.9 is nonplanar.

Figure 25.9: The twisted cube graph

Since K_5 is nonplanar, so is any graph that contains it, such as K_6 or K_7.

25.5.4 Show that K_5, K_6, K_7 may each be drawn on a torus without edges crossing. (Hint: It may be helpful to represent the torus as a square with identified edges, as in Figure 22.11.)

25.5.5 Show also that $K_{3,3}$ may be drawn on the torus without edges crossing.

25.6   The Kőnig Infinity Lemma

The Kőnig infinity lemma first appeared in Kőnig (1926), where it played
a minor role in set theory. Kőnig (1927) noticed that it is quite generally
useful as a "means of reasoning from the finite to the infinite," and finally
Kőnig (1936) placed it firmly in the graph theory setting where it seems
most natural. Today, it is usually stated as a property of infinite trees.

The lemma itself is quite easy. Its proof runs along similar lines to
those of the Bolzano–Weierstrass theorem of Section 25.3. (But, like the
Bolzano–Weierstrass theorem, it is surprisingly powerful.) To state it concisely we say that a tree has *finite branching* if each vertex belongs to only
finitely many edges, and to further indulge the tree metaphor we call an
infinite path in a tree an *infinite branch*.

Kőnig infinity lemma. *If T is an infinite tree with finite branching, then T
has an infinite branch.*

To obtain an infinite branch in T we start at any vertex v. Since T is
infinite it contains infinitely many vertices, so at least one of the finitely
many edges out of v leads into an infinite subtree T_1 of T (by the infinite
pigeonhole principle).

We choose such an edge as the initial edge of a path into T_1, with v_1 as
its first vertex in T_1. Since T_1 is also infinite, there is at least one edge out
of v_1 that leads into an infinite subtree T_2 of T_1, and so on.

Thus by repeatedly choosing an edge that leads into an infinite subtree,
we obtain an infinite path in T. This is our infinite branch.   □

It should be clear that this is essentially the same argument as that for
the Bolzano–Weierstrass theorem in Section 25.3. In the proof of Bolzano–
Weierstrass there is implicitly a *tree of subintervals* of $[0, 1]$, branching
occurs each time we split a subinterval in two, and an infinite branch of
this tree gives the limit point we seek.

The Kőnig infinity lemma allows us to obtain many kinds of limit object by constructing a "tree of finite approximations" to the object we seek.
Typically, we cannot foresee which finite approximations will extend to infinity, but the Kőnig infinity lemma allows us not to worry. If we construct
the tree of *all* finite approximations, the limit object is sure to occur as an
infinite branch.

Application to map coloring

One of the most famous theorems of combinatorics is the *four color theorem* of Appel and Haken (1976). It states that any map (in the sense of geography) can be validly colored with four colors, where valid coloring satisfies the condition that adjacent regions (countries, states, or counties say) are given different colors. Despite its fame, the four color theorem occupies a lonely place in mathematics, because its known proofs are inordinately long and seemingly unrelated to other parts of the subject.

However, the theorem has one corollary that is interesting and appropriate here: an *infinite* version that follows elegantly from the usual finite version by the Kőnig infinity lemma. Indeed, the passage from coloring finite maps to coloring infinite maps was one of the first applications of the lemma, pointed out by Kőnig (1927).

Given an infinite map M, with countries $C_1, C_2, C_3, \ldots$, one constructs a *tree of valid 4-colorings* of finite submaps (namely, of maps M_n consisting of countries $C_1, C_2, \ldots, C_n$). The four color theorem ensures that this tree is infinite, so it has an infinite branch. And it is clear from the nature of the tree that an infinite branch represents a valid 4-coloring of M.

To illustrate the construction of the tree, consider the map M whose first few regions $C_1, C_2, C_3, C_4, \ldots$ are shown in Figure 25.10 The tree of

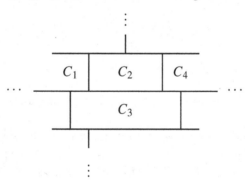

Figure 25.10: Part of an infinite map M

valid 4-colorings is now constructed from M so that each branch of length n, starting at the top vertex, represents a valid 4-coloring of the submap M_n whose regions are $C_1, C_2, \ldots, C_n$.

- The top vertex (level 0) has four outgoing edges to vertices at level 1. These four vertices represent the four possible colorings of C_1 and we give them four different colors (shades of gray in Figure 25.11, to make life easier for the printer).

- Each vertex at level 1 has up to four downward edges, leading to vertices at level 2 that are given the allowable colorings of C_2 (colors different from those of the vertex at level 1 on the same branch when C_2 is adjacent to C_1, as here).

- Each vertex at level 2 likewise has up to four downward edges, leading to vertices at level 3 that are given the allowable colorings of C_3 (colors different from any of those on the same branch assigned to regions adjacent to C_3, as C_1 and C_2 are here), and so on.

Figure 25.11 shows the complete levels 1 and 2 of the tree of colorings for the map M given in Figure 25.10, and parts of levels 3 and 4.

Each vertex at level n is the endpoint of a unique path from level 0 that assigns colors validly to $C_1, C_2, \ldots, C_n$. For example, the vertex v represents the coloring (which is valid because C_4 is not adjacent to C_1)

$$C_1 = \bullet, \quad C_2 = \bullet, \quad C_3 = \bullet, \quad C_4 = \bullet.$$

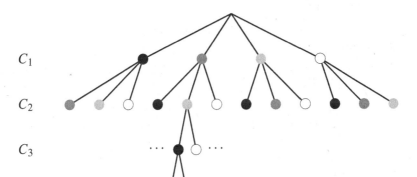

Figure 25.11: Part of the tree of 4-colorings for M

By the four color theorem, there are paths to every level. Thus the tree is infinite, and so it has an infinite branch, by the Kőnig infinity lemma. Clearly, an infinite branch assigns valid colors to all the regions $C_1, C_2, C_3, \ldots$ in M; hence M can be 4-colored.

EXERCISES

A remarkably simple proof that finite 4-coloring implies countable 4-coloring, due to G. Galperin, may be found in the book Soifer (2009). This proof depends on the Bolzano–Weierstrass theorem. Given a map with countries $C_1, C_2, C_3, \ldots$, we represent each valid coloring of the submap with countries $C_1, C_2, \ldots, C_n$ by a finite decimal number $0.a_1 a_2 \ldots a_n$, where a_i is 1, 2, 3, or 4 according as C_i is colored with the first, second, third, or fourth color.

25.6.1 It follows from the 4-color theorem that this set of numbers is infinite. Explain why it has a limit point.

25.6.2 Explain how the limit point gives a valid coloring of the map with countries $C_1, C_2, C_3, \ldots$.

This proof suggests a way to replace the Kőnig infinity lemma by the Bolzano-Weierstrass theorem in other proofs. However, it is often more natural to construct a tree than a set of real numbers. Here is an example: proving that a countable graph G is planar if and only if all its finite subgraphs G_n (involving the first n vertices of G) are planar. We call a graph "planar" if it can be realized as a plane graph.

25.6.3 Assuming that each finite planar graph has only finitely many topologically distinct planar realizations, describe a suitable tree of planar realizations of finite subgraphs $G_1, G_2, G_3, \ldots$ of G.

25.6.4 Deduce from Exercise 25.6.3 and the Kőnig infinity lemma that G is planar if all of $G_1, G_2, G_3, \ldots$ are planar.

25.6.5 Deduce from Exercise 25.6.4 that G is planar if it contains no subdivision of K_5 or $K_{3,3}$.

As another application of the Kőnig infinity lemma/Bolzano–Weierstrass, we now prove a famous theorem of topology: the two-dimensional *Brouwer fixed point theorem*. It states that *any continuous map f of the triangle into itself has a fixed point*, that is, a point P such that $f(P) = P$. The key to the following proof is Sperner's lemma, from the exercises for Section 25.4.

The idea of the proof applies to any triangle, but for convenience we take our triangle to be the equilateral triangle T in $\mathbb{R}^3$ with vertices

$$V_1 = (1, 0, 0), \quad V_2 = (0, 1, 0), \quad V_3 = (0, 0, 1).$$

The triangle T is the part of the plane $x_1 + x_2 + x_3 = 1$ for which $x_1, x_2, x_3 \geq 0$. Hence

$$0 \leq x_1, x_2, x_3 \leq 1$$

for each point (x_1, x_2, x_3) in T.

Now suppose that $f : T \to T$ is a continuous map. For any point $\mathbf{x} = (x_1, x_2, x_3)$ we will write $f(\mathbf{x})_i$ for the ith coordinate of $f(\mathbf{x})$.

25.6.6 If $f(\mathbf{x}) \neq \mathbf{x}$ for each point $\mathbf{x}$ in T, show that $f(\mathbf{x})_i < x_i$ for some i.

We now "color" each point $\mathbf{x}$ in T with the *least* i such that $f(\mathbf{x})_i < x_i$. In particular, we apply this coloring to the vertices in an infinite sequence of triangulations of T, the first three of which are shown in Figure 25.12. These triangulations are obtained by repeating the subdivision of an equilateral triangle into four equilateral subtriangles.

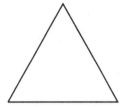 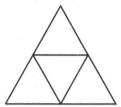 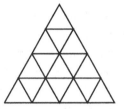

Figure 25.12: Triangulations of an equilateral triangle

25.6.7 Explain why V_1, V_2, V_3 get colors 1, 2, 3 respectively and

- Vertices on $V_1 V_2$ do not get color 3.
- Vertices on $V_2 V_3$ do not get color 1.
- Vertices on $V_3 V_1$ do not get color 2.

Thus the coloring of each triangulation in the sequence satisfies the conditions of Sperner's lemma, so each triangulation contains a subtriangle with vertices of all three colors (a "3-colored subtriangle").

25.6.8 With the help of the Bolzano–Weierstrass theorem, show that there is a convergent sequence of 3-colored triangles, that is, a sequence of 3-colored triangles whose vertices all tend to the same point $\mathbf{y} = (y_1, y_2, y_3)$.

Now, since we assume that f has no fixed point, $\mathbf{y}$ has a certain color: 1, 2, or 3.

25.6.9 Observing that $f(\mathbf{y})_i = y_i$ for at most *one* coordinate of $\mathbf{y}$ (why?), deduce from the continuity of f that any sufficiently small triangle, sufficiently close to $\mathbf{y}$, has at least two vertices of the same color.

This contradicts the definition of $\mathbf{y}$, so the assumption that f has no fixed points is false.

25.7 Ramsey Theory

Ramsey theory takes its name from the English mathematician and logician Frank Plumpton Ramsey. Ramsey (1929) laid the foundations of the subject with two important theorems, now known as the *finite Ramsey theorem* and the *infinite Ramsey theorem*. Ramsey introduced these theorems in a paper on mathematical logic, so they were not noticed by many mathematicians. His influence grew with the paper of Erdős and Szekeres (1935), which gave a simpler approach to Ramsey theory and introduced it to a wider audience. Erdős and Szekeres gave an elegant proof of the finite Ramsey theorem and also observed that it follows from the infinite Ramsey theorem by the Kőnig infinity lemma—an important connection that Ramsey himself had not noticed.

To motivate the finite Ramsey theorem we note, as many have done, the following curious fact: *in any group of six people there are either three mutual acquaintances or three mutual nonacquaintances.* We translate this fact into the language of graph theory by letting the six people be vertices of a graph, drawing a red edge between any two who are acquainted, and a blue edge between any two who are not acquainted. Then the fact about acquaintances translates into the following fact about graphs:

Baby Ramsey theorem. *A K_6 whose edges are colored with two colors always contains a monochromatic triangle (that is, a triangle whose edges are all of the same color).*

To see why this is so, notice first that each vertex v of a K_6 belongs to five edges. It follows that at least three of the edges out of v have the same color (shown as black in Figure 25.13). So, the only way to avoid a triangle of this color is to join the ends s, t, u of these three edges by edges of the *other* color (shown as gray).

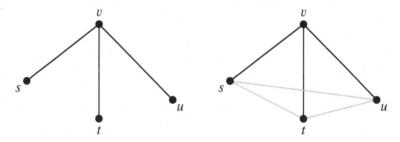

Figure 25.13: Monochrome edges and monochrome triangle

Thus there is always a monochromatic triangle.   □

Notice also that 6 is the smallest value of n for which a 2-colored K_n always contains a monochromatic triangle (or a monochromatic K_3, as we will say for short). This is because there is a 2-colored K_5 that does *not* contain a monochromatic K_3, as Figure 25.14 shows.

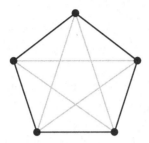

Figure 25.14: A 2-colored K_5 with no monochromatic K_3

This raises the following question: is there an n such that any 2-colored K_n contains a monochromatic K_4? The answer is yes, and 18 is the least such value of n. (So, in any group of 18 people there are either four mutual acquaintances or four mutual nonacquaintances.) More generally, we have the following "finite Ramsey theorem":

Ramsey theorem for 2-colorings of K_n. *For any m there is an n such that any 2-coloring of K_n contains a monochromatic K_m.*

This is not the most general finite Ramsey theorem. Instead of coloring edges of K_n, which are 2-element subsets of the n-element set of vertices, we can color k-element subsets of an n-element set. And instead of two colors we can use any finite number. However, we will stick to 2-coloring of edges of graphs, since the ideas of Ramsey theory can be illustrated perfectly well in this setting.

We will not give a direct proof of the Ramsey theorem for 2-colorings of K_n, since the elegant Erdős–Szekeres approach is outlined in the exercises. Instead, we will give a proof of a countably infinite Ramsey theorem, from which the finite Ramsey theorem follows easily.

The countably infinite Ramsey theorem

To state this theorem we consider the *countably infinite complete graph K_ω*, which has a countable infinity of vertices $v_1, v_2, v_3, \ldots$ and an edge between any two of them. Thus the edges of K_ω can be enumerated as follows.

- The edges from v_1 to $v_2, v_3, v_4, \ldots$.

- The edges from v_2 to $v_3, v_4, v_5, \ldots$.

- The edges from v_3 to $v_4, v_5, v_6, \ldots$.

- And so on.

Given a 2-coloring of these edges, we find a monochromatic K_ω on a countably infinite subset of the vertices by "thinning out" the enumeration above. The thinning process involves infinitely many applications of the infinite pigeonhole principle, rather like the proof of the Bolzano–Weierstrass theorem in Section 25.3.

Step 1. By the infinite pigeonhole principle, infinitely many edges out of v_1 have the same color. We let W_1 be the set of vertices to which these monochromatic edges lead, and let w_1 be the first member of W_1 in the ordering of vertices.

Step 2. By the infinite pigeonhole principle again, infinitely many of the edges from w_1 to other vertices in W_1 have the same color. Let W_2 be the set of vertices in W_1 to which these edges lead, and let w_2 be the first member of W_2 in the ordering of vertices.

Step n. By the infinite pigeonhole principle (for the nth time), infinitely many of the edges from w_{n-1} to other vertices in W_{n-1} have the same color. Let W_n be the set of vertices in W_{n-1} to which these edges lead, and let w_n be the first of them in the ordering of vertices.

By induction, W_n is infinite for each n, and hence we get an infinite subsequence $w_1, w_2, w_3, \ldots$ of the original vertices. Also the edges from w_{n-1} to $w_n, w_{n+1}, w_{n+2}, \ldots$ are all of the same color, since they are a subset of the edges from w_{n-1} to the members of W_n, and the latter edges have the same color by construction.

It is *not* necessarily the case that the color of the edges from w_1 to $w_2, w_3, \ldots$ is the same as the color of the edges from w_2 to $w_3, w_4 \ldots$, and so on. However, by the infinite pigeonhole principle again, the *same color occurs infinitely often*. Therefore (**Step ω**) we can choose an infinite subsequence $x_1, x_2, x_3, \ldots$ of $w_1, w_2, w_3, \ldots$ such that

$$\text{color of edges from } x_1 \text{ to } x_2, x_3, x_4 \ldots$$
$$= \text{color of edges from } x_2 \text{ to } x_3, x_4, x_5 \ldots$$
$$= \text{color of edges from } x_3 \text{ to } x_4, x_5, x_6, \ldots$$

and so on. That is, the complete graph with the vertices $x_1, x_2, x_3, \ldots$ is monochromatic.   □

This proves the countably infinite Ramsey theorem. To deduce the finite Ramsey theorem we suppose, for the sake of contradiction, that the finite Ramsey theorem is false. That is, there is an m such that, for any n, there is a *rogue 2-coloring* of the edges of K_n, namely, a 2-coloring that contains no monochromatic K_m.

Now construct a *tree of rogue 2-colorings* as follows. At level 0 put a dummy vertex, connected to all vertices at level 1. At each level $n > 0$ put a vertex for each rogue 2-coloring of K_n. Also connect each rogue 2-coloring of K_n to each rogue 2-coloring of K_{n+1} that extends it. Each rogue 2-coloring of K_{n+1} necessarily extends a unique rogue 2-coloring of K_n (obtained by deleting all edges that include the vertex $n+1$). Hence the graph of all rogue 2-colorings is indeed a tree.

Also, the tree is infinite, by the assumption that rogue 2-colorings of K_n exist for all n. Thus the tree has an infinite branch, by the Kőnig infinity lemma. But an infinite branch defines a 2-coloring of K_ω, obtained by the series of coloring extensions from K_n to K_{n+1} along the branch.

It follows, by the infinite Ramsey theorem, that there is a monochromatic complete graph on an infinite set of vertices $x_1, x_2, x_3, \ldots$. In particular, there is a monochromatic K_m on the vertices $x_1, x_2, \ldots, x_m$. This contradicts our assumption that *no* 2-coloring in the tree contains a monochromatic K_m. Hence the finite Ramsey theorem is true. $\qquad\square$

EXERCISES

An elegant inductive proof of the finite Ramsey theorem was given by Erdős and Szekeres (1935). Assuming that edges of graphs are colored red or blue (for the sake of argument), they define the *Ramsey number*

$R(p, q) = $ minimum m such that a 2-colored K_m contains a red K_p or a blue K_q.

Then they prove that $R(p, q)$ exists for all $p, q \geq 2$ by induction on $p+q$ as follows.

- Clearly $R(2, 2)$ exists and equals 2.

- If $R(p - 1, q)$ and $R(p, q - 1)$ exist then so does $R(p, q)$, and in fact

$$R(p, q) \leq R(p - 1, q) + R(p, q - 1).$$

The hard part is to prove the inequality $R(p, q) \leq R(p - 1, q) + R(p, q - 1)$, which is done by considering a 2-coloring of K_m, where

$$m = R(p - 1, q) + R(p, q - 1).$$

25.7.1 Explain why one of the following exists in the 2-colored K_m:

Case 1. A vertex u with at least $R(p, q - 1)$ blue edges attached to it.

Case 2. A vertex v with at least $R(p - 1, q)$ red edges attached to it.

25.7.2 In Case 1, show (by considering the $K_{R(p,q-1)}$ at the ends of the blue edges out of u) that K_m contains either a red K_p or a blue K_q.

25.7.3 In Case 2, show (by considering the $K_{R(p-1,q)}$ at the ends of the red edges out of v) that K_m contains either a red K_p or a blue K_q.

25.7.4 Explain why $R(2, 3) = R(3, 2) = 3$.

25.7.5 Look again at the proof of the "Baby Ramsey theorem" and compare it with the $p = 3$, $q = 3$ case of the proof above.

In Section 25.3 we mentioned that there is a proof of the monotonic sequence theorem that does not appeal to limit points (via the Bolzano–Weierstrass theorem). There is a purely combinatorial proof, using the infinite Ramsey theorem.

Given an infinite sequence $x_1, x_2, x_3, \ldots$ of distinct real numbers, take the x_i as vertices of a graph, with

- A red edge connecting x_i and x_j if $i < j$ and $x_i < x_j$.

- A blue edge connecting x_i and x_j if $i < j$ and $x_i > x_j$.

25.7.6 Conclude, from the infinite Ramsey theorem, that $x_1, x_2, x_3, \ldots$ contains an infinite monotonic subsequence.

We reiterate that the infinite pigeonhole principle, which underlies both Bolzano–Weierstrass and the infinite Ramsey theorem, is used infinitely often in both proofs. Thus the two proofs of the monotonic subsequence theorem appear to be roughly equal in complexity.

25.8   Hard Theorems of Combinatorics

It may seem like excessive use of force to employ the infinite Ramsey theorem to prove the finite Ramsey theorem, when reasoning about finite sets will suffice. However, the proof is a model for others, some of which have *no* finite alternative. The most historically important of these is the Paris–Harrington theorem mentioned in Section 24.8. As we said there, the Paris–Harrington theorem is not provable in Peano arithmetic (PA), and hence it is not provable in finite set theory either. Yet it follows easily from the infinite Ramsey theorem!

The reason is that the Paris–Harrington theorem is simply the finite Ramsey theorem with an additional "largeness" condition imposed on the monochromatic subset. Paris and Harrington call a finite set S of natural numbers *large* if the number of members of S is greater than the smallest member of S. It is this condition that makes their theorem unprovable in PA, but it presents no obstacle to a proof from the infinite Ramsey theorem.

The proof begins, as in Section 25.7, by supposing that the theorem is false for some size m of monochromatic set. This means that we can construct a tree of "rogue colorings" not containing a large monochromatic subset of size m or more. But then the Kőnig infinity lemma gives an *infinite* monochromatic subset, from which we can extract a *finite* set S for which the number of members is at least m and greater than the smallest member of S. Thus S is large and we have a contradiction.

The Paris–Harrington theorem is certainly a natural variation of the classical finite Ramsey theorem. Nevertheless, it was engineered by logicians for the express purpose of being unprovable in PA, and hence unprovable in finite set theory. Are there any theorems of combinatorics, of independent interest, that happen to be unprovable in finite set theory? Well, of course, the infinite Ramsey theorem is one such, because it implies the Paris–Harrington theorem. However, the infinite Ramsey theorem cannot even be *stated* in finite set theory, because it is not a sentence about finite sets. It would be better to say that the infinite Ramsey theorem is *stronger* than finite set theory, or that it is an "axiom of infinity." Other theorems that originate in combinatorics are also axioms of infinity in this sense. Two of the most famous of them are *Kruskal's theorem* and the *Robertson–Seymour theorem*.

The Kruskal and Robertson–Seymour theorems

The theorem of Kruskal (1960) concerns the ordering of finite trees by the embedding relation. We say that a tree T_1 *embeds in* a tree T_2, written $T_1 \leq T_2$, if there is a continuous one-to-one map of T_1 into T_2. The relation $\leq$ is an example of a *partial order*, which means that it satisfies

$$T \leq T \qquad \text{(reflexivity)}$$
$$T_1 \leq T_2 \quad \text{and} \quad T_2 \leq T_3 \quad \text{implies} \quad T_1 \leq T_3 \qquad \text{(transitivity)}$$

It is *not* a linear order, like the ordering of real numbers, because there are trees T_1 and T_2 such that neither $T_1 \leq T_2$ nor $T_2 \leq T_1$. Nevertheless, "complete disorder is impossible" in the following sense:

Kruskal's theorem. *Any infinite sequence of finite trees contains an infinite monotonic subsequence: $T_1 \preceq T_2 \preceq T_3 \preceq \cdots$.*

Kruskal's theorem does not extend to arbitrary finite graphs, because one can give an infinite sequence of graphs, none of which embeds in any other. An example is the sequence of polygon graphs, shown in Figure 25.15.

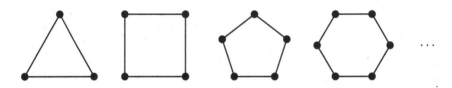

Figure 25.15: The polygon graphs

However, a similar theorem holds for a slight relaxation of the embedding relation called the *graph minor relation*. We say that G_1 is a *minor* of graph G_2 if some *blowup* of G_1 embeds in G_2. We "blow up" G_1 by finitely often replacing a vertex by two vertices connected by an edge. Thus any tree is a blowup of a single vertex, and each polygon is a blowup of any polygon with a smaller number of edges. The graph minor theorem is the following:

Robertson–Seymour theorem. *Any infinite sequence of finite graphs contains an infinite monotonic sequence under the graph minor relation.*

The proof of Kruskal's theorem is too difficult to include here. An accessible proof may be found in the last chapter of Diestel (2005). The proof of the Robertson–Seymour theorem is barely accessible to anybody, since it occupies a series of about 20 papers, published between 1983 and 2004.

Finite consequences of Kruskal and Robertson–Seymour

The apparent difficulty of these two theorems is matched by their logical strength. Like the infinite Ramsey theorem, they imply theorems that can be stated, but not proved, in finite set theory. In 1981, the American logician Harvey Friedman discovered variants of Kruskal's theorem, concerning *finite* sequences of trees, that follow from it but are not provable in finite set theory. In fact, these variants of Kruskal's theorem lie at a "higher level of unprovability" than the Paris–Harrington theorem, in a sense that can be made precise by ordinal numbers. Friedman also discovered finite variants of the Robertson–Seymour theorem that follow from it but are likewise highly unprovable in finite set theory.

These discoveries suggest that some kind of convergence between combinatorics, logic, and set theory is underway. Combinatorics seems to be the most fruitful source of easy-to-understand but hard-to-prove theorems, and it is also seems to be the place where insights from the infinite world most clearly illuminate the finite world. The importance of principles connecting the finite world to the infinite world has been emphasized by the Australian mathematician Terry Tao:

> These principles allow one to tap the power of the infinitary
> world (for instance, the ability to take limits and perform completions or closures of objects) in order to establish results in
> the finitary world, or at least to take the intuition gained in the
> infinitary world and transfer it to a finitary setting.
>
> Tao (2009), p. 165

Another spectacular finite consequence of the Robertson–Seymour theorem is the following generalization of Kuratowski's theorem on planar graphs: *for any surface S there is a finite set of "forbidden minors" for graphs on S*. That is, if a graph G cannot be drawn on S without edges crossing, then one of the forbidden minors for S is a minor of G. This follows very easily from the graph minor theorem because, in a minimal set of forbidden minors, one graph cannot be a minor of another. So the set of forbidden minors cannot be infinite, by the Robertson–Seymour theorem.

We call this a generalization of Kuratowski's theorem because Wagner (1937) showed that K_5 and $K_{3,3}$ make up a minimal set of forbidden minors for the plane. It is not clear whether the generalization to arbitrary surfaces can be proved in finite set theory, but it is certainly a difficult theorem. Apart from the plane (or sphere), the only surface for which the set of forbidden minors is known is the projective plane, for which the minimal set has 35 members. For the torus we do not even know the *number* of forbidden minors, but it is known to be at least 16000.

EXERCISES

An interesting example of a graph that is *not* forbidden for the projective plane is the Petersen graph, mentioned in the exercises to Section 25.5. In fact, when the projective plane is constructed by identifying antipodal points on the sphere, the Petersen graph arises as the image of the *dodecahedron graph* on the sphere.

25.8.1 By labeling antipodal vertices of a dodecahedron A and A', B and B', C and C', ... show that the vertex pairs $\{A, A'\}$, $\{B, B'\}$, $\{C, C'\}$, ... and the corresponding edge pairs form a Petersen graph.

Other graphs that are not forbidden for the projective plane are K_5 and K_6.

25.8.2 Show that K_6, and hence K_5, may be drawn on the projective plane without edges crossing. (Hint: Draw K_6 on a Möbius band.)

On the other hand, K_7 *is* forbidden on the projective plane.

25.8.3 A graph with V vertices, E edges, and F faces on the projective plane corresponds to a graph with $2V$ vertices, $2E$ edges, and $2F$ faces on the sphere. Why? Conclude that $V - E + F = 1$ for graphs on the projective plane.

25.8.4 Use the formula in Exercise 25.8.3 to show that K_7 cannot be drawn on the projective plane without edges crossing.

25.9   Biographical Notes: Erdős

Paul Erdős (or Erdős Pál as he was known in Hungary) was born on March 26, 1913, in Budapest, to a middle-class Jewish family. His parents, Lajos and Anna, were mathematics teachers, so Paul was immersed in mathematics from childhood. He took to it immediately, so much so that it became almost his whole world for the rest of his life. His single-minded devotion to mathematics—remarkable even for a mathematician—undoubtedly had a lot to do with the shocks experienced by his family when he was a child.

Just as he was born, his two older sisters were carried off by scarlet fever. Then in 1914, with the onset of World War I, Lajos was drafted into the Austro-Hungarian army. Shortly thereafter, he was captured by the Russians and imprisoned for six years in Siberia. Before his return, the Austro-Hungarian empire vanished, and Hungary experienced a postwar upheaval with the Hungarian Commune of Béla Kún. In 1919, the Commune collapsed after 132 days, and was followed by a wave of "White Terror" with attacks on Communists and Jews.

Understandably, Anna became a very protective mother to her only remaining child, but she took protectiveness to bizarre extremes. Paul did not enter school until age 11 (about the same time that he first learned to tie his shoes) and he did not butter his own bread until 1934, when he made his first overseas trip. There were some things he never learned to do and, after Anna died in 1971, colleagues had to step in to make his travel arrangements, manage his bank account, and so on.

But always, Paul was learning and discovering mathematics. At the age of four he discovered negative numbers by himself. At the age of 10, Lajos showed him Euclid's proof that there are infinitely many primes, and at the age of 13 he had his first publication—the solution to a problem in a journal for secondary school students. At the age of 18, in his first year at the Science University of Budapest, he came to the attention of serious mathematicians with a new and elementary proof of a theorem first proved by the Russian mathematician Pafnuty Chebyshev in 1850. As Erdős himself might have said (in words penned by the mathematician Nathan Fine):

> Chebyshev said it, and I'll say it again,
> There is always a prime between N and $2N$.

Chebyshev's theorem is a new demonstration that there are infinitely many primes, and it also gives an idea of how *dense* they are. As such, it can be viewed as the first step towards a proof of the *prime number theorem*: *the number of primes less than N is asymptotically equal to $N/\log N$.* That is, if $\pi(N)$ is the number of primes less than N, then the ratio of $\pi(N)$ to $N/\log N$ tends to 1 as N tends to infinity.

The prime number theorem was conjectured around 1800 by Legendre and Gauss, but not proved for almost 100 years, independently by Hadamard (1896) and de la Vallée Poussin (1896). Their proofs made heavy use of analysis, particularly the zeta function of Euler and Riemann studied in Section 10.7, and for a long time a proof by elementary methods was thought impossible. Erdős had something to say about this later, as we will see.

Number theory was Erdős's first love, but he discovered the joys of graph theory through a course given by Dénes Kőnig while he was at university. In 1935 he made his mark on the infant field of combinatorics with the Ramsey theory paper Erdős and Szekeres (1935). This joint paper with George Szekeres had a seemingly frivolous origin, but it was typical of the way that Erdős liked to work: start with a simple problem that contains the germ of a new idea, and generalize.

Szekeres and his future wife, Esther Klein, belonged to a circle of student friends around Erdős. In the early 1930s they used to gather for mathematical conversations at the Erdős home, the city park, or on hikes in the hills around Budapest. On one of these excursions, Esther Klein raised a problem in elementary geometry: does every set of five points in the plane, no three of which are in a line, include four that are the vertices of a convex quadrilateral? Klein proved that the answer is yes, by considering the various possible cases, but a more general problem beckoned: for any integer n, is there an N such that any N points in the plane (again, with no three in a line) include n that are the vertices of convex n-gon? It was this apparently simple problem that led Erdős and Szekeres to Ramsey theory.

In later years, Erdős came to call the problem of convex subsets the "happy ending" problem; not only because it put Ramsey theory on the map, but because George and Esther got married in 1936 and had a long and happy life together. (They both died on August 28, 2005, within an hour of each other.)

The 1930s were a time of mathematical ferment for Erdős and his friends, but also a time of great anxiety. Most of them were Jewish, so it was clear to them that they had to leave Europe to survive. George and Esther Szekeres went to Shanghai and then Australia, and Erdős first to the UK, then to the US. This was the beginning of his restless search for what he called "another roof, another proof"—never with a permanent job, or a home, or more than a suitcase full of possessions. However, his travels began auspiciously enough, with a one-year fellowship at the Institute for Advanced Study (IAS) at Princeton in 1938-1939.

The institute was set up in the 1930s as a haven for top mathematicians, physicists, and other scholars. Its first permanent member was Albert Einstein, and other stellar refugees from Europe soon followed, such as John von Neumann and Kurt Gödel. The idea of the IAS was to allow its members complete freedom to pursue their researches, without any teaching or administrative responsibilities. This was fine for a dynamo like Erdős, who spent a very productive year there, but not necessarily for more contemplative types. Gödel, in particular, published only a handful of papers in his 40 years at the Institute, and spent long periods studying the philosophical writings of Leibniz and Russell. Once, in exasperation, Erdős told Gödel "you became a mathematician so that people should study you, not that you should study Leibniz!"

The IAS may in fact have had some reservations about Gödel, because he was not made a full professor there until 1953. However, at least they renewed his position every year until he became a permanent member. Erdős became the only person ever "fired" from the IAS: not having his one-year fellowship renewed, and being allowed to stay for one further year only when outside funding was found for him. It is not clear what Erdős did wrong. Certainly, he was eccentric, but so was Gödel. Perhaps his peers at the IAS thought his elementary methods were immature and shallow. If so, they had to think again, because in 1949 Erdős discovered the most famous elementary proof of all: an elementary proof of the prime number theorem.

Alas, the Erdős proof was tarnished by its entanglement with another elementary proof of the same theorem, by the Norwegian mathematician Atle Selberg. Selberg set the ball rolling in 1948 with an elementary proof of a famous theorem of Dirichlet (1837), according to which *there are infinitely many primes in any arithmetic progression $a+b, 2a+b, 3a+b, \ldots$, where* $\gcd(a, b) = 1$. Dirichlet's theorem was the oldest example of an arithmetic theorem with apparently unavoidable use of analysis in its proof, so Selberg's proof was already a sensation. But Selberg (and Erdős, as soon as he heard about Selberg's proof) thought he could do better, with an elementary proof of the prime number theorem.

What followed in 1949 was a proof by Selberg and a proof by Erdős, as Selberg preferred, and not a joint paper, as Erdős would have preferred. For his proof, Selberg was awarded the highest honor in mathematics in 1950, the Fields Medal, and a position at the IAS. Erdős was awarded the Cole Prize in 1952 (a high honor, but not as high as the Fields Medal) and the offer of a permanent position at Notre Dame. He spent a happy year at Notre Dame, but his restless spirit could not be contained there. Indeed, with the the coming of McCarthyism in the 1950s, Erdős found himself shut out of the the US until 1959. Even then, it took the intervention of senator Hubert Humphrey to obtain permission for a brief visit.

In 1963, the US finally became hospitable to Erdős again. His visa problems ceased, and he met the mathematician Ron Graham, who became his chief facilitator and protector. For the next three decades, Graham managed Erdős's affairs, eventually going so far as to build a special room in his house where Erdős stayed on his frequent visits. Graham also championed the mathematics of Erdős, particularly graph theory and Ramsey theory. He coauthored the definitive book on Ramsey theory, Graham et al. (1990). During these three decades, Erdős followed a chaotic whirlwind path from roof to roof, proof to proof. Wherever he was, he drew people into mathematical conversations, assessed their abilities, and set them working on problems appropriate to their mathematical strength. From these collaborations, hundreds of joint papers followed, and with them the concept of

Erdős number. Persons have Erdős number 1 if they have written a joint paper with Erdős (there are currently 511 such persons), and Erdős number $n + 1$ if they do not have Erdős number n but have coauthored a paper with someone having Erdős number n.

Erdős used to say that he wished to emulate Euler and die while doing mathematics (see Schechter (1998), p. 201). He made virtually sure of it, by sleeping very little and doing mathematics almost constantly while awake. He even continued talking to mathematician friends in the operating theater during a cornea transplant and insertion of a heart pacemaker in 1995 and 1996. And when he finally did die, on September 20, 1996, it was during a mathematics conference in Warsaw. He got his wish.

Figure 25.16: Paul Erdős

Most mathematicians today remember Erdős as an old man (Figure 25.16), and indeed he liked to joke that he was a billion years old: when he was a child the earth was said to be one billion years old; when he was an adult the earth was said to be two billion years old. But in a sense he is forever young—the Peter Pan of mathematics as his old friend Marta Sved called him (see Soifer (2009), p. 235). He never lost the ability to inspire young people to do mathematics, or to revive the mathematical energy of his older colleagues. His legacy of awarding prize money for the solution of hard problems has been maintained by Ron Graham, who continues to pose Erdős-style problems and to pay out when they are solved. And Erdős even continues to publish. Over 70 papers have appeared since his death. He is as old, and young, as mathematics.

Bibliography

Abel, N. H. (1826). Démonstration de l'impossibilité de la résolution algébrique des équations générales qui passent le quatrième degré. *J. reine und angew. Math. 1*, 65–84. *Oeuvres Complètes* 1: 66–87.

Abel, N. H. (1827). Recherches sur les fonctions elliptiques. *J. reine und angew. Math. 2*, 101–181. *3*, 160–190. In his *Oeuvres Complètes* 1: 263–388.

Abel, N. H. (1829). Mémoire sur une classe particulière d'equations résolubles algébriquement. *J. reine und angew. Math. 4*, 131–156. *Œuvres Complètes* 1: 478–507.

Ackermann, W. F. (1937). Der Widerspruchsfreiheit der allgemeine Mengenlehre. *Math. Ann. 112*, 305–315.

Adyan, S. I. (1957). Unsolvability of some algorithmic problems in the theory of groups (Russian). *Trudy Moskov. Mat. Obshch. 6*, 231–298.

Akivis, M. A. and B. A. Rosenfeld (1993). *Élie Cartan (1869–1951)*. Providence, RI: American Mathematical Society. Translated from the Russian manuscript by V. V. Goldberg.

Alberti, L. B. (1436). *Trattato della pittura*. Reprinted in *Il trattato della pittura e i cinque ordine architettonici*, R. Carabba, 1913.

Alexander, J. W. (1919). Note on two three-dimensional manifolds with the same group. *Trans. Amer. Math. Soc. 20*, 339–342.

Apéry, R. (1981). Interpolation de fractions continues et irrationalité de certaines constantes. In *Mathematics*, pp. 37–53. Paris: Bib. Nat.

Appel, K. and W. Haken (1976). Every planar map is four colorable. *Bull. Amer. Math. Soc. 82*, 711–712.

Argand, J. R. (1806). *Essai sur une manière de représenter les quantités imaginaires dans les constructions géométriques*. Paris.

589

Artin, M. (1991). *Algebra*. Englewood Cliffs, NJ: Prentice-Hall Inc.

Assmus, Jr., E. F. and H. F. Mattson (1966). Perfect codes and the Mathieu groups. *Arch. Math. (Basel) 17*, 121–135.

Ayoub, R. (1984). The lemniscate and Fagnano's contributions to elliptic integrals. *Arch. Hist. Exact Sci. 29*(2), 131–149.

Bachet de Méziriac, C. G. (1621). *Diophanti Alexandrini libri sex*. Toulouse.

Baer, R. (1928). Zur Axiomatik der Kardinalarithmetik. *Math. Zeit. 29*, 381–396.

Baillet, A. (1691). *La vie des Monsieur Des-Cartes*. Paris: Daniel Horthemels.

Ball, W. W. R. (1890). Newton's classification of cubic curves. *Proc. London Math. Soc. 22*, 104–143.

Baltrušaitis, J. (1977). *Anamorphic Art*. New York: Harry Abrams.

Banach, S. and A. Tarski (1924). Sur la décomposition des ensembles de points en parties respectivement congruentes. *Fund. Math. 6*, 244–277.

Banville, J. (1981). *Kepler: A Novel*. London: Secker and Warburg.

Baron, M. E. (1969). *The Origins of the Infinitesimal Calculus*. Oxford: Pergamon Press.

Bashmakova, I. G. (1981). Arithmetic of algebraic curves from Diophantus to Poincaré. *Historia Math. 8*(4), 393–416.

Beeckman, I. (1628). Journal. Beeckman (1634), quoted in *Œuvres de Descartes*, volume 10, pp. 344–346.

Beeckman, I. (1634). *Journal tenu par Isaac Beeckman de 1604 à 1634*. The Hague: Nijhoff. Edited by C. de Waard, 4 vols.

Beltrami, E. (1865). Risoluzione del problema: Riportare i punti di una superficie sopra un piano in modo che le linee geodetiche vengano rappresentate da linee rette. *Ann. Mat. pura appl., ser. 1 7*, 185–204. In his *Opere Matematiche* 1: 262–280.

Beltrami, E. (1868a). Saggio di interpretazione della geometria non-euclidea. *Giorn. Mat. 6*, 284–312. In his *Opere Matematiche* 1: 262–280, English translation in Stillwell (1996).

Beltrami, E. (1868b). Teoria fondamentale degli spazii di curvatura costante. *Ann. Mat. pura appl., ser. 2 2*, 232–255. In his *Opere Matematiche* 1: 406–429, English translation in Stillwell (1996).

Bernoulli, D. (1728). Observationes de seriebus. *Comment. Acad. Sci. Petrop. 3*, 85–100. In Bernoulli (1982), pp. 49–64.

Bernoulli, D. (1743). Letter to Euler, 4 September 1743. In Eneström (1906).

Bernoulli, D. (1753). Réflexions et éclaircissemens sur les nouvelles vibrations des cordes exposées dans les mémoires de l'académie de 1747 & 1748. *Hist. Acad. Sci. Berlin 9*, 147–172.

Bernoulli, D. (1982). *Die Werke von Daniel Bernoulli, Band 2*. Basel: Birkhäuser.

Bernoulli, Jakob (1690). Quaestiones nonnullae de usuris cum solutione problematis de sorte alearum propositi in Ephem. Gall. A. 1685. *Acta Erud. 11*, 219–233.

Bernoulli, Jakob (1692). Lineae cycloidales, evolutae, ant-evolutae ... Spira mirabilis. *Acta. Erud. 11*, 207–213.

Bernoulli, Jakob (1694). Curvatura laminae elasticae. *Acta. Erud. 13*, 262–276.

Bernoulli, Jakob (1697). Solutio problematis fraternorum. *Acta Erud. 16*, 211–217.

Bernoulli, Jakob (1713). Ars conjectandi. *Opera* 3: 107–286.

Bernoulli, Jakob and Johann (1704). *Über unendliche Reihen*. Ostwald's *Klassiker*, vol. 171. Engelmann, Leipzig, 1909.

Bernoulli, Johann (1691). Solutio problematis funicularii. *Acta Erud. 10*, 274–276. In his *Opera Omnia* 1: 48–51.

Bernoulli, Johann (1696). Problema novum ad cujus solutionem mathematici invitantur. *Acta Erud. 15*, 270. In his *Opera Omnia* 1: 161.

Bernoulli, Johann (1697). Principia calculi exponentialum. *Acta Erud. 16*, 125–133. In his *Opera Omnia* 1: 179–187.

Bernoulli, Johann (1699). *Disputatio medico-physica de nutritione*. Groningen.

Bernoulli, Johann (1702). Solution d'un problème concernant le calcul intégral, avec quelques abrégés par raport à ce calcul. *Mém. Acad. Roy. Soc. Paris*, 289–297. In his *Opera Omnia* 1: 393–400.

Bernoulli, Johann (1712). Angulorum arcuumque sectio indefinita. *Acta Erud. 31*, 274–277. In his *Opera Omnia* 1: 511–514.

Bertrand, J. (1860). Remarque à l'occasion de la note précédente. *Comp. Rend. 50*, 781–782.

Bézout, E. (1779). *Théorie générale des équations algébriques*. Paris: Ph.-D. Pierres. English translation: *General Theory of Algebraic Equations*, by Eric Feron, Princeton University Press, Princeton, 2006.

Biggs, N. L., E. K. Lloyd, and R. J. Wilson (1976). *Graph Theory: 1736–1936*. Oxford: Oxford University Press.

Birkhoff, G. (Ed.) (1973). *A Source Book in Classical Analysis*. Cambridge, MA.: Harvard University Press. With the assistance of Uta Merzbach.

Boltyansky, V. G. (1978). *Hilbert's Third Problem*. Washington, DC: V. H. Winston & Sons. Translated from the Russian by Richard A. Silverman, with a foreword by Albert B. J. Novikoff, Scripta Series in Mathematics.

Bolyai, F. (1832a). *Tentamen juventutem studiosam in elementa matheseos purae, elementaris ac sublimioris, methodo intuitiva, evidentiaque huic propria, introducendi*. Marosvásárhely.

Bolyai, J. (1832b). Scientiam spatii absolute veram exhibens: a veritate aut falsitate Axiomatis XI Euclidei (a priori haud unquam decidanda) independentem. Appendix to Bolyai (1832a), English translation in Bonola (1912).

Bolzano, B. (1817). *Rein analytischer Beweis des Lehrsatzes dass zwischen je zwey Werthen, die ein entgegengesetzes Resultat gewähren, wenigstens eine reelle Wurzel der Gleichung liege*. Ostwald's Klassiker, vol. 153. Engelmann, Leipzig, 1905. English translation in Russ (2004) pp. 251–277.

Bombelli, R. (1572). *L'algebra. Prima edizione integrale. Introduzione di U. Forti. Prefazione di E. Bortolotti*. Reprint by Biblioteca scientifica Feltrinelli. 13. Milano: Giangiacomo Feltrinelli Editore. LXIII (1966).

Bonnet, O. (1848). Mémoire sur la théorie générale des surfaces. *J. Éc. Polytech. 19*, 1–146.

Bonola, R. (1912). *Noneuclidean Geometry*. Chicago: Open Court. Reprinted by Dover, New York, 1955.

Boole, G. (1847). *Mathematical Analysis of Logic*. Reprinted by Basil Blackwell, London, 1948.

Borcherds, R. E. (1994). Sporadic groups and string theory. In *First European Congress of Mathematics, Vol. I (Paris, 1992)*, Volume 119 of *Progr. Math.*, pp. 411–421. Basel: Birkhäuser.

Borel, E. (1898). *Leçons sur la théorie des fonctions*. Paris: Gauthier-Villars.

Bos, H. J. M. (1981). On the representation of curves in Descartes' *Géométrie*. *Arch. Hist. Exact Sci.* 24(4), 295–338.

Bos, H. J. M. (1984). Arguments on motivation in the rise and decline of a mathematical theory; the "construction of equations," 1637–ca. 1750. *Arch. Hist. Exact Sci.* 30(3-4), 331–380.

Bosse, A. (1648). *Manière universelle de Mr Desargues*. Paris: P. Des-Hayes.

Bourgne, R. and J.-P. Azra (1962). *Ecrits et mémoires mathématiques d'Évariste Galois: Édition critique intégrale de ses manuscrits et publications*. Gauthier-Villars & Cie, Imprimeur-Éditeur-Libraire, Paris. Préface de J. Dieudonné.

Boyer, C. B. (1956). *History of Analytic Geometry*. Scripta Mathematica, New York.

Boyer, C. B. (1959). *The History of the Calculus and Its Conceptual Development*. New York: Dover Publications Inc.

Boyer, C. B. (1968). *A History of Mathematics*. New York: John Wiley & Sons Inc.

Brahana, H. R. (1921). Systems of circuits on 2-dimensional manifolds. *Ann. Math. 23*, 144–168.

Brahmagupta (628). *Brâhma-sphuṭa-siddhânta*. Partial English translation in Colebrooke (1817).

Brieskorn, E. and H. Knörrer (1981). *Ebene algebraische Kurven*. Basel: Birkhäuser Verlag. English translation: *Plane Algebraic Curves*, by John Stillwell, Birkhäuser Verlag, 1986.

Briggs, H. (1624). *Arithmetica logarithmica*. London: William Jones.

Bring, E. S. (1786). *Meletemata quaedam mathematica circa transformationem aequationum algebraicarum*. Lund University. Promotionschrift.

Burnside, W. (1911). *The Theory of Groups of Finite Order*. Cambridge: Cambridge University Press. Second edition, reprinted by Dover, New York, 1955.

Burton, D. M. (1985). *The History of Mathematics*. Boston, MA.: Allyn and Bacon Inc.

Cajori, F. (1913). History of the exponential and logarithmic concepts. *Amer. Math. Monthly 20*, 5–14, 35–47, 75–84, 107–117, 148–151, 173–182, 205–210.

Cantor, G. (1872). Über die Ausdehnung eines Satzes aus der Theorie der trigonometrischen Reihen. *Math. Ann. 5*, 123–132. In his *Gesammelte Abhandlungen*, 92–102.

Cantor, G. (1874). Über eine Eigenschaft des Inbegriffes aller reellen algebraischen Zahlen. *J. reine und angew. Math. 77*, 258–262. In his *Gesammelte Abhandlungen*, 145–148. English translation by W. Ewald in Ewald (1996), Vol. II, pp. 840–843.

Cantor, G. (1880). Über unendlich lineare Punktmannigfaltigkeiten, 2. *Math. Ann. 17*, 355–358. In his *Gesammelte Abhandlungen*, 145–148.

Cantor, G. (1883). *Grundlagen einer allgemeinen Mannigfaltigkeitslehre*. Leizig: Teubner. In his *Gesammelte Abhandlungen*, 165–204. English translation by W. Ewald in Ewald (1996), Vol. II, pp. 878–919.

Cantor, G. (1891). Über eine elementare Frage der Mannigfaltigkeitslehre. *Jahresber. deutsch. Math. Verein. 1*, 75–78. English translation by W. Ewald in Ewald (1996), Vol. II, pp. 920–922.

Cardano, G. (1545). *Ars magna*. 1968 translation *The great art or the rules of algebra* by T. Richard Witmer, with a foreword by Oystein Ore. The M.I.T. Press, Cambridge, MA-London.

Cardano, G. (1575). *De Vita Propria Liber*. English translation *The Book of My Life*, Dover, New York 1962.

Cartan, E. (1894). *Sur la structure des groupes de transformations finis et continus*. Paris: Nony et Co.

Cartan, E. (1908). Nombres complexes. In *Encyclopédie des sciences mathématiques, I 5*, pp. 329–468. Paris: Jacques Gabay.

Cartan, E. (1936). La topologie des espaces représentatives des groupes de Lie. *L'Enseignement Math. 35*, 177–200.

Cauchy, A.-L. (1813a). Démonstration du théorème général de Fermat sur les nombres polygones. *Mém. Sci. Math. Phys. Inst. France, ser. 1 14*, 177–220. In his *Œuvres*, ser. 2, 6: 320–353.

Cauchy, A.-L. (1813b). Recherches sur les polyèdres – premier mémoir. *J. de l'École Polytechnique 9*, 68–86. Partial English translation in Biggs et al. (1976), pp. 81–83.

Cauchy, A.-L. (1815). Mémoire sur le nombre des valeurs qu'une fonction peut acquerir, lorsqu'on y permute de toutes les manières possibles les quantités qu'elle renferme. *J. Éc. Polytech. 18*, 1–28. In his *Œuvres*, ser. 2, 1: 62–90.

Cauchy, A.-L. (1825). *Mémoire sur les intégrales définies prises entre des limites imaginaires*. Paris.

Cauchy, A.-L. (1837). Letter to Coriolis, 29 January 1837. *Comp. Rend. 4*, 214–218. In his *Œuvres*, ser. 1, 4: 38–42.

Cauchy, A.-L. (1844). Mémoire sur les arrangements que l'on peut former avec des lettres données, et sur les permutations ou substitutions à l'aide desquelles on passe d'un arrangement à un autre. *Ex. anal. phys. math. 3*, 151–252. In his *Œuvres*, ser. 2, 13: 171–282.

Cauchy, A.-L. (1846). Sur les intégrales qui s'étendent à tous les points d'une courbe fermée. *Comp. Rend. 23*, 251–255. In his *Œuvres*, ser. 1, 10: 70–74.

Cavalieri, B. (1635). *Geometria indivisibilibus continuorum nova quadam ratione promota*. Bononi: Clement Ferroni.

Cayley, A. (1845a). On certain results relating to quaternions. *Phil. Mag. XXXVI*, 141–145. In his *Collected Mathematical Papers* 1: 123–126.

Cayley, A. (1845b). On Jacobi's elliptic functions and on quaternions. *Phil. Mag. XXXVI*, 208–211. In his *Collected Mathematical Papers*, p. 127. The part relevant to octonions is in Hamilton's *Mathematical Papers* 3: 650–651.

Cayley, A. (1854). On the theory of groups, as depending on the symbolic equation $\theta^n = 1$. *Phil. Mag. 7*, 40–47. In his *Collected Mathematical Papers* 2: 123–130.

Cayley, A. (1855). Recherches sur les matrices dont les termes des fonctions linéaires d'une seule indéterminée. *J. reine und angew. Math. 50*, 313–317. In his *Collected Mathematical Papers* 2: 216–220.

Cayley, A. (1858). A memoir on the theory of matrices. *Phil. Trans. Roy. Soc. London 148*, 17–37. In his *Collected Mathematical Papers* 2: 475–496.

Cayley, A. (1859). A sixth memoir on quantics. *Phil. Trans. Roy. Soc. 149*, 61–90. In his *Collected Mathematical Papers* 2: 561–592.

Cayley, A. (1878). The theory of groups. *Amer. J. Math. 1*, 50–52. In his *Collected Mathematical Papers* 10: 401–403.

Chaitin, G. J. (1970). Computational complexity and Gödel's incompleteness theorem. *Notices Amer. Math. Soc. 17*, 672.

Chandler, B. and W. Magnus (1982). *The History of Combinatorial Group Theory*. New York: Springer-Verlag.

Chevalley, C. (1955). Sur certains groupes simples. *Tôhoku Math. J. (2) 7*, 14–66.

Church, A. (1936). An unsolvable problem in elementary number theory. *Amer. J. Math. 58*, 345–363.

Church, A. (1938). Review. *J. Symb. Logic 3*, 46.

Clagett, M. (1959). *The Science of Mechanics in the Middle Ages*. The University of Wisconsin Press, Madison. Publications in Medieval Science, 4.

Clagett, M. (1968). *Nicole Oresme and the Medieval Geometry of Qualities and Motions*. Madison: University of Wisconsin Press.

Clairaut, A.-C. (1740). Sur l'intégration ou la construction des équations différentialles du premier ordre. *Mém. Acad. Sci. Paris*, 294.

Clairaut, A.-C. (1743). *Théorie de la figure de la Terre tirée des principes de l'hydrodynamique*. Paris: Durand.

Clebsch, A. (1864). Über einen Satz von Steiner und einige Punkte der Theorie der Curven dritter Ordnung. *J. reine und angew. Math. 63*, 94–121.

Cohen, M. R. and I. E. Drabkin (1958). *Source Book in Greek Science*. Cambridge, MA.: Harvard University Press.

Cohen, P. (1963). The independence of the continuum hypothesis I, II. *Proc. Nat. Acad. Sci. 50, 51*, 1143–1148, 105–110.

Cohn, H. and A. Kumar (2004). The densest lattice in twenty-four dimensions. *Electron. Res. Announc. Amer. Math. Soc. 10*, 58–67 (electronic). The full proof is in *Annals of Mathematics 170* (2009), 1003–1050.

Cole, F. N. (1893). Simple groups as far as order 660. *Amer. J. Math. 15*, 305–315.

Colebrooke, H. T. (1817). *Algebra, with Arithmetic and Mensuration, from the Sanscrit of Brahmegupta and Bháscara*. London: John Murray. Reprinted by Martin Sandig, Wiesbaden, 1973.

Coleman, A. J. (1989). The greatest mathematical paper of all time. *Math. Intelligencer 11*(3), 29–38.

Connelly, R. (1977). A counterexample to the rigidity conjecture for polyhedra. *Inst. Hautes Études Sci. Publ. Math. (47)*, 333–338.

Coolidge, J. L. (1945). *A History of the Conic Sections and Quadric Surfaces*. Oxford University Press.

Copernicus, N. (1543). De revolutionibus orbium coelestium. English translation *On the revolutions*, Polish Science Publishers, Warsaw, 1978.

Cotes, R. (1714). Logometria. *Phil. Trans. 29*, 5–45.

Cotes, R. (1722). *Harmonia mensurarum*. Cambridge: Robert Smith.

Cox, D. A. (1984). The arithmetic-geometric mean of Gauss. *Enseign. Math. (2) 30*(3-4), 275–330.

Cox, D. A. (1989). *Primes of the Form $x^2 + ny^2$*. New York: John Wiley & Sons Inc.

Cramer, G. (1750). *Introduction à l'analyse des lignes courbes algébriques*. Geneva.

Crossley, J. N. (1987). *The Emergence of Number*, 2nd ed. Singapore: World Scientific Publishing Co.

Crowe, M. J. (1967). *A History of Vector Analysis*. Notre Dame, IN.: University of Notre Dame Press.

d'Alembert, J. le. R. (1746). Recherches sur le calcul intégral. *Hist. Acad. Sci. Berlin 2*, 182–224.

d'Alembert, J. le. R. (1747). Recherches sur la courbe que forme une corde tendue mise en vibration. *Hist. Acad. Sci. Berlin 3*, 214–219.

d'Alembert, J. le. R. (1752). *Essai d'une nouvelle théorie de la résistance des fluides*. Paris: David.

Davenport, J. H. (1981). *On the Integration of Algebraic Functions*. Berlin: Springer-Verlag.

David, F. N. (1962). *Games, Gods and Gambling*. London: Charles Griffin.

Davis, M. (Ed.) (1965). *The Undecidable. Basic papers on undecidable propositions, unsolvable problems and computable functions*. Raven Press, Hewlett, NY.

Davis, M. (1973). Hilbert's tenth problem is unsolvable. *Amer. Math. Monthly 80*, 233–269.

de la Hire, P. (1673). *Nouvelle méthode en géométrie*. Paris.

de la Vallée Poussin, C. J. (1896). Recherches analytiques sur la théorie des nombres premiers. *Ann. Soc. Sci. Bruxelles 20*, 183–256.

de Moivre, A. (1698). A method of extracting the root of an infinite equation. *Phil. Trans. 20*, 190–193.

de Moivre, A. (1707). Æquationem quarundum potestatis tertiae, quintae septimae, nonae & superiorum, ad infinitum usque pergendo, in terminis finitis, ad instar regularum pro cubicus que vocantur Cardani, resolutio analytica. *Phil. Trans. 25*, 2368–2371.

de Moivre, A. (1730). *Miscellanea analytica de seriebus et quadraturis*. London: J. Tonson and J. Watts.

Dedekind, R. (1871). Supplement X. In Dirichlet's *Vorlesungen über Zahlentheorie*, 2nd ed., Vieweg 1871.

Dedekind, R. (1872). *Stetigkeit und irrationale Zahlen*. Braunschweig: Vieweg und Sohn. English translation in: *Essays on the Theory of Numbers*, Dover, New York, 1963.

Dedekind, R. (1876). Bernhard Riemann's Lebenslauf. In Riemann's *Werke*, 2nd ed. pp. 539–558.

Dedekind, R. (1877). *Theory of Algebraic Integers*. Cambridge: Cambridge University Press. Translated from the 1877 French original and with an introduction by John Stillwell.

Dedron, P. and J. Itard (1973). *Mathematics and Mathematicians, Vol. 1*. Milton Keynes: Open University Press.

Degen, C. F. (1822). Adumbratio demonstrationis theorematis arithmeticae maxime generalis. *Mém. l'Acad. Imp. Sci. St. Petersbourg VIII*, 207–219.

Dehn, M. (1900). Über raumgleiche Polyeder. *Gött. Nachr. 1900*, 345–354..

Dehn, M. (1910). Über die Topologie des dreidimensionalen Raumes. *Math. Ann. 69*, 137–168.

Dehn, M. (1912). Über unendliche diskontinuierliche Gruppen. *Math. Ann. 71*, 116–144.

Dehn, M. and P. Heegaard (1907). Analysis situs. *Enzyklopädie der Mathematischen Wissenschaften*, vol. IIAB3, 153–220, Teubner, Leipzig.

Desargues, G. (1639). *Brouillon projet d'une atteinte aux évènements des rencontres du cône avec un plan*. In Taton (1951), pp. 99–180.

Descartes, R. (1637). *The geometry of René Descartes. (With a facsimile of the first edition, 1637.)*. New York, NY: Dover Publications Inc. Translated by David Eugene Smith and Marcia L. Latham, 1954.

Descartes, R. (1638). Letter to Mersenne, 18 January 1638. *Œuvres* 1, 490.

Diacu, F. and P. Holmes (1996). *Celestial Encounters*. Princeton, NJ: Princeton University Press.

Dickson, L. E. (1901a). *Linear Groups with an Exposition of the Galois Field Theory*. Leipzig: Teubner.

Dickson, L. E. (1901b). Theory of linear groups in an arbitrary field. *Trans. Amer. Math. Soc* 2, 363–394.

Dickson, L. E. (1903). *Introduction to the Theory of Algebraic Equations*. New York: Wiley.

Dickson, L. E. (1914). *Linear Algebras*. Cambridge: Cambridge University Press.

Dickson, L. E. (1920). *History of the Theory of Numbers. Vol. II: Diophantine Analysis*. New York: Chelsea Publishing Co. 1966 reprint of Carnegie Institute, Washington, edition.

Diestel, R. (2005). *Graph Theory* (Third ed.). Berlin: Springer-Verlag.

Dirichlet, P. G. L. (1829). Sur la convergence des séries trigonométriques qui servent à représenter une fonction arbitraire entre des limites données. *J. reine und angew. Math 4*, 157–169. In his *Werke* 1: 117–132.

Dirichlet, P. G. L. (1837). Beweis des Satzes, dass jede unbegrentze arithmetische Progression, deren erstes Glied und Differenz ganze Zahlen ohne gemeinschaftlichen Factor sind, unendliche viele Primzahlen enthält. *Abh. Akad. Wiss. Berlin*, 45–81. In his *Werke* 1: 315–342.

Dirichlet, P. G. L. (1863). *Vorlesungen über Zahlentheorie*. Braunschweig: F. Vieweg und Sohn. English translation *Lectures on Number Theory*, with Supplements by R. Dedekind, translated from the German and with an introduction by John Stillwell, American Mathematical Society, Providence, RI, 1999.

Dombrowski, P. (1979). *150 Years after Gauss' "Disquisitiones generales circa superficies curvas"*. Paris: Société Mathématique de France. With the original text of Gauss.

Donaldson, S. K. (1983). An application of gauge theory to four-dimensional topology. *J. Differential Geom. 18*(2), 279–315.

Dostrovsky, S. (1975). Early vibration theory: physics and music in the seventeenth century. *Arch. History Exact Sci. 14*(3), 169–218.

du Bois-Reymond, P. (1875). Über asymptotische Werte, infinitäre Approximationen und infinitäre Auflösung von Gleichungen. *Math. Ann. 8*, 363–414.

Dugas, R. (1957). *A History of Mechanics*. Editions du Griffon, Neuchâtel, Switzerland. Foreword by Louis de Broglie. Translated into English by J. R. Maddox.

Dugas, R. (1958). *Mechanics in the Seventeenth Century*. Editions du Griffon, Neuchâtel, Switzerland.

Dunnington, G. W. (1955/2004). *Carl Friedrich Gauss*. Washington, DC: Mathematical Association of America. Reprint of the 1955 original [Exposition Press, New York], with an introduction and commentary by Jeremy Gray, with a brief biography of the author by Fritz-Egbert Dohse.

Dürer, A. (1525). *Underweysung der Messung*. Facsimile of 1525 edition by Collegium Graphicum, Portland, Oregon, 1972. English translation: *The Painter's Manual*, Albaris Books, New York, 1977.

Dyck, W. (1882). Gruppentheoretische Studien. *Math. Ann. 20*, 1–44.

Dyck, W. (1883). Gruppentheoretische Studien II. *Math. Ann. 22*, 70–108.

Edwards, Jr., C. H. (1979). *The Historical Development of the Calculus*. New York: Springer-Verlag.

Edwards, H. M. (1974). *Riemann's Zeta Function*. Academic Press, New York-London. Pure and Applied Mathematics, Vol. 58.

Edwards, H. M. (1977). *Fermat's Last Theorem*. New York: Springer-Verlag.

Edwards, H. M. (1984). *Galois Theory*. New York: Springer-Verlag.

Eisenstein, G. (1847). Beiträge zur Theorie der elliptische Functionen. *J. reine und angew. Math. 35*, 137–274.

Eisenstein, G. (1850). Über einige allgemeine Eigenschaften der Gleichung, von welcher die Theorie der ganzen Lemniscate abhängt. *J. reine und angew. Math. 39*, 556–619.

Eneström, G. (1906). Der Briefwechsel zwischen Leonhard Euler und Daniel Bernoulli. *Bibl. Math. ser. 3 7*, 126–156.

Engelsman, S. B. (1984). *Families of Curves and the Origins of Partial Differentiation*. Amsterdam: North-Holland Publishing Co.

Erdős, P. and G. Szekeres (1935). A combinatorial problem in geometry. *Compositio Math. 2*, 463–470.

Euler, L. (1728a). De linea brevissima in superficie quacunque duo quaelibet puncta iungente. *Comm. Acad. Sci. Petrop. 3*, 110–124. In his *Opera Omnia*, series 1, 25: 1–12.

Euler, L. (1728b). Letter to John Bernoulli, 10 December 1728. *Bibl. Math.*, ser. 3, **4**, 352–354.

Euler, L. (1734). De summis serierum reciprocarum. *Comm. Acad. Sci. Petrop. 7*. In his *Opera Omnia*, ser. 1, 14: 73–86.

Euler, L. (1736). Theorematum quorundam ad numeros primos spectantium demonstratio. *Comm. Acad. Sci. Petrop. 8*, 141–146. In his *Opera Omnia*, ser. 1, 2: 33–37.

Euler, L. (1743). *Addimentum I de curvis elasticis. Opera Omnia*, ser. 1, 24: 231–297, English translation in *Isis* **20** (1933), 72–160.

Euler, L. (1746). Letter to Goldbach, 14 June 1746. Briefwechsel *Opera Omnia*, ser. quarta A, 1, 52.

Euler, L. (1748a). *Introductio in analysin infinitorum, I.* Volume 8 of his *Opera Omnia*, series 1. English translation, *Introduction to the Analysis of the Infinite. Book I*, Springer-Verlag, 1988.

Euler, L. (1748b). *Introductio in analysin infinitorum, II.* Volume 9 of his *Opera Omnia*, series 1. English translation, *Introduction to the Analysis of the Infinite. Book II*, Springer-Verlag, 1988.

Euler, L. (1748c). Letter to Goldbach, 4 May 1748. In Fuss (1968), **1**, 450–455.

Euler, L. (1749). Letter to Goldbach, 12 April 1749. In Fuss (1968), **1**, 493–495.

Euler, L. (1750). Letter to Goldbach, 9 June 1750. In Fuss (1968), **I**, 521–524.

Euler, L. (1752). Elementa doctrinae solidorum. *Novi Comm. Acad. Sci. Petrop. 4*, 109–140. In his *Opera Omnia*, ser. 1, 26: 71–93.

Euler, L. (1758). Theoremata arithmetica nova methodo demonstrata. *Novi Comm. Acad. Sci. Petrop. 8*, 74–104. In his *Opera Omnia*, ser. 1, 2: 531–555.

Euler, L. (1760). Recherches sur la courbure des surfaces. *Mém. Acad. Sci. Berlin 16*, 119–143. In his *Opera Omnia*, ser. 1, 28: 1–22.

Euler, L. (1768). *Institutiones calculi integralis. Opera Omnia*, ser. 1, 11.

Euler, L. (1770a). De summis serierum numeros bernoullianos involventium. *Novi Comm. Acad. Sci. Petrop. 14*, 129–167.

Euler, L. (1770b). *Elements of Algebra*. Translated from the German by John Hewlett. Reprint of the 1840 edition, with an introduction by C. Truesdell, Springer-Verlag, New York, 1984.

Euler, L. (1776). Formulae generales pro translatione quacunque corporum rigidorum. *Novi Comm. Acad. Sci. Petrop.* 20, 189–207.

Euler, L. (1777). De repraesentatione superficiei sphaericae super plano. *Acta Acad. Sci. Imper. Petrop.* *1*, 107–132.

Euler, L. (1849). De numeris amicabilibus. *Comm. Arith.* 2, 627–636. In his *Opera Omnia*, ser. 1, 5: 353–365.

Ewald, W. (1996). *From Kant to Hilbert: A Source Book in the Foundations of Mathematics. Vol. I, II.* New York: The Clarendon Press, Oxford University Press.

Fagnano, G. C. T. (1718). Metodo per misurare la lemniscata. *Giorn. lett. d'Italia 29*. In his *Opere Matematiche*, 2: 293–313.

Faltings, G. (1983). Endlichkeitssätze für abelsche Varietäten über Zahlkörpern. *Invent. Math.* *73*(3), 349–366.

Fauvel, J. and J. Gray (Eds.) (1988). *The History of Mathematics: A Reader*. Basingstoke: Macmillan Press Ltd. Reprint of the 1987 edition.

Federico, P. J. (1982). *Descartes on Polyhedra*. New York: Springer-Verlag. A study of the *De solidorum elementis*.

Feit, W. and J. G. Thompson (1963). Solvability of groups of odd order. *Pacific J. Math.* *13*, 775–1029.

Fermat, P. (1629). Ad locos planos et solidos isagoge. *Œuvres* 1, 92–103. English translation in Smith (1959), 389–396.

Fermat, P. (1640a). Letter to Frenicle, 18 October 1640. *Œuvres* 2: 209.

Fermat, P. (1640b). Letter to Mersenne, 25 December 1640. *Œuvres* 2: 212.

Fermat, P. (1654). Letter to Pascal, 25 September 1654. *Œuvres* 2: 310–314.

Fermat, P. (1657). Letter to Frenicle, February 1657. *Œuvres* 2: 333–334.

Fermat, P. (1670). Observations sur Diophante. *Œuvres* 3: 241–276.

Fibonacci (1202). *Liber abaci*. In *Scritti di Leonardo Pisano*, edited by Baldassarre Boncompagni, Rome 1857–1862. English translation *Fibonacci's Liber abaci*, by L.E. Sigler, Springer, New York, 2002.

Fibonacci (1225). *Flos Leonardo Bigolli Pisani super solutionibus quarundam quaestionum ad numerum et ad geometriam pertinentium.*

Field, J. V. and J. J. Gray (1987). *The Geometrical Work of Girard Desargues.* New York: Springer-Verlag.

Fourier, J. (1822). *La théorie analytique de la chaleur.* Paris: Didot. English translation, *The Analytical Theory of Heat*, Dover, New York, 1955.

Fowler, D. H. (1980). Book II of Euclid's *Elements* and a pre-Eudoxan theory of ratio. *Arch. Hist. Exact Sci. 22*(1-2), 5–36.

Fowler, D. H. (1982). Book II of Euclid's *Elements* and a pre-Eudoxan theory of ratio. II. Sides and diameters. *Arch. Hist. Exact Sci. 26*(3), 193–209.

Freedman, M. H. (1982). The topology of four-dimensional manifolds. *J. Differential Geom. 17*, 357–453.

Frege, G. (1879). *Begriffschrift.* English translation in van Heijenoort (1967).

Freudenthal, H. (1951). *Oktaven, Ausnahmegruppen und Oktavengeometrie.* Mathematisch Instituut der Rijksuniversiteit te Utrecht, Utrecht.

Frey, G. (1986). Links between stable elliptic curves and certain Diophantine equations. *Ann. Univ. Sarav. Ser. Math. 1*(1), iv+40.

Fritsch, R. (1984). The transcendence of π has been known for about a century— but who was the man who discovered it? *Resultate Math. 7*(2), 164–183.

Frobenius, G. (1878). Über lineare Substitutionen und bilineare Formen. *J. reine und angew. Math. 84*, 1–63. In his *Gesammelte Abhandlungen* 1: 343–405.

Fuss, P.-H. (1968). *Correspondance mathématique et physique de quelques célèbres géomètres du XVIIIème siècle. Tomes I, II.* New York: Johnson Reprint Corp. Reprint of the Euler correspondence originally published by l'Académie Impériale des Sciences de Saint-Pétersbourg. The Sources of Science, No. 35.

Galileo Galilei (1604). Letter to Paolo Scarpi, 16 October 1604. In the *Works of Galileo* 10: 115.

Galileo Galilei (1638). *Dialogues Concerning Two New Sciences.* English translation reprinted by Dover, New York, 1952.

Galois, E. (1831a). Analyse d'un mémoire sur la résolution algébrique des équations. In Bourgne and Azra (1962), pp. 163–165.

Galois, E. (1831b). Mémoire sur les conditions de résolubilité des équations par radicaux. In Bourgne and Azra (1962), pp. 43–71.

Gannon, T. (2006). *Moonshine beyond the Monster*. Cambridge Monographs on Mathematical Physics. Cambridge: Cambridge University Press.

Gauss, C. F. (1799). Demonstratio nova theorematis omnem functionem algebraicum rationalem integram unius variabilis in factores reales primi vel secundi gradus resolvi posse. Helmstedt dissertation, in his *Werke* 3: 1–30.

Gauss, C. F. (1801). *Disquisitiones arithmeticae*. Translated and with a preface by Arthur A. Clarke. Revised by William C. Waterhouse, Cornelius Greither and A. W. Grootendorst and with a preface by Waterhouse, Springer-Verlag, New York, 1986.

Gauss, C. F. (1811). Letter to Bessel, 18 December 1811. *Briefwechsel mit F. W. Bessel*, Georg Olms Verlag, Hildesheim, 1975, pp. 155–160. English translation in Birkhoff (1973).

Gauss, C. F. (1816). Demonstratio nova altera theorematis omnem functionem algebraicum rationalem integram unius variabilis in factores reales primi vel secundi gradus resolvi posse. *Comm. Recentiores (Gottingae) 3*, 107–142. In his *Werke* 3: 31–56.

Gauss, C. F. (1818). Determinatio attractionis quam in punctum quodvis positionis datae exerceret planeta si eius massa per totam orbitam ratione temporis quo singulae partes describuntur uniformiter esset dispertita. *Comm. Soc. Reg. Sci. Gottingensis Rec. 4*. In his *Werke* 3: 331–355.

Gauss, C. F. (1819). Die Kugel. *Werke* 8: 351–356.

Gauss, C. F. (1822). Allgemeine Auflösung der Aufgabe; die Theile einer gegebenen Fläche so abzubilden, dass die Abbildung dem Abgebildeten in den kleinsten Theilen ähnlich wird. *Astr. Abh. 3*, 1–30. In his *Werke* 4: 189–216. English translation, *Phil. Mag.*, new ser., **4** (1828), 104–113, 206–215.

Gauss, C. F. (1825). Die Seitenkrümmung. *Werke* 8: 386–395.

Gauss, C. F. (1827). *Disquisitiones generales circa superficies curvas*. Göttingen: König. Ges. Wiss. Göttingen. English translation in Dombrowski (1979).

Gauss, C. F. (1828). Letter to Bessel, 30 March 1828. *Briefwechsel mit F. W. Bessel*, Georg Olms Verlag, Hildesheim, 1975, 477–478.

Gauss, C. F. (1831). Letter to Schumacher, 12 July 1831. *Werke* 8: 215–218.

Gauss, C. F. (1832a). Cubirung der Tetraeder. *Werke* 8: 228–229.

Gauss, C. F. (1832b). Letter to W. Bolyai, 6 March 1832. *Briefwechsel zwischen C. F. Gauss und Wolfgang Bolyai*, eds. F. Schmidt and P. Stäckel. Leipzig, 1899. Also in his *Werke* 8: 220–224.

Gauss, C. F. (1832c). Theoria residuorum biquadraticorum. *Comm. Sòc. Reg. Sci. Gött. Rec. 4*. In his *Werke* 2: 67–148.

Gauss, C. F. (1846a). Letter to Gerling, 2 October 1846. *Briefwechsel mit Chr. L. Gerling*, Georg Olms Verlag, Hildesheim, 1975, pp. 738–741.

Gauss, C. F. (1846b). Letter to Schumacher, 28 November 1846. Excerpt translated in Kaufmann-Bühler (1981), p. 50.

Gelfond, A. O. (1961). *The Solution of Equations in Integers*. San Francisco, CA.: W. H. Freeman and Co. Translated from the Russian and edited by J. B. Roberts.

Gödel, K. (1931). Über formal unentscheidbare Sätze der Principia Mathematica und verwandter Systeme. I. *Monatsh. Math. Phys. 38*, 173–198.

Gödel, K. (1938). The consistency of the axiom of choice and the generalized continuum hypothesis. *Proc. Nat. Acad. Sci 25*, 220–224.

Gödel, K. (1946). Remarks before the Princeton bicentennial conference on problems in mathematics. In Davis (1965).

Gödel, K. (1949). An example of a new type of cosmological solutions of Einstein's field equations of gravitation. *Rev. Modern Physics 21*, 447–450.

Golay, M. (1949). Notes on digital encoding. *Proc. IRE 37*, 657.

Goldstine, H. H. (1977). *A History of Numerical Analysis from the 16th through the 19th Century*. New York: Springer-Verlag. Studies in the History of Mathematics and Physical Sciences, Vol. 2.

Gomes Teixeira, F. (1995a). *Traité des courbes spéciales remarquables planes et gauches. Tome I*. Paris: Éditions Jacques Gabay. Translated from the Spanish, revised and augmented. Reprint of the 1908 translation.

Gomes Teixeira, F. (1995b). *Traité des courbes spéciales remarquables planes et gauches. Tome II*. Paris: Éditions Jacques Gabay. Translated from the Spanish, revised and augmented. Reprint of the 1909 translation.

Gomes Teixeira, F. (1995c). *Traité des courbes spéciales remarquables planes et gauches. Tome III*. Paris: Éditions Jacques Gabay. Reprint of the 1915 original.

Goursat, E. (1900). Sur la définition générale des fonctions analytiques, d'après Cauchy. *Trans. Amer. Math. Soc. 1*, 14–16.

Graham, R. L., B. L. Rothschild, and J. H. Spencer (1990). *Ramsey Theory* (Second ed.). New York: John Wiley & Sons Inc.

Grandi, G. (1723). Florum geometricorum manipulus. *Phil. Trans. 32*, 355–371.

Graves, J. T. (1844). Letter to Hamilton, 22 January 1844. In Hamilton's *Mathematical Papers* 3: 649.

Graves, R. P. (1975). *Life of Sir William Rowan Hamilton.* New York: Arno Press. Reprint of the edition published by Hodges, Figgis, Dublin, 1882–1889.

Gray, J. (1982). From the history of a simple group. *Math. Intelligencer 4*(2), 59–67.

Green, G. (1828). An essay on the application of mathematical analysis to the theories of electricity and magnetism. In his *Papers*, 1–115.

Gregory, J. (1667). *Vera circuli et hyperbolae quadratura.* Padua: Jacobus de Cadorinius.

Gregory, J. (1668). *Geometriae pars universalis.* Padua: Paolo Frambotto.

Gregory, J. (1670). Letter to Collins, 23 November 1670. In Turnbull (1939), pp. 118–133.

Gregory, J. (1671). Letter to Gideon Shaw, 29 January 1671. In Turnbull (1939), pp. 356–357.

Griess, Jr., R. L. (1982). The friendly giant. *Invent. Math. 69*(1), 1–102.

Grünbaum, B. (1985). Geometry strikes again. *Math. Mag. 58*(1), 12–17.

Hadamard, J. (1896). Sur la distribution des zéros de la fonction $\zeta(s)$ et ses conséquences arithmétiques. *Bull. Soc. Math. France 24*, 199–220.

Hall, Jr., M. (1967). *Combinatorial Theory.* Blaisdell Publishing Co. Ginn and Co., Waltham, MA–Toronto, Ont.–London.

Hamilton, R. S. (1982). Three-manifolds with positive Ricci curvature. *J. Differential Geom. 17*, 255–306.

Hamilton, W. R. (1835). Theory of conjugate functions, or algebraic couples. Communicated to the Royal Irish Academy, 1 June 1835. In his *Mathematical Papers* 3: 76–96.

Hamilton, W. R. (1853). Preface to *Lectures on Quaternions.* In his *Mathematical Papers* 3: 117–155.

Hamilton, W. R. (1856). Memorandum respecting a new system of roots of unity. *Phil. Mag. 12*, 496. In his *Mathematical Papers* 3: 610.

Hamilton, W. R. (1865). Letter to his son Archibald, 5 August 1865. In Graves (1975), vol. II, Ch. XXIX, 434–435.

Hamming, R. W. (1950). Error detecting and error correcting codes. *Bell System Tech. J.* 29, 147–160.

Hankins, T. L. (1980). *Sir William Rowan Hamilton*. Baltimore, MD.: Johns Hopkins University Press.

Harnack, A. (1885). Über den Inhalt von Punktmengen. *Math. Ann.* 25, 241–250.

Hausdorff, F. (1914). *Grundzüge der Mengenlehre*. Leipzig: Von Veit.

Hawkins, T. (2000). *Emergence of the Theory of Lie Groups*. New York: Springer-Verlag.

Heath, T. L. (1897). *The Works of Archimedes*. Cambridge: Cambridge University Press. Reprinted by Dover, New York, 1953.

Heath, T. L. (1910). *Diophantus of Alexandria: A Study in the History of Greek Algebra*. New York: Dover Publications Inc. 1964 reprint of the Cambridge University Press 2nd ed.

Heath, T. L. (1921). *A History of Greek Mathematics*. Oxford: Clarendon Press. Reprinted by Dover, New York, 1981.

Heath, T. L. (1925). *The Thirteen Books of Euclid's Elements*. Cambridge: Cambridge University Press. Reprinted by Dover, New York, 1956.

Hermite, C. (1858). Sur la résolution de l'équation du cinquième degré. *Comp. Rend.* 46, 508–515. In his *Œuvres*, 2, 5–12.

Hermite, C. (1859). Sur la théorie des équations modulaires. *Comp. Rend.* 48,49, 48: 940–947, 940–947, 1079–1084, 1095–1102; 49: 16–24, 110–118, 141–144. In his *Œuvres*, 2, 38–82.

Hermite, C. (1873). Sur la fonction exponentielle. *C. R. LXXVII.* 18–24, 74–49, 226–233, 285–293.. In his *Œuvres* 3, 150–181.

Higman, G. (1961). Subgroups of finitely presented groups. *Proc. Roy. Soc. Lond., ser. A* 262, 455–475.

Hilbert, D. (1897). *The Theory of Algebraic Number Fields*. Translated from the German and with a preface by Iain T. Adamson. With an introduction by Franz Lemmermeyer and Norbert Schappacher. Springer-Verlag, Berlin, 1998.

Hilbert, D. (1899). *Grundlagen der Geometrie*. Leipzig: Teubner. English translation: *Foundations of Geometry,* Open Court, Chicago, 1971.

Hilbert, D. (1900a). Mathematische Probleme. Vortrag, gehalten auf dem internationalen Mathematiker-Congress zu Paris 1900. *Gött. Nachr.* 1900, 253–297.

Hilbert, D. (1900b). Über das Dirichlet'sche Princip. *Jahresber. Deutschen Math. Ver.* 8, 184–188.

Hilbert, D. (1901). Über Flächen von constanter Gaussscher Krümmung. *Trans. Amer. Math. Soc. 2,* 87–89. In his *Gesammelte Abhandlungen* 2: 437–438.

Hilbert, D. and P. Bernays (1936). *Grundlagen der Mathematik I.* Berlin: Springer.

Hilbert, D. and S. Cohn-Vossen (1932). *Anschauliche Geometrie.* Berlin: Julius Springer. English translation: *Geometry and the Imagination,* Chelsea, New York, 1952.

Hobbes, T. (1656). Six lessons to the professors of mathematics. *The English Works of Thomas Hobbes,* vol. 7, 181–356, Scientia Aalen, Aalen, West Germany, 1962.

Hobbes, T. (1672). Considerations upon the answer of Doctor Wallis. *The English Works of Thomas Hobbes,* vol. 7, 443–448, Scientia Aalen, Aalen, West Germany, 1962.

Hoe, J. (1977). *Les systèmes d'équations polynômes dans le Siyuan yujian (1303) par Chu Shih-chieh.* Institut des Hautes Études Chinoises, Collège de France, Paris. Mémoires de l'Institut des Hautes Études Chinoises, Vol. VI.

Hofmann, J. E. (1974). *Leibniz in Paris, 1672–1676.* London: Cambridge University Press. His growth to mathematical maturity, Revised and translated from the German with the assistance of A. Prag and D. T. Whiteside.

Hölder, O. (1896). Über den Casus Irreducibilis bei der Gleichung dritten Grades. *Math. Ann. 38,* 307–312.

Hooke, R. (1675). A description of helioscopes, and some other instruments. In R. T. Gunther, *Early Science in Oxford,* vol. 8, Oxford, 1931.

Hurwitz, A. (1898). Über die komposition der quadratischen Formen von beliebig vielen Variablen. *Göttinger Nachrichten,* 309–316. In his *Mathematische Werke* 2: 565–571.

Huygens, C. (1646). Letters to Mersenne, November 1646. In his *Œuvres Complètes* 1: 34–40.

Huygens, C. (1659a). Fourth part of a treatise on quadrature. *Œuvres Complètes* 14: 337.

Huygens, C. (1659b). Piece on the cycloid, 1 December 1659. *Œuvres Complètes* 16: 392–413.

Huygens, C. (1659c). Recherches sur la théorie des développées. *Œuvres Complètes* 14: 387–405.

Huygens, C. (1671). Letter to Lodewijk Huygens, 29 October 1671. *Œuvres Complètes* 7: 112–113.

Huygens, C. (1673). *Horologium oscillatorium.* In his *Œuvres Complètes* 18: 69–368, English translation *The Pendulum Clock*, Iowa State University Press, Ames, IA, 1986.

Huygens, C. (1691). Christianii Hugenii, dynastae in Zülechem, solutio ejusdem problematis. *Acta Erud. 10*, 281–282. In his *Œuvres Complètes* 10: 95–98.

Huygens, C. (1692). Letter to the Marquis de l'Hôpital, 29 December 1692. *Œuvres Complètes* 10: 348–355.

Huygens, C. (1693a). Appendix to Huygens (1693b). *Œuvres Complètes* 10: 481–422.

Huygens, C. (1693b). Letter to H. Basnage de Beauval, February 1693. *Œuvres Complètes* 10: 407–417.

Jacobi, C. G. J. (1829). *Fundamenta nova theoriae functionum ellipticarum.* Königsberg: Bornträger. In his *Werke* 1: 49–239.

Jacobi, C. G. J. (1834). De usu theoriae integralium ellipticorum et integralium abelianorum in analysi diophantea. *J. reine und angew. Math. 13*, 353–355. In his *Werke* 2: 53–55.

Jones, J. P. and Y. V. Matiyasevich (1991). Proof of recursive unsolvability of Hilbert's tenth problem. *Amer. Math. Monthly 98*(8), 689–709.

Jordan, C. (1866). Sur la déformation des surfaces. *J. Math., ser. 2 11*, 105–109.

Jordan, C. (1870). *Traité des substitutions et des équations algébriques.* Sceaux: Éditions Jacques Gabay. 1989 Reprint of the 1870 original.

Jordan, C. (1887). *Cours de Analyse de l'École Polytechnique.* Paris: Gauthier-Villars.

Jordan, C. (1892). Remarques sur les intégrales définies. *J. Math., ser. 4 8*, 69–99.

Kac, M. (1984). How I became a mathematician. *American Scientist 72*, 498–499.

Kaestner, A. G. (1761). *Anfangsgründe der Analysis der Unendlichen—Die mathematischen Anfangsgründe.* Göttingen. 3. Teil, 2. Abteilung.

Kahn, D. (1967). *The Codebreakers.* London: Weidenfeld and Nicholson.

Kanamori, A. (1994). *The Higher Infinite.* Berlin: Springer-Verlag.

Kaufmann-Bühler, W. (1981). *Gauss. A Biographical Study.* Berlin: Springer-Verlag.

Kepler, J. (1596). *Mysterium cosmographicum.* English translation of 1621 edition, *The Secret of the Universe*, Abaris, New York, 1981.

Kepler, J. (1604). *Ad vitellionem paralipomena, quibus astronomiae pars optica traditur.* Frankfurt: Marnium & Aubrii.

Kepler, J. (1609). *Astronomia nova.* English translation *New Astronomy*, Cambridge University Press, Cambridge, 1992.

Kepler, J. (1619). *Harmonice mundi.* English translation *The Harmony of the World*, American Philosophical Society, 1997.

Kőnig, D. (1926). Sur les correspondances multivoques des ensembles. *Fundamenta Mathematicae 8*, 114–134.

Kőnig, D. (1927). Über eine Schlussweise aus dem Endlichen ins Unendliche. *Acta Litterarum ac Scientiarum 3*, 121–130.

Kőnig, D. (1936). *Theorie der endlichen und unendlichen Graphen.* Leipzig: Akademische Verlagsgesellschaft. English translation by Richard McCoart, *Theory of Finite and Infinite Graphs*, Birkhäuser Boston 1990.

Killing, W. (1888). Die Zusammensetztung der stetigen endlichen Transformationsgruppen. *Math. Ann. 31*, 252–290.

Klein, F. (1871). Über die sogenannte Nicht-Euklidische Geometrie. *Math. Ann. 4*, 573–625. In his *Gesammelte Mathematische Abhandlungen* 1: 254–305. English translation in Stillwell (1996).

Klein, F. (1872). *Vergleichende Betrachtungen über neuere geometrische Forschungen (Erlanger Programm).* Leipzig: Akademische Verlagsgesellschaft. In his *Gesammelte Mathematischen Abhandlungen* 1: 460–497.

Klein, F. (1874). Bemerkungen über den Zusammenhang der Flächen. *Math. Ann. 7*, 549–557.

Klein, F. (1876). Über binäre Formen mit lineare Transformation in sich selbst. *Math. Ann. 9*, 183–208. In his *Gesammellte Mathematische Abhandlungen* 2: 275–301.

Klein, F. (1882a). Letter to Poincaré, 14 May 1882. *Gesammelte Mathematische Abhandlungen* 3: 615–616.

Klein, F. (1882b). Neue Beiträge zur Riemannschen Funktionentheorie. *Math. Ann. 21*, 141–218. In his *Gesammellte Mathematische Abhandlungen* 3: 630–710.

Klein, F. (1884). *Vorlesungen über das Ikosaeder und die Auflösung der Gleichungen vom fünften Grade*. Stuttgart: Teubner. Reprinted in 1993 by Birkhäuser Verlag, with an introduction and commentary by Peter Slodowy. English translation *Lectures on the Icosahedron* by Dover, 1956.

Klein, F. (1924). *Elementarmathematik vom höheren Standpunkte aus. Erster Band: Arithmetik-Algebra-Analysis*. Berlin: Springer. English translation *Elementary mathematics from an advanced standpoint. Arithmetic-algebra-analysis*. Reprinted by Dover Publications Inc., New York, 1953.

Klein, F. (1928). *Vorlesungen über Nicht-Euklidische Geometrie*. Berlin: Springer.

Kline, M. (1972). *Mathematical Thought from Ancient to Modern Times*. New York: Oxford University Press.

Koblitz, N. (1985). *Introduction to Elliptic Curves and Modular Forms*. New York: Springer-Verlag.

Koebe, P. (1907). Über die Uniformisierung beliebiger analytischer Kurven. *Göttinger Nachrichten*, 191–210.

Koestler, A. (1959). *The Sleepwalkers*. London: Hutchinson.

Kolmogorov, A. N. (1933). *Grundbegriffe der Wahrscheinlichkeitsrechnung*. Berlin: Springer. English translation, *Foundations of the Theory of Probability*, Chelsea, New York, 1956.

Kowal, C. T. and S. Drake (1980). Galileo's observations of Neptune. *Nature 287*, 311.

Kreisel, G. (1980). Kurt Gödel. *Biog. Mem. Fellows Roy. Soc. 26*, 149–224.

Kronecker, L. (1857). Über die elliptischen Functionen für welche complexe Multiplication stattfindet. Read to the Prussian Academy of Sciences, 29 October 1857. In his *Werke* 4: 179–183.

Kronecker, L. (1881). Zur Theorie der Elimination einer Variablen aus zwei algebraischen Gleichungen. *Monatsber. König. Preuss. Akad. Wiss. Berlin*, 535–600. In his *Werke* 2: 113–192.

Krummbiegel, B. and A. Amthor (1880). Das Problema bovinum des Archimedes. *Schlömilch Z. XXV. III. A.* 121–136, 153–171.

Kruskal, J. B. (1960). Well-quasi-ordering, the tree theorem, and Vazsonyi's conjecture. *Trans. Amer. Math. Soc.* 95, 210–225.

Kummer, E. E. (1844). De numeris complexis, qui radicibus unitatis et numeris realibus constant. *Gratulationschrift der Univ. Breslau zur Jubelfeier der Univ. Königsberg.* Also in Kummer (1975), vol. 1, 165–192.

Kummer, E. E. (1975). *Collected Papers.* Berlin: Springer-Verlag. Volume I: Contributions to Number Theory, edited and with an introduction by André Weil.

Kuratowski, K. (1930). Sur le problème des courbes gauches en topologie. *Fundamenta Mathematicae* 15, 271–283.

Lagrange, J. L. (1768). Solution d'un problème d'arithmétique. *Miscellanea Taurinensia 4*, 19ff. In his *Œuvres* 1: 671–731.

Lagrange, J. L. (1770). Demonstration d'un théorème d'arithmétique. *Nouv. Mém. Acad. Berlin.* In his *Œuvres* 3: 189–201.

Lagrange, J. L. (1771). Réflexions sur la résolution algébrique des équations. *Nouv. Mém. Acad. Berlin.* In his *Œuvres* 3: 205–421.

Lagrange, J. L. (1772). Recherches sur la manière de former des tables des planètes d'après les seules observations. *Mém. Acad. Roy. Sci. Paris.* In his *Œuvres* 6: 507–627.

Lagrange, J. L. (1773a). Recherches d'arithmétique. *Nouv. Mém. Acad. Berlin*, 265ff. In his *Œuvres* 3: 695–795.

Lagrange, J. L. (1773b). Solutions analytiques de quelques problèmes sur les pyramides triangulaires. *Nouv. Mém. Acad. Berlin.* Also *Œuvres* 3, 658–692.

Lagrange, J. L. (1779). Sur la construction des cartes géographiques. *Nouv. Mém. Acad. Berlin.* In his *Œuvres* 4: 637–692.

Lagrange, J. L. (1785). Sur une nouvelle méthode de calcul intégral. *Mém. Acad. Roy. Soc. Turin 2.* In his *Œuvres* 2: 253–312,

Lam, L. Y. and T. S. Ang (1992). *Fleeting Footsteps.* River Edge, NJ: World Scientific Publishing Co. Inc. With an English translation of *The Mathematical Classic* of Sun Zi.

Lambert, J. H. (1766). Die Theorie der Parallellinien. *Mag. reine und angew. Math. (1786),* 137–164, 325–358.

Lambert, J. H. (1772). *Anmerkungen und Zusätze zur Entwerfung der Land- und Himmelscharten.* English translation by Waldo R. Tobler, Michigan Geographical Publication No. 8, Department of Geography, University of Michigan, 1972.

Lamé, G. (1847). Démonstration générale du théorème de Fermat. *Comp. rend. 24,* 310–315.

Laplace, P. S. (1787). Mémoire sur les inégalités séculaires des planètes et des satellites. *Mém. Acad. Roy. Sci. Paris,* 1–50. In his *Oeuvres Complètes* 11: 49–92.

Laurent, P.-A. (1843). Extension du théorème de M. Cauchy relatif à la convergence du développement d'une fonction suivant les puissances ascendantes de la variable. *Comp. Rend. 17,* 348–349.

Lebesgue, H. (1902). Intégrale, longueur, aire. *Ann. Mat., ser. 3, 7,* 231–359.

Leech, J. (1967). Notes on sphere packings. *Canad. J. Math. 19,* 251–267.

Legendre, A.-M. (1794). *Élements de géométrie.* Paris: F. Didot.

Legendre, A.-M. (1825). *Traité des fonctions elliptiques.* Paris: Huzard-Courcier.

Leibniz, G. W. (1666). *Dissertatio de arte combinatoria.* In Leibniz's *Mathematische Schriften* 5, 7–79.

Leibniz, G. W. (1675). De bisectione laterum. See Schneider (1968).

Leibniz, G. W. (1684). Nova methodus pro maximis et minimis. *Acta Erud. 3,* 467–473. In his *Mathematische Schriften* 5, 220–226. English translation in Struik (1969).

Leibniz, G. W. (1686). De geometria recondita et analysi indivisibilium atque infinitorum. *Acta Erud. 5,* 292–300. Also in Leibniz's *Mathematische Schriften* 5, 226–233.

Leibniz, G. W. (1691). De linea in quam flexile se pondere proprio curvat, ejusque usu insigni ad inveniendas quotcunque medias proportionales et logarithmos. *Acta Erud. 10,* 277–281. In his *Mathematische Schriften* 5: 243–247.

Leibniz, G. W. (1697). Communicatio suae pariter duarumque alienarum ad edendum sibi primum a Dn. Joh. Bernoullio. *Acta Erud. 16*, 205–210. In his *Mathematische Schriften* 5: 331–336.

Leibniz, G. W. (1702). Specimen novum analyseos pro scientia infiniti circa summas et quadraturas. *Acta Erud. 21*, 210–219. In his *Mathematische Schriften* 5: 350–361.

Lenstra, H. W. (2002). Solving the Pell equation. *Notices Amer. Math. Soc. 49*, 182–192.

Levi ben Gershon (1321). *Maaser Hoshev*. German translation by Gerson Lange: *Sefer Maasei Choscheb*, Frankfurt 1909.

l'Hôpital, G. F. A. d. (1696). *Analyse des infiniment petits*. English translation *The Method of Fluxions both Direct and Inverse*, William Ynnis, London 1730.

l'Hôpital, G. F. A. d. (1697). Solutio problematis de linea celerrimi descensus. *Acta Erud. 16*, 217–220.

Li, Y. and S. R. Du (1987). *Chinese Mathematics: A Concise History*. New York: The Clarendon Press Oxford University Press. Translated from the Chinese and with a preface by John N. Crossley and Anthony W.-C. Lun. With a foreword by Joseph Needham.

Libbrecht, U. (1973). *Chinese Mathematics in the Thirteenth Century*. Cambridge, MA.: M.I.T. Press. The *Šhu-shu chiu-chang* of Ch'in Chiu-shao, MIT East Asian Science Series, 1.

Lindemann, F. (1882). Über die Zahl π. *Math. Ann. 20*, 213–225.

Liouville, J. (1833). Mémoire sur les transcendantes elliptiques de première et de seconde espèce considérées comme fonctions de leur amplitude. *J. Éc. Polytech. 23*, 37–83.

Liouville, J. (1850). Note IV to Monge's *Application de l'analyse à la géometrie*, 5th ed. Bachelier, Paris.

Lobachevsky, N. I. (1829). *On the foundations of geometry*. Kazansky Vestnik. (Russian).

Lobachevsky, N. I. (1836). Application of imaginary geometry to some integrals. *Zap. Kazan Univ. 1*, 3–166. (Russian).

Lohne, J. A. (1965). Thomas Harriot als Mathematiker. *Centaurus 11*(1), 19–45.

Lohne, J. A. (1979). Essays on Thomas Harriot. *Arch. Hist. Exact Sci.* 20(3-4), 189–312. I. Billiard balls and laws of collision, II. Ballistic parabolas, III. A survey of Harriot's scientific writings.

Lyusternik, L. A. (1966). *Convex Figures and Polyhedra.* D. C. Heath and Co., Boston, MA. Translated and adapted from the first Russian edition (1956) by Donald L. Barnett.

Maclaurin, C. (1720). *Geometrica organiza sive descriptio linearum curvarum universalis.* London: G. and J. Innys.

Magnus, W. (1930). Über diskontinuierliche Gruppen mit einer definierenden Relation (der Freiheitssatz). *J. reine und angew. Math. 163*, 141–165.

Magnus, W. (1974). *Noneuclidean Tesselations and Their Groups.* Academic Press, New York–London. Pure and Applied Mathematics, Vol. 61.

Mahoney, M. J. (1973). *The Mathematical Career of Pierre de Fermat.* Princeton, NJ: Princeton University Press.

Markov, A. (1958). The insolubility of the problem of homeomorphy (Russian). *Dokl. Akad. Nauk SSSR 121*, 218–220.

Masotti, A. (1960). Sui "Cartelli di matematica disfida" scambiati fra Lodovico Ferrari e Niccolò Tartaglia. *Ist. Lombardo Accad. Sci. Lett. Rend. A 94*, 31–41. (1 plate).

Mathieu, E. (1861). Mémoire sur l'étude des fonctions des plusieurs quantités, sur le manière de les former et sur les substitutions qui les laissent invariables. *J. Math. Pures Appl. 6*, 241–323.

Mathieu, E. (1873). Sur la fonction cinq fois transitive de 24 quantités. *J. Math. Pures Appl. 18*, 25–46.

Matiyasevich, Y. V. (1970). The Diophantineness of enumerable sets (russian). *Dokl. Akad. Nauk SSSR 191*, 279–282.

McKean, H. and V. Moll (1997). *Elliptic Curves.* Cambridge: Cambridge University Press.

Melzak, Z. A. (1976). *Companion to Concrete Mathematics. Vol. II. Mathematical Ideas, Modeling and Applications.* New York: Wiley-Interscience (John Wiley & Sons). Foreword by Wilhelm Magnus.

Mengoli, P. (1650). *Novae quadraturae arithmeticae seu de additione fractionum.* Bononi: Iacob Montij.

Mercator, N. (1668). *Logarithmotechnia.* London: William Godbid and Moses Pitt.

Mersenne, M. (1625). *La vérité des sciences.* Paris: Toussainct du Bray.

Mersenne, M. (1636). *Harmonie Universelle.* Facsimile published by CNRS, Paris, 1963.

Milnor, J. (2003). Towards the Poincaré conjecture and the classification of 3-manifolds. *Notices Amer. Math. Soc. 50,* 1226–1233.

Minding, F. (1839). Wie sich entscheiden lässt, ob zwei gegebene krumme Flächen auf einander abwickelbar sind oder nicht; nebst Bemerkungen über die Flächen von unveränderlichem Krümmungsmasse. *J. reine und angew. Math. 19,* 370–387.

Minding, F. (1840). Beiträge zur Theorie der kürzesten Linien auf krummen Flächen. *J. reine und angew. Math. 20,* 323–327.

Möbius, A. F. (1827). Der barycentrische Calcul. *Werke* 1, 1–388.

Möbius, A. F. (1863). Theorie der Elementaren Verwandtschaft. *Werke* 2: 433–471.

Moise, E. E. (1963). *Elementary Geometry from an Advanced Standpoint.* Addison-Wesley Publishing Co., Inc., Reading, MA-Palo Alto, CA-London.

Moore, E. H. (1893). A doubly infinite system of simple groups. *Bull. New York Math. Soc. 3,* 73–78.

Mordell, L. J. (1922). On the rational solutions of the indeterminate equations of the third and fourth degrees. *Cambr. Phil. Soc. Proc. 21, 179–192.*

Morgan, J. and G. Tian (2007). *Ricci Flow and the Poincaré Conjecture.* Providence, RI: American Mathematical Society.

Nasar, S. and D. Gruber (2006). Manifold Destiny. *The New Yorker.* August 28: 44–57.

Nathanson, M. B. (1987). A short proof of Cauchy's polygonal number theorem. *Proc. Amer. Math. Soc. 99*(1), 22–24.

Needham, T. (1997). *Visual Complex Analysis.* Oxford: Clarendon Press.

Neugebauer, O. and A. Sachs (1945). *Mathematical Cuneiform Texts.* New Haven, CT: Yale University Press.

Neumann, C. (1865). *Vorlesungen über Riemann's Theorie der Abelschen Integralen*. Leipzig: Teubner.

Neumann, C. (1870). Zur Theorie des logarithmischen und des Newtonschen Potentiales, zweite Mitteilung. *Ber. König. Sächs. Ges. Wiss., math.-phys. Cl.*, 264–321.

Newton, I. (1665a). Annotations on Wallis. *Mathematical Papers* 1, 96–111.

Newton, I. (1665b). The geometrical construction of equations. *Mathematical Papers* 1, 492–516.

Newton, I. (1665c). Normals, curvature and the resolution of the general problem of tangents. *Mathematical Papers* 1: 245–297.

Newton, I. (1667). Enumeratio curvarum trium dimensionum. *Mathematical Papers* 12, 10–89.

Newton, I. (1669). De analysi. *Mathematical Papers*, 2, 206–247.

Newton, I. (1670s). De resolutione quaestionum circa numeros. *Mathematical Papers*, 4: 110–115.

Newton, I. (1671). De methodis serierum et fluxionum. *Mathematical Papers*, 3, 32–353.

Newton, I. (1676a). Letter to Oldenburg, 13 June 1676. In Turnbull (1960), pp. 20–47.

Newton, I. (1676b). Letter to Oldenburg, 24 October 1676. In Turnbull (1960), pp. 110–149.

Newton, I. (1687). *Philosophiae naturalis principia mathematica*. London: William Dawson & Sons, Ltd. Facsimile of first edition of 1687.

Newton, I. (1695). Enumeratio linearum tertii ordinis. *Mathematical Papers*, 7, 588–645.

Newton, I. (1697). The twin problems of Johann Bernoulli's "Programma" solved. *Phil. Trans. 17*, 388–389. In his *Mathematical Papers* 8: 72–79.

Nicéron, F. (1638). *La perspective curieuse*. Paris: P. Billaine.

Nielsen, J. (1927). Untersuchungen zur Topologie der geschlossenen zweiseitigen Flächen. *Acta Math. 50*, 189–358.

Novikov, P. S. (1955). On the algorithmic unsolvability of the word problem in group theory (Russian). *Dokl. Akad. Nauk SSSR Mat. Inst. Tr. 44*. English translation in *Amer. Math. Soc. Transl.* ser. 2, **9**, 1–122.

O'Donnell, S. (1983). *William Rowan Hamilton*. Dún Laoghaire: Boole Press. With a foreword by A. J. McConnell.

Ore, O. (1953). *Cardano, the gambling scholar. With a translation from the Latin of Cardano's "Book on games of chance," by S. H. Gould*. Princeton, NJ.: Princeton University Press.

Ore, O. (1957). *Niels Henrik Abel: Mathematician Extraordinary*. Minneapolis, MN.: University of Minnesota Press.

Oresme, N. (1350a). *Quaestiones super geometriam Euclidis*. Edited by H. L. L. Busard. Janus, suppléments, Vol. III, E. J. Brill, Leiden, 1961.

Oresme, N. (1350b). *Tractatus de configurationibus qualitatum et motuum*. English translation in Clagett (1968).

Ostrogradsky, M. (1828). Démonstration d'un théorème du calcul integral. *Mém. Acad. Sci. St. Petersburg, ser. 6 1*, 39–53.

Ostrowski, A. (1920). Über den ersten und vierten Gaussschen Beweis des Fundamentalsatzes der Algebra. *Gauss Werke* 10, part 2, 1–18.

Pacioli, L. (1509). *De divina proportione*. Venice: Paganius Paganinus.

Paige, L. J. (1957). A note on the Mathieu groups. *Canad. J. Math. 9*, 15–18.

Paris, J. and L. Harrington (1977). A mathematical incompleteness in Peano arithmetic. In *Handbook of Mathematical Logic*, ed. J. Barwise, North-Holland, Amsterdam.

Pascal, B. (1640). *Essay pour les coniques*. Paris.

Pascal, B. (1654). Traité du triangle arithmétique, avec quelques autres petits traités sur la même manière. English translation in *Great Books of the Western World*, Encyclopedia Britannica, London, 1952, 447–473.

Pearson, K. (1978). *The History of Statistics in the 17th and 18th Centuries*. New York: Macmillan Co. Lectures given at University College, London, during the academic sessions 1921–1933. Edited and with a preface by Egon S. Pearson.

Pierpont, J. (1895). Zur Geschichte der Gleichung des V. Grades (bis 1858). *Monatsh. f. Math. VI. 15-68.*.

Plofker, K. (2009). *Mathematics in India*. Princeton, NJ: Princeton University Press.

Plücker, J. (1830). Über ein neues Coordinatensystem. *J. reine angew. Math. 5*, 1–36. *Gesammelte Mathematische Abhandlungen* 124–158.

Plücker, J. (1847). Note sur le théorème de Pascal. *J. reine angew. Math. 34*, 337–340. *Gesammelte Mathematische Abhandlungen* 413–416.

Poincaré, H. (1882). Théorie des groupes fuchsiens. *Acta Math. 1*, 1–62. In his *Œuvres* 2: 108–168. English translation in Poincaré (1985), 55–127.

Poincaré, H. (1883). Mémoire sur les groupes Kleinéens. *Acta Math. 3*, 49–92. English translation in Poincaré (1985), 255–304.

Poincaré, H. (1892). *New Methods of Celestial Mechanics. Vol. 1*. Periodic and asymptotic solutions, translated from the French, revised reprint of the 1967 English translation, with endnotes by V. I. Arnol'd, edited and with an introduction by Daniel L. Goroff, American Institute of Physics, New York, 1993.

Poincaré, H. (1893). *New Methods of Celestial Mechanics. Vol. 2*. Approximations by series, translated from the French, revised reprint of the 1967 English translation, with endnotes by V. M. Alekseev, edited and with an introduction by Daniel L. Goroff, American Institute of Physics, New York, 1993.

Poincaré, H. (1895). Analysis situs. *J. Éc. Polytech., ser. 2 1*, 1–121. In his *Œuvres* 6: 193–288.

Poincaré, H. (1899). *New Methods of Celestial Mechanics. Vol. 3*. Integral invariants and asymptotic properties of certain solutions, translated from the French, revised reprint of the 1967 English translation, with endnotes by G. A. Merman, edited and with an introduction by Daniel L. Goroff, American Institute of Physics, New York, 1993.

Poincaré, H. (1901). Sur les propriétés arithmétiques des courbes algébriques. *J. Math. 7*, 161–233. In his *Œuvres* 5: 483–548.

Poincaré, H. (1904). Cinquième complément à l'analysis situs. *Palermo Rend. 18*, 45–110. In his *Œuvres* 6: 435–498.

Poincaré, H. (1907). Sur l'uniformisation des fonctions analytiques. *Acta Math. 31*, 1–63. In his *Œuvres* 4: 70–139.

Poincaré, H. (1918). *Science et Méthode*. Paris: Flammarion. English translation in *The Foundations of Science*, Science Press, New York, 1929, 357–553.

Poincaré, H. (1955). Le Livre du Centenaire de la Naissance de Henri Poincaré. *Œuvres* 11.

Poincaré, H. (1985). *Papers on Fuchsian Functions*. New York: Springer-Verlag. Translated from the French and with an introduction by John Stillwell.

Pólya, G. (1954a). An elementary analogue to the Gauss–Bonnet theorem. *Amer. Math. Monthly 61*, 601–603.

Pólya, G. (1954b). *Induction and Analogy in Mathematics. Mathematics and Plausible Reasoning, Vol. I.* Princeton, NJ.: Princeton University Press.

Poncelet, J. V. (1822). *Traité des propriétés projectives des figures.* Paris: Bachelier.

Post, E. L. (1936). Finite combinatory processes. Formulation 1. *J. Symb. Logic 1*, 103–105.

Post, E. L. (1941). Absolutely unsolvable problems and relatively undecidable propositions. Account of an anticipation. In Davis (1965), pp. 340–433.

Post, E. L. (1944). Recursively enumerable sets of positive integers and their decision problems. *Bull. Amer. Math. Soc. 50*, 284–316.

Prouhet, E. (1860). Remarques sur un passage des œuvres inédits de Descartes. *Comp. Rend. 50*, 779–781.

Puiseux, V.-A. (1850). Recherches sur les fonctions algébriques. *J. Math. 15*, 365–480.

Rabinovitch, N. L. (1970). Rabbi Levi ben Gershon and the origins of mathematical induction. *Arch. Hist. Exact Sci. 6*, 237–248.

Rajagopal, C. T. and M. S. Rangachari (1977). On an untapped source of medieval Keralese mathematics. *Arch. History Exact Sci. 18*(2), 89–102.

Rajagopal, C. T. and M. S. Rangachari (1986). On medieval Kerala mathematics. *Arch. Hist. Exact Sci. 35*(2), 91–99.

Ramsey, F. P. (1929). On a problem of formal logic. *Proc. Lond. Math. Soc. 30*, 291–310.

Raspail, F. V. (1839). *Lettres sur les Prisons de Paris, Vol. 2.* Paris.

Ribet, K. A. (1990). On modular representations of $\mathrm{Gal}(\overline{\mathbf{Q}}/\mathbf{Q})$ arising from modular forms. *Invent. Math. 100*(2), 431–476.

Richeson, D. S. (2008). *Euler's Gem.* Princeton: Princeton University Press.

Riemann, G. F. B. (1851). Grundlagen für eine allgemeine Theorie der Functionen einer veränderlichen complexen Grösse. *Werke*, 2nd ed., 3–48.

Riemann, G. F. B. (1854a). Über die Darstellbarkeit einer Function durch eine trigonometrische Reihe. *Werke*, 2nd ed., 227–264.

Riemann, G. F. B. (1854b). Über die Hypothesen, welche der Geometrie zu Grunde liegen. *Werke*, 2nd ed., 272–287.

Riemann, G. F. B. (1857). Theorie der Abel'schen Functionen. *J. reine und angew. Math.* 54, 115–155. *Werke*, 2nd ed., 82–142.

Riemann, G. F. B. (1858a). *Elliptische Funktionen*. Ed. H. Stahl, Leipzig, 1899.

Riemann, G. F. B. (1858b). Vorlesungen über die hypergeometrische Reihe. *Werke*, 2nd ed., Dover, New York, 1953.

Riemann, G. F. B. (1859). Über die Anzahl der Primzahlen unter einer gegebenen Grösse. *Werke*, 2nd ed., 145–153. English translation in Edwards (1974), 299–305.

Robert, A. (1973). *Elliptic Curves*. Berlin: Springer-Verlag. Notes from postgraduate lectures given in Lausanne 1971/72, Lecture Notes in Mathematics, Vol. 326.

Robinson, A. (1966). *Non-standard Analysis*. Amsterdam: North-Holland Publishing Co.

Rodrigues, O. (1840). Des lois géométriques qui régissent les déplacements d'un système solide dans l'espace, et de la variation des coordonnées provenant de ces déplacements considérés indépendamment des causes qui peuvent les produire. *J. de Math. Pures et Appliquées, ser. 1* 5, 380–440.

Ronan, M. (2006). *Symmetry and the Monster*. Oxford: Oxford University Press.

Rose, P. L. (1976). *The Italian Renaissance of Mathematics*. Geneva: Librairie Droz. Studies on humanists and mathematicians from Petrarch to Galileo, Travaux de l'Humanisme et Renaissance, 145.

Rosen, M. (1981). Abel's theorem on the lemniscate. *Amer. Math. Monthly* 88(6), 387–395.

Rothman, T. (1982). Genius and biographers: the fictionalization of Évariste Galois. *Amer. Math. Monthly* 89(2), 84–106.

Ruffini, P. (1799). *Teoria generale delle equazioni in cui si dimostra impossibile la soluzione algebraica delle equazioni generale di grade superiore al quarto*. Bologna.

Russ, S. (2004). *The Mathematical Works of Bernard Bolzano*. Oxford: Oxford University Press.

Saccheri, G. (1733). *Euclides ab omni naevo vindicatus*. Milan: Pauli Antoni Montani. English translation, Open Court, Chicago, 1920.

Salmon, G. (1851). Théorèmes sur les courbes de troisième degré. *J. reine und angew. Math. 42*, 274–276.

Schechter, B. (1998). *My Brain Is Open*. New York: Simon & Schuster.

Schneider, I. (1968). Der Mathematiker Abraham de Moivre (1667–1754). *Arch. Hist. Exact Sci. 5*, 177–317.

Schooten, F. v. (1659). *Geometria à Renato Des Cartes*. Amsterdam: Louis and Daniel Elzevir.

Schwarz, H. A. (1870). Über einen Grenzübergang durch alternirendes verfahren. *Vierteljahrsch. Natur. Ges. Zürich 15*, 272–286. In his *Mathematische Abhandlungen* 2: 133–143.

Schwarz, H. A. (1872). Über diejenigen Fälle, in welchen die Gaussische hypergeometrische Reihe eine algebraische Function ihres vierten Elementes darstellt. *J. reine und angew. Math. 75*, 292–335. In his *Mathematische Abhandlungen* 2: 211–259.

Scott, J. F. (1952). *The Scientific Work of René Descartes (1596–1650)*. Taylor and Francis, Ltd., London.

Seifert, H. and W. Threlfall (1934). *Lehrbuch der Topologie*. Leipzig: Teubner. English translation *A Textbook of Topology*, Academic Press, New York, 1980.

Shelah, S. (1984). Can you take Solovay's inaccessible away? *Israel J. Math. 48*(1), 1–47.

Shen, K.-S., J. N. Crossley, and W.-C. Lun (1999). *The Nine Chapters on the Mathematical Art. Companion and Commentary*. Oxford: Oxford University Press.

Shirley, J. W. (1983). *Thomas Harriot: A Biography*. New York: The Clarendon Press Oxford University Press.

Siegel, C. L. (1969). *Topics in Complex Function Theory. Vol. I: Elliptic Functions and Uniformization Theory*. Wiley-Interscience (a Division of John Wiley & Sons), New York-London-Sydney. Translated from the original German by A. Shenitzer and D. Solitar. Interscience Tracts in Pure and Applied Mathematics, no. 25.

Sitnikov, K. (1960). The existence of oscillatory motion in the three-body problem. *Soviet Physics Dokl. 5*, 647–650.

Sluse, R. F. (1673). A method of drawing tangents to all geometrical curves. *Phil. Trans. 7*, 5143–5147.

Smale, S. (1961). Generalized Poincaré conjecture in dimensions greater than four. *Ann. of Math. 74*, 391–406.

Smith, D. E. (1959). *A Source Book in Mathematics*. New York: Dover Publications Inc. 2 vols.

Soifer, A. (2009). *The Mathematical Coloring Book*. New York: Springer.

Solovay, R. M. (1970). A model of set-theory in which every set of reals is Lebesgue measurable. *Ann. of Math. (2) 92*, 1–56.

Sperner, E. (1928). Neuer Beweis für die Invarianz der Dimensionzahl und des Gebietes. *Abh. Math. Sem. Univ. Hamburg 6*, 265–272.

Srinivasiengar, C. N. (1967). *The History of Ancient Indian Mathematics*. The World Press Private, Ltd., Calcutta.

Stäckel, P. (1901). Die Entdeckung der nichteuklidischen Geometrie durch Johann Bolyai. *Mat.-natur. ber. Ungarn. Budapest 17*, 1–19.

Stäckel, P. (1913). *Wolfgang und Johann Bolyai. Geometrische Untersuchungen*. Leipzig: Teubner.

Stedall, J. (2003). *The Greate Invention of Algebra*. Oxford: Oxford University Press.

Sternberg, S. (1969). *Celestial Mechanics. Part I*. New York-Amsterdam: W. A. Benjamin, Inc. Mathematics Lecture Note Series: XXII.

Stevin, S. (1586). *De Weeghdaet*. Leyden: Christoffel Plantijn.

Stillwell, J. (1982). The word problem and the isomorphism problem for groups. *Bull. Amer. Math. Soc. (N.S.) 6*, 33–56.

Stillwell, J. (1993). *Classical Topology and Combinatorial Group Theory, 2nd ed.* New York, NY: Springer-Verlag.

Stillwell, J. (1996). *Sources of Hyperbolic Geometry*. Providence, RI: American Mathematical Society.

Stillwell, J. (2003). *Elements of Number Theory*. New York, NY: Springer-Verlag.

Stillwell, J. (2008). *Naive Lie Theory*. New York, NY: Springer-Verlag.

Stirling, J. (1717). *Lineae tertii ordinis Neutonianae*. Oxford: Edward Whistler.

Strubecker, K. (1964). *Differentialgeometrie I, II, III*. Berlin: Walter de Gruyter.

Struik, D. (1969). *A Source Book of Mathematics 1200–1800*. Cambridge: Harvard University Press.

Stubhaug, A. (2000). *Niels Henrik Abel and His Times*. Berlin: Springer-Verlag. Translated from the 1996 Norwegian original by Richard H. Daly.

Stubhaug, A. (2002). *The Mathematician Sophus Lie*. Berlin: Springer-Verlag. Translated from the 2000 Norwegian original by Richard H. Daly.

Szabó, I. (1977). *Geschichte der mechanischen Prinzipien und ihrer wichtigsten Anwendungen*. Basel: Birkhäuser Verlag. Wissenschaft und Kultur, 32.

Tao, T. (2009). *Structure and Randomness: Pages from Year One of a Mathematical Blog*. American Mathematical Society.

Tartaglia, N. (1546). *Quesiti et Inventioni Diverse*. Facsimile of 1554 edition, edited by A. Masotti, by Ateneo di Brescia, Brescia.

Taton, R. (1951). *L'œuvre mathématique de G. Desargues*. Paris: Presses universitaires de France.

Taurinus, F. A. (1826). *Geometriae prima elementa*. Cologne.

Taylor, B. (1713). De motu nervi tensi. *Phil. Trans 28*, 26–32.

Taylor, B. (1715). *Methodus incrementorum directa et inversa*. London: William Innys.

Thompson, T. M. (1983). *From Error-Correcting Codes through Sphere Packings to Simple Groups*. Washington, DC: Mathematical Association of America.

Thurston, W. P. (1997). *Three-Dimensional Geometry and Topology. Vol. 1*. Princeton, NJ: Princeton University Press. Edited by Silvio Levy.

Tietze, H. (1908). Über die topologische Invarianten mehrdimensionaler Mannigfaltigkeiten. *Monatsh. Math. Phys. 19*, 1–118.

Tits, J. (1956). Les groupes de Lie exceptionnels et leur interprétation géométrique. *Bull. Soc. Math. Belg. 8*, 48–81.

Torricelli, E. (1643). *De solido hyperbolico acuto*. Partial English translation in Struik (1969).

Torricelli, E. (1644). *De dimensione parabolae*.

Torricelli, E. (1645). *De infinitis spirabilus*. Reprint edited by E. Carruccio, Domus Galiaeana, Pisa 1955.

Truesdell, C. (1954). *Rational fluid mechanics, 1687–1765.* Orell Füssli, Zürich. Leonhardi Euleri Opera Omnia, Series secunda, Vol. XII: IV–CXXV.

Truesdell, C. (1960). *The rational mechanics of flexible or elastic bodies, 1638–1788.* Orell Füssli, Zürich. Leonhardi Euleri Opera Omnia, Series secunda, Vol. XI, sectio secunda.

Turing, A. (1936). On computable numbers, with an application to the Entscheidungsproblem. *Proc. Lond. Math. Soc., ser. 2 42,* 230–265.

Turnbull, H. W. (1939). *James Gregory (1638–1675).* University of St. Andrews James Gregory Tercentenary, St. Andrews. G. Bell and Sons, London.

Turnbull, H. W. (1960). *The Correspondence of Isaac Newton, Vol. II: 1676–1687.* New York: Cambridge University Press.

Ulam, S. (1930). Zur Masstheorie in der allgemeinen Mengenlehre. *Fund. Math. 15,* 140–150.

Van Brummelen, G. (2009). *The Mathematics of the Heavens and the Earth.* Princeton, NJ: Princeton University Press.

van Dalen, D. and A. Monna (1972). *Sets and Integration. An Outline of the Development.* Groningen: Wolters-Noordhoff Publishing.

van der Waerden, B. (1976). Pell's equation in Greek and Hindu mathematics. *Russ. Math. Surveys 31*(5), 210–225.

van der Waerden, B. L. (1949). *Modern Algebra.* New York: Frederick Ungar.

van der Waerden, B. L. (1954). *Science Awakening.* Groningen: P. Noordhoff Ltd. English translation by Arnold Dresden.

van der Waerden, B. L. (1983). *Geometry and Algebra in Ancient Civilizations.* Berlin: Springer-Verlag.

van Heijenoort, J. (1967). *From Frege to Gödel. A Source Book in Mathematical Logic, 1879–1931.* Cambridge, Mass.: Harvard University Press.

Vandermonde, A.-T. (1771). Mémoire sur la résolution des équations. *Hist. Acad. Roy. Sci..*

Veblen, O. (1905). Theory of plane curves in nonmetrical analysis situs. *Trans. Amer. Math. Soc. 6,* 83–98.

Veblen, O. and W. Bussey (1906). Finite Projective Geometries. *Trans. Amer. Math. Soc. 7,* 241–259.

Viète, F. (1579). *Universalium inspectionium ad canonem mathematicum liber singularis.*

Viète, F. (1591). De aequationum recognitione et emendatione. In his *Opera*, 82–162. English translation in Viète (1983).

Viète, F. (1593). Variorum de rebus mathematicis responsorum libri octo. In his *Opera*, 347–435.

Viète, F. (1615). Ad angularium sectionum analyticen theoremata. In his *Opera*, 287–304.

Viète, F. (1983). *The Analytic Art.* Kent, OH: The Kent State University Press. Nine studies in algebra, geometry and trigonometry from the *Opus Restitutae Mathematicae Analyseos, seu Algebra Nova*, translated by T. Richard Witmer.

Vitali, G. (1905). *Sul problema della misura dei gruppi di punti di una retta.* Bologna.

von Neumann, J. (1923). Zur Einführung der transfiniten Zahlen. *Acta lit. acad. sci. Reg. U. Hungar. Fran. Jos. Sec. Sci. 1*, 199–208. English translation in van Heijenoort (1967) 347–354.

von Staudt, K. G. C. (1847). *Geometrie der Lage.* Nurnberg: Bauer und Raspe.

Vrooman, J. R. (1970). *René Descartes. A Biography.* New York: Putman.

Wagner, K. W. (1936). Bemerkungen zum Vierfarbenproblem. *Jahresber. Deutsch. Math.-Ver. 46*, 26–32.

Wagner, K. W. (1937). Über eine Eigenschaft der ebenen Komplexe. *Math. Ann. 114*, 570–590.

Wagon, S. (1985). *The Banach-Tarski Paradox.* Cambridge: Cambridge University Press. With a foreword by Jan Mycielski.

Wallis, J. (1655a). Arithmetica infinitorum. *Opera* 1: 355–478. English translation *The Arithmetic of Infinitesimals* by Jacqueline Stedall, Springer, New York, 2004.

Wallis, J. (1655b). De sectionibus conicis. *Opera* 1: 291–354.

Wallis, J. (1657). Mathesis universalis. *Opera* 1: 11–228.

Wallis, J. (1659). Tractatus duo. Prior, de cycloide. Posterior, de cissoid. *Opera* 1: 489–569.

Wallis, J. (1663). De postulato quinto; et definitione quinta Lib. 6 Euclidis. *Opera* 2: 669–678.

Wallis, J. (1673). On imaginary numbers. From his *Algebra*, Vol. 2. In Smith (1959) 1: 46–54.

Wallis, J. (1696). Autobiography. *Notes and Records, Roy. Soc. London*, **25**, (1970), 17–46.

Wantzel, P. L. (1837). Recherches sur les moyens de reconnaitre si un problème de géométrie peut se resoudre avec la règle et le compas. *J. Math.* 2, 366–372.

Weber, H. (1892). Leopold von Kronecker. *Jahresber. Deutsch. Math. Verein.* 2, 19.

Weeks, J. R. (1985). *The Shape of Space*. New York: Marcel Dekker Inc.

Weierstrass, K. (1863). Vorlesungen über die Theorie der elliptischen Funktionen. *Mathematische Werke* 5.

Weierstrass, K. (1874). *Einleitung in die Theorie der analytischen Funktionen.* Summer Semester 1874. Notes by G. Hettner. Mathematische Institut der Universität Göttingen.

Weierstrass, K. (1884). Zur Theorie der aus *n* Haupteinheiten gebildeten complexen Grössen. *Göttingen Nachrichten*, 395–414. In his *Mathematische Werke* 2: 311–332.

Weil, A. (1975). Introduction to Kummer (1975).

Weil, A. (1976). *Elliptic Functions According to Eisenstein and Kronecker.* Berlin: Springer-Verlag. Ergebnisse der Mathematik und ihrer Grenzgebiete, Band 88.

Weil, A. (1984). *Number Theory. An Approach through History, from Hammurapi to Legendre.* Boston, MA.: Birkhäuser Boston Inc.

Wessel, C. (1797). Om Directionens analytiske Betegning, et Forsøg anvendt fornemmelig til plane og sphæriske Polygoners Opløsning. *Danske Selsk. Skr. N. Samml. 5.* English translation in Smith (1959), vol. 1, 55–66.

Westfall, R. S. (1980). *Never at Rest.* Cambridge: Cambridge University Press. A biography of Isaac Newton.

Whitehead, A. N. and B. Russell (1910). *Principia Mathematica.* Cambridge: Cambridge University Press. 3 vols. 1910, 1912, 1913.

Whitehead, J. H. C. (1935). A certain open manifold whose group is unity. *Quart. J. Math. 6*, 268–279.

Whiteside, D. T. (1961). Patterns of mathematical thought in the later seventeenth century. *Arch. History Exact Sci. 1*, 179–388 (1961).

Whiteside, D. T. (1964). Introduction to *The Mathematical Works of Isaac Newton*. Vol. I. Johnson Reprint Corp., New York, 1964.

Whiteside, D. T. (1966). Newton's marvellous year: 1666 and all that. *Notes and Records, Roy. Soc. Lond. 21*, 32–41.

Wiles, A. (1995). Modular elliptic curves and Fermat's last theorem. *Ann. of Math. (2) 141*(3), 443–551.

Woodin, W. H. (1999). *The Axiom of Determinacy, Forcing Axioms, and the Nonstationary Ideal*. Berlin: Walter de Gruyter & Co.

Wright, L. (1983). *Perspective in Perspective*. London: Routledge and Kegan Paul.

Wussing, H. (1984). *The Genesis of the Abstract Group Concept*. Cambridge, MA.: MIT Press. Translated from the German by Abe Shenitzer.

Xia, Z. (1992). The existence of noncollision singularities in Newtonian systems. *Ann. of Math. (2) 135*(3), 411–468.

Yáng Huí (1261). *Compendium of analyzed mathematical methods in the "Nine Chapters"*.

Zermelo, E. (1904). Beweis dass jede Menge wohlgeordnet werden kann. *Math. Ann. 59*, 514–516. English translation in van Heijenoort (1967).

Zeuthen, H. G. (1903). *Geschichte der Mathematik im 16. und 17. Jahrhundert*. Leipzig: Teubner. Johnson Reprint Corp., New York, 1977.

Zhū Shijié (1303). *Siyuan yujian*. French translation in Hoe (1977).

Index